S0-AZA-204

Leptin and Reproduction

Leptin and Reproduction

Edited by

Michael C. Henson

Tulane University Health Sciences Center
New Orleans, Louisiana

and

V. Daniel Castracane

Texas Tech Health Sciences Center
Amarillo, Texas
and
Diagnostic Systems Laboratories, Inc.
Webster, Texas

Kluwer Academic / Plenum Publishers
New York ● Boston ● Dordrecht ● London ● Moscow

Library of Congress Cataloging-in-Publication Data

Leptin and reproduction/edited by Michael C. Henson and V. Daniel Castracane.
    p.   ; cm.
  Includes bibliographical references and index.
  ISBN 0-306-47488-3 (alk. paper)
    1. Leptin—Physiological effect. 2. Human reproduction. I. Henson, Michael Chris. II.
  Castracane, V. Daniel.
    [DNLM: 1. Reproduction—physiology. 2. Leptin—physiology. WQ 205 L611 2003]
  QP572.L48 L468 2003
  612.6—dc21

                                                                            2002040786

ISBN 0-306-47488-3

©2003 Kluwer Academic/Plenum Publishers, New York
233 Spring Street, New York, New York 10013

http://www.wkap.nl/

10 9 8 7 6 5 4 3 2 1

A C.I.P. record for this book is available from the Library of Congress

Contributors

Paul Bischof, Ph.D., Department of Obstetrics and Gynaecology, University Hospital of Geneva

Susann Blüher, M.D., Division of Endocrinology, Department of Medicine, Beth Israel Deaconess Medical Center, Harvard Medical School

John D. Brannian, Ph.D., Department of Obstetrics and Gynecology, University of South Dakota School of Medicine

V. Daniel Castracane, Ph.D., Department of Obstetrics and Gynecology and the Women's Health Research Institute of Amarillo, Texas Tech University Health Sciences Center—Amarillo; Diagnositc Systems Laboratories, Inc., Webster, Texas

Farid F. Chehab, Ph.D., Department of Laboratory Medicine, University of California—San Francisco

Helen Christou, M.D., Division of Newborn Medicine, Children's Hospital, Harvard Medical School

Robert V. Considine, Ph.D., Department of Medicine, Indiana University School of Medicine

Ekaterini Domali, M.D., Department of Obstetrics and Gynecology, University of Thessalia Medical School

Jennifer E. Dominguez, B.A., Department of Medicine and Obstetrics and Gynecology, Columbia University

Priya S. Duggal, Ph.D., Reproductive Medicine Unit, Department of Obstetrics and Gynaecology, The Queen Elizabeth Hospital, University of Adelaide

Deborah E. Edwards, Ph.D., Interdisciplinary Program in Molecular and Cellular Biology, Tulane University Health Sciences Center

Amanda Ewart-Toland, Ph.D., Department of Laboratory Medicine, University of California—San Francisco

I. Sadaf Farooqi, M.D., Ph.D., University Departments of Medicine and Clinical Biochemistry, Addenbrooke's Hospital—Cambridge

Douglas L. Foster, Ph.D., Department of Obstetrics and Gynecology and Department of Ecology and Evolutionary Biology, University of Michigan

Keith A. Hansen, M.D., Department of Obstetrics and Gynecology, University of South Dakota School of Medicine

Michael C. Henson, Ph.D., Departments of Obstetrics and Gynecology, Physiology, and Structural and Cellular Biology, the Tulane National Primate Research Center and Interdisciplinary Program in Molecular and Cellular Biology, Tulane University Health Sciences Center

Dorina Islami, M.D., Department of Obstetrics and Gynaecology, University Hospital of Geneva

Leslie M. Jackson, Ph.D., Department of Obstetrics and Gynecology and Department of Ecology and Evolutionary Biology, University of Michigan

Sharada Karanth, Ph.D., Pennington Biomedical Research Center, Louisiana State University

Joaquin Lado-Abeal, M.D., Ph.D., Department of Medicine, The University of Chicago Medical Center

Denis A. Magoffin, Ph.D., Cedars-Sinai Burns and Allen Research Institute, The David Geffen School of Medicine at UCLA

David R. Mann, Ph.D., Department of Physiology and the Cooperative Reproductive Science Research Center, Morehouse School of Medicine

Christos S. Mantzoros, M.D., D.Sc., Division of Endocrinology, Department of Internal Medicine, Beth Israel Deaconess Medical Center, Harvard Medical School

Claudio A. Mastronardi, M.S., Pennington Biomedical Research Center, Louisiana State University

Samuel M. McCann, M.D., Pennington Biomedical Research Center, Louisiana State University

Ioannis E. Messinis, M.D., Ph.D., Department of Obstetrics and Gynecology, University of Thessalia Medical School

Khalid Mounzih, Ph.D., Department of Laboratory Medicine, University of California—San Francisco

Reid L. Norman, Ph.D., Department of Pharmacology, Texas Tech University Health Sciences Center—Lubbock

Robert J. Norman, M.D., Reproductive Medicine Unit, Department of Obstetrics and Gynaecology, The Queen Elizabeth Hospital, University of Adelaide

Scott Ogus, B.S., Department of Laboratory Medicine, University of California—San Francisco

Tony M. Plant, Ph.D., Department of Cell Biology and Physiology and the Specialized Cooperative Center Program of Reproduction Research, University of Pittsburgh Medical School

Lucilla Poston, Ph.D., Department of Women's Health, Guy's Kings and St. Thomas' School of Medicine, King's College, London

Jun Qiu, M.D., Ph.D., Department of Laboratory Medicine, University of California—San Francisco

Shanti Serdy M.D., Division of Endocrinology, Department of Internal Medicine, Beth Israel Deaconess Medical Center, Harvard Medical School

Jeremy T. Smith, Ph.D., School of Anatomy and Human Biology, The University of Western Australia

Leon J. Spicer, Ph.D., Department of Animal Science, Oklahoma State University

Brendan J. Waddell, Ph.D., School of Anatomy and Human Biology, The University of Western Australia

Michelle P. Warren, M.D., Departments of Medicine and Obstetrics and Gynecology, Columbia University

Wen H. Yu, Ph.D., Pennington Biomedical Research Center, Louisiana State University

Preface

The isolation of leptin in 1994 and its characterization as a factor influencing appetite, energy balance, and adiposity, immediately thrust the polypeptide into the rapidly growing body of literature centered on the physiology of obesity. The growing clinical awareness of obesity as a major health risk in developed societies dovetailed perfectly with any of a number of roles that leptin might play in this aberrant physiological condition. Almost unnoticed amidst the excitement generated by early leptin publications was the suggestion that the "fat hormone" might also regulate a wide range of systems and events important to reproduction, including pubertal development, gonadal endocrinology, fertility, and pregnancy. Recognizing this potential, a relatively small cadre of researchers began to examine leptin specifically as a reproductive hormone, thus creating a new and fertile field of investigation. Interest in this area has since gained momentum and an increased number of participants have now made significant contributions to our understanding of many leptin-related mechanisms that are relevant to reproductive biology.

Leptin and Reproduction is the first major volume to specifically address leptin as a reproductive hormone and closely examines the advances made in the short time since this field of interest developed. Preeminent researchers from many of the subdisciplines working within this area present a welcomed compendium of the wealth of related literature and voice novel interpretations of current advances. These interpretations evolved, in many cases, from ongoing work in their own laboratories and provide fresh and valuable perspectives. We trust that this volume will be of great interest to both basic scientists and clinicians in many areas of reproductive biology, and that original insights gained from our excellent team of contributors will help to inspire the next generation of exciting new advances in the field.

MICHAEL C. HENSON
V. DANIEL CASTRACANE
Editors

Contents

I INTRODUCTION

1 Leptin: From Satiety Signal to Reproductive Regulator **3**
Michael C. Henson, V. Daniel Castracane, and Deborah E. Edwards

1.1 Leptin's Past 3
1.2 Leptin's Present 5
    1.2.1 General Reproductive Function 5
    1.2.2 Puberty 6
    1.2.3 Pregnancy 7
    1.2.4 Genetics 7
    1.2.5 Clinical Challenges 8
1.3 Leptin's Future? 8

II GENERAL REPRODUCTIVE FUNCTION

**2 Neuroendocrine Regulation of Leptin Secretion and Its
Role in Nitric Oxide Secretion** **15**
S. M. McCann, S. Karanth, W. H. Yu, and C. A. Mastronardi

2.1 Abstract 15
2.2 Hypothalamic Control of Gonadotropin Secretion 16
    2.2.1 FSHRH 17
    2.2.2 Localization of lGnRH in the Brain of Rats by
           Immunocytochemistry 18
    2.2.3 Evidence for a Specific FSHRH Receptor 19
    2.2.4 Mechanism of Action of FSHRH, LHRH, and
           Leptin on Gonadotropin Secretion 20
2.3 Leptin in Reproduction 21
    2.3.1 Effect of Leptin on LH Release 22

2.3.2 Effect of Leptin on FSH Release 22
2.3.3 Effect of Leptin on LHRH Release 22
2.3.4 The Effect of Intraventricularly Injected Leptin on Plasma Gonadotropin Concentrations in Ovariectomized, Estrogen-Primed Rats 23
2.3.5 Mechanism of Action of Leptin on the Hypothalamic Pituitary Axis 23
2.4 Role of Leptin in the Response to Stress and Infection 26
2.5 Control of NO Release by Leptin 28
2.6 Results and Discussion 30

3 Endocrine Regulation of Leptin Production **39**
Robert V. Considine

3.1 Introduction 39
3.2 Serum Leptin is Proportional to Body Fat Mass 39
3.3 Gender Dimorphism in Serum Leptin 40
3.3.1 Role of Fat Distribution 40
3.3.2 Role of Gonadal Steroids 41
3.4 Effect of Energy Intake on Serum Leptin 41
3.4.1 Role of Insulin and Glucose 41
3.4.2 Hexosamines Link Insulin/Glucose to Leptin Synthesis 42
3.5 Inhibition of Leptin Release by the Sympathetic Nervous System ... 42
3.6 Other Hormones and Factors that Influence Leptin Release 43
3.6.1 Glucocorticoids 43
3.6.2 Cytokines 43
3.6.3 Activation of PPARγ 44
3.7 Transcriptional Regulation of the *LEP* Gene 44
3.7.1 Hexosamines 44
3.7.2 Glucocorticoids 44
3.7.3 PPARγ Agonists 45
3.7.4 β-Adrenergics and cAMP 45
3.7.5 Non-Transcriptional Regulation of Leptin Release 45
3.8 Leptin Synthesis in Tissues Other than Fat 46
3.9 Summary and Conclusions 46

4 Regulation and Roles of Leptin during the Menstrual Cycle and in Menopause **53**
Ioannis E. Messinis and Ekaterini Domali

4.1 Secretion of GnRH and Gonadotropins 54
4.2 Ovary 56
4.2.1 Regulation of Leptin Production by Ovarian Steroids 56
4.2.2 Effects of Leptin on Ovarian Function 61

4.2.3 Leptin Levels during the Normal Menstrual Cycle 62
4.2.4 Leptin Values in Superovulated Cycles 65
4.3 Endometrium 66
4.4 Menopause 67
4.5 Conclusions 69

**5 Impact of Leptin on Ovarian Folliculogenesis and
Assisted Reproduction** **77**
John D. Brannian and Keith A. Hansen

5.1 Introduction 77
5.2 Leptin and Leptin Receptors in the Ovary 78
    5.2.1 Leptin Receptors 78
    5.2.2 Leptin in Follicular Fluid 78
    5.2.3 Evidence for Ovarian Secretion of Leptin 78
5.3 Leptin and Gonadotropin Stimulation of the Ovary 79
    5.3.1 Leptin Dynamics during Exogenous Gonadotropin
          Stimulation 79
    5.3.2 Impact of Leptin on Ovarian Responsiveness 80
5.4 Leptin and ART Outcomes 81
    5.4.1 Obesity and Fertility 81
    5.4.2 Leptin and IVF Outcomes 81
5.5 Leptin, Folliculogenesis, and Oocyte Maturation 83
    5.5.1 Leptin and Steroidogenesis 83
    5.5.2 Leptin and Follicle Growth 83
    5.5.3 Leptin and Ovulation 84
    5.5.4 Leptin and Perifollicular Blood Flow and
          Follicular Metabolism 84
    5.5.5 Leptin Effects on Oocytes and Preimplantation Embryos 85
    5.5.6 Leptin Signaling Interactions 86
5.6 Leptin Resistance 87
    5.6.1 Central and Peripheral Leptin Resistance 87
    5.6.2 Evidence of Leptin Resistance in the Ovary 88
5.7 Conclusions 88

6 The Effect of Leptin on Ovarian Steroidogenesis **97**
Leon J. Spicer

6.1 Introduction 97
6.2 Leptin Levels During Female Reproductive Cycles 98
6.3 Role of Leptin in Ovarian Function 98
    6.3.1 Leptin and GC Steroidogenesis 99
    6.3.2 TC Steroidogenesis 103
    6.3.3 Luteal Cell Steroidogenesis 104
6.4 Summary and Conclusions 104

7 Regulation of Leptin and Leptin Receptor in the Human Uterus: Possible Roles in Implantation and Uterine Pathology **111**
V. Daniel Castracane and Michael C. Henson

7.1 Leptin and Leptin Receptor in the Endometrium 112
7.2 Leptin and Implantation 112
7.3 Leptin and Endometrial Pathology 113

8 Leptin and Reproduction in the Male **117**
Joaquin Lado-Abeal and Reid L. Norman

8.1 Introduction 117
8.2 Fatness and Fertility 118
    8.2.1 Fat and Fertile—Native Leptin 118
    8.2.2 Fat and Infertile—Inactive Leptin 118
8.3 Central Actions of Leptin 119
8.4 Leptin and Fertility 119
    8.4.1 Human Leptin Mutations 119
    8.4.2 Lessons from Lipodystrophies (Life without Fat) 120
    8.4.3 Non-Primate Studies 120
    8.4.4 Non-Human Primate Studies 120
        8.4.4.1 Leptin and Puberty 120
        8.4.4.2 Leptin and Adult Reproductive Function 120
8.5 Energy Availability, Leptin, and Fertility 123
    8.5.1 Energy Availability and the HPA 123
    8.5.2 Energy Availability and Gonadotropin Secretion 123
8.6 Leptin and Gonadal Steroids 125
    8.6.1 Does Leptin Affect Testosterone Secretion? 125
    8.6.2 Does Testosterone Affect Leptin Levels? 125
8.7 Summary 126

III PUBERTY

9 Leptin and Pubertal Development in Higher Primates **133**
David R. Mann and Tony M. Plant

9.1 Introduction 133
9.2 Development of the Hypothalamic–Pituitary–Gonadal Axis 134
9.3 Leptin Receptors and Circulating Forms of Leptin 135
9.4 Potential Sites of Action of Leptin Within the Hypothalamic–Pituitary–Gonadal Axis 137
9.5 Patterns of Circulating Leptin During Pubertal Development 138
    9.5.1 Nonhuman Primate 138
    9.5.2 Human 141
9.6 Experimental Evidence (Including those of Nature) that Leptin is Involved in Pubertal Development in Higher Primates 144

9.6.1 Nonhuman Primate 144
9.6.2 Human 144
9.7 Conclusions 146

10 Leptin and Pubertal Development in Humans **151**
Susann Blüher and Christos S. Mantzoros

10.1 Role of Leptin in the Regulation of the
Hypothalamic–Pituitary–Gonadal Axis 151
10.1.1 Production of Leptin and Expression of the
Leptin Receptor 151
10.1.2 Leptin is Secreted in a Pulsatile Fashion, and
Circulating Leptin Levels are Associated with
those of Other Hormones 152
10.1.3 Central Role of Leptin in Regulating Reproductive
Function—Leptin and the Hypothalamic–Pituitary Axis .. 152
10.1.4 Peripheral Role of Leptin in Regulating Reproductive
Function: Leptin and the Gonads 153
10.2 Normal Pubertal Maturation 154
10.2.1 Physiological Aspects 154
10.2.2 Neuroendocrine Aspects 155
10.3 Leptin and Normal Pubertal Development 155
10.3.1 Leptin in Childhood 155
10.3.1.1 Circulating Leptin Levels in Neonates 155
10.3.1.2 Circulating Leptin Levels during Childhood .. 155
10.3.2 Leptin in Puberty 156
10.3.2.1 Circulating Leptin Levels and the
Onset of Puberty 156
10.3.2.2 Leptin in Male Puberty 156
10.3.2.3 Leptin in Female Puberty 157
10.3.3 Leptin's Role in Activating the HPG Axis
during Human Puberty 157
10.3.4 Leptin and Sexual Maturation in Nonhuman Primates ... 158
10.4 Leptin and Aberrant Pubertal Development 159
10.4.1 Obesity and Puberty 159
10.4.2 Eating Disorders and Puberty 159
10.4.3 Strenuous Exercise and Puberty 160
10.4.4 Precocious and Delayed Puberty 160
10.5 Conclusions and Future Directions 161

**11 Integration of Leptin with Other Signals Regulating the
Timing of Puberty: Lessons Learned from the Sheep Model** **169**
Douglas L. Foster and Leslie M. Jackson

11.1 Overview 169
11.2 The Many Facets of Puberty 170

11.2.1 Event versus Process 170
11.2.2 Level of Inquiry 171
11.3 Growth Cues 172
11.3.1 Energetics of Puberty 172
11.3.2 Criteria for a Blood-Borne Metabolic Signal Timing Puberty 172
11.3.3 Studies in the Sheep Model 172
11.3.3.1 Exogenous Leptin Prevents Fasting-Induced Hypogonadotropism 172
11.3.3.2 Exogenous Leptin Stimulates LH during Fasting-Induced Hypogonadotropism 173
11.3.3.3 Exogenous Leptin Stimulates LH in the Hypogonadotropic Sexually Immature Male .. 174
11.3.3.4 Exogenous Leptin Stimulates LH in the Hypogonadotropic Growth-Retarded Male Lamb 176
11.3.3.5 Central versus Peripheral Action of Leptin in the Sheep 177
11.4 Integration of External and Internal Information to Time Puberty 179
11.4.1 Multiple Cues for Puberty 179
11.4.2 Growth and Seasonal Cues Timing Initiation of Ovulations in Female Sheep 180
11.4.3 Prenatal Programming of the Type of Cues Used for Puberty in the Sheep 181
11.5 Consideration of Species Differences in the Role of Leptin in Timing Puberty 183
11.6 Conclusions 184

IV PREGNANCY

12 Leptin and Fetal Growth and Development **189**
Helen Christou, Shanti Serdy, and Christos S. Mantzoros

12.1 Introduction 189
12.2 Leptin is Produced by the Placenta and Leptin Receptor is Expressed by Fetal Tissues 190
12.3 Role of Leptin in Normal Reproductive Function 190
12.4 Role of Leptin in Normal Pregnancy and in Pathologic Conditions during Pregnancy 191
12.5 Role of Leptin in Early Embryonic Development 192
12.6 Role of Leptin in Fetal Growth and Development 192
12.7 Leptin and Angiogenesis 194
12.8 Role of Leptin in Hematopoiesis and Immune Function 195
12.9 Role of Leptin in Ossification and Bone Development 195

12.10 Leptin and Brain Development 196
12.11 Conclusions and Future Directions 196

13 Leptin in the Placenta **201**
Dorina Islami and Paul Bischof

13.1 Abstract 201
13.2 Introduction 202
13.3 Placental Leptin Secretion 202
13.4 Leptin in Rodent Pregnancy 204
13.5 Leptin Secretion and Regulation during Human Pregnancy 204
   13.5.1 Leptin in the Mother 204
   13.5.2 Leptin in the Fetus and Neonate 207
13.6 Role of Leptin in Pregnancy 210
13.7 Leptin and Trophoblast Invasion 212

14 Leptin in Rodent Pregnancy **221**
Brendan J. Waddell and Jeremy T. Smith

14.1 Introduction 221
14.2 Pregnancy in the Rat and Mouse 221
14.3 Plasma Leptin Levels Throughout Gestation 222
14.4 Leptin Receptor Expression in Pregnancy 223
   14.4.1 Leptin Receptor Isoforms and their Functional Properties . 223
   14.4.2 Hypothalamus 224
   14.4.3 Uterus and Ovary 225
   14.4.4 Placenta 226
14.5 Leptin Transport and Metabolism in Pregnancy 228
   14.5.1 Effects of Placental Ob-Re on Maternal Leptin 228
   14.5.2 Placental Ob-Ra and Transplacental Passage of
          Maternal Leptin to the Fetus 228
14.6 Biological Actions of Leptin in Pregnancy and Lactation 230
   14.6.1 Implantation and Placentation 230
   14.6.2 Fetal and Placental Growth 231
   14.6.3 Maternal Metabolism 232
   14.6.4 Lactation 233
14.7 Summary and Conclusions 233

15 Leptin in Primate Pregnancy **239**
Michael C. Henson and V. Daniel Castracane

15.1 Introduction 239
15.2 Leptin Dynamics in Mammalian Pregnancy 240
15.3 Leptin Regulation by Steroid Hormones 243
15.4 Roles of Leptin in Pregnancy 250

15.5 Summary of Hypotheses Concerning the Regulation of
Leptin and its Roles in Primate Pregnancy 253

V GENETICS

16 Genetics and Physiology Link Leptin to the Reproductive System .. **267**
Farid F. Chehab, Amanda Ewart-Toland, Khalid Mounzih, Jun Qiu, and Scott Ogus

16.1 Introduction 267
16.2 Adipose Tissue and Reproduction 268
16.3 Role of Fat in Reproduction and in Signaling Puberty 268
16.4 Sterility of *ob/ob* and *db/db* Mice 269
16.5 Leptin and its Receptor 270
16.6 Correction of the Sterility of *ob/ob* Mice 271
    16.6.1 *ob/ob* Females 271
    16.6.2 *ob/ob* Males 273
16.7 Correction of the Sterility in *db/db* Mice 275
16.8 Impact of Modifier Genes on the *ob/ob* Phenotype 275
16.9 Leptin and Puberty 279
16.10 Leptin in Pregnancy 281
16.11 Concluding Remarks 283

17 Leptin and the Onset of Puberty: Insights from Rodent and Human Genetics **287**
I. Sadaf Farooqi

17.1 Abstract 287
17.2 Introduction 287
17.3 *Ob/ob* Mice and the Discovery of Leptin 288
    17.3.1 Responses to Leptin Administration 288
17.4 Transgenic Models 289
17.5 Leptin Action in Humans 290
    17.5.1 Heterozygotes for Leptin Mutation 290
    17.5.2 Response to Leptin Administration 291
17.6 Downstream Mediators of Leptin Action 292
    17.6.1 Melanocortin Pathways 293
    17.6.2 Neuropeptide Y 293

VI CLINICAL CHALLENGES

18 Does Leptin Play a Role in Preeclampsia? **299**
Lucilla Poston

18.1 The Syndrome of Preeclampsia 299

18.2 Leptin in Normal Pregnancy 300
18.3 Why Investigate Leptin in Preeclampsia? 300
18.4 Leptin in Maternal Blood in Preeclampsia 300
18.5 The Possible Origins of Raised Maternal Blood Leptin
     Concentrations in Preeclampsia 303
     18.5.1 The Placenta as a Source of Leptin 303
     18.5.2 The Role of Placental Hypoxia 304
     18.5.3 Interactions with Inflammatory Mediators 304
18.6 Potential Benefits of Raised Leptin in Preeclamptic Pregnancies . 305
18.7 Potential Disadvantages of Raised Leptin in
     Preeclamptic Pregnancies 305
18.8 Leptin Concentrations in Umbilical Cord Blood in Preeclampsia . 306
18.9 Conclusions 307

19 Leptin and Hypothalamic Amenorrhea 311
   Michelle P. Warren and Jennifer E. Dominguez

19.1 Definitions 311
19.2 The Hypothalamus and Amenorrhea 312
     19.2.1 The Reproductive Axis 312
            19.2.1.1 GnRH Pulsatility 312
19.3 Phases of Dysfunction in HA 312
     19.3.1 Exercise-Induced 313
     19.3.2 Eating Disorder-Induced 314
     19.3.3 Idiopathic/Functional HA 314
19.4 Etiology of HA 314
     19.4.1 Athletic 315
     19.4.2 Eating Disorder-Induced 316
     19.4.3 Functional HA 317
19.5 The Mechanism of GnRH Inhibition 318
     19.5.1 Leptin 319
19.6 HA and Osteoporosis 319
     19.6.1 Mechanisms of Bone Loss in HA 321
            19.6.1.1 Leptin and Bone Accrual 323
19.7 Treatment of HA 324

20 The Role of Leptin in Polycystic Ovary Syndrome 333
   Denis A. Magoffin, Priya S. Duggal, and Robert J. Norman

20.1 The Polycystic Ovary Syndrome 333
20.2 Insulin and Obesity in PCOS 334
20.3 Leptin and Obesity in PCOS 335
20.4 PCOS and Leptin in the Hypothalamus 336
20.5 The Role of Leptin in Ovarian Hyperandrogenism 337
     20.5.1 PCOS and Ovarian Leptin 337
     20.5.2 Leptin Receptors in PCOS 337

20.5.3 Direct Effects of Leptin on
Human Ovarian Steroidogenesis 337
20.6 Leptin and Insulin Resistance in PCOS 338
20.7 Leptin and Ovulation 339
20.8 Leptin and Assisted Reproduction 340
20.8.1 PCOS, Leptin, and Ovulation Induction 340
20.8.2 PCOS, Leptin, and IVF 340
20.9 Summary 341

Index 347

1

Introduction

1

Leptin: From Satiety Signal to Reproductive Regulator

Michael C. Henson, V. Daniel Castracane, and
Deborah E. Edwards

1.1. LEPTIN'S PAST

The hypothalamus was identified many years ago as an important site in the regulation of body weight (Hetherington and Ranson, 1940). Although the metabolic messenger(s) responsible for transmitting information regarding adipose energy stores to this site was unknown, numerous investigations suggested that a circulating product of adipose tissue was the agent involved. This unidentified entity was made more tangible in the late 1950s by classic parabiosis experiments, which surgically linked rodents bearing induced lesions of the ventromedial nucleus (Hervey, 1959) to intact controls. In these rats that featured conjoined circulatory systems, the lesioned animals rapidly became obese as a result of hyperphagia, while intact control animals lost body weight due to decreased food intake. Although these studies directly supported the prevailing theories of lipostasis, the identity of the putative "satiety factor" remained a mystery until quite recently.

Leptin, the 16 kD product of the *obese/Lep* gene was originally described by Zhang and co-workers in 1994. The obesity of *ob/ob* mice was attributed to an inability to synthesize this hormone, which implied a potential use for leptin in the treatment of human obesity and initiated worldwide interest in the polypeptide (Friedman and Halaas, 1998).

Michael C. Henson • Departments of Obstetrics and Gynecology, Physiology, and Structural and Cellular Biology, and the Tulane National Primate Research Center and Interdisciplinary Program in Molecular and Cellular Biology, Tulane University Health Sciences Center.

V. Daniel Castracane • Department of Obstetrics and Gynecology and the Women's Health Research Institute of Amarillo, Texas Tech University Health Sciences Center—Amarillo; Diagnostic Systems Laboratories, Inc., Webster, Texas.

Deborah E. Edwards • Interdisciplinary Program in Molecular and Cellular Biology, Tulane University Health Science Center.

Leptin and Reproduction. Edited by Henson and Castracane, Kluwer Academic/Plenum Publishers, 2003.

3

The name of the hormone is taken from the Greek "leptos," meaning "thin" or "lean," and is descriptive of the body weight-reducing quality of the hormone. The gene is highly conserved between species, and in the human, it is located at chromosomal position 7q31.3, spans 20 kb and contains 3 exons, which are separated by 2 introns (Green et al., 1995; Isse et al., 1995). The gene encodes a 167 amino acid protein including a 21 amino acid secretory sequence. An intrachain disulfide bond maintains stability of the native state folded protein, and is essential for bioactivity (Grasso et al., 1997). The mouse protein exhibits 84 percent homology with human leptin (Zhang et al., 1994), and both share many structural similarities to other helical cytokines, including interleukin-6 (IL-6) and growth hormone (Madej et al., 1995; Zhang et al., 1997). Although adipose tissue is a primary source of leptin, its production has more recently been attributed to a variety of other tissues, which include the stomach (Sobhani et al., 2000), fetal cartilage (Hoggard et al., 1998), pituitary (Jin et al., 1999), mammary tissue (Smith-Kirwin et al., 1998), and placenta (Masuzaki et al., 1997). Leptin is found in the circulation in a free form or complexed with leptin-binding proteins. The half-life of free leptin is about 30 min (Trayhurn et al., 1999), with the kidneys being responsible for approximately 80 percent of leptin clearance from the peripheral circulation (Meyer et al., 1997).

Use of an expression cloning strategy assisted in the identification of a receptor for leptin, which was later localized to human chromosome 1q31 (Tartaglia et al., 1995). Analysis of this receptor revealed a single-spanning transmembrane protein belonging to the class I cytokine receptor superfamily. Discovery of multiple forms of the receptor has led to the recognition of at least five leptin receptor isoforms, including four membrane bound receptor isoforms and one circulating, soluble isoform. The membrane bound isoforms consist of three distinct domains: (1) an extracellular, leptin binding domain, (2) a transmembrane domain, (3) and an intracellular, cytoplasmic tail. The leptin receptor is present in many tissues. In fact, upon analysis, there are very few tissues that do not express mRNA for at least one isoform of the leptin receptor. This suggests that many tissues and organs are primed for leptin action, and that the polypeptide may exert physiological influences via paracrine mechanisms.

The intracellular, cytoplasmic domain of the leptin receptor varies in length with each isoform. Membrane bound isoforms originate from a single mRNA transcript and are produced as a result of alternative splicing. The 29th amino acid of the intracellular domain provides the point of divergence, as isoforms remain identical to this point, but manifest the results of alternative splicing after it (Chen et al., 1996; Lee et al., 1996). Three isoforms terminate between amino acids 32 and 40, giving rise to a category of short leptin receptor isoforms ($LepR_S$). Roles for these isoforms remain undefined, but it is speculated that one might facilitate the transport of leptin across the blood–brain barrier (Kastin et al., 1999), or activate the Mitogen Activating Protein (MAP) kinase pathway (Bjorbaek et al., 1997). The most notable isoform with respect to signaling ability also has the longest intracellular domain ($LepR_L$). It contains a cytoplasmic tail of 303 amino acids and is the only isoform to include all the domains necessary for signaling through the Janus-Activated Kinase/Signal Tranducers and Activators of Transcription (JAK/Stat) pathway (Tartaglia, 1997). In most tissues, $LepR_S$ is expressed at much greater levels than $LepR_L$. However, the hypothalamus constitutes a notable exception to this rule (Ghilardi et al., 1996). This high expression of $LepR_L$ in a neuronal center recognized in the regulation of appetite and satiety is in accord with its role as the main signaling isoform of the receptor. In this capacity, evidence suggests (as reviewed by Ahima and Flier, 2000; Wauters et al., 2000) that in addition to observations linking leptin (via hypothalamic $LepR_L$) with the inhibition of neuropeptide Y (NPY), the polypeptide also acts to up-regulate corticotrophin-releasing hormone (CRH) in the paraventricular nucleus. In addition, the glucagon-like peptide-1 (GLP-1), proopiomelanocortin (POMC), and cocaine and amphetamine regulated transcript (CART) systems are involved in mechanisms linking leptin with energy balance via satiety. Other neuroendocrine messengers potentially involved in leptin mechanisms include melanin concentrating hormone (MCH), cholecystokinin (CCK), orexins, bombesin, serotonin and urocortin. Interactions with glucocorticoids and insulin probably modulate the effects of leptin in brain regions dedicated to appetite.

Although originally hailed as the "anti-obesity" hormone, leptin's ability to affect adiposity is counteracted in humans by a natural resistance that is characterized by hyperleptinemia. Mechanisms thought to mediate this relationship include rare receptor defects, equally rare deficiencies in leptin production or secretion, a quantitative imbalance between leptin and its binding protein in the peripheral circulation, and/or the relative inability of leptin to breach the blood–brain barrier. In this capacity, suppressors of cytokine signaling-3 (SOCS-3) protein is an excellent candidate as a molecular mediator of leptin resistance in the brain (Bjorbaek et al., 1998), as it effectively blocks leptin signaling. Similarly, the soluble leptin receptor, a potential product of proteolytic cleavage of membrane-bound isoforms in the human (Maamra et al., 2001), can bind leptin in the circulation, thereby rendering it biologically inactive, augmenting its half life and perhaps contributing to leptin resistance.

1.2. LEPTIN'S PRESENT

In addition to its role as the "adipostatic" hormone, leptin has been linked as an endocrine regulator to processes as widely divergent as hematopoiesis, angiogenesis, bone formation, the immune response, diabetes, and general fertility; an array of associations that have combined with the leptin-obesity interaction to fuel almost 6,000 entries in the peer-reviewed literature by the end of 2002. None of these leptin-associated fields has engendered greater interest than those oriented toward reproduction (Castracane and Henson, 2002). In this respect, a steady stream of informative, well-conceived reviews have reported on the reproductive implications of leptin, covering the myriad of probable interactions with gonadotropins, steroids, the ovary, the testis, puberty, the menstrual cycle, pregnancy, and the menopause (Hoggard et al., 1998; Clarke and Henry, 1999; Gonzalez et al., 2000; Baldelli et al., 2002; Brann et al., 2002; Moschos et al., 2002). In the present volume, many of the most pressing questions linking leptin with both normal and aberrant reproductive function receive a focused, in-depth treatment for the first time. This is obtained through the collective contributions of experts representing the range of subdisciplines important to reproductive biology. Both basic and clinical researchers address leptin and reproduction with respect to endocrine and neuroendocrine regulatory mechanisms, as well as the direct interactions with gonadal function, gestation, and related pathologies.

1.2.1. General Reproductive Function

In the current volume, readers are presented in Chapter 2 with a detailed explanation by McCann et al. of pathways regulating leptin action in the brain and exciting new evidence of the polypeptide's role in regulating nitric oxide production, via nitric oxide synthase, in adipocytes (Mastronardi et al., 2002). In Chapter 3, Considine expands on the theme of central neuroendocrine regulation to include an examination of peripheral mechanisms and energy availability, as well as the unique effect of gender on leptin synthesis. New findings that demonstrate roles for glucose and hexosamines as transcriptional regulators (Zhang et al., 2002) are summarized. In light of the unique mechanisms evidenced by leptin-producing tissues and linked to gender, regulation of leptin biosynthesis during the menstrual cycle and following the onset of the menopause are worthy of special consideration. Therefore, important pathways are discussed by Messinis and Domali (Chapter 4) and evidence is presented of progesterone working in tandem with estrogen to enhance leptin production (Messinis et al., 2001), an association that may explain the increase in serum leptin concentrations commonly observed in the luteal phase of the menstrual cycle.

Endocrine interactions in the ovary are further examined by Brannian and Hansen, in Chapter 5, who develops a hypothetical model of how leptin and leptin resistance might impact follicular development and oocyte maturation. He proposes that elevations in leptin concentrations, related to obesity, directly impair oocyte quality and embryonic development (Brannian et al., 2001). Therefore, women availing themselves of assisted reproductive technologies might improve their chances of becoming pregnant by first lowering their serum leptin levels through weight loss. Specific effects of leptin on the ovary are pursued further by Spicer (Chapter 6), who describes the polypeptide's direct inhibitory effect on steroidogenesis in granulosa and thecal cells, an effect that results in an overall decrease in estradiol secretion (Spicer, 2001). It is clear, however, that further studies will be required to fully elucidate the actions of leptin in regulating ovarian steroidogenesis, as roles in these cell types may be species-dependent. In Chapter 7, we examine tissue-specific interactions, focusing on leptin and the uterine endometrium. Recent investigations in this area link the polypeptide with mechanisms important to implantation, and to pathologies such as endometriosis. Interactions of leptin and steroid hormones have also been proposed in the male, with respect to a negative effect of testosterone. Lado-Abeal and Norman propose, however, that the primary effects of leptin on reproduction in the male appear to be at the hypothalamic level (Lado-Abeal et al., 2002). They conclude, in Chapter 8, that the maintenance of available energy is more important than a direct effect of leptin in regulating male reproductive function.

1.2.2. Puberty

Be it causal or permissive in nature, the role of leptin in regulating puberty has been a subject of controversy since it was first alluded to by Jeffrey Flier during a symposium (Flier, 1996). The vast body of work that followed is throughly reviewed in the current volume. These concepts are first examined by Mann and Plant in Chapter 9, with special focus on the nonhuman primate. Due, in large measure, to results of their own investigations (Mann et al., 2000; Plant, 2001), they conclude that leptin does not appear to trigger the initiation of puberty, but rather that once leptin reaches a preset threshold, puberty may proceed if other critical control mechanisms are in place and functional. Blüher and Mantzoros further explore the effects of leptin on human pubertal development in Chapter 10 and agree that leptin communicates vital information concerning the body's energy stores to the brain at this critical period of development. However they also implicate leptin as a trigger of puberty onset in females, because female leptin levels, unlike levels in the male, continue to rise in late puberty. They conclude that although substantial evidence suggests that leptin can act to regulate gonadotropin levels (Chan and Mantzoros, 2001), currently ongoing interventional studies will need to be completed before conclusive evidence of leptin's effect on puberty in humans is known. In Chapter 11, Foster and Jackson review the preponderance of studies dedicated to the role of leptin as a permissive regulator of GnRH/LH (Smith et al., 2002). Their discussion of past findings is augmented by an explanation of their ongoing work in the sheep, which brings new focus on leptin's integration with other endocrine signals that are important for the initiation of reproductive activity and provide new information on yet another animal model.

1.2.3. Pregnancy

In response to the seminal observation by Masuzaki et al. (1997) of leptin production in the human placenta and the more recent report of Lepercq et al. (2001) that human fetal adipose tissues also produce the polypeptide, many investigations have attempted to elucidate the roles and regulatory mechanisms involved during gestation (as reviewed, Henson and Castracane 2000, 2002; Bajoria et al., 2002; Sagawa et al., 2002). In Chapter 12, Christou and colleagues from Mantzoros' group review the extensive literature dedicated to leptin as a modulator of conceptus growth and development. They concede that although leptin may, at least in part, act on fetal growth via insulin and GH/IGF-I, cumulative evidence indicates a direct effect that is independent of those axes (Christou et al., 2001). They further conclude that the way in which leptin's influence on growth is affected by conditions such as diabetes, intrauterine growth restriction, and pre-eclampsia must be a priority for future research.

In addition to its effects on growth, in Chapter 13 Islami and Bischof propose a novel role for the polypeptide as an autocrine/paracrine regulator of the invasion phase of human implantation (Castellucci et al., 2000). In this scenario, leptin originating in the placental trophoblast induces metalloproteinases and regulates integrin expression, which in turn induces an invasive phenotype in cytotrophoblasts. This effect of leptin, as a proponent of pregnancy maintenance, then continues with the up-regulation of human chorionic gonadotropin (hCG), another trophoblastic product. Similarly, as reported by Waddell and Smith (Chapter 14), leptin is required for the establishment of rodent pregnancy with likely effects on decidualization, implantation, and placental formation occurring through signaling-competent isoforms of the leptin receptor. Significantly, pregnancy in the rodent exemplifies a state of relative leptin resistance in the maternal compartment due largely to increased leptin binding activity, which is potentiated by a soluble form of the receptor (Seeber et al., 2002). In Chapter 15, we continue this theme by reporting that in primate pregnancy, maternal and placental leptin production may be ultimately dependent on fetal and placental steroids to regulate synthesis in a tissue-specific manner (O'Neil et al., 2001). In addition, leptin produced via these steroid-associated mechanisms in the placenta may be directed, in some measure, to both the maternal and fetal circulations and may feed back to regulate its own production, as well as the synthesis or release of other trophoblastic hormones.

1.2.4. Genetics

Observations in the *ob/ob* mouse provided some of the first genetically based evidence linking leptin with reproduction (Chehab et al., 1996). In this ground-breaking work, the sterility defect was corrected in homozygous females by the administration of human recombinant leptin. In the present volume, Chehab and colleagues (Chapter 16) expand on their early investigations to suggest that anomalies precipitated by leptin deficiency may be under the discrete control of multiple modifier genes (Qiu et al., 2001). In Chapter 17, Farooqi reviews the current thinking about rodent genetics and the onset of puberty, and expands on work with this model to include his laboratory's studies of genetically leptin-deficient humans (Farooqi et al., 2001). In these cases, administration of recombinant human leptin resulted in the onset of normal reproductive function at an appropriate developmental age, suggesting that the polypeptide serves as a "metabolic gate" for the initiation of puberty in humans. The author concludes that leptin's actions are mediated both through the central hypothalamic pathway and by direct action on peripheral organs.

1.2.5. Clinical Challenges

Recent investigations have begun to address the gap between basic research and the development of therapeutic strategies designed to alleviate leptin-associated reproductive anomalies. Much of this research has focused on pre-eclampsia, hypothalamic amenorrhea, and polycystic ovary syndrome. Pre-eclampsia is the focus of Chapter 18, in which Poston discusses the elevated maternal leptin levels common to the condition, which is a major cause of premature birth, and the attendant problems of immaturity. Evidence may suggest that because the condition is associated with oxidative stress (Poston and Chappell, 2001), and peripheral leptin levels are often reported to be enhanced under hypoxic conditions, elevated leptin synthesis may constitute a placental attempt to improve nutrient availability in the face of such adversity. Additionally, increased leptin levels could signify an attempt to: (1) increase placental growth, in light of the polypeptide's perceived role as an angiogenic factor; (2) blunt responses to apoptotic stimuli; and/or (3) counteract endothelial dysfunction by stimulating nitric oxide synthesis.

Hypothalamic amenorrhea is described by Warren and Dominguez, in Chapter 19, as a lack of regular menstrual periods resulting from a deficiency in the secretion of hypothalamic GnRH. Although a number of questions concerning regulatory mechanisms remain to be answered (Warren and Fried, 2001), the causal role of energy deficit in hypothalamic amenorrhea and the observation that the condition is reversible with weight gain, provide compelling evidence for leptin involvement. Finally, in Chapter 20, Magoffin and colleagues characterize polycystic ovary disease as the most common reproductive endocrine disease in women of reproductive age. Approximately three quarters of all women exhibiting anovulatory infertility are afflicted, which accounts for about one third of those women with secondary amenorrhea and about nine tenths of those women with oligomenorrhea. Obesity and elevated leptin concentrations are associated with the incidence of polycystic ovary disease, but current data do not support a direct causal role for leptin. In vitro studies do, however, support the potential for leptin to significantly affect ovarian steroidogenesis (Agarwal et al., 1999), but the clinical relevance of this in relation to polycystic ovary disease has yet to be established.

1.3. LEPTIN'S FUTURE?

In light of the contributions of so many researchers to this volume, leptin's place in the list of hormones important to reproduction seems secure. Intriguingly, organs influenced by the polypeptide represent the breadth of reproductive biology and include the hypothalamus, pituitary, gonads and the placental trophoblast. Potential roles for leptin range from being a detached modulator of the energy reserves needed for reproduction to take place to a direct promoter of puberty onset; from an inhibitor of ovarian steroidogenesis to a regulator of fetal development, placental invasion and trophoblast endocrinology. It seems likely that such varied roles must have direct clinical importance, as they are intimately related to the success of various assisted reproductive technologies, as well as to an array of common pathologies that are associated with delayed reproductive maturity, general ovarian dysfunction, intrauterine growth restriction, and placental insufficiency.

Studies dedicated to the further elucidation of leptin's roles and regulation, which in many cases appear specific with regard to the tissue, physiological state, or species affected, will doubtless continue. Such extensive research effort will probably uncover new and even more exciting interactions important to reproductive biology. In this regard, envisioning effective clinical intervention therapies, based on what is learned, will likely be the scientific community's greatest challenge in the years to come.

REFERENCES

Agarwal, S. K., Vogel, K., Weitsman, S. R., and Magoffin, D. A. (1999). Leptin antagonizes the insulin-like growth factor-I augmentation of steroidogenesis in granulosa and theca cells of the human ovary. *J. Clin. Endocrinol. Metab., 84*, 1072–1076.

Ahima, R. S. and Flier, J. S. (2000). Leptin. *Ann. Rev. Physiol., 62*, 413–437.

Bajoria, R., Sooranna, S. R., Ward, B. S., and Chatterjee, R. (2002). Prospective function of placental leptin at maternal-fetal interface. *Placenta, 13*, 103–115.

Baldelli, R., Dieguez, C., and Casanueva, F. F. (2002). The role of leptin in reproduction: experimental and clinical aspects. *Ann. Med., 34*, 5–18.

Bjorbaek, C., Elmquist, J. K., Franz, J. D., Shoelson, S. E., and Flier, J. S. (1998). Identification of SOCS-3 as a potential mediator of central leptin resistance. *Mol. Cell, 1*, 619–625.

Bjorbaek, C., Uotani, S., da Silva, B., and Flier, J. S. (1997). Divergent signaling capacities of the long and short isoforms of the leptin receptor. *J. Biol. Chem., 272*, 32686–32695.

Brann, D. W., Wade, M. F., Dhandapani, K., Mahesh, V. B., and Buchanan, C. D. (2002). Leptin and reproduction. *Steroids, 67*, 95–104.

Brannian, J. D., Schmidt, S. M., Kreger, D. O., and Hansen, K. A. (2001). Baseline non-fasting serum leptin concentration to body mass index ratio is predictive of IVF outcomes. *Hum. Reprod., 16*, 1819–1826.

Castellucci, M., De Matteis, R., Meisser, A., Cancello, R., Monsurro, V., Islami, D., Sarzani, R., Marzioni, D., Cinti, S., and Bischof, P. (2000). Leptin modulates extracellular matrix molecules and metalloproteinases: possible implications for trophoblast invasion. *Mol. Hum. Reprod., 6*, 951–958.

Castracane, V. D. and Henson, M. C. (2002). When did leptin become a reproductive hormone? *Sem. Reprod. Med., 20*, 89–92.

Chan, J. L. and Mantzoros, C. S. (2001). Leptin and the hypothalamic-pituitary regulation of the gonadotropin–gonadal axis. *Pituitary, 4*, 89–92.

Chehab, F. F., Lim, M. E., and Lu, E. (1996). Correction of the sterility defect in homozygous obese female mice by treatment with the human recombinant leptin. *Nat. Genet., 12*, 318–320.

Chen, H., Charlat, O., Tartaglia, L. A., Woolf, E. A., Weng, X., Ellis, S. J., Lakey, N. D., Culpepper, J., Moore, K. J., Breitbart, R. E., Duyk, G. M., Tepper, R. I., and Morgenstern, J. P. (1996). Evidence that the diabetes gene encodes the leptin receptor: identification of a mutation in the leptin receptor gene in *db/db* mice. *Cell, 84*, 491–495.

Christou, H., Connors, J. M., Ziotopoulou, M., Hatzidakis, V., Papathanassoglou, E., Ringer, S. A., and Mantzoros, C. S. (2001). Cord blood leptin and insulin-like growth factor levels are independent predictors of fetal growth. *J. Clin. Endocrinol. Metab., 86*, 935–938.

Clarke, I. J. and Henry, B. A. (1999). Leptin and reproduction. *Rev. Reprod., 4*, 48–55.

Farooqi, I. S., Keogh, J. M., Kamath, S., Jones, S., Gibson, W. T., Trussell, R., Jebb, S. A., Lip, G. Y., and O'Rahilly, S. (2001). Partial leptin deficiency and human adiposity. *Nature, 414*, 34–35.

Flier, J. S. (1996). Symposium. Neurobiology of *OB* protein (Leptin): a peripheral signal acting on control neural networks to regulate body energy balance. Abstract 469.2, p. 1193. Society for Neuroscience, 26th Annual Meeting. Washington, DC.

Friedman, J. M. and Halaas, J. L. (1998). Leptin and the regulation of body weight in mammals. *Nature, 395*, 763–770.

Ghilardi, N., Zeigler, S., Wiestner, A., Stoffel, R., Heim, M. H., and Skoda, R. C. (1996). Defective STAT signaling by the leptin receptor in diabetic mice. *Proc. Natl. Acad. Sci. USA, 93*, 6231–6235.

Gonzalez, R. R., Simon, C., Caballero-Campo, P., Norman, R., Chardonnens, D., Devoto, L., and Bischof, P. (2000). Leptin and reproduction. *Hum. Reprod. Update, 6*, 290–300.

Grasso, P., Leinung, M. C., Ingher, S. P., and Lee, D. W. (1997). *In vivo* effects of leptin related synthetic peptides on body weight and food intake in female *ob/ob* mice: localization of leptin activity to domains between amino acid residues 106–140. *Endocrinology, 138*, 1413–1418.

Green, E. D., Maffei, M., Braden, V. V., Proenca, R., DeSilva, U., Zhang, Y., Chua, S. C., Leibel, R., Weissanbach, J., and Friedman, J. M. (1995). The human *obese* (*OB*) gene: RNA expression pattern and mapping on the physical, cytogenetic and genetic maps of chromosome 7. *Genome Res., 5*, 5–12.

Henson, M. C. and Castracane, V. D. (2000). Leptin in pregnancy. *Biol. Reprod., 63*, 1219–1228.

Henson, M. C. and Castracane, V. D. (2002). Leptin: roles and regulation in primate pregnancy. *Sem. Reprod. Med., 20*, 113–122.

Hervey, G. R. (1959). The effects of lesions in the hypothalamus in parabiotic rats. *J. Physiol., 145*, 336–352

Hetherington, A. W. and Ranson, S. W. (1940). Hypothalamic lesions and adiposity in the rat. *Anat. Res., 78*, 149–172.

Hoggard, N., Hunter, L., Trayhurn, P., Williams, L. M., and Mercer, J. G. (1998). Leptin and reproduction. *Proc. Nutr. Soc., 57*, 421–427.

Isse, N., Ogawa, Y., Tamura, N., Masuzaki, H., Mori, K., Okazaki, T., Satoh, N., Shigemoto, M., Yoshimasa, Y., Nishi, S., Hosoda, K., Inazawa, J., and Nakao, K. (1995). Structural organization and chromosomal assignment of the human *obese* gene. *J. Biol. Chem., 270*, 27728–27733.

Jin, L., Burguera, B. G., Couce, M. E., Scheithauer, B. W., Lamsan, J., Eberhardt, N. L., Kulig, E., and Lloyd, R. V. (1999). Leptin and leptin receptor expression in normal and neoplastic human pituitary: evidence of a regulatory role for leptin on pituitary cell proliferation. *J. Clin. Endocrinol. Metab., 84*, 2903–2911.

Kastin, A. J., Weihong, P., Maness, L. M., Koletsky, R. J., and Ernsberger, P. (1999). Decreased transport of leptin across the blood–brain barrier in rats lacking the short form of the leptin receptor. *Peptides, 20*, 1449–1453.

Lado-Abeal, J., Veldhuis, J. D., and Norman, R. L. (2002). Glucose relays information regarding nutritional status to the neural circuits that control somatotropic, corticotropic, and gonadotropic axes in adult male rhesus macaques. *Endocrinology, 143*, 403–410.

Lee, G. H., Proenca, R., Montez, J. M., Carroll, K. M., Darishzadeh, J. G., Lee, J. I., and Friedman, J. M. (1996). Abnormal splicing of the leptin receptor in diabetic mice. *Nature, 379*, 632–635.

Lepercq, J., Challier, J. C., Guerre-Millo, M., Cauzac, M., Vidal, H., and Hauguel-de Mouzon, S. (2001). Prenatal leptin production: evidence that fetal adipose tissue produces leptin. *J. Clin. Endocrinol. Metab., 86*, 2409–2413.

Maamra, M., Bidlingmaier, M., Postel-Vinay, M. C., Wu, Z., Strasburger, C. J., and Ross, R. J. M. (2001). Generation of human soluble leptin receptor by proteolytic cleavage of membrane-anchored receptors. *Endocrinology, 142*, 4389–4393.

Madej, T., Boguski, M. S., and Bryant, S. H. (1995). Threading analysis suggests that the *obese* gene product may be a helical cytokine. *FEBS Lett., 373*, 13–18.

Mann, D. R., Akinbami, M. A., Gould, K. G., and Castracane, V. D. (2000). A longitudinal study of leptin during development in the male rhesus monkey: the effect of body composition and season on circulating leptin levels. *Biol. Reprod., 62*, 285–291.

Mastronardi, C. A., Yu, W. H., and McCann, S. M. (2002). Resting and circadian release of nitric oxide is controlled by leptin in male rats. *Proc. Natl. Acad. Sci. USA, 99*, 5721–5726.

Masuzaki, H., Ogawa, Y., Sagawa, N., Hosoda, K., Matsumoto, T., Mise, H., Nishimura, H., Yoshimasa, Y., Tanaka, I., Mori, T., and Nakao, K. (1997). Nonadipose tissue production of leptin: leptin as a novel placenta-derived hormone in humans. *Nat. Med., 3*, 1029–1033.

Messinis, I. E., Papageorgiou, I., Milingos, S., Kollios, G., and Seferiadis, K. (2001). Oestradiol plus progesterone treatment increases serum leptin concentrations in normal women. *Hum. Reprod., 16,* 1827–1832.

Meyer, C., Robson, D., Rackovsky, N., Nadkarni, V., and Gerich, J. (1997). Role of the kidney in human leptin metabolism. *Am. J. Physiol., 273,* E903–E907.

Morrison, C. D., Wood, R., McFadin, E. L., Whitley, N. C., and Keisler, D. H. (2002). Effect of intravenous infusion of recombinant ovine leptin on feed intake and serum concentrations of GH, LH, insulin, IGF-1, cortisol, and thyroxine in growing prepubertal ewe lambs. *Domest. Anim. Endocrinol., 22,* 103–112.

Moschos, S., Chan, J. L., and Mantzoros, C. S. (2002). Leptin and reproduction: a review. *Fertil. Steril., 77,* 433–444.

O'Neil, J. S., Green, A. E., Edwards, D. E., Swan, K. F., Gimpel, T., Castracane, V. D., and Henson, M. C. (2001). Regulation of leptin and leptin receptor in baboon pregnancy: effects of advancing gestation and fetectomy. *J. Clin. Endocrinol. Metab., 86,* 2518–2524.

Plant, T. M. (2001). Neurobiological bases underlying the control of the onset of puberty in the rhesus monkey: a representative higher primate. *Front. Neuroendocrinol., 22,* 107–139.

Poston, L. and Chappell, L. C. (2001). Is oxidative stress involved in the aetiology of pre-eclampsia? *Acta Paediatr. Suppl., 90,* 3–5.

Qiu, J., Ogus, S., Mounzih, K., Ewart-Toland, A., and Chehab, F. F. (2001). Leptin-deficient mice backcrossed to the *BALB/cJ* genetic background have reduced adiposity, enhanced fertility, normal body temperature, and severe diabetes. *Endocrinology, 142,* 3421–3425.

Sagawa, N., Yura, S., Itoh, H., Mise, H., Kakui, K., Korita, D., Takemura, M., Nuamah, M. A., Ogawa, Y., Masuzaki, H., Nakao, K., and Fujii, S. (2002). Role of leptin in pregnancy—a review. *Placenta 23, Suppl. A, Trophoblast Res., 16,* S80–S86.

Seeber, R. M., Smith, J. T., and Waddell, B. J. (2002). Plasma leptin-binding activity and hypothalamic leptin receptor expression during pregnancy and lactation in the rat. *Biol. Reprod., 66,* 1762–1767.

Smith, G. D., Jackson, L. M., and Foster, D. L. (2002). Leptin regulation of reproductive function and fertility. *Theriogenology, 57,* 73–86.

Smith-Kirwin, S. M., O'Connor, D. M., De Johnston, J., Lancey, E. D., Hassink, S. G., and Funanage, V. L. (1998). Leptin expression in human mammary epithelial cells and breast milk. *J. Clin. Endocrinol. Metab., 83,* 1810–1813.

Sobhani, I., Bado, A., Vissuzaine, C., Buyse, M., Kermorgant, S., Laigneau, J. P., Attoub, S., Lehy, T., Henin, D., Mignon, M., and Lewin, M. J. (2000). Leptin secretion and leptin receptor in the human stomach. *Gut, 47,* 178–183.

Spicer, L. J. (2001). Leptin: A possible metabolic signal affecting reproduction. *Domest. Anim. Endocrinol., 21,* 251–270.

Tartaglia, L. A., Dembski, M., Weng, X., Deng, N., Culpepper, J., Devos, R., Richards, G. J., Campfield, L. A., Clark, F. T., Deeds, J., Muir, C., Sanker, S., Moriarity, A., Moore, K. J., Smutko, J. S., Mays, G. G., Wolf, E. A., Monroe, C. A., and Tepper, R. I. (1995). Identification and expression cloning of a leptin receptor, OB-R. *Cell, 83,* 1263–1271.

Tartaglia, L. A. (1997). The leptin receptor. *J. Biol. Chem., 272,* 6093–6096.

Trayhurn, P., Hoggard, N., Mercer, J. G., and Rayner, D. V. (1999). Leptin: fundamental aspects. *Int. J. Obes., 23,* 22–28.

Warren, M. P. and Fried, J. L. (2001). Hypothalamic amenorrhea. The effects of environmental stresses on the reproductive system: a central effect on the central nervous system. *Endocrinol. Metab. Clin. North Am., 30,* 611–629.

Wauters, M., Considine, R. V., and Van Gaal, L. F. (2000). Human leptin: from an adipocyte hormone to an endocrine mediator. *Eur. J. Endocrinol., 143,* 293–311.

Zhang, F., Basinksi, M. B., Beals, J. M., Briggs, S. L., Churgay, L. M., Clawson, D. K., DiMarchi, R. D., Furman, T. C., Hale, J. E., Hsiung, H. M., Schoner, B. E., Smith, D. P., Zhang, X. Y., Wery, J. P., and Schevitz, R. W. (1997). Crystal structure of the obese protein leptin-E100. *Nature, 387,* 206–209.

Zhang, P., Klenk, E. S., Lazzaro, M. A., Williams, L. B., and Considine, R. V. (2002). Hexosamines regulate leptin production in 3T3-L1 adipocytes through transcriptional mechanisms. *Endocrinology, 143,* 99–106.

Zhang, Y., Proenca, R., Maffei, M., Barone, M., Leopold, L., and Friedman, J. M. (1994). Positional cloning of the mouse *obese* gene and its human homologue. *Nature, 372,* 425–432.

II

General Reproductive Function

2

Neuroendocrine Regulation of Leptin Secretion and Its Role in Nitric Oxide Secretion

S. M. McCann, S. Karanth, W. H. Yu, and C. A. Mastronardi

2.1. ABSTRACT

In mammals hypothalamic control of gonadotropin secretion is mediated by mammalian luteinizing hormone-releasing hormone (mLHRH), or gonadotropin-releasing hormone (GnRH) and a follicle-stimulating hormone (FSH) RH that is lGnRH-III or a closely related polypeptide. We believe that the differential pulsatile release of FSH and LH and its differential secretion at different times of the estrous cycle are accounted for by differential secretion of FSHRH and LHRH. Both FSHRH and LHRH act on the gonadotropes to increase intracellular free calcium that combines with calmodulin to activate neural nitric oxide synthase (nNOS). The NO released activates guanylyl cyclase that converts guanosine triphosphate (GTP) into cyclic guanosine monophosphate (cGMP) that induces exocytosis of gonadotropin-containing secretory granules. The adipocyte hormone leptin has equal potency with LHRH to activate the release of FSH and LH from male rat pituitaries by NO. It also releases LHRH from arcuate-nuclear median eminence explants and LH from ovariectomized, estrogen-blocked rats, whereas it inhibits FSH release from the same preparation and these actions are brought about via NO, which also controls LHRH release stimulated by other means. Just prior to proestrus, sensitivity to LH-releasing action of leptin peaks suggesting that it plays a role in induction of puberty.

S. M. McCann, S. Karanth, W. H. Yu, and C. A. Mastronardi • Pennington Biomedical Research Center, Louisiana State University.

Leptin and Reproduction. Edited by Henson and Castracane, Kluwer Academic/Plenum Publishers, 2003.

Our hypothesis is that leptin release is neurohormonally controlled. Conscious, male rats bearing indwelling external jugular catheters were injected with the test drug or 0.9 percent NaCl (saline) and blood samples were drawn thereafter to measure plasma leptin. Anesthesia decreased plasma leptin concentrations within 10 min to a minimum at 120 min, followed by a rebound at 360 min. Administration of lipopolysaccharide (LPS) (iv) increased plasma leptin to almost twice baseline by 120 min that remained on a plateau for 360 min, accompanied by increased adipocyte leptin mRNA. Anesthesia largely blunted the LPS-induced leptin release at 120 min. Isoproterenol (β-adrenergic agonist) failed to alter plasma leptin but reduced LPS-induced leptin release significantly. Propranolol (β-receptor antagonist) produced a significant increase in plasma leptin but had no effect on the response to LPS. Phentolamine (α-adrenergic receptor blocker) not only increased plasma leptin ($P < 0.001$), but also augmented the LPS-induced increase ($P < 0.001$). α-Bromoergocryptine (dopamine D-2 receptor agonist) decreased plasma leptin ($P < 0.01$) and blunted the LPS-induced rise in plasma leptin release ($P < 0.001$). We conclude that leptin is at least in part neurally controlled since anesthesia decreased plasma leptin and blocked its response to LPS. The findings that phentolamine and propranolol increased plasma leptin concentrations suggest that leptin release is inhibited by the sympathetic nervous system mediated principally by α-adrenergic receptors since phentolamine but not propranolol augmented the response to LPS. Since α-bromoergocryptine decreased basal and LPS-induced leptin release, dopaminergic neurons may inhibit basal and LPS-induced leptin release by suppression of release of prolactin from the adenohypophysis.

Since leptin stimulates nitric oxide (NO) release from the hypothalamus and anterior pituitary gland, we hypothesized that it might also release NO from adipocytes, the principal source of leptin. Consequently, plasma concentrations of leptin and NO, estimated from its metabolites NO_3 and NO_2 (NO_3–NO_2), were measured in adult male rats. There was a linear increase of both leptin and NO_3–NO_2 with body weight that was associated with a parallel rise in fat mass. These findings indicate that release of leptin and NO is directly related to adipocyte mass. Furthermore, there was a parallelism in circadian rhythm of both substances, with peaks at 01:30 hr and nadirs at 07:30 hr. Measurement of both leptin and NO_3–NO_2 in plasma from individual rats revealed that NO_3–NO_2 increased linearly with leptin. Incubation of epididymal fat pads (EFP) with leptin or its intravenous injection in conscious rats increased NO_3–NO_2 release. The release of NO_3–NO_2 in vivo and in vitro exceeded that of leptin by many fold indicating that leptin activates NO synthase (NOS). Leptin increased TNF-α release at a 100-fold lower dose than required for NO release in vitro and in vivo, suggesting that it may also participate in leptin-induced NO release. However, since many molecules of leptin were required to release a molecule of TNF-α in vivo and in vitro, we believe that leptin-induced TNF-α release is an associated phenomenon not involved in NO production. The results support the hypothesis that adipocytes play a major role in NO release by activating NOS in the adipocytes and the adjacent capillary endothelium.

2.2. HYPOTHALAMIC CONTROL OF GONADOTROPIN SECRETION

The control of gonadotropin secretion is extremely complex as revealed by the research of the past 40 years since the discovery of LHRH (McCann et al., 1960), now commonly called GnRH (McCann and Ojeda, 1996). This was the second of the hypothalamic-releasing hormones characterized. It stimulates FSH release, albeit in smaller amounts than LH. For this reason, it was renamed GnRH (Reichlin, 1992; McCann and Ojeda, 1996). Overwhelming evidence indicates that there must be a separate FSHRH because pulsatile release of LH and FSH can be dissociated. In the castrated male rat, roughly half of the FSH pulses occur in the absence of LH pulses and only a small fraction of the pulses of both gonadotropins are coincident. LHRH antisera or antagonists can suppress pulsatile release of LH without altering FSH pulses (Culler and Negro-Vilar, 1987; McCann et al., 1993). LH, but not FSH pulses can be suppressed by alcohol (Dees et al., 1985), delta-9-tetrahydrocannabinol, and cytokines, such as interleukin-1 alpha (IL-1α) (Rettori et al., 1994). In addition, a number of peptides inhibit LH, but not FSH release and a few stimulate FSH without affecting LH (McCann and Krulich, 1989; McCann et al., 1993).

The hypothalamic areas controlling LH and FSH are separable. Stimulation in the dorsal medial anterior hypothalamic area causes selective FSH release, whereas lesions in this area selectively suppress the pulses of FSH and not LH (Lumpkin et al., 1989). Conversely, stimulations or lesions in the medial preoptic region can augment or suppress LH release, respectively, without affecting FSH release. Electrical stimulation in the preoptic region releases only LH, whereas lesions in this area inhibit LH release without inhibiting FSH release. The medial preoptic area contains most of the perikarya of LHRH neurons. The axons of these neurons project from the preoptic region to the anterior and mid-portions of the median eminence. Extracts of the anterior–mid-median eminence contain LH-releasing activity commensurate with the content of immunoassayable LHRH, whereas extracts of the caudal median eminence and organum vasculosum lamina terminalis (OVLT) contain more FSH-releasing activity than can be accounted for by the content of LHRH (McCann et al., 1993).

Lesions confined to the rostral and mid-median eminence can selectively inhibit pulsatile LH release without altering FSH pulsations, whereas lesions that destroy the caudal and mid-median eminence can selectively block FSH pulses in castrated male rats (Lumpkin et al., 1989; McCann et al., 1993; Marubayashi et al., 1999). Therefore, the putative FSHRH may be synthesized in neurons with perikarya in the dorsal anterior hypothalamic area with axons that project to the mid- and caudal median eminence to control FSH release selectively.

2.2.1. FSHRH

In the search for FSHRH, we first believed that it might be an analog of LHRH (Dhariwal et al., 1967) and later had many such analogs synthesized. We tested the forms of GnRH that were known to exist in lower species (Yu et al., in preparation). We had not tested lamprey GnRH-III, but when we realized that an antiserum that crossed reacted with lGnRH-I and lGnRH-III, immunostained neural fibers in the arcuate nucleus proceeding to the median eminence of human brains obtained at autopsy, it occurred to us that lGnRH-III could be the FSHRH since lGnRH-I had little activity to release either LH or FSH (Yu et al., 1997a). Indeed, lGnRH-III is a potent FSH-releasing factor with little or no LH-releasing activity both in vitro when incubated with hemipituitaries of male rats and in vivo when injected into conscious, ovariectomized, estrogen–progesterone-blocked rats (Yu et al., 1997a). The lowest dose tested in that preparation

(10 pmol) produced a highly significant increase in plasma FSH with no increase in LH, a 10-fold higher dose (100 pmol) increased plasma FSH similarly and had no effect on LH release. LGnRH-III 10^{-9} M produced highly significant FSH release in vitro, whereas, LH-releasing activity did not appear until 10^{-6} M (Yu et al., 1997a).

We fractionated 1,000 rat hypothalami by gel filtration on Sephadex G-25 and determined the FSH- and LH-releasing activity of the various fractions by bioassay on male rat hemipituitaries and compared the localization of these activities with that of LHRH and lGnRH determined by RIA. LGnRH was assayed by RIA using a specific antiserum for lGnRH, that recognized lGnRH I and III equally but did not cross-react with LHRH or cGnRH-II. A peak of LHRH immunoreactivity was found as well as three peaks of lGnRH immunoreactivity that preceded the peak of LHRH. The first peak eluted was quite small. The second peak was much larger and the third peak was of the greatest magnitude. Only the second peak altered gonadotropin release and it produced selective FSH release.

To determine whether this activity was caused by lGnRH, anterior hemipituitaries were incubated with normal rabbit serum or the lGnRH antiserum (1 : 1,000) and the effect on FSH and LH-releasing activity of the FSH-releasing fraction and the LH-releasing activity of LHRH was determined. The antiserum had no effect on basal release of either FSH or LH, but eliminated the FSH-releasing activity of the active fraction without altering the LH-releasing activity of LHRH. Since lGnRH-1 has minimal nonselective activity to release FSH or LH, whereas, previous experiments had shown that lGnRH-III highly selectively releases FSH with a potency equal to that of LHRH to release LH, these results support the hypothesis that the FSH-releasing activity observed in the FSHRH fraction was caused by lGnRH-III or a very closely related peptide (Yu et al., 2000).

2.2.2. Localization of lGnRH in the Brain of Rats by Immunocytochemistry

Using the same antiserum that we employed to find the location of lGnRH after gel-filtration of rat hypothalami, we attempted to localize lGnRH neurons by immunocyto-chemistry. Immunoreactive lGnRH-like cell bodies were found in the ventromedial preoptic area with axons projecting to the rostral wall of the third ventricle (3V) and OVLT. Another population of lGnRH-like cell bodies was located in the dorsomedial preoptic area with axons projecting caudally and ventrally to the external layer of the median eminence. On the other hand, using a highly specific, monoclonal antiserum against mGnRH to localize the mGnRH neurons, so that their localization could be compared with that of lGnRH neurons, we found that there were no mGnRH cells or fibers in the dorsomedial preoptic area that contained perikarya and fibers of lGnRH neurons (Dees et al., 1999).

Furthermore, immunoabsorption studies indicated that the cell bodies of the lGnRH neurons were eliminated by lGnRH-III, but not by mGnRH whereas the axons in the median eminence were eliminated by lGnRH-III but only slightly reduced by absorption with mGnRH. Using an antiserum against cGnRH-II that visualized cGnRH-II neurons in the chicken hypothalamus, no such neurons could be visualized in the rat hypothalamus (Dees et al., 1999).

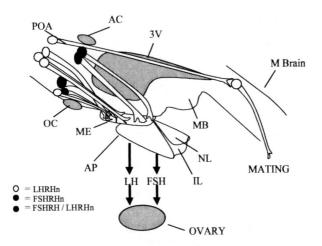

Figure 2.1. Parasuggital section of preoptic and hypothalamic region of the rat brain illustrating the distribution of FSHRH and LHRH neurons. Abbreviations: POA = preoptic area, AC = anterior commissure, OC = optic chiasm, 3V = third ventricle, ME = median eminence, MB = mammilliary body, M Brain = midbrain, AP = anterior pituitary, IL = intermediate lobe, NL = neural lobe.

Since the lGnRH antiserum (#3952) recognized lGnRH-I and lGnRH-III equally, a specific antiserum against lGnRH-III without cross-reactivity with lGnRH-I was needed to prove that the lGnRH neurons visualized in the rat brain were indeed lGnRH-III neurons. Our recent studies indicate that the specific lGnRH-III antiserum (#39-82-78-3) visualizes the same population of neurons seen with the less specific lGnRH antiserum (3952). Also, the staining of cells and fibers could be eliminated by lGnRH-III but was not affected by lGnRH-I or mGnRH. Consequently, the results strongly support the hypothesis that the lGnRH-III neurons that are located in the areas of the brain responsible for control of FSH are the FSHRH neurons (Hiney et al., 2002) (Figure 2.1). The lGnRH-III neurons whose cell bodies are located in the caudal dorsal medial preoptic area with axons projecting to the median eminence are in the very regions that have been shown to be selectively involved in the control of FSH release by lesion and stimulation studies described earlier (pp. 17, 18).

2.2.3. Evidence for a Specific FSHRH Receptor

The selectivity of release of FSH by FSHRH and lGnRH-III suggests the probability of a specific FSHRH receptor on pituitary gonadotropes. We hypothesized that we might demonstrate the presence of lGnRH-III receptors on gonadotropes using biotinylated lGnRH-III and mLHRH. Biotinylated lGnRH-III (10^{-9} M) bound to 80 percent of FSH gonadotropes and only 50 percent of LH gonadotropes of acutely dispersed pituitary cells, a finding that indicates that there are receptors on gonadotropes that bind this peptide (Childs et al., 2001). The binding of biotinylated lGnRH-III was not displaced with LHRH, which indicates that it is highly specific for the putative FSHRH receptors. It appears that in this situation as with monolayer cultured pituitary cells, the

FSHRH receptors disappear with time in culture since with 24 hr in culture, the binding of the biotinylated lGnRH-III was significantly decreased. We believe, on the basis of our recent studies with the biotinylated lGnRH-III, that ultimately a specific receptor for this peptide will be found in the pituitary and we believe it is in all probability the FSHRH.

Previously, cGnRH-II has been reported in various tissues in monkeys and humans, but the peptide was found in the hypothalamus of only fetal monkeys and also in the midbrain central gray (Lescheid et al., 1997). In adult monkeys, there were only a few cells and fibers in the caudal hypothalamus suggesting that this peptide would not reach the pituitary via the portal vessels, but must have other actions perhaps on mating behavior or in regulation of cell division. We found no cGnRH-II perikarya and only scant fibers in the hypothalamus of the rat in contrast with the readily observed lGnRH-III neurons and terminals in the median eminence (Hiney et al., 2002). Furthermore, cGnRH-II, as indicated above, has little or no selective FSH-releasing activity; however, its receptor has been cloned and found in the anterior pituitary gland (Neill et al., 2001). There is little doubt that it is an ancient GnRH existing from fish to mammals, but its role in control of gonadotropin secretion remains to be clarified.

2.2.4. Mechanism of Action of FSHRH, LHRH and Leptin on Gonadotropin Secretion

It is well known that FSH and LH release is controlled by calcium ions (Ca^{++}) (Wakabayashi et al., 1969; Stojilkovic, 1998) and that interaction of LHRH with its receptor causes an increase in intracellular free calcium and also activates the phosphatidyl inositol cycle that mobilizes internal calcium. The resulting increase in intracellular free calcium mediates the releasing action of LHRH (for a review, see Stojikovic, 1998); however, we earlier showed a role for a cGMP and not cyclic adenosine monophosphate (cAMP) in controlling the release of LH and FSH mediated by LHRH (Nakano et al., 1978; Naor et al., 1978; Snyder et al., 1980). This was before it was accepted that NO is a physiologically significant, gaseous transmitter that acts by activation of guanylyl cyclase that converts GTP to cGMP. cGMP activates protein kinase G that causes exocytosis of gonadotropin secretory granules.

To test the hypothesis that the FSH-releasing activity of lGnRH-III (or FSHRH) is mediated by calcium and NO, calcium was removed from the medium and a chelating agent (ethylene glycol-N-N-N′-N′-tetraacetic acid) that would remove any residual Ca^{++} was added. The action of purified FSHRH and lGnRH-III was blocked in the absence of Ca^{++}. N^G-monomethyl-L-arginine (NMMA), a competitive inhibitor of NOS, was added to the medium in other experiments. We found that this competitive inhibitor of NOS, NMMA, completely blocked the FSH-releasing activity of not only purified FSHRH but also of lGnRH-III. Furthermore, sodium nitroprusside (NP), a releaser of NO, stimulated both LH and FSH release and the activity of LHRH to release both LH and FSH was also blocked by NMMA. These data indicate that FSHRH (or lGnRH-III) acts on its putative receptor via a calcium-dependent, nitric oxide pathway to release FSH specifically, whereas LHRH acts on its receptor similarly to increase intracellular Ca^{++} that activates NOS in the gonadotrophs to cause release of LH and to a lesser extent FSH (Yu et al., 1997c; Yu et al., 2002) (Figure 2.2).

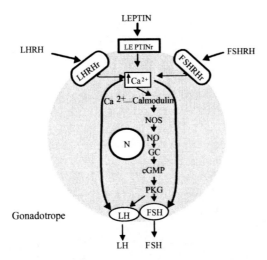

Figure 2.2. Schematic diagram illustrating the mechanism of the gonadotropin-releasing action of FSHRH, LHRH and leptin, N = nucleus, other abbreviations are in list of abbreviations. The principal pathway is via NO, cGMP and PKG, but Ca^{++} may have an independent action.

2.3. LEPTIN IN REPRODUCTION

The hypothesis that leptin may play an important role in reproduction stems from several findings. First, the Ob/Ob mouse, lacking the leptin gene, is infertile and has atrophic reproductive organs. Gonadotropin secretion is impaired and very sensitive to negative feedback by gonadal steroids as is the case for prepubertal animals (Swerdloff et al., 1976; Swerdloff et al., 1978). It has now been shown that treatment with leptin can recover the reproductive system in the Ob/Ob mouse by leading to growth and function of the reproductive organs and fertility (Chehab et al., 1996) via secretion of gonadotropins (Barash et al., 1996).

The critical weight hypothesis of the development of puberty states that when body fat stores have reached a certain point puberty occurs (Frisch and McArthur, 1974). This hypothesis in its original form does not hold because if animals are underfed, puberty is delayed, but with access to food, rapid weight gain leads to the onset of puberty at weights well below the critical weight under normal nutritional conditions (Ronnekleiv et al., 1978). We hypothesized that during this period of refeeding, or at the time of the critical weight in the normally fed animals, as the fat stores increase, there is increased release of leptin from the adipocytes into the blood stream and this acts on the hypothalamus to stimulate the release of LHRH with resultant induction of puberty. Indeed, leptin has been found recently to induce puberty (Chehab et al., 1997).

We initiated studies on its possible effects on hypothalamic–pituitary function. We anticipated that it would also be active in adult rats and, therefore, studied its effect on the release of FSH and LH from hemipituitaries, and also its possible action to release LHRH from MBH explants in vitro. To determine if it was active in vivo, we used a model that we have often used to evaluate stimulatory effects of peptides on LH release; namely, the ovariectomized, estrogen-primed rat. Because our supply of leptin was limited, we began by microinjecting it into the 3V in conscious animals bearing implanted third ventricular cannulae, and also catheters in the external jugular vein extending to the right atrium, so that we could draw blood samples before and after the injection of leptin and measure the effect on plasma FSH and LH (Yu et al., 1997b).

2.3.1. Effect of Leptin on LH Release

We found that under our conditions, leptin had a bell-shaped dose–response curve to release LH from anterior pituitaries incubated in vitro. There was no consistent stimulation of LH release with a concentration of 10^{-5} M. Results became significant with 10^{-7} M and remained on a plateau through 10^{-11} M with reduced release at a concentration of 10^{-12} M that was no longer significant statistically. The release was not significantly less than that achieved with LHRH (4×10^{-8} M). Under these conditions, there was no additional release of LH when leptin (10^{-7} M) was incubated together with LHRH (4×10^{-8} M). In certain other experiments, there was an additive effect when leptin was incubated with LHRH; however, this effect was not uniformly seen. The results indicate that leptin was only slightly less effective to release LH, than LHRH itself (Yu et al., 1997b).

2.3.2. Effect of Leptin on FSH Release

In the incubates from these same glands, we also measured FSH release and found that it showed a similar pattern as that of LH, except that the sensitivity in terms of FSH release was much less than that for LH. The minimal effective dose for FSH was 10^{-9} M, whereas it was 10^{-11} M for LH. The responses were roughly of the same magnitude at the effective concentrations as obtained with LH and the responses were clearly equivalent to those observed with 4×10^{-9} M LHRH. Combination of LHRH with a concentration of leptin that was just below significance gave a clear additive effect (Yu et al., 1997b).

The action of leptin to stimulate both LH and FSH release was mediated by the long form of the leptin receptor that is located on the cell surface of the gonadotropes. The mechanism of action is the same as that of FSHRH and LHRH; the only difference is that the action is mediated by leptin receptors that increase intracellular free Ca^{++} activating nNOS that generates NO that activates GC followed by PKG leading to extrusion of FSH and LH secretory granules from the gonadotropes (Figure 2.2).

2.3.3. Effect of Leptin on LHRH Release

There was no significant effect of leptin in a concentration range of 10^{-6}–10^{-12} M on LHRH release during the first 30 min of incubation; however, during the second 30 min, the highest concentration produced a borderline significant decrease in LHRH release with 10^{-6} M, followed by a tendency to increase with lower concentrations and a significant ($P < 0.01$), plateaued increase with the lowest concentrations tested (10^{-10} and 10^{-12} M). Both the FSH- and LH-releasing actions of leptin were blocked by NMMA indicating that NO mediates its action (Yu et al., 1997c).

2.3.4. The Effect of Intraventricularly Injected Leptin on Plasma Gonadotropin Concentrations in Ovariectomized, Estrogen-Primed Rats

The injection of the diluent for leptin into the 3V (Krebs-Ringer Bicarbonate, 5 μl) had no effect on pulsatile FSH or LH release, but the injection of leptin (10 μg) uniformly produced an increase in plasma LH with a variable time-lag ranging from 10 to 50 min, so that the maximal increase in LH from the starting value was highly significant $P < 0.01$) and constituted a mean increase of 60 percent above the initial concentration. In contrast, leptin inhibited FSH release on comparison with the results with the diluent, but the effect was delayed and occurred mostly in the second hour. Therefore, at this dose of estrogen (10 μg estradiol benzoates, 72 hr before experiments), leptin stimulates the release of LHRH and inhibits the release of FSHRH (Walczewska et al., 1999).

2.3.5. Mechanism of Action of Leptin on the Hypothalamic Pituitary Axis

We have shown that leptin exerts its action at both hypothalamic and pituitary levels by activating NOS, since its effect to release LHRH, FSH, and LH in vitro (Yu et al., 1997b) is blocked by NMMA. Leptin, in essence, is a cytokine secreted by the adipocytes. It, like the cytokines, seems to reach the brain via a transport mechanism mediated by the Ob/Ob$_a$ receptors (Cioffi et al., 1996) in the choroid plexus (Schwartz et al., 1996). These receptors have an extensive extracellular domain, but a greatly truncated intracellular domain (Cioffi et al., 1996) and mediate transport of the cytokine by a saturable mechanism (Banks et al., 1996). Following uptake into the cerebrospinal fluid (CSF) through the choroid plexus, leptin is carried by the flow of CSF to the 3V, where it either diffuses into the hypothalamus through the ependymal layer lining the ventricle or combines with Ob/Ob$_a$ (Cioffi et al., 1996) receptors on terminals of responsive neurons that extend to the ventricular wall.

The Ob/Ob$_b$ receptor has a large intracelluar domain that presumably mediates the action of the protein (Schwartz et al., 1996). These receptors are wide-spread throughout the brain (Schwartz et al., 1996), but particularly localized in the region of the paraventricular (PVN) and arcuate nuclei (AN). Leptin activates stat 3 within 30 min after its intraventricular injection (Vaisse et al., 1996). Stat 3 is a protein that is important in conveying information to the nucleus to initiate DNA-directed messenger ribonucleic acid (mRNA) synthesis. Following injection of bacterial LPS, it is also activated, but in this case, the time delay is 90 min presumably because LPS has been shown to induce IL-1 beta (β) mRNA in the same areas—namely, the PVN and AN (Schwartz et al., 1996)—IL-1β mRNA would then cause production of IL-1β, which would activate stat 3. On entrance into the nucleus, stat 3 would activate or inhibit DNA-directed mRNA synthesis. In the case of leptin, it activates corticotropin-releasing hormone (CRH) mRNA in the PVN, whereas in the AN, it inhibits neuropeptide Y (NPY) mRNA resulting in increased CRH synthesis and presumably release in the PVN and decreased NPY synthesis and release in the AN (Schwartz et al., 1996). Presumably, the combination of leptin with these transducing receptors also either increases or decreases the firing rate of that particular neuron. In the case of the AN-median eminence area, leptin may enter the median eminence by diffusion between the tanycytes or alternatively by combining with its receptors on terminals of neurons projecting to the tanycytes. Activation of these neurons would induce LHRH release.

The complete pathway of leptin action in the MBH to stimulate LHRH release is not yet elucidated. Arcuate neurons bearing Ob/Ob receptors may project to the median eminence to the tanycyte/portal capillary junction. Leptin would either combine with its receptors on the terminals that transmit information to the cell bodies in the AN or diffuse to the AN to combine with its receptors on the perikarya of AN neurons. Because leptin decreases NPY mRNA, and presumably NPY biosynthesis in NPY neurons in the AN (Chehab et al., 1997), we postulate that leptin causes a decrease in NPY release. Because NPY inhibited LH release in intact and castrated male rats (Reznikov and McCann, 1993), we hypothesize that NPY decreases the release of LHRH by inhibiting the noradrenergic neurons which mediate pulsatile release of LHRH. Therefore, when the release of NPY is inhibited by leptin, noradrenergic impulses are generated, which act on the α_1 receptor on the NOergic neurons causing the release of NO which diffuses to the LHRH terminals and activates LHRH release by activating guanylate cyclase and cyclooxygenase$_1$ as shown in our prior experiments reviewed above. Leptin acts to activate NOS as indicated because its release of LHRH is blocked by inhibition of NOS (Yu et al., 1997c). The LHRH enters the portal vessels and is carried to the anterior pituitary gland where it acts to stimulate FSH and particularly LH release by combining with its receptors on the gonadotropes. The release of LH and to a lesser extent FSH is further increased by the direct action of leptin on its receptors in the pituitary gland (Cioffi et al., 1996; Naivar et al., 1996; Yu et al., 1997c) (Figure 2.3).

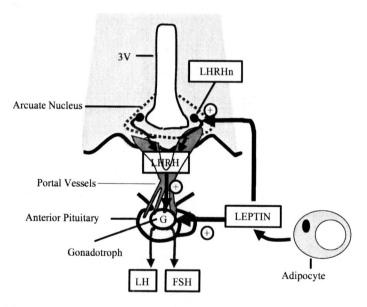

Figure 2.3. Schematic diagram illustrating the dual action of leptin to stimulate LHRH release from the hypothalamus and gonadotropin release from the anterior pituitary gland. 3V = third ventricle.

Recent experiments suggest the possibility that serotonin may play a role in the actions of leptin on LHRH release since the transducing leptin receptor has been localized to serotonergic neurons in the raphè nuclei (Finn et al., 2001). Previously, serotonin has been found to have either stimulatory or inhibitory action on LHRH release (Vitale and Chiocchio, 1993). Therefore, it is possible, that leptin could act by this indirect pathway to activate serotonergic input to the LHRH neurons in the hypothalamus to influence reproduction. Further work is necessary to substantiate this hypothesis.

Galanin is an important gastrointestinal peptide that is found in the central nervous system, and within the hypothalamus of mammalian species including man and primates (Bartfai et al., 1993). Galanin releases growth hormone by hypothalamic action in the rat (Ottlecz et al., 1986) and in humans (Giustina et al., 1995). In the rat, Rhesus macaque, and human, galanin increases LH secretion, suggesting that it stimulates LHRH release (Kordower et al., 1992; Evans et al., 1993; Finn et al., 2000; Cunningham et al., 2002).

Galanin and its three receptors are expressed in the brain and hypothalamus. Galanin, a 30 amino acid peptide, is highly conserved and the N-terminable 13 amino acids activate the galanin receptors (Habert-Ortoli et al., 1994; Fathi et al., 1998; Kolakowski et al., 1998).

Screening purified hypothalamic extracts for binding to galanin receptors revealed an active fraction that was isolated and characterized as a 60 amino acid peptide named galanin-like peptide (GALP) (Ohtaki et al., 1999). Amino acids 9–21 of GALP are identical to amino acids 1–13 of galanin. GALP mRNA was shown to be expressed in the hypothalamic arcuate nucleus and median eminence, as well as within the posterior pituitary (Jureus et al., 2000; Kerr et al., 2000; Larm and Gundlach, 2000). In the Rhesus macaque, GALP mRNA is confined to the Arc nucleus, median eminence and neurohypophysis (Cunningham et al., 2002). Nearly all GALP mRNA-expressing cells in the arcuate of nucleus also express mRNA for the long form of the leptin receptor.

Furthermore in the Ob/Ob mice lacking leptin, the expression of GALP mRNA in the brain is remarkably reduced. The ICV injection of leptin into Ob/Ob mice increased both the number of GALP mRNA expressing neurons and their content of GALP mRNA, indicating that leptin stimulates the synthesis of GALP. Furthermore, in fasted rats leptin levels are decreased and GALP mRNA is also reduced. Leptin injected ICV restored GALP mRNA levels to normal in these animals (Juréus et al., 2000; Takatsu et al., 2001).

Intraventricular injection of GALP stimulated both C-Fos expression in GnRH neurons and LH secretion (Matsumoto et al., 2001). LH secretion induced by GALP was blocked by concurrent treatment with the GnRH receptor antagonist, Cetrorelix, indicating that GALP had stimulated the release of LHRH (Matsumoto et al., 2001). These recent findings raise the possibility that leptin may act not only by directly releasing LHRH and LH (Yu et al., 1997b), but also by stimulating the release of GALP, which would further stimulate the release of LHRH into the portal vessels that would augment the stimulation of LH release.

We hypothesize that leptin may be a critical factor in induction of puberty as the animal nears the so-called critical weight. Either metabolic signals reaching the adipocytes, or signals related to their content of fat cause the release of leptin, which increases LHRH and gonadotropin release, thereby initiating puberty and finally ovulation and onset of menstrual cycles. Indeed, in prepubertal rats, sensitivity of LH release from intraventricularly injected leptin peaks a short time before the first preovulatory release of LH (Dearth et al., 2000). Since this enhanced sensitivity is prior to the first preovulatory LH surge, leptin may prime the system leading to increased sensitivity of LHRH release to the rising estradiol levels. In the male, the system would work similarly; however, there is no preovulatory LH surge brought about by the positive feedback of estradiol. Sensitivity to leptin is undoubtedly under steroid control and we are actively working to elucidate this problem.

During fasting, the leptin signal is removed and LH pulsatility and reproductive function decline quite rapidly. In women with anorexia nervosa, this causes a reversion to the prepubertal state, which can be reversed by feeding. Thus, leptin would have a powerful influence on reproduction throughout the reproductive lifespan of the individual. The consequences to gonadotropin secretion of overproduction of leptin, as has already been demonstrated in human obesity, are not clear. There are often reproductive abnormalities in this circumstance and whether they are due to excess leptin production or other factors, remains to be determined. In conclusion, it is now clear that leptin plays an important role in control of reproduction by actions on the hypothalamus and pituitary.

2.4. ROLE OF LEPTIN IN THE RESPONSE TO STRESS AND INFECTION

In inflammatory stress induced by peripheral or central administration of LPS in rodents, there are a number of changes in the immune system such as stimulation of release of acute phase proteins, induction of NOS activity and synthesis, and increased release of cytokines (Gottschall et al., 1992; Rettori et al., 1994; Wong et al., 1996; McCann et al., 1998; Turnbull et al., 1998; McCann et al., 2000). Proinflammatory cytokines, such as tumor necrosis factor-alpha (TNF-α) and interleukin-1 (IL-1) released from immune cells reach the central nervous system (CNS) and increase release of corticotrophin-releasing hormone (CRH) that activates the pituitary–adrenal axis (Kakizaki et al., 1999). LPS also alters the release of other pituitary hormones in rats, increasing prolactin release and decreasing growth hormone (GH) and thyroid-stimulating hormone (TSH) and luteinizing hormone (LH) release (Rettori et al., 1994). There are a variety of other changes in the CNS caused by LPS, as well, many of which are probably mediated by LPS-induced cytokine release, such as alterations of central and peripheral catecholamine levels (Song, et al., 1999; Wang and White, 1999), and alteration of neurotransmitter release in different areas of the brain (MohanKumar et al., 1999). These result in induction of fever, loss of appetite, libido, and somnolence and many metabolic changes.

Leptin is a cytokine (Madej et al., 1995; Rock et al., 1996; Kline et al., 1997; Zhang et al., 1997) and like other proinflammatory cytokines, such us TNF-α, IL-1 and IL-6, leptin is also secreted in response to the inflammatory stress induced by LPS (Mastronardi et al., 2000; McCann et al., 2001). In previous research we have determined that IV injection of LPS in conscious, freely moving rats bearing indwelling jugular catheters increased plasma leptin concentrations 100 percent over baseline concentrations within 2 hr (Mastronardi et al., 2000b). Similar results were found also in experiments in mice (Finck et al., 1998). Recent research in humans showed that plasma leptin levels were increased in survivors of acute sepsis (Bornstein et al., 1998) and it was shown in another in vivo paradigm in mice that leptin exerted a protective effect on TNF-α-induced sepsis (Takahashi et al., 1999), suggesting that leptin might be an anti-stress cytokine.

Plasma leptin concentrations have a circadian rhythm with values peaking in the middle of the night in both humans (Licinio et al., 1998) and rats (Mastronardi et al., 2000a). There is also an ultradian rhythm of plasma leptin in humans (Licinio et al., 1998). These studies suggest that leptin release is controlled by the CNS.

The fact that plasma leptin concentration gradually declined in anesthetized male rats during surgery led us to hypothesize that leptin secretion was under neural control (Mastronardi et al., 2001a). In order to assess the neural control of leptin we used another type of stress, namely, the inflammatory stress induced by LPS that is known to increase leptin release (Mastronardi et al., 2000b). Furthermore, it was shown that LPS increased catecholamine levels in the central nervous system (MohanKumar et al., 1999; Wang and White, 1999), in the periphery (Song et al., 1999) and also activated the hypothalamic–pituitary–adrenal axis (HPA) (Rettori et al., 1994; Turnbull et al., 1998). Therefore, we considered the present paradigm a suitable model to assess the neural control of leptin.

Our previous results had shown that LPS evoked a rapid and long-lasting increase in plasma leptin concentrations with the first increase obtained within 10 min and a plateau existing from 2 to 6 hr (Mastronardi et al., 2000b). The current results show that at 6 hr there was a highly significant increase in leptin mRNA in EFP induced by LPS. The early release of leptin within 10 min could not have occurred by new synthesis of leptin, and instead, must be due to the release of stored leptin that has been found in pinocytotic vesicles in adipocytes (Bornstein et al., 2000). Therefore, presumably, LPS acts on its receptors within the brain or receptors on afferent neurons, such as vagal afferents, to activate neural or hormonal mechanisms that evoke exocytosis of leptin-containing vesicles, which accounts for the initial elevation of plasma leptin. This is followed by induction of leptin mRNA stimulating leptin synthesis that contributes to the release that occurs later, along with that of preformed leptin.

As in our previous results with placement of jugular catheters in anesthetized rats (Mastronardi et al., 2001a), plasma leptin decreased in the current experiments. Moreover, anesthesia also decreased LPS-induced plasma leptin release in the same period of time, providing further support for our concept that leptin is under neural control. Surprisingly, ketamine anesthesia, either alone or in the presence of LPS, provoked a rebound in plasma leptin levels after 120 min that reached concentrations similar to or even greater than those present in the LPS-treated rats. We hypothesize that this rebound may be caused by decreased negative feedback of the depressed plasma leptin concentrations during anesthesia, which at the termination of anesthesia act centrally to stimulate leptin release.

To further understand the possible role of the sympathetic nervous system in the LPS-induced leptin release, we studied the effects of α-adrenergic and β-adrenergic agonists and antagonists on the response to LPS. Injection of the β-adrenergic agonist, isoproterenol, slightly but significantly decreased plasma leptin concentrations and in the presence of LPS largely blunted the LPS-induced increase in plasma leptin concentrations. Recently, it has been shown that there is noradrenergic innervation not only of brown but also of white fat (Bartness and Bamshad, 1998). The fact that propranolol increased baseline concentrations of plasma leptin is consistent with the hypothesis that there is β-adrenergic inhibitory tone depressing leptin release during resting conditions. However, in the presence of LPS, propranolol slightly but not significantly decreased LPS-induced leptin release, suggesting that the inhibitory β tone under resting conditions was not present after injection of LPS. Therefore, our data suggest that either circulating epinephrine and/or norepinephrine, or norepinephrine released from noradrenergic terminals may inhibit leptin release by acting on β-adrenergic receptors present on cell membranes of the adipocytes (Scriba et al., 2000). In future researches we plan to study the effect on leptin release of specific antagonists of β receptor subtypes, such as β-3 adrenergic receptors known to be present on adipocytes.

Phentolamine, the α-adrenergic antagonist, induced a rapid and highly significant increase in plasma leptin either alone or in the presence of LPS, suggesting that there is a strong inhibitory tone acting through the α-adrenergic receptors to inhibit not only basal but LPS-stimulated leptin release.

Recently, we found that prolactin stimulated leptin release and that α-bromoergocryptine, a D-2 receptor agonist that inhibits prolactin release from the anterior pituitary gland decreased plasma leptin concentrations (Mastronardi et al., 2000), results that were independently obtained by Gualillo et al. (1999). Moreover, we have shown previously that LPS increased prolactin release (Rettori et al., 1994) in a similar experimental paradigm as that used here. Thus, it is likely that LPS-induced prolactin release increases leptin release by activation of prolactin receptors on adipocytes (McAveney et al., 1996). Therefore, we studied the effect of α-bromoergocryptine, an inhibitor of prolactin release, alone or in the presence of LPS. As in the case of anesthesia, α-bromoergocryptine alone or in the presence of LPS, decreased plasma leptin concentrations initially, but this decrease was followed by a later rebound of a lesser extent than that observed in the ketamine experiments. The rebound in both cases may be related to the initial lowering of plasma leptin reducing its negative feedback. Then, as the drug or anesthesia dissipates, there follows an increase in leptin release leading to the rebound in plasma levels, presumably caused by increased prolactin release.

The decline in leptin in the bromocryptine-injected animals was much less than in the anesthetized animals and the rebound was also less, perhaps because the lesser decrease in plasma leptin on comparison with that of anesthetized rats resulted in a lesser negative feedback of leptin in these animals (Mastronardi et al., 2001c).

We hypothesize that LPS acts on the CNS to inhibit the secretion of dopamine (DA), removing the inhibition exerted by tuberoinfundibullar dopaminergic neurons on the secretion of prolactin (PRL) (see summary diagram, Figure 2.4). Therefore, the secretion of PRL is increased from the lactotropes (L). Thereafter, PRL circulates to the adipose tissue and acting on its receptors (PRL_r) on the adipocytes increases the release of leptin that is stored in pinocytotic vesicles in the cytoplasma adjacent to the cell membrane. The sympathetic nervous system exerts a tonic inhibitory effect on leptin release mediated predominantly by α- and to a lesser extent by β-adrenergic receptors that are still present and may be augmented by LPS (Mastronardi et al., 2001c).

2.5. CONTROL OF NO RELEASE BY LEPTIN

Our previous studies have shown that leptin controls the release of NO by activating NOS, both in the basal hypothalamus and also in the anterior pituitary gland (Yu et al., 1997a,b). Although leptin has been localized to the hypothalamic arcuate nucleus, in the very region where it controls NO release (Morash et al., 1999), the major storehouse of leptin is in the adipocytes (Bornstein et al., 2000). Therefore, it occurred to us that leptin released from the adipocytes might also activate NOS and cause release of NO.

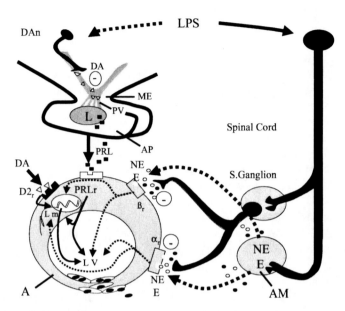

Figure 2.4. The neurohormonal control of leptin release. For description see text, p. 25. Open arrows indicate inhibition and closed arrows indicate stimulation. Abbreviations: tuberoinfundibullar dopaminergic neurons (DAn), dopamine (DA), median eminence (ME), portal vessels (PV), lactotropes (L), anterior pituitary (AP), prolactin (PRL), PRL receptors (PRL$_r$), DA2 receptors (D2$_r$), leptin mRNA (Lm), Leptin vesicles (LV), adipocyte (A), epinephrine (E), norepinephrine (NE), α-adrenergic receptors (A-α_r), beta adrenergic receptors (A-β_r), and adrenal medulla (AM).

It has been shown that endothelial (e) NOS is present on the cell membrane of adipocytes (Ribiere et al., 1996), that leptin receptors are on their cell surface (Bornstein et al., 2000), and that they contain a significant storehouse of leptin in small pinocytotic vesicles beneath the cell membrane (Bornstein et al., 2000). We hypothesized that leptin is extruded by exocytosis from these vesicles and acts on its receptors on the surface of the adipocytes, and on adjacent capillary endothelial cells to activate eNOS. The activation of eNOS increases NO that is released into the interstitial fluid and diffuses into the capillaries, adjacent arterioles and venules. The released NO would diffuse into the smooth muscle of these small vessels and activate guanylyl cyclase causing generation of cGMP from guanosine triphosphate (GTP). cGMP acts by protein kinase G to relax vascular smooth muscle (McCann et al., 1998). Thus, if leptin releases NO as hypothesized, it would dilate arterioles and venules, increasing blood flow to the adipocytes.

Vasodilatation is also produced by release of acetyl choline from parasympathetic terminals that activate eNOS in vascular endothelium. The NO released dilates vascular smooth muscle thereby lowering blood pressure. Under resting conditions there is tonic NOergic vasodilatory tone since blockade of NOS by inhibition of the enzyme elevates blood pressure (Vargas et al., 1991). NO also plays a role in renal function and the metabolites of NO produced in the kidney would also reach the circulation and diffuse into the urine (Soares et al., 1999). In fact, NO plays a role in the function of nearly all organs in the body. It is rapidly metabolized to its principal metabolites nitrite [NO$_2$] and nitrate [NO$_3$] that diffuse into the circulation and serve as an index of NO production.

Consequently, plasma NO_3–NO_2 concentration [NO_3–NO_2] would be an index of the sum total of NO released from all the organs of the body that would include brain, pituitary, heart, vascular system, kidney, adipose tissue, and immune cells.

Proinflammatory cytokines, such as IL-1, IL-6, and TNF-α, that are secreted in response to infection also increase NO production by inducing the expression of inducible (i) NOS not only from monocytes and macrophages but also from parenchymal cells in many other organs (Oswald et al., 1999; McCann et al., 2000; McCann et al., 2001). Leptin is a cytokine (Zhang et al., 1997) that is synthesized and released principally by adipocytes in adipose tissue that constitutes 10 to 25 percent of body weight (Klinger et al., 1996). We hypothesize that release of leptin from the adipocytes stimulates significant release of NO that would make an important contribution to the plasma levels of NO_3–NO_2. Moreover, unlike the classical cytokines that are hardly measurable during baseline conditions and reach concentrations in the picomolar range only during stress, such as inflammatory stress provoked by LPS (Mastronardi et al., 2001b, c) or surgical stress (Mastronardi et al., 2001a), baseline concentrations of leptin in plasma are in the nM range suggesting that leptin might be a major factor to control NO release and consequently plasma NO_3–NO_2 concentrations.

Therefore, both plasma leptin and NO_3–NO_2 in blood samples from individual rats were measured to determine if there was any correlation between the plasma levels of leptin and [NO_3–NO_2] under a variety of conditions. The effect of leptin on release of NO from EFP incubated in vitro and on plasma NO_3–NO_2 in conscious male rats was also determined. In addition, TNF-α release from EFP and plasma was determined since we recently discovered that adipose tissue produces significant amounts of TNF-α together with leptin during the inflammatory stress produced by LPS (Mastronardi, Yu and McCann, 2001 unpublished data), and TNF-α is also an important stimulator of NO production (Oswald et al., 1999). The results support the hypothesis that leptin stimulates the release of NO from adipose tissue and that this NO is a major contributor to plasma NO_3–NO_2 and accounts for the parallel circadian rhythm of both substances.

2.7. RESULTS AND DISCUSSION

Since leptin increased the release of LHRH from the hypothalamus and the gonadotropins, LH and FSH, from the pituitary by activating NOS and releasing NO (Yu et al., 1997a,b), we hypothesized that leptin might also control the release of NO into the blood stream. Therefore, our first experiment was to determine if a correlation existed between plasma levels of leptin and NO, measuring NO_3–NO_2 as an index of NO production. As the body weight of the rats increased, the concentrations of both leptin and NO_3–NO_2 linearly increased in plasma; however, the rate of increase of NO_3–NO_2 was 75,000-fold greater than that of leptin. This remarkably greater increase in NO_3–NO_2 than leptin can be explained by hypothesizing that leptin directly or indirectly activates NOS. If this is the case, one molecule of leptin might activate one molecule of NOS that could by its enzymatic action convert 75,000 molecules of arginine into 75,000 molecules of NO metabolized to the same number of molecules of NO_3–NO_2. The plasma concentration of NO_3–NO_2 is an estimate of NO production since it is metabolized stoichiometrically to NO_3–NO_2. However, the plasma concentration of NO_3-NO_2 is the balance of production minus excretion in the urine and further metabolism of NO_3-NO_2. There have been no studies of the clearance of NO_3-NO_2 from plasma; however, physical stress decreases plasma NO_3-NO_2 with a half-life of 10 min (Mastronardi et al., 2001a), indicating that its removal by further metabolism and renal excretion is very rapid. We hypothesize that the concomitant increase in leptin with body weight is related to increased size of the fat stores of the animals as they grew. Indeed, it has recently been reported that the fat stores increase in proportion to weight gain even in male rats of the size range employed herein that are quite young (Iossa et al., 1999).

Since there was a positive linear correlation between plasma leptin and NO_3-NO_2 and we showed that leptin fluctuates in a circadian rhythm in the rat (Mastronardi et al., 2000a), as in humans (Licinio et al., 1998) we hypothesized that NO_3-NO_2 would also fluctuate in parallel with leptin. Indeed, there was a circadian variation of plasma NO_3-NO_2 that correlated throughout the 24 hr with that of plasma leptin. Since we have shown that LPS-induced leptin release is not mediated by NO (Finck et al., 1998), it is unlikely that NO increases leptin release. On the contrary, on the basis of our hypothesis, leptin would stimulate NO secretion. Therefore, we determined the effect of leptin on NO_3-NO_2 by performing in vitro and in vivo experiments. Since the main source of leptin is the adipocytes, which have leptin receptors on their cell membranes (Bornstein et al., 2000) and express both the inducible NOS (iNOS) and the constitutive endothelial NOS (eNOS) isoforms (Ribiere et al., 1996), we hypothesized that leptin would increase NO release from incubated EFP.

The highest dose of leptin tested (10^{-5} M) increased NO_3-NO_2 release into the medium indicating, that leptin directly or indirectly activates NOS causing release of NO. Although the concentration required was quite high, such concentrations probably exist adjacent to adipocytes after exocytosis of pinocytotic vesicles containing leptin, which is sufficient for autocrine activation of NOS by leptin receptors on adipocytes and paracrine activation of eNOS by leptin receptors on vascular endothelium adjacent to the adipocytes. Therefore, leptin release from adipocytes would increase NO production leading to increased plasma NO_3-NO_2. The released NO would diffuse to adjacent arterioles and venules relaxing their smooth muscle via activation of guanylyl cyclase and generation of cGMP from GTP. Relaxation would be induced by protein kinase G, resulting in dilatation of vessels and increased blood flow to the adipocytes. As was the case in vivo, each molecule of leptin added to the incubation medium, once the threshold concentration was reached, released 12 molecules/hr of NO suggesting that leptin activated NOS. Surprisingly, the in vivo experiments suggested that the number of molecules of NO released by a single leptin molecule was approximately 1030 molecules/hr. This suggests that leptin was much more effective to activate NOS in vivo than in vitro for some unexplained reasons. In a recent paper published while our research was in progress, it was also shown in anesthetized rats that IV administration of leptin into the femoral vein increased plasma NO_3-NO_2 (Fruhbeck, 1999). Similar to our results the increase in plasma NO_3-NO_2 in their experimental paradigm was nearly 100 percent (Fruhbeck, 1999).

Furthermore, since in recent research we have determined that the EFP is an important source of TNF-α (Mastronardi, Yu and McCann, unpublished data), we also studied the possible role of leptin to alter TNF-α secretion. Surprisingly, we determined that a 100-fold lower concentration of leptin (10^{-7} M) than the one that elicited the increase in NO_3-NO_2 was effective to increase TNF-α by 4-fold, and the highest concentration of leptin (10^{-5} M) increased TNF-α 60-fold, suggesting that leptin-induced TNF-α might directly activate NOS resulting in release of NO that was metabolized to NO_3–NO_2. TNF-α has been demonstrated to be a potent stimulator of NO (Oswald et al., 1999) by inducing the expression of iNOS that is present in adipose tissue (Ribiere et al., 1996).

The in vitro results were confirmed in vivo since IV administration of leptin into conscious rats bearing a jugular catheter also increased plasma NO_3–NO_2 and TNF-α. As was the case in vitro, the dose of leptin to increase plasma NO_3–NO_2 was higher (10-fold) than that which released TNF-α supporting the hypothesis that leptin released TNF-α that in turn activates NOS to release NO. However, if this is the case, why does the dose of leptin have to be increased 100-fold before there is significant release of NO in vitro? Perhaps the two act together to activate NOS inducing NO release.

Interestingly, in contrast to NO where each molecule of leptin released many NO molecules, more than 700 molecules of leptin were required to release a molecule of TNF-α in vivo and approximately 10^7 molecules were required in vitro. These findings suggest that the release of TNF-α by leptin does not contribute to leptin-induced NO secretion, but is an associated event.

In summary, we hypothesize (Figure 2.5) that prolactin stimulates leptin release from the adipocytes (Mastronardi et al., 2000a). Leptin combines with its transducing leptin receptor (OB-Rb) present on the cell membrane of both adipocytes and endothelial cells to increase the activity of eNOS by increasing intracellular free calcium that combines with calmodulin and activates eNOS. The NO released into the interstitium diffuses into the smooth muscle cells of the adjacent arterioles and venules. There, it activates GC that converts GTP into cGMP that relaxes the myocytes, thereby dilating the vessels and increasing the blood flow to adipose tissue required for metabolism.

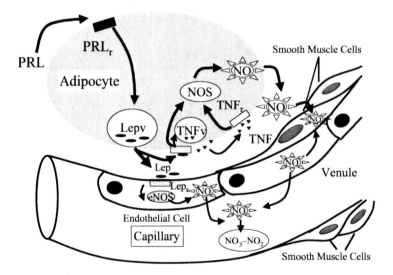

Figure 2.5. Leptin controls NO release from adipocytes. For description see last paragraph of Discussion on p. 32. Closed arrows indicate stimulation. Prolactin (PRL), PRL receptor (PRL_r), leptin (Lep), Lep vesicles (Lepv), Lep receptor (Lep_r), TNF vesicle (TNFv), and TNF receptor (TNF_r).

ACKNOWLEDGMENTS

This work was supported by NIH Grant MH51853. We would like to thank Judy Scott and Nicole Mestayer for their excellent secretarial support.

REFERENCES

Banks, W. A., Kastin, A. J., Huang, W., Jaspan, J. B., and Maness, L. M. (1996). Leptin enters the brain by a saturable system independent of insulin. *Peptides, 17*, 305–311.

Barash, I. A., Cheung, C. C., Weigle, D. S., Ren, H., Kabigting, E. B., Kuijper, J. L., Clifton, D. K., and Steiner, R. A. (1996). Leptin is a metabolic signal to the reproductive system. *Endocrinology, 137*, 3144–3147.

Bartfai, T., Hökfelt, T., and Langel, Ü. (1993). Galanin—a neuroendocrine peptide. *Crit. Rev. Neurobiol., 7*, 229–274.

Bartness, T. J., and Bamashad, M. (1998). Innervation of mammalian white adipose tissue: implications for regulation of total body fat. *Am. J. Physiol, 275*, 1399–1411.

Bornstein, S. R., Abu-Asab, M., Glasow, A., Path, G., Hauner, H., Tsokos, M., Chrousos, G. P., and Scherbaum, W. A. (2000). Immunohistochemical and ultrastructural localization of leptin and leptin receptor in human white adipose tissue and differentiating human adipose cells in primary culture. *Diabetes, 49*, 532–538.

Bornstein, S. R., Licinio, J., Tauchnitz, R., Engelmann, L., Negrao, A. B., Gold, P., and Chrousos, G. P. (1998). Lipopolysaccharide-induced changes in monoamines in specific areas of the brain: blockade by interleukin-1 receptor antagonist. *J. Clin. Endocrinol. Metab., 83*, 280–283.

Branchek, T. A., Smith, K. E., Gerald, C., and Walker, M. W. (2000). Galanin receptor subtypes. *Trends Pharmacol Sci., 21*, 109–117.

Chehab, F. F., Lim, M. E., and Ronghua, L. (1996). Correction of the sterility defect in homozygous obese female mice by treatment with the human recombinant leptin. *Nature Genetics, 12*, 318–320.

Chehab, F. F., Mounzih, K., Lu, R., and Lim, M. E. (1997). Early onset of reproductive function in normal female mice treated with leptin. *Science, 275*, 88–90.

Childs, G. V., Miller, B. T., Chico, D. E., Unabia, G. C., Yu, W. H., and McCann, S. M. (2001). Preferential expression of receptors for lamprey gonadotropin releasing hormone-III (GnRH-III) by FSH cells: support for its function as an FSH-RF. *The Endocrine Society's 83rd Annual Meeting*, Denver, CO, June 20–23, pp. 506, Abstract # P3-268.

Cioffi, J., Shafer, A., Zupancic, T., Smith-Gbur, J., Mikhail, A., Platika, D., and Snodgrass, H. (1996). Novel B219/ob receptor isoforms: possible role of leptin in hematopoiesis and reproduction. *Nature Med., 2*, 585–588.

Culler, M. D. and Negro-Vilar, A. (1987). Pulsatile follicle-stimulating hormone secretion is independent of luteinizing hormone-releasing hormone (LHRH): Pulsatile replacement of LHRH bioactivity in LHRH-immunoneutralized rats. *Endocrinology, 120*, 2011–2021.

Cunningham, M. J., Scarlett, J. M., and Steiner, R. A. (2002). Cloning and distribution of galanin-like peptide mRNA in the hypothalamus and pituitary of the macaque. *Endocrinology, 143*, 755–763.

Dearth, R. K., Hiney, J. K., and Dees, W. L. (2000) Leptin acts centrally to induce prepubertal secretion of LH in the female rat. *Peptides, 21*, 387–392.

Dees, W. L., Hiney, J. K., Sower, S. A., Yu, W. H., and McCann, S. M. (1999). Localization of immunoreactive lamprey gonadotropin-releasing hormone in the rat brain. *Peptides, 20*, 1503–1511.

Dees, W. L., Dearth, R. K., Hooper, R. N., Brinsko, S. P., Romano, J. E., Rahe, C. H., Yu, W. H., and McCann, S. M. (2001). Lamprey gonadotropin-releasing hormone-III selectively releases follicle stimulating hormone in the bovine. *Domest. Anim. Endocrinol., 20*, 279–288.

Dees, W. L., Rettori, V., Kozlowski, G., and McCann, S. M. (1985). Ethanol and the pulsatile release of luteinizing hormone, follicle-stimulating hormone and prolactin in ovariectomized rats. *Alcohol, 2*, 641–646.

Dhariwal, A. P. S., Watanabe, S., Antunes-Rodrigues, J., and McCann, S. M (1967). Chromatographic behavior of follicle stimulating hormone-releasing factor on Sephadex and carboxy methyl cellulose. *Neuroendocrinology, 2*, 294–303.

Evans, H. F. and Shine, J. (1991). Human galanin: molecular cloning reveals a unique structure. *Endocrinology, 129*, 1682–1684.

Evans, H. F., Huntley, G. W., Morrison, J. H., and Shine, J. (1993). Localisation of mRNA encoding the protein precursor of galanin in the monkey hypothalamus and basal forebrain. *J. Comp. Neurol., 328*, 203–212.

Fathi, Z., Battaglino, P. M., Iben, L. G., Li, H., Baker, E., Zhang, D., McGovern, R., Mahle, C. D., Sutherland, G. R., Iismaa, T. P., Dickinson, K. E., and Zimanyi, I. A. (1998). Molecular characterization, pharmacological properties and chromosomal localization of the human GALR2 galanin receptor. *Brain Res. Mol. Brain Res., 58*, 156–169.

Fattori, E., Cappelletti, M., Costa P., Sellitto, C., Cantoni, L., Carelli, M., Faggioni, R., Fantuzzi, G., Ghezzi, P., and Poli, V. (1994). Defective inflammatory response in interleukin 6-deficient mice. *J. Exp. Med., 180*, 1243–1250.

Finck, B. N., Kelley, K. W., Dantzer, R., and Johnson, R. W. (1998). Lipopolysaccharide-induced changes in monoamines in specific areas of the brain: blockade by interleukin-1 receptor antagonist. *Endocrinology, 139*, 2278–2283.

Finn, P. D., Pau, K.-Y. F., Spies, H. G., Cunningham, M. J., Clifton, D. K., and Steiner, R. A. (2000). *Neuroendocrinology, 71*, 16–26.

Finn, P. D., Cunningham, M. J., Rickard, D. G., Clifton, D. K., and Steiner, R. A. (2001). Serotonergic neurons are targets for leptin in the monkey. *Endocrinology, 86*, 422–426.

Frisch, R. E. and McArthur, J. W. (1974). Menstrual cycles: Fatness as a determinant of minimum weight for height necessary for their maintenance or onset. *Science, 185*, 949–951.

Fruhbeck, G. (1999). Pivotal role of nitric oxide in the control of blood pressure after leptin administration. *Diabetes, 48*, 903–908.

Giustina, A., Gastaldi, C., Bugari, G., Chiesa, L., Loda, G., Tironi, C., and Negro-Vilar, A. (1995). Role of galanin in the regulation of somatotrope and gondotrope function in young ovulatory women. *Metabolism, 44*, 1028–1032.

Gottschall, P. E., Komaki, G., and Arimura, A. (1992). Increased circulating interleukin-1 and interleukin-6 after intracerebroventricular injection of lipopolysaccharide. *Neuroendocrinology, 6*, 935–938.

Gualillo, O., Lago, F., Garcia, M., Menendez, C., Senaris, R., Casanueva, F. F., and Dieguez, C. (1999). Prolactin stimulates leptin secretion by rat white adipose tissue. *Endocrinology, 11*, 5149–5153.

Habert-Ortoli, E., Amiranoff, B., Loquet, I., Laburthe, M., and Mayaux, J.-F. (1994). Molecular cloning of a functional human galanin receptor. *Proc. Natl. Acad. Sci., 91*, 9780–9783.

Iossa, S., Lionetti, L., Mollica, M. P., Barletta, A., and Liverini, G. (1999). Energy intake and utilization vary during development in rats. *J. Nutr., 129*, 1593–1596.

Juréus, A., Cunningham, M. J., McClain, M. E., Clifton, D. K., and Steiner, R. A. (2000). Galanin-like peptide (GALP) is a target for regulation by leptin in the hypothalamus of the rat. *Endocrinology, 141*, 2703–2706.

Kakizaki, Y., Watanobe, H., Kohsaka, A., and Suda, T. (1999). Temporal profiles of interleukin-1beta, interleukin-6, and tumor necrosis factor-alpha in the plasma and hypothalamic paraventricular nucleus after intravenous or intraperitoneal administration of lipopolysaccharide in the rat: estimation by push–pull perfusion. *Endocr. J., 46*, 487–496.

Kerr, N. C. H., Holmes, F. E., and Wynick, D. (2000). Galanin like-peptide (GALP) is expressed in rat hypothalamus and pituitary, but not in DRG. *Neuroreport., 11*, 3909–3913.

Kline, A. D., Becker, G. W., Churgay, L. M., Landen, B. E., Martin, D. K., Muth, W. L., Rathnachalam, R., Richardson, J. M., Schoner, B., Ulmer, M., and Hale, J. E. (1997). Leptin is a four-helix bundle: secondary structure by NMR. *FEBS Lett., 407*, 239–242.

Klinger, M. M., MacCarter, G. D., and Boozer, C. N. (1996). Body weight and composition in the Sprague Dawley rat: comparison of three outbred sources. *Lab. Anim. Sci., 46*, 67–70.

Kolakowski Jr., L. F., O'Neill, G. P., Howard, A. D., Broussard, S. R., Sullivan, K. A., Feighner, D., Sawzdargo, M., Nguyen, T., Kargman, S., Shiao, L. L., Hreniuk, D. L., Tan, C. P., Evans, J., Abramovitz, M., Chanteauneuf, A., Coulombe, N., Ng, G., Johnson, M. P., Tharain, A., Khoshbouei, H., George, S. R., Smith, R. G., and O'Dowd, B. F. (1998). Molecular characteristics and expression of cloned human galanin receptors GALR2 and GALR3. *Neurochem., 71*, 2239–2251.

Kordower, J. H., Le, H. K., and Mufson, R. J. (1992). Galanin immunoreactivity in the primate central nervous system. *J. Comp. Neurol., 319*, 479–500.

Larm, J. A. and Gundlach, A. L. (2000). Galanin-like peptide (GALP) mRNA expression is restricted to accurate nucleus of hypothalamus in adult male rat brain. *Neuroendocrinology, 72*, 67–71.

Lescheid, D. W., Terasawa, E., Abler, L. A., Urbanski, H. F., Warby, C. M., Millar, R. P., and Sherwood, N. M. (1997). A second form of gonadotropin-releasing hormone (GnRH) with characteristics of chicken GnRH-II is present in the primate brain. *Endocrinology, 138*, 5618–5629.

Licinio, J., A. B. Negrão, C. Mantzoros, V. Kaklamani, M.-L. Wong, P. B. Bongiorno, A. Mulla, L. Cearnal, J. D. Veldhuis, J. S. Flier, S. M. McCann, and P. W. Gold. (1998). Synchronicity of frequently sampled, 24-h concentrations of circulating leptin, luteinizing hormone, and estradiol in healthy women. *Proc Natl. Acad. Sci., 95*, 2541–2546.

Lumpkin, M. D. and McCann, S. M. (1984). Effect of destruction of the dorsal anterior hypothalamus on selective follicle stimulating hormone secretion. *Endocrinology, 115*, 2473–2480.

Lumpkin, M. D., Moltz, J. H., Yu, W. H., Samson, W. K., and McCann, S. M. (1987). Purification of FSH-releasing factor: its dissimilarity from LHRH of mammalian, avian, and piscian origin. *Brain Res. Bull., 18*, 175–178.

Lumpkin, M. D., McDonald, J. K., Samson, W. K., and McCann, S. M. (1989). Destruction of the dorsal anterior hypothalamic region suppresses pulsatile release of follicle stimulating hormone but not luteinizing hormone. *Neuroendocrinology, 50*, 229–235.

Madej, T., Boguski, M. S., and Bryant, S. H. (1995). Lipopolysaccharide-induced changes in monoamines in specific areas of the brain: blockade by interleukin-1 receptor antagonist. *FEBS Lett., 373*, 13–18.

Marubayashi, U., Yu, W. H., and McCann, S. M. (1999). Median eminence lesions reveal separate hypothalamic control of pulsatile follicle-stimulating hormone and luteinizing hormone release. *Proc. Soc. Exp. Biol. Med., 220*, 139–146.

Mastronardi, C. A., Walczewska, A. Yu, W. H., Karanth, S., Parlow, A. F., and McCann, S. M. (2000a). The possible role of prolactin in the circadian rhythm of leptin secretion in male rats. *Proc. Exp. Biol. Med., 224*, 152–158.

Mastronardi, C. A., Yu, W. H., Rettori, V., and McCann, S. M. (2000b). Lipopolysaccharide-induced leptin release is not mediated by nitric oxide, but is blocked by dexamethasone. *Neuroimmunomodulation, 8*, 91–97.

Mastronardi, C. A., Yu, W. H., and McCann, S. M. (2001a). Comparisons of the effects of anesthesia and stress on release of tumor necrosis factor-α, leptin, and nitric oxide in adult male rats. *Exp. Biol. Med., 226*, 296–300.

Mastronardi, C. A., Yu, W. H., and McCann, S. M. (2001b). Lipopolysaccharide-induced tumor necrosis factor alpha release is controlled by the central nervous system. *Neuroimmunomodulation, 9*, 148–156.

Mastronardi, C. A., Yu, W. H., Srivastava, V. K., Dees, W. L., and McCann, S. M. (2001c). Lipopolysaccharide-induced leptin release is neurally controlled. *Proc. Natl. Acad. Sci., 98*, 14720–14725.

Mastronardi, C. A., Yu, W. H., and McCann, S. M. (2002). Resting and circadian release of nitric oxide is controlled by leptin in male rats. *Proc. Natl. Acad. Sci., 99*, 5721–5726, 2000.

Matsumoto, H., Noguchi, J., Takatsu, Y., Horikoshi, Y., Kumano, S., Ohtaki, T., Kitada, C., Itoh, T., Onada, H., Nishimura, O., and Fujino, M. (2001). Stimulation effect of galanin-like peptide (GALP) on luteinizing hormone-releasing hormone (LH) secretion in male rats. *Endocrinology, 142*, 3693–3696.

McAvery, K. M., Gimble, J. M., and Yu-Lee, L. (1996). Prolactin receptor expression during adipocyte differentiation of bone marrow stroma. *Endocrinology, 137*, 5723–5726

McCann, S. M., Kimura, M., Karanth, S., Yu, W. H., and Rettori, V. (1998). Role of nitric oxide in the neuroendocrine responses to cytokines. *Ann. N.Y. Acad. Sci., 840*, 174–184.

McCann, S. M., Kimura, M., Karanth, S., Yu, W. H., Mastronardi, C. A., and Rettori, V. (2000). The mechanism of action of cytokines to control the release of hypothalamic and pituitary hormones in infection. *Ann. N. Y. Acad. Sci., 917*, 4–18.

McCann, S. M., Kimura, M., Yu, W. H., Mastronardi, C. A., Rettori, V., and Karanth, S. (2001). Cytokines and pituitary hormone secretion. In G. Litwack (Ed.) *Vitamins and Hormone* (Vol 63). Academic Press, New York, pp. 29–62.

McCann, S. M. and Krulich, L. (1989). Role of neurotransmitters in control of anterior pituitary hormone release. *Endocrinology* (2nd ed.). WB Saunders, Philadelphia, pp. 117–130.

McCann, S. M., Licinio, J., Wong, M. L., Yu, W. H., Karanth, S., and Rettori, V. (1998). The nitric oxide hypothesis of aging. *Exp. Gerontol., 33*, 813–826.

McCann, S. M., Marubayashi, U., Sun, H-Q., and Yu, W. H. (1993). Control of follicle stimulating hormone and luteinizing hormone release by hypothalamic peptides. In G. P. Chrousos and G. Tolis (Eds.) *Intraovarian Regulators and Polycystic Ovarian Syndrome: Recent Progress on Clinical and Therapeutic Aspects* (Vol. 687). Ann. New York Acad. Sci., pp. 55–59.

McCann, S. M. and Ojeda, S. R. (1996). The anterior pituitary and hypothalamus. In J. Griffin and S. R. Ojeda (Eds.) *Textbook of Endocrine Physiology* (3rd ed.). Oxford University Press, New York, pp. 101–133.

McCann, S. M. and Rettori, V. (1997). The role of nitric oxide in reproduction. *Proc. Natl. Acad. Sci., 94,* 2735–2740.

McCann, S. M., Taleisnik, S., and Friedman, H. M. (1960). LH-releasing activity in hypothalarmic extracts. *Proc. Soc. Exp. Biol. Med, 104,* 432–434.

Mizunuma, H., Samson, W. K., Lumpkin, M. D., Moltz, J. H., Fawcett, C. P., and McCann, S. M. (1983). Purification of a bioactive FSH-releasing factor (FSHRH). *Brain Res. Bull., 10,* 623–629.

MohanKumar, S. M., MohanKumar, P. S., and Quadri, S. K. (1999). Lipopolysaccharide-induced changes in monoamines in specific areas of the brain: blockade by interleukin-1 receptor antagonist. *Brain Res., 824,* 232–237.

Morash, B., Li, A., Murphy, P. R., Wilkinson, M., and Ur, E. (1999). Leptin gene expression in the brain and pituitary gland. *Endocrinology, 140,* 5995–5998.

Naivar, J. S., Dyer, C. J., Matteri, R. L., and Keisler, D. H. (1996). Expression of leptin and its receptor in sheep tissues. *Proc. Soc. Study Reprod., 391,* 154 (abstract).

Nakano, H., Fawcett, C. P., Kimura, F., and McCann, S. M. (1978). Evidence for the involvement of guanosine 3',5'-cyclic monophosphate in the regulation of gonadotropin release. *Endocrinology, 103,* 1527–1533.

Naor, Z., Fawcett, C. P., and McCann, S. M. (1978). The involvement of cGMP in LHRH stimulated gonadotropin release. *Amer. J. Physiol., 235,* 586–590.

Neill, J. D., Duck, L. W., Sellers, J. C., and Musgrove, L. C. (2001). A gonadotropin-releasing hormone (GnRH) receptor specific for GnRH II in primates. *Biochem. & Biophys. Res. Comm., 282,* 1012–1018.

Ohtaki, T., Kumaano, S., Ishibashi, Y., Ogi, K., Matsui, H., Harada, M., Kitada, C., Kurokawa, T., Onda, H., and Fujino, M. (1999). Isolation and cDNA cloning of a novel galanin-like peptide (GALP) from porcine hypothalamus. *J. Biol. Chem., 274,* 37041–37045.

Ojeda, S. R., Jameson, H. E., and McCann, S. M. (1977). Hypothalamic areas involved in prostaglandin (PG)-induced gonadotropin release. II: Effect of PGE2 and PGF2alpha implants on follicle stimulating hormone release. *Endocrinology, 100,* 1595–1603.

Oswald, I. P., Dozois, C. M., Fournout, S., Petit, J. F., and Lemaire, G. (1999). Tumor necrosis factor is required for the priming of peritoneal macrophages by trehalose dimycolate. *Eur. Cytokine Netw., 10,* 533–540.

Ottlecz, A., Samson, W. K., and McCann, S. M. (1986). Galanin: evidence for a hypothalamic site of action to release growth hormone. *Peptides, 7,* 51–53.

Reichlin, S. (1992). In J. D. Wilson and D. W. Foster (Eds.) *Williams' Textbook of Endocrinology, Neuroendocrinology.* WB Saunders, Philadelphia, pp. 135–219.

Rettori, V., Dees, W. L., Hiney, J. K., Lyson, K., and McCann, S. M. (1994). An interleukin-1-alpha-like neuronal system in the preoptic-hypothalamic region and its induction by bacterial lipopolysaccharide in concentrations which alter pituitary hormone release. *Neuroimmunomodulation, 1,* 251–258.

Reznikov, A. G. and McCann, S. M. (1993). Effects of neuropeptide Y on gonadotropin and prolactin release in normal, castrated or flutamide-treated male rats. *Neuroendocrinology, 57,* 1148–1154.

Ribiere, C., Jaubert, A. M., Gaudiot, N., Sabourault, D., Marcus, M. L., Boucher, J. L., Denis-Henriot, D., and Giudicelli Y. (1996). White adipose tissue nitric oxide synthase: a potential source for NO production. *Biochem. Biophys. Res. Commun., 222,* 706–712.

Rock, F. L., Altmann, S. W., van Heek, M., Kastelein, R. A., and Bazan, J. F. (1996). The leptin haemopoietic cytokine fold is stabilized by an intrachain disulfide bond. *Horm. Metab. Res., 12,* 649–652.

Ronnekleiv, O. K., Ojeda, S. R., and McCann, S. M. (1978). Undernutrition, puberty and the development of estrogen positive feedback in the female rat. *Biol. Reprod., 19,* 414–424.

Schally, A. V., Saito, T., Arimura, A., Muller, E. E., Bowers, C. Y., and White, W. F. (1966). Purification of follicle-stimulating hormone-releasing factor (FSH-RF) from bovine hypothalamus. *Endocrinology, 79,* 1087–1094.

Schwartz, M. W., Seeley, R. J., Campfield, L. A., Burn, P., and Baskin, D. G. (1996). Identification of targets of leptin action in rat hypothalamus. *J. Clin. Invest., 98,* 1101–1106.

Snyder, G., Naor, Z., Fawcett, C. P., and McCann, S. M. (1980). Gonadotropin release and cyclic nucleotides: Evidence for LHRH-induced elevation of cyclic GMP levels in gonadotrophs. *Endocrinology, 107,* 1627–1633.

Soares, T. J., Coimbra, T. M., Martins, A. R., Pereira, A. G., Carnio, E. C., Branco, L. G., Albuquerque-Araujo, W. I., de Nucci, G., Favaretto, A. L., Gutkowska, J., McCann, S. M., and Antunes-Rodrigues, J. (1999). Atrial natriuretic peptide and oxytocin induce natriuresis by release of cGMP. *Proc. Natl. Acad. Sci. USA, 96*, 278–283.

Song, D. K., Im, Y. B., Jung, J. S., Suh, H. W., Huh, S. O., Park, S. W., Wie, M. B., and Kim, Y. H. (1999). Differential involvement of central and peripheral norepinephrine in the central lipopolysaccharide-induced interleukin-6 responses in mice. *J. Neurochem., 72*, 1625–1633.

Stojilkovic, S. S. (1998). Calcium signaling systems. In M.P. Conn and H.M. Goodman (Eds.) *Handbook of Physiology, Section 7 The Endocrine System, Volume I: Cellular Endocrinology*. Oxford Press, New York, pp. 177–224.

Swerdloff, R., Batt, R., and Bray, G. (1976). Reproductive hormonal function in the genetically obese (ob/ob) mouse. *Endocrinology, 98*, 1359–1364.

Swerdloff, R., Peterson, M., Vera, A., Batt, R., Heber, D., and Bray, G. (1978). The hypothalamic–pituitary axis in genetically obese (ob/ob) mice: response to luteinizing hormone-releasing hormone. *Endocrinology, 103*, 542–547.

Takahashi, N., Waelput, W., and Guisez, Y. (1999). Lipopolysaccharide-induced changes in monoamines in specific areas of the brain: blockade by interleukin-1 receptor antagonist. *J. Exp. Med., 189*, 207–212.

Takatsu, Y., Matsumoto, H., Ohtaki, T., Kumano, S., Kitada, C., Onda, H., Nishimura, O., and Fujino, M. (2001). Distribution of galanin-like peptide in the rat brain. *Endocrinology, 142*, 1626–1634.

Turnbull, A. V., Lee, S., and Rivier, C. (1998). Mechanisms of hypothalamic–pituitary–adrenal axis stimulation by immune signals in the adult rat. *Ann. N. Y. Acad. Sci., 840*, 434–443.

Vaisse, C., Halaas, J. L., Horvath, C. M., Darnell, J. E. Jr., Stoffel, M., and Friedman, J. M. (1996). Leptin activation of stat3 in the hypothalamus of wildtype and Ob/Ob mice but not Db/Db mice. *Nat. Genet., 14*, 95–97.

Vargas, H. M., Cuevas, J. M., Ignarro, L. J., and Chaudhuri, G. (1991). Comparison of the inhibitory potencies of N(G)-methyl-, N(G)-nitro- and N(G)-amino-L-arginine on EDRF function in the rat: evidence for continuous basal EDRF release. *J. Pharmacol. Exp. Ther., 257*, 1208–1215.

Vitale, M. L. and Chiocchio, S. R. (1993). Serotonin, a neurotransmitter involved in the regulation of luteinizing hormone release. *Endocrinology, 14*, 480–493.

Wakabayashi, K., Kamberi, I. A., and McCann, S. M. (1969). In vitro responses of the rat pituitary to gonadotrophin-releasing factors and to ions. *Endocrinology, 85*, 1046–1056.

Walczewska, A., Yu, W. H., Karanth, S., and McCann, S. M. (1999). Estrogen and leptin have differential effects on FSH and LH release in female rats. *Proc. Soc. Exp. Biol. Med., 222*, 70–77.

Wang, Y. S. and White, T. D. (1999). The bacterial endotoxin lipopolysaccharide causes rapid inappropriate excitation in rat cortex. *J. Neurochem., 72*, 652–660.

Wong, M. L., Rettori, V., Al-Shekhlee, A., Bongiorno, P. B., Canteros, G., McCann, S. M., Gold, P. W., and Licinio, J. (1996). Inducible nitric oxide synthase gene expression in the brain during systemic inflammation. *Nat. Med., 5*, 581–584.

Yu, W. H., Karanth, S., Walczewska, A., Sower, S. A., and McCann, S. M. (1997a). A hypothalamic follicle-stimulating hormone-releasing decapeptide in the rat. *Proc. Natl. Acad. Sci., 94*, 9499–9503.

Yu, W. H., Kimura, M., Walczewska, A., Karanth, S., and McCann, S. M. (1997b). Role of leptin in hypothalamic–pituitary function. *Proc. Natl. Acad. Sci., 94*, 1023–1028.

Yu, W. H., Walczewska, A., Karanth, S., and McCann, S. M. (1997c). Nitric oxide mediates leptin-induced luteinizing hormone-releasing hormone (LHRH) and LHRH and leptin-induced LH release from the pituitary gland. *Endocrinology, 138*, 5055–5058.

Yu, W. H., Karanth, S., Sower, S. A., Parlow, A. F., and McCann, S. M. (2000). The similarity of FSH-releasing factor to lamprey gonadotropin-releasing hormone III (lGnRH-III). *Proc. Soc. Exp. Biol. Med., 224*, 87–92.

Yu, W. H., Karanth, S., Mastronardi, C. A., Sealfon, S., Dean, C., Dees, W. L., and McCann, S. M. (2002). Lamprey GnRH-III acts on its putative receptor via nitric oxide to release follicle-stimulating hormone specifically. *Expl. Biol. Med.*, accepted.

Zhang, F., Basinski, M. B., Beals, J. M., Briggs, S. L., Churgay, L. M., Clawson, D. K., DiMarchi, R. D., Furman, T. C., Hale, J. E., Hsiung, H. M., Schoner, B. E., Smith, D. P., Zhang, X. Y., Wery, J. P., and Schevitz, R. W. (1997). Lipopolysaccharide-induced changes in monoamines in specific areas of the brain: blockade by interleukin-1 receptor antagonist. *Nature, 387*, 206–209.

Endocrine Regulation of Leptin Production

ROBERT V. CONSIDINE

3.1. INTRODUCTION

The *LEP* gene (originally termed *ob* gene in rodents) was discovered through a molecular genetics search for the defective gene that resulted in morbid obesity in *ob/ob* mice (Zhang et al., 1994). Early observations demonstrated that leptin, the protein product of the *LEP* gene, was synthesized by adipose tissue and released into the bloodstream to provide a signal to the central nervous system of the size of the adipose tissue mass, thus linking energy stores to mechanisms regulating energy intake and expenditure (reviewed in Matson et al., 1996; Considine and Caro, 1997; Friedman, 1998). It has since been recognized that leptin can be synthesized by cells other than adipocytes, and that this hormone participates in other physiologic processes (Considine and Caro, 1999). A substantial amount of work has provided insight into the mechanisms that regulate leptin synthesis and secretion from the adipose tissue, although many details remain to be elucidated. This chapter will summarize the endocrine regulation of leptin production in adipose tissue, focusing on findings in humans and human tissue manipulated in vitro. Regulation of leptin synthesis in tissues other than adipose tissue, and the possible regulatory function(s) of the hormone produced by these tissues will be discussed.

3.2. SERUM LEPTIN IS PROPORTIONAL TO BODY FAT MASS

Before considering the effect of various hormones and factors to regulate leptin release, it is important to recognize that circulating leptin concentrations are proportional

ROBERT V. CONSIDINE • Department of Medicine, Indiana University School of Medicine.

Leptin and Reproduction. Edited by Henson and Castracane, Kluwer Academic/Plenum Publishers, 2003.

to the amount of body fat under states of consistent food intake. Leptin is highly correlated with fat mass in adults, children, and newborns (Considine and Caro, 1997; Blum et al., 1998). Therefore, serum leptin is greater in subjects with larger body fat stores. In contrast, leptin is significantly reduced in lipodystrophic states compared to healthy lean subjects (Oral et al., 2002). In addition to the amount of body fat, serum leptin is influenced by the distribution of fat to specific depots, and the size of the adipocyte itself. Leptin release from abdominal subcutaneous adipocytes is greater than that from abdominal omental (visceral) adipocytes (Van Harmelen et al., 1998; Williams et al., 2000), an effect that is likely related to adipocyte size. *LEP* gene expression is greater in larger adipocytes than smaller adipocytes isolated from the same piece of adipose tissue (Hamilton et al., 1995) and leptin secretion is strongly correlated with fat cell volume (Lonnqvist et al., 1997). Adipocytes from the subcutaneous adipose tissue are larger than those from the omental depot (Fried and Kral, 1987). The link between adipocyte size and regulation of leptin synthesis has not been completely elucidated. However, recent work has suggested that UDP-N-acetylglucosamine (UDP-GlcNAc), an end product of hexosamine biosynthesis, may link cell size to leptin release (Considine et al., 2000). UDP-GlcNAc in human subcutaneous adipose tissue is positively correlated with BMI in lean and obese subjects. Further, increased synthesis of UDP-GlcNAc results in increased leptin release and inhibition of hexosamine biosynthesis reduces leptin production in vitro. These observations suggest that UDP-GlcNAc may provide a signal of adipocyte size that determines leptin production.

Alterations in the amount of adipose tissue with weight loss or weight gain are reflected by serum leptin. A decrease in adipose tissue results in a decrease in leptin while an increase in adipose tissue significantly increases circulating leptin concentrations (reviewed in Matson et al., 1996; Considine and Caro, 1997; Friedman, 1998). The mechanism linking leptin synthesis and release with weight change likely involves alterations in both the number and size of adipocytes within the tissue.

3.3. GENDER DIMORPHISM IN SERUM LEPTIN

Serum leptin is greater in women than in men. Two separate explanations have been put forth to explain this observation.

3.3.1. Role of Fat Distribution

In the first studies to examine serum leptin concentrations it was recognized that leptin was significantly greater in women than in men (Maffei et al., 1995; Considine et al., 1996). The most straightforward explanation for this finding was that, in general, women have a greater amount of body fat compared to men of equivalent weight or body mass index (BMI), and this explanation was sufficient for many early studies in which fat mass was not quantitated. However, in men and women of equivalent fat mass leptin is significantly greater in women (Rosenbaum and Leibel, 1999). One explanation for this finding is that women have significantly more subcutaneous adipose tissue mass relative to visceral adipose mass than men. Several studies utilizing adipose tissue obtained primarily or entirely from females have observed that *LEP* gene expression/leptin secretion is greater in subcutaneous than omental adipocytes from the same individual (Montague et al., 1997;

Lefebve et al., 1998; Van Harmelen et al., 1998; Williams et al., 2000). As discussed above, cell size may explain in part, the greater leptin production in subcutaneous versus omental adipose tissue and the gender effect on serum leptin. However, Rosenbaum et al. (2001) recently examined serum leptin in healthy lean men and women tightly matched for body fat mass and subcutaneous adipose tissue (SAT) volume. These investigators conclude that visceral adipose tissue (VAT) volume is small compared to that of SAT, and that differences in the relative amounts of VAT and SAT do not explain the gender dimorphism in serum leptin. Rosenbaum et al. suggest that gonadal steroids and/or primary genetic differences are greater determinants of serum leptin than adipose tissue distribution.

3.3.2. Role of Gonadal Steroids

Several observations, both in vivo and in vitro, suggest that gonadal steroids regulate leptin synthesis and release from the adipose tissue. Administration of testosterone therapy to hypogonadal men reduces serum leptin, although the concommitant reduction in fat mass with therapy precludes the assignment of a direct effect of testosterone on leptin secretion (Sih et al., 1997; Jockenhovel et al., 1997). As boys progress through puberty, leptin levels are reduced as serum testosterone increases (Mantzoros et al., 1997b; Horlick et al., 2000). Testosterone administration in female to male transsexuals decreases leptin, and estrogens, in combination with anti-androgens, increase leptin in male to female transsexuals, independent of changes in adipose tissue mass (Elbers et al., 1997). In vitro, estradiol stimulates, and dihydrotestosterone inhibits leptin release from cultured human adipose tissue pieces or differentiated preadipocytes (Wabitsch et al., 1997; Casabiell et al., 1998; Pineiro et al., 1999). These in vitro observations suggest that the gonadal steroids directly regulate leptin synthesis and release from adipose tissue in vivo, although the mechanism(s) through which this effect occurs has not yet been elucidated.

3.4. EFFECT OF ENERGY INTAKE ON SERUM LEPTIN

Serum leptin concentrations reflect caloric intake independent of changes in adipose tissue mass. However, leptin is not a true post-prandial hormone such as CCK, GLP-1, or insulin. Studies both in vivo and in vitro suggest that glucose and insulin link energy intake to leptin production in the adipose tissue.

3.4.1. Role of Insulin and Glucose

Serum leptin declines during short-term fasting (24 hr) in the absence of any change in adipose tissue mass. A slow infusion of glucose to maintain blood concentrations at 90 mg/dl prevents the fasting-induced fall in leptin, suggesting a role for insulin or glucose as the nutritional signal that is recognized by the adipocyte for leptin synthesis (Boden et al., 1996; Kolaczynski et al., 1996). During more prolonged energy restriction (moderate and severe) reductions in serum leptin were best correlated with the reduction in glycemia (Wisse et al., 1999).

Serum leptin exhibits a diurnal profile that is entrained to food intake (Sinha et al., 1996; Schoeller et al., 1997). The peak in serum leptin occurs at ~02.00 hr, in both lean and obese subjects under normal living conditions. Day/night reversal shifts the peak in serum leptin by 12 hr. A meal shift of 6.5 hr without changing light or sleep cycles shifts the leptin peak 5–7 hr. These observations suggest that the nocturnal rise in leptin may represent a delayed postprandial response and are in agreement with observations that glucose and insulin are important regulators of leptin synthesis.

The macronutrient content of the food ingested can also influence serum leptin. Consumption of high fat/low carbohydrate meals (60/20 percent) leads to a reduction in leptin levels (Havel et al., 1999). High fat/low carbohydrate meals induce smaller postprandial excursions in insulin and glucose than meals of standard fat/carbohydrate content, implicating insulin and glucose in the nutritional regulation of leptin production.

Direct in vivo evidence that glucose and insulin regulate serum leptin has been obtained with hyperinsulinemic–euglycemic clamp techniques. Serum leptin is elevated by the end of prolonged (9 hr) hyperinsulinemic–euglycemic clamps using physiologic insulin concentrations over that at the start of clamp (Saad et al., 1998). Clamps utilizing supraphysiologic insulin concentrations increase serum leptin in 4–8 hr (Utriainen et al., 1996; Saad et al., 1998).

3.4.2. Hexosamines Link Insulin/Glucose to Leptin Synthesis

One mechanism by which glucose metabolism regulates leptin synthesis is through hexosamine biosynthesis (Wells et al., 2001). The endproduct of this pathway, UDP-GlcNAc, is utilized in O-linked glycosylation reactions to post-translationally modify proteins such as transcription factors. O-glycosylation has been demonstrated to regulate the activity of Sp1, a transcription factor with binding sites in the *LEP* gene promoter. In rodents, infusion of glucosamine, uridine or free fatty acids during a 3 hr hyperinsulinemic–euglycemic clamp increased tissue UDP-GlcNAc, muscle *LEP* gene expression and serum leptin, compared to saline-infused controls clamped under identical conditions (Wang et al., 1998). In a transgenic mouse model overexpressing the rate-limiting enzyme in hexosamine biosynthesis (GFAT) in skeletal muscle and adipose tissue, serum leptin is increased (McClain et al., 2000).

In humans, UDP-GlcNAc is increased 3.2-fold in the subcutaneous adipose tissue of obese compared to lean subjects, and tissue UDP-GlcNAc content is positively associated with BMI (Considine et al., 2000). Glucosamine, an intermediate in UDP-GlcNAc biosynthesis, increased leptin release from cultured human subcutaneous adipocytes, and inhibition of GFAT activity with 6-diazo-5-oxo-norleucine reduced glucose-stimulated leptin release. These observations suggest that hexosamine biosynthesis links glucose metabolism to leptin production. In addition, glucose metabolism to endproducts other than hexosamines, such as ATP, may also influence leptin synthesis and release (Levy et al., 2000).

3.5. INHIBITION OF LEPTIN RELEASE BY THE SYMPATHETIC NERVOUS SYSTEM

Leptin binding to its receptor in the hypothalamus activates neural pathways to reduce food intake, and activates the sympathetic nervous system to increase energy expenditure (Haynes et al., 1997; Trayhurn et al., 1998). Activation of sympathetic nerves also appears to be the negative feedback signal to the adipose tissue to reduce leptin release (Rayner and Trayhurn, 2001). Administration of isoproterenol to human subjects acutely reduces serum leptin within 20–60 min of the start of the infusion (Donahoo et al., 1997; Pinkney et al., 1998; Stumvoll et al., 2000). A similar reduction in leptin was found with epinephrine infusion for 3 hr (Carulli et al., 1999). The effects of catecholamines on leptin release are postulated to occur through β-adrenergic receptors and increased cAMP in the adipocytes. However, in one study isoproterenol reduced leptin release when co-infused with acipimox, an inhibitor of adenylate cyclase (Stumvoll et al., 2000). This observation suggests that catecholamines may also regulate leptin synthesis and release through non-traditional mechanisms not activated by cAMP. In vitro, catecholamines and cAMP have been shown to reduce *LEP* mRNA and leptin synthesis in human adipose tissue pieces and adipocytes differentiated in vitro (Halleux et al., 1998; Ricci and Fried, 1999; Scriba et al., 2000). Similar observations have been made with various rodent models and in 3T3-L1 cells (Rayner and Trayhurn, 2001).

3.6. OTHER HORMONES AND FACTORS THAT INFLUENCE LEPTIN RELEASE

3.6.1. Glucocorticoids

The synthetic glucocorticoid dexamethasone is a potent stimulus for leptin secretion in vivo and in vitro and several studies suggest that glucocorticoids synergize with insulin to increase leptin production (Fried et al., 2000). However a physiologic relevance for glucocorticoid-induced leptin release has been questioned. Patients with adrenal insufficiency in whom hydrocortisone was withdrawn for 72 hr have a normal diurnal pattern of serum leptin, indicating that cortisol is not an absolute requirement for leptin synthesis (Purnell and Samuels, 1999). Further, adreno-chorticophin hormone (ACTH) and cortisol levels are not elevated in obese humans (Drent, 1998); therefore cortisol does not appear to have a direct role in the increase in serum leptin in obese subjects. However, it has recently been recognized that 11β-hydroxysteroid dehydrogenase in the adipose tissue plays a major role in the modulation of local glucocorticoid concentrations in the tissue by reactivating glucocorticoids from inactive metabolites (Sandeep and Walker, 2001). Local glucocorticoid concentrations may therefore play an important role in the regulation of leptin synthesis and release. However, as discussed below, the mechanism through which glucocorticoids regulate leptin production is incompletely understood.

3.6.2. Cytokines

The effect of cytokines on leptin in humans is complicated. In vivo, TNF-α (Zumbach et al., 1997) and interleukin 1α (Janik et al., 1997) increase serum leptin and survivors of acute sepsis exhibit elevated leptin levels (Borstein et al., 1998). In vitro, TNFα produces significant time- and dose-dependent inhibition of leptin production from cultured primary adipocytes and preadipocytes differentiated in vitro (Gottschling-Zeller et al., 1999; Fawcett et al., 2000; Zhang et al., 2000). It is, therefore, possible that the cytokine-induced increase in serum leptin in vivo may not result from direct effects of the cytokine on the adipose tissue. The mechanism through which TNFα regulates leptin production in mature adipocytes is not yet known.

3.6.3. Activation of PPARγ

Thiazolidinediones are potent activators of the transcription factor PPARγ (Berger and Moller, 2002). Thiazolidinediones and 15-deoxy-prostaglandin J_2, a putative naturally occuring ligand for PPARγ, inhibit *LEP* gene expression in isolated human adipocytes (Williams et al., 2000) and cultured 3T3-L1 cells (Sinha et al., 1999; Berger et al., 2001). In vivo in humans, troglitazone has been reported to have no effect on serum leptin in two studies (Nolan et al., 1996; Mantzoros et al., 1997a) and reduce leptin in a third study (Shimizu et al., 1998). The observation that thiazolidinediones do not change serum leptin in vivo may be the result of a dual effect of the drug to improve insulin sensitivity, which should increase leptin synthesis, and to directly decrease *LEP* gene expression. As discussed below, activated PPARγ interacts with the *LEP* gene promoter to reduce transcription through a novel mechansim.

3.7. TRANSCRIPTIONAL REGULATION OF THE *LEP* GENE

Changes in serum leptin in vivo, and leptin secretion from cultured adipocytes in vitro, are associated with parallel changes in the amount of *LEP* gene expression in the adipose tissue/adipocyte. Leptin also does not appear to be stored in secretory granules to any great extent but rather is released in a constitutive fashion. These observations suggest that *LEP* transcription is a major point of regulation of leptin synthesis and release.

3.7.1. Hexosamines

Approximately 5 kb of the human *LEP* gene promoter has been mapped and many *cis*-acting regulatory elements are present (Gong et al., 1996; Hollenberg et al., 1997). The proximal promoter region ($\sim$200 bp), which is sufficient for adipose expression of the gene, contains a TATA box, C/EBP (CCAAT/enhancer binding protein) interaction sequence and three GC boxes, which bind Sp1 (Gong et al., 1996; Miller et al., 1996). The C/EBP binding motif is a consistent finding for adipocyte-specific genes (Lane et al., 1999) and mutation of this site eliminates all *LEP* promoter activity (Miller et al., 1996). Mutation of the GC boxes also significantly reduces promoter activity (Mason et al., 1998; Zhang et al., 2002). One mechanism through which glucose and insulin appear to regulate the *LEP* promoter is the synthesis of hexosamines. As discussed above, O-glycosylation of Sp1 alters its transcriptional activity, either by preventing degradation of the transcription factor or by enhancing its DNA binding activity at these GC boxes (Wells et al., 2001).

3.7.2. Glucocorticoids

Glucocorticoids strongly increase LEP mRNA in adipose tissue in vivo or in adipocytes treated in culture. A traditional glucocorticoid response element is located at -1283 to -1276 bp of the LEP promoter sequence (Gong et al., 1996). However, De Vos et al. (1998) observed that dexamethasone-stimulated LEP promoter activity did not require this GRE as truncated promoter constructs lacking the site were activated as well as constructs containing the site. Deletion mapping of the human promoter indicated that a cis-element for glucocorticoid interaction is located between -55 and $+31$ bp. This promoter sequence does not bind the glucocorticoid receptor and therefore the mechanism through which glucocorticoids activate transcription at this site is unresolved.

3.7.3. PPARγ Agonists

Thiazolidinediones reduce LEP gene expression (De Vos et al., 1998), and a PPARγ binding sequence is located at -3951 to -3939 bp (Hollenberg et al., 1997). However, as is the case for dexamethasone and its traditional GRE on the leptin promoter, troglitazone-activated PPARγ does not inhibit LEP gene promoter activity through its recognized consensus sequence. Rather, deletion mapping identified a cis-element for activated PPARγ between -65 to $+9$ bp. PPARγ does not bind to any sequence within this region but it has been proposed that PPARγ may antagonize the binding of C/EBPα to down regulate LEP gene transcription.

3.7.4. β-Adrenergics and cAMP

β-Adrenergic agonists and cAMP inhibit LEP gene expression in vivo and in vitro in cultured adipocytes (Considine and Caro, 1999). Two CREB (cAMP response) elements have been identified at -1384 to -1380 bp and -2329 to -2323 bp of the promoter sequence (Gong et al., 1996), but the functionality of these sites has not been demonstrated. Cyclic AMP regulates the transcription of many genes through the activation of protein kinase-A (PKA) and the subsequent phosphorylation of the transcription factor CREB (Shaywitz and Greenberg, 1999). For most genes studied, CREB is an activator of gene expression, not an inhibitor, as it would need to be in the case of LEP gene transcription. This fact points to the possiblity of a novel mechanism for cAMP/CREB to regulate LEP promoter activity.

3.7.5. Non-Transcriptional Regulation of Leptin Release

Very little work has been done on the regulation of leptin synthesis and release that occurs via non-transcriptional means. Two studies in rat adipocytes suggest that insulin may acutely increase leptin secretion without increasing LEP transcription (Barr et al., 1997; Bradley and Cheatham, 1999). In the study of Barr et al. (1997) insulin acutely increased the movement of leptin from the endoplasmic reticulum to the plasma membrane for release. However, acute insulin-induced changes in serum leptin have not been observed in human subjects, possibly because the amount of leptin released into the circulation via this mechanism is small relative to the circulating concentration.

3.8. LEPTIN SYNTHESIS IN TISSUES OTHER THAN FAT

The adipose tissue is the primary site of leptin synthesis and release into the circulation; however, other tissues appear to be able to make the hormone. Leptin synthesized in tissues other than fat does not appear to significantly influence the serum concentration of the hormone but likely acts in a paracrine or autocrine manner to regulate metabolism in the tissue.

Of particular interest to readers of this text is the fact that leptin is synthesized by human placenta from the same gene as that in the adipose tissue. Contained within a medium reiteration frequency repeat element (MER11) of the placental *LEP* promoter is an enhancer, which permits the expression of leptin by novel placental-specific transcription factors (Bi et al., 1997). These factors have not yet been identified. Placental leptin production and its physiologic effects are covered in greater detail in other chapters.

Leptin mRNA and protein have been discovered in the fundic epithelium of the stomach and its synthesis appears to be regulated by the gastrointestinal hormone CCK (Bado et al., 1998). CCK-8 induced a time-dependent increase in leptin release from rat stomach perfused in vitro, and in vivo CCK-8 increased leptin release from rat stomach following an 18 hr fast. Bado et al. (1998) propose that stomach-derived leptin synergizes with CCK to exert satiety effects through local actions on vagus afferent neurons. Leptin-responsive gastric vagal afferent terminals have been identified (Wang et al., 1997).

LEP gene expression in rat muscle has been observed following infusion of the hexosamine biosynthetic intermediate glucosamine or feeding (Wang et al., 1998, 1999). Evidence that muscle cells rather than adipocytes within the skeletal muscle synthesized leptin was provided by immunohistochemistry and the fact that RNA for the adipose specific gene AdipoQ was not increased. The authors propose that skeletal muscle leptin may play an important paracrine role in intermediary metabolism.

Bone marrow contains adipocytes that appear to be functionally similar to adipocytes derived from the fat pads (Gimble et al., 1996). Human bone marrow adipocytes differentiated in vitro synthesize and secrete leptin (Laharrague et al., 1998). Leptin receptors are present on many different hematopoietic cells such as blast cells, promyelocytes, promonocytes, macrophages, and others. Leptin increases the proliferation of many of these cell types. Leptin also prevents the starvation-induced reduction in immune function and stimulates the proliferation of naive and memory T cells (Lord et al., 1998). These observations suggest that leptin synthesized by bone marrow adipocytes has a regulatory role in hematopoiesis and immune function.

3.9. SUMMARY AND CONCLUSIONS

Serum leptin is synthesized and released into the circulation in proportion to the amount of body fat. However, additional factors regulate leptin synthesis (Figure 3.1). After accounting for body fat, the single most important determinant of serum leptin is gender, resulting in more circulating leptin in women than men of equivalent body fat. The gonadal steroids are responsible for this gender dichotomy in serum leptin, influencing the site of fat deposition and directly acting on leptin synthesis in the adipose tissue. Glucose metabolism

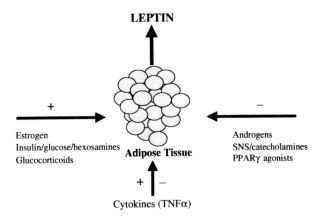

Figure 3.1. Several hormones and factors influence leptin synthesis and release from adipose tissue. The amount of adipose tissue and gender (through the action of the gonadal steriods) are major determinants of leptin release SNS: sympathetic nervous system.

to hexosamines appears to link caloric intake to leptin synthesis in the adipose tissue and the sympathetic nervous system, acting as the negative feedback arm of the leptin pathway, inhibits leptin release from the adipose tissue. Other regulators of leptin synthesis and release include glucocorticoids, cytokines and agonists of PPARγ, although the physiologic role of these factors is not fully understood. Leptin can be synthesized in several tissues other than adipose tissue and likely acts in a paracrine/autocrine fashion at these sites.

The major function of serum leptin is to inform the central nervous system of the size of energy stores in the body. With this information the brain can modulate energy intake and expenditure to maintain the appropriate energy balance needed for survival and successful reproduction. Therefore, leptin has an important, central role in the physiology of reproduction.

ACKNOWLEDGMENT

Work of the author cited in this chapter was supported in part by grants from the National Institutes of Health (DK51140), the American Diabetes Association and the Showalter Trust.

REFERENCES

Bado, A., Levasseur, S., Attoub, S., Kermorgant, S., Laigneau, J.-P., Bortoluzzi, M.-N., Moizo L., Lehy, T., Guerre-Millo, M., Le Marchand-Brustel, Y., and Lewin, M. J. M. (1998). The stomach is a source of leptin. *Nature, 394,* 790–793.

Barr, V. A., Malide, D., Zarnowski, M. J., Taylor, S. I., and Cushman, S. W. (1997). Insulin stimulates both leptin secretion and production by white adipose tissue. *Endocrinology, 138,* 4463–4472.

Berger, J. and Moller, D. E. (2002). The mechanisms of action of PPARs. *Annu. Rev. Med., 53*, 409–435.

Berger, J., Tanen, M., Elbrecht, A., Hermanowski-Vosatka, A., Moller, D. E., Wright, S. D., and Thieringer, R. (2001). Peroxisome proliferator-activated receptor-gamma ligands inhibit adipocyte 11beta -hydroxysteroid dehydrogenase type 1 expression and activity. *J. Biol. Chem., 276*, 12629–12635.

Bi, S., Gavrilova, O., Gong, D-W., Mason, M. M., and Reitman, M. (1997). Identification of a placental enhancer for the human leptin gene. *J. Biol. Chem., 272*, 30583–30588.

Blum, W. F., Englaro, P., Attanasio, A. M., Kiess, W., and Rascher, W. (1998). Human and clinical perspectives on leptin. *Proc. Nutr. Soc., 57*, 477–485.

Boden, G., Chen, X., Mozzoli, M., and Ryan, I. (1996). Effect of fasting on serum leptin in normal human subjects. *J. Clin. Endocrinol. Metab., 81*, 3419–3423.

Bornstein, S. R., Licinio, J., Tauchnitz, R., Engelmann, L., Negrao, A. B., Gold, P., and Chrousos, G. P. (1998). Plasma leptin levels are increased in survivors of acute sepsis: associated loss of diurnal rhythm, in cortisol and leptin secretion. *J. Clin. Endocrinol. Metab., 83*, 280–283.

Bradley, R. L. and Cheatham, B. (1999). Regulation of ob gene expression and leptin secretion by insulin and dexamethasone in rat adipocytes. *Diabetes, 48*, 272–278.

Carulli, L., Ferrari, S., Bertolini, M., Tagliafico, E., and Del Rio, G. (1999). Regulation of ob gene expression: evidence for epinephrine-induced suppression in human obesity. *J. Clin. Endocrinol. Metab., 84*, 3309–3312.

Casabiell, X., Pineiro, V., Peino, R., Lage, M., Camina, J. P., Gallego, R., Vallejo, L. G., Dieguez, C., and Casanueva, F. F. (1998). Gender differences in both spontaneous and stimulated leptin secretion by human omental adipose tissue in vitro: dexamethasone and estradiol stimulate leptin release in women, but not in men. *J. Clin. Endocrinol. Metab., 83*, 2149–2155.

Considine, R. V. and Caro, J. F. (1997). Leptin and the regulation of body weight. *Int. J. Biochem. Cell. Biol., 29*, 1255–1272.

Considine, R. V. and Caro, J. F. (1999). Pleotropic Cellular Effects of Leptin. *Curr. Opin. Endocrinol. Diabetes, 6*, 163–169.

Considine, R. V., Cooksey, R. C., Williams, L. B., Fawcett, R. L., Zhang, P., Ambrosius, W. T., Whitfield, R. M., Jones, R. M., Inman, M., Huse, J., and McClain, D. A. (2000). Hexosamines regulate leptin production in human subcutaneous adipocytes. *J. Clin. Endocrinol Metab., 85*, 3551–3556.

Considine, R. V., Sinha, M. K., Heiman, M. L., Kriauciunas, A., Stephens, T. W., Nyce, M. R., Ohannesian, J. P., Marco, C. C., McKee, L. J., Bauer, T. L., and Caro, J. F. (1996). Serum immunoreactive-leptin concentrations in normal-weight and obese humans. *N. Engl. J. Med., 334*, 292–295.

De Vos, P., Lefebvre, A. M., Shrivo, I., Fruchart, J. C., and Auwerx, J. (1998). Glucocorticoids induce the expression of the leptin gene through a non-classical mechanism of transcriptional activation. *Eur. J. Biochem., 253*, 619–626.

Donahoo, W. T., Jensen, D. R., Yost, T. J., and Eckel, R. H. (1997). Isoproterenol and somatostatin decrease plasma leptin in humans: a novel mechanism regulating leptin secretion. *J. Clin. Endocrinol. Metab., 82*, 4139–4143.

Drent, M. L. (1998). Effects of obesity on endocrine function. In G. A. Bray, C. Bouchard, and W. P. T. James (Eds.), *Handbook of Obesity*. Marcel Dekker, New York, pp. 753–773.

Elbers, J. M., Asscheman, H., Seidell, J. C., Frolich, M., Meinders, A. E., and Gooren, L. J. (1997). Reversal of the sex difference in serum leptin levels upon cross-sex hormone administration in transsexuals. *J. Clin. Endocrinol. Metab., 82*, 3267–3270.

Fawcett, R. L., Waechter, A. S., Williams, L. B., Zhang, P., Louie, R., Jones, R. M., Inman, M., Huse, J., and Considine, R. V. (2000). TNFα inhibits leptin production in subcutaneous and omental adipocytes from morbidly obese humans. *J. Clin. Endocrinol. Metab., 85*, 530–535.

Fried, S. K. and Kral, J. G. (1987). Sex differences in regional distribution of fat cell size and lipoprotein lipase activity in morbidly obese patients. *Int. J. Obesity, 11*, 129–140.

Fried, S. K., Ricci, M. R., Russell, C. D., and Laferrere, B. (2000). Regulation of leptin production in humans. *J. Nutr., 130*, 3127S–3131S.

Friedman, J. M. (1998). Leptin, leptin receptors, and the control of body weight. *Nutr. Rev., 56*, S38–S46.

Gimble, J. M., Robinson, C. E., Wu, X., and Kelly, K. A. (1996). The function of adipocytes in the bone marrow stroma: an update. *Bone, 19*, 421–428.

Gong, D. W., Bi, S., Pratley, R. E., and Weintraub, B. D. (1996). Genomic structure and promoter analysis of the human *obese* gene. *J. Biol. Chem., 271*, 3971–3974.

Gottschling-Zeller, H., Birgel, M., Scriba, D., Blum, W. F., and Hauner, H. (1999). Depot-specific release of leptin from subcutaneous and omental adipocytes in suspension culture: effect of tumor necrosis factor-alpha and transforming growth factor-beta1. *Eur. J. Endocrinol., 141*, 436–442.

Halleux, C. M., Servais, I., Reul, B. A., Detry, R., and Brichard, S. M. (1998). Multihormonal control of ob gene expression and leptin secretion from cultured human visceral adipose tissue: increased responsiveness to glucocorticoids in obesity. *J. Clin. Endocrinol. Metab., 83*, 902–910.

Hamilton, B. S., Paglia, D., Kwan, A. Y. M., and Deitel, M. (1995). Increased *obese* mRNA expression in omental fat cells from massively obese humans. *Nat. Med., 1*, 953–956.

Havel, P. J., Townsend, R., Chaump, L., and Teff, K. (1999). High fat meals reduce 24 h circulating leptin concentrations in women. *Diabetes, 48*, 334–341.

Haynes, W. G., Sivitz, W. I., Morgan, D. A., Walsh, S. A., and Mark, A. L. (1997). Sympathetic and cardiorenal actions of leptin. *Hypertension, 30*, 619–623.

Hollenberg, A. N., Susulic, V. S., Madura, J. P., Zhang, B., Moller, D. E., Tontonoz, P., Sarraf, P., Spiegelman, B. M., and Lowell, B. B. (1997). Functional antagonism between CCAAT/enhancer binding protein-α and peroxisome proliferator-activated receptor-γ on the leptin promoter. *J. Biol. Chem., 272*, 5283–5290.

Horlick, M. B., Rosenbaum, M., Nicolson, M., Levine, L. S., Fedun, B., Wang, J., Pierson, R. N. Jr, and Leibel, R. L. (2000). Effect of puberty on the relationship between circulating leptin and body composition. *J. Clin. Endocrinol. Metab., 85*, 2509–2518.

Janik, J. E., Curti, B. D., Considine, R. V., Rager, H. C., Powers, G. C., Alvord, W. G., Smith, J. W. 2nd, Gause, B. L., and Kopp, W. C. (1997). Interleukin 1 alpha increases serum leptin concentrations in humans. *J. Clin. Endocrinol. Metab., 82*, 3084–3086.

Jockenhovel, F., Blum, W. F., Vogel, E., Englaro, P., Muller-Wieland, D., Reinwein, D., Rascher, W., and Krone, W. (1997). Testosterone substitution normalizes elevated serum leptin levels in hypogonadal men. *J. Clin. Endocrinol. Metab., 82*, 2510–2513.

Kolaczynski, J. W., Considine, R. V., Ohannesian, J., Marco, C., Opentanova, I., Nyce, M. R., Myint, M., and Caro, J. F. (1996). Responses of leptin to short-term fasting and refeeding in humans: a link with ketogenesis but not ketones themselves. *Diabetes, 45*, 1511–1515.

Laharrague, P., Larrouy, D., Fontanilles, A.-M., Truel, N., Campfield, A., Tenenbaum, R., Galitzky, J., Corberand, J. X., Penicaud, L., and Casteilla, L. (1998). High expression of leptin by human bone marrow adipocytes in primary culture. *FASEB J., 12*, 747–752.

Lane, M. D., Tang, Q. Q., and Jiang, M. S. (1999). Role of the CCAAT enhancer binding proteins (C/EBPs) in adipocyte differentiation. *Biochem. Biophys. Res. Commun., 266*, 677–683.

Lefebvre, A. M., Laville, M., Veg, N., Riou, J. P., van Gaal, L., Auwerx, J., and Vidal, H. (1998). Depot-specific differences in adipose tissue gene expression in lean and obese subjects. *Diabetes, 47*, 98–103.

Levy, J. R., Gyarmati, J., Lesko, J. M., Adler, R. A., and Stevens, W. (2000). Dual regulation of leptin secretion: intracellular energy and calcium dependence of regulated pathway. *Am. J. Physiol. Endocrinol. Metab., 278*, E892–E901.

Lonnqvist, F., Nordfors, L., Jansson, M., Thorne, A., Schalling, M., and Arner, P. (1997). Leptin secretion from adipose tissue of women. Relatioship to plasma levels and gene expression. *J. Clin. Invest., 99*, 2398–2404.

Lord, G. M., Matarese, G., Howard, J. K., Baker, R. J., Bloom, S. R., and Lechler, R. I. (1998). Leptin modulates the T-cell immune response and reverses starvation-induced immunosuppression. *Nature, 394*, 897–901.

Maffei, M., Halaas, J., Ravussin, E., Pratley, R. E., Lee, G. H., Zhang, Y., Fei, H., Kim, S., Lallone, R., Ranganathan, S., Kern, P. A., and Friedman, J. M. (1995). Leptin levels in human and rodent: Measurement of plasma leptin and *ob* RNA in obese and weight-reduced subjects. *Nat. Med., 1*, 1155–1161.

Mantzoros, C. S., Dunaif, A., and Flier, J. S. (1997a). Leptin concentrations in the polycystic ovary syndrome. *J. Clin. Endocrinol. Metab., 82*, 1687–1691.

Mantzoros, C. S., Flier, J. S., and Rogol, A. D. (1997b). A longitudinal assessment of hormonal and physical alterations during normal puberty in boys. V. Rising leptin levels may signal the onset of puberty. *J. Clin. Endocrinol. Metab., 82*, 1066–1070.

Mason, M. M., He, Y., Chen, H., Quon, M. J., and Reitman, M. (1998). Regulation of leptin promoter function by Sp1, C/EBP, and a novel factor. *Endocrinology, 139*, 1013–1022.

Matson, C. A., Wiater, M., and Weigle, D. S. (1996). Leptin and the regulation of body adiposity. *Diabetes Rev., 4*, 488–508.

McClain, D. A., Alexander, T., Cooksey, R. C., and Considine, R. V. (2000). Hexosamines stimulate leptin production in transgenic mice. *Endocrinology, 141,* 1999–2002.

Miller, S. G., De Vos, P., Guerre-Millo, M., Wong, K., Hermann, T., Staels, B., Briggs, M. R., and Auwerx, J. (1996). The adipocyte specific transcription factor C/EBPα modulates human *ob* gene expression. *Proc. Natl. Acad. Sci. USA, 93,* 5507–5511.

Montague, C. T., Prins, J. B., Sanders, L., Digby, J. E., and O'Rahilly, S. (1997). Depot and sex-specific differences in human leptin mRNA expression. *Diabetes, 46,* 342–347.

Nolan, J. J., Olefsky, J. M., Nyce, M. R., Considine, R. V., and Caro, J. F. (1996). Effect of troglitazone on leptin production. Studies in vitro and in human subjects. *Diabetes, 45,* 1276–1278.

Oral, E. A., Simha, V., Ruiz, E., Andewelt, A., Premkumar, A., Snell, P., Wagner, A. J., DePaoli, A. M., Reitman, M. L., Taylor, S. I., Gorden, P., and Garg, A. (2002). Leptin-replacement therapy for lipodystrophy. *N. Engl. J. Med., 346,* 570–578.

Pineiro, V., Casabiell, X., Peino, R., Lage, M., Camina, J. P., Menendez, C., Baltar, J., Dieguez, C., and Casanueva, F. F. (1999). Dihydrotestosterone, stanozol, androstenedione and dehydroepiandrosterone sulphate inhibit leptin secretion in female but not male samples of omental adipose tissue in vitro: lack of effect of testosterone. *J. Endocrinol., 160,* 425–432.

Pinkney, J. H., Coppack, S. W., and Mohamed Ali, V. (1998). Effect of isoproterenol on plasma leptin and lipolysis in humans. *Clin. Endocrinol. (Oxf.), 48,* 407–411.

Purnell, J. Q. and Samuels, M. H. (1999). Levels of leptin during hydrocortisone infusions that mimic normal and reversed diurnal cortisol levels in subjects with adrenal insufficiency. *J. Clin. Endocrinol. Metab., 84,* 3125–3128.

Rayner, D. V. and Trayhurn, P. (2001). Regulation of leptin production: sympathetic nervous system interactions. *J. Mol. Med., 79,* 8–20.

Ricci, M. R. and Fried, S. K. (1999). Isoproterenol decreases leptin expression in adipose tissue of obese humans. *Obes. Res., 7,* 233–240.

Rosenbaum, M. and Leibel, R. L. (1999). Clinical Review 107. Role of gonadal steroids in the sexual dimorphisms in body composition and circulating concentrations of leptin. *J. Clin. Endocrinol. Metab., 84,* 1784–1789.

Rosenbaum, M., Pietrobelli, A., Vasselli, J. R., Heymsfield, S. B., and Leibel, R. L. (2001). Sexual dimorphism in circulating leptin concentrations is not accounted for by differences in adipose tissue distribution. *Int. J. Obes. Relat. Metab. Disord., 25,* 1365–1371.

Saad, M. F., Khan, A., Sharma, A., Michael, R., Road-Gabriel, M. G., Boyadjian, R., Jinagouda, S. D., Steil, G. M., and Kamdar, V. (1998). Physiological insulinemia acutely modulates plasma leptin. *Diabetes, 47,* 544–549.

Sandeep, T. C. and Walker, B. R. (2001). Pathophysiology of modulation of local glucocorticoid levels by 11beta-hydroxysteroid dehydrogenases. *Trends Endocrinol. Metab., 12,* 446–453.

Schoeller, D. A., Cella, L. K., Sinha, M. K., and Caro, J. F. (1997). Entrainment of the diurnal rhythm of plasma leptin to meal timing. *J. Clin. Invest., 100,* 1882–1887.

Scriba, D., Aprath-Husmann, I., Blum, W. F., and Hauner, H. (2000). Catecholamines suppress leptin release from in vitro differentiated subcutaneous human adipocytes in primary culture via beta1- and beta2-adrenergic receptors. *Eur. J. Endocrinol., 143,* 439–445.

Shaywitz, A. J. and Greenberg, M. E. (1999). CREB: a stimulus-induced transcription factor activated by a diverse array of extracellular signals. *Ann. Rev. Biochem., 68,* 821–861.

Shimizu, H., Tsuchiya, T., Sato, N., Shimomura, Y., Kobayashi, I., and Mori, M. (1998). Troglitazone reduces plasma leptin concentration but increases hunger in NIDDM patients. *Diabetes Care, 21,* 1470–1474.

Sih, R., Morley, J. E., Kaiser, F. E., Perry, H. M. 3rd, Patrick, P., and Ross, C. (1997). Testosterone replacement in older hypogondal men: a 12-month randomized controlled trial. *J. Clin. Endocrinol. Metab., 82,* 1661–1667.

Sinha, D., Addya, S., Murer, E., and Boden, G. (1999). 15-Deoxy-delta (12,14) prostaglandin J2: a putative endogenous promoter of adipogenesis suppresses the ob gene. *Metabolism, 48,* 786–791.

Sinha, M., Ohannesian, J. P., Heiman, M. L., Kriauciunas, A., Stephens, T. W., Magosin, S., Marco, C., and Caro, J. F. (1996). Nocturnal rise of leptin in lean, obese, and non-insulin-dependent diabetes mellitus subjects. *J. Clin. Invest., 97,* 1344–1347.

Stumvoll, M., Fritsche, A., Tschritter, O., Lehmann, R., Wahl, H. G., Renn, W., and Haring, H. (2000). Leptin levels in humans are acutely suppressed by isoproterenol despite Acipomox-induced inhibition of lipolysis, but not by free fatty acids. *Metabolism, 49,* 335–339.

Trayhurn, P., Duncan, J. S., Hoggard, N., and Rayner, D. V. (1998). Regulation of leptin production: a dominant role for the sympathetic nervous system? *Proc. Nutr. Soc., 57*, 413–419.

Utriainen, T., Malmstrom, R., Makimattila, S., and Yki-Jarvinen, H. (1996). Supraphysiological hyperinsuline-mia increases plasma leptin concentrations after 4 h in normal subjects. *Diabetes, 45*, 1364–1366.

Van Harmelen, V., Reynisdotir, S., Eriksson, P., Thorne, A., Hoffstedt, J., Lonnqvist, F., and Arner, P. (1998). Leptin secretion from subcutaneous and visceral adipose tissue of women. *Diabetes, 47*, 913–917.

Wabitsch, M., Blum, W. F., Muche, R., Braun, M., Hube, F., Rascher, W., Heinze, E., Teller, W., and Hauner, H. (1997). Contribution of androgens to the gender difference in leptin production in obese children and adolescents. *J. Clin. Invest., 100*, 808–813.

Wang, J., Liu, R., Hawkins, M., Barzilai, N., and Rossetti, L. (1998). A nutrient-sensing pathway regulates leptin gene expression in muscle and fat. *Nature, 393*, 684–688.

Wang, J., Liu, R., Liu, L., Chowdhury, R., Barzilai, N., Tan, J., and Rossetti, L. (1999). The effect of leptin on Lep expression is tissue-specific and nutritionally regulated. *Nat. Med., 5*, 895–899.

Wang, Y. H., Tache, Y., Sheibel, A. B., Go, V. L. W., and Wei, J. Y. (1997). Two types of leptin responsive gastric vagal afferent terminals: an in vitro single unit study in rats. *Am. J. Physiol., 272*, R833–R837.

Wells, L., Vosseller, K., and Hart, G. W. (2001). Glycosylation of nucleocytoplasmic proteins: signal transduction and O-GlcNAc. *Science, 291*, 2376–2378.

Williams, L. B., Fawcett, R. L., Waechter, A. S., Zhang, P., Kogon, B. E., Jones, R. M., Inman, M., Huse, J., and Considine, R. V. (2000). Leptin production in adipocytes from morbidly obese subjects: Stimulation by dexamethasone, inhibition with troglitazone and influence of gender. *J. Clin. Endocrinol. Metab., 85*, 2678–2684.

Wisse, B. E., Campfield, L. A., Marliss, E. B., Morais, J. A., Tenenbaum, R., and Gougeon, R. (1999). Effect of prolonged moderate and severe energy restriction and refeeding on plasma leptin concentrations in obese women. *Am. J. Clin. Nutr., 70*, 321–330.

Zhang, H. H., Kumar, S., Barnett, A. H., and Eggo, M. C. (2000). Tumour necrosis factor-alpha exerts dual effects on human adipose leptin synthesis and release. *Mol. Cell. Endocrinol., 159*, 79–88.

Zhang, P., Klenk, E. S., Lazzaro, M. A., Williams, L. B., and Considine, R. V. (2002). Hexosamines regulate leptin production in 3T3-L1 adipocytes through transcriptional mechanisms. *Endocrinology, 143*, 99–106.

Zhang, Y., Proenca, R., Maffei, M., Barone, M., Leopold, L., and Friedman, J. M. (1994). Positional cloning of the mouse *obese* gene and its human homologue. *Nature, 372*, 425–432.

Zumbach, M. S., Boehme, M. W., Wahl, P., Stremmel, W., Ziegler, R., and Nawroth, P. P. (1997). Tumor necrosis factor increases serum leptin levels in humans. *J. Clin. Endocrinol. Metab., 82*, 4080–4082.

4

Regulation and Roles of Leptin during the Menstrual Cycle and in Menopause

Ioannis E. Messinis and Ekaterini Domali

Regular menstruation in women is indicative of the normal function of the hypothalamic–pituitary–ovarian axis. Although this means that the negative and positive feedback mechanisms are intact, it also indicates that the various autocrine and paracrine mechanisms involved are effective. Accumulated evidence has indicated that leptin, a protein produced by the adipose tissue (Zhang et al., 1994), may affect human reproductive function through various pathways (Mantzoros and Moschos, 1998; Messinis and Milingos, 1999; Wauters et al., 2000; Brann et al., 2002). Leptin receptors have been identified throughout the body including the hypothalamic–pituitary–ovarian system (Tartaglia et al., 1995; Karlsson et al., 1997; Zamorano et al., 1997; Hakansson et al., 1998; Loffler et al., 2001). Especially, mRNA of leptin receptor and protein were expressed in immortalized neurons of gonadotropin releasing hormone (GnRH) (Magni et al., 1999). It can be assumed, therefore, that leptin affects gonadotropin secretion and steroid production. Nevertheless, data available at the moment have been derived mainly from experimental studies and may not directly apply to in vivo mechanisms regulating menstrual cyclicity. In many respects, these data are interesting and provide the basis for further research. This chapter will discuss the possible physiological role of leptin during the normal menstrual cycle with particular reference to interference of this protein with the function of the various parts of the reproductive axis from the hypothalamus to the uterus.

Ioannis E. Messinis and Ekaterini Domali • Department of Obstetrics and Gynecology, University of Thessalia Medical School.

Leptin and Reproduction. Edited by Henson and Castracane, Kluwer Academic/Plenum Publishers, 2003.

4.1. SECRETION OF GnRH AND GONADOTROPINS

It has been known for some years that mice carrying a mutation of the *ob* gene (ob/ob mice) are obese and infertile and suffer from diabetes with insulin resistance (Halaas et al., 1995; Pelleymounter et al., 1995; Chehab et al., 1996). Treatment of these animals with leptin results in the activation of their reproductive system and in normal fertility (Halaas et al., 1995; Pelleymounter et al., 1995; Mounzih et al., 1997).

Experiments in rats have demonstrated the ability of leptin to stimulate in vitro the secretion of GnRH from hypothalamic explants and the release of follicle stimulating hormone (FSH) and luteinizing hormone (LH) from anterior pituitaries, but the sensitivity of FSH release was much less than that of LH (Yu et al., 1997a). These effects of leptin are mediated at least in part by nitric oxide both at the hypothalamic and the pituitary level (Yu et al., 1997b). Leptin is able to stimulate LH pulse frequency and pulse amplitude in rats fasted for 5 days (Gonzalez et al., 1999). It is interesting that leptin also stimulated prolactin secretion in these animals. In these in vitro experiments adult male rats were used; however, recent studies using female rats have also reported that pituitary cells in culture can secrete LH and FSH under the influence of leptin (De Biasi et al., 2001; Ogura et al., 2001). The stimulating action of leptin on LH secretion has also been demonstrated after the intraventricular injection of this protein into ovariectomized rats pretreated with estradiol (Yu et al., 1997a). Also, in estrogen treated ovariectomized rats, a decrease in LH pulse frequency was observed during fasting for 48 hr, but normal pulsatility of LH was restored after treatment with leptin regardless of the presence of estrogens (Nagatani et al., 1998). Similar data were obtained in male monkeys, in which during the second day of fasting LH pulse frequency and amplitude as well as basal LH and FSH plasma levels were significantly higher in leptin treated as compared to saline treated animals (Finn et al., 1998). It is evident from these data that leptin provides information about food intake to central mechanisms regulating pulsatile gonadotropin secretion (Figure 4.1). A certain period of time is probably required for leptin to be effective, since intravenous administration of this substance to pubertal male rhesus macaques after fasting for only 12 hr had no effect on plasma LH levels (Lado-Abeal et al., 1999).

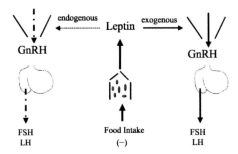

Figure 4.1. Leptin as a signal providing information about food intake to central mechanisms regulating pulsatile GnRH secretion. During fasting leptin secretion is reduced resulting in decreased secretion of GnRH and gonadotropins, while exogenous leptin restores normal FSH and LH secretion (based on the findings in rats of Finn et al., 1998; Nagatani et al., 1998).

Whether leptin can affect gonadotropin secretion in humans is not known. However, in women with anorexia nervosa, decreased serum levels of leptin and gonadotropins, which increased during weight gain, have been found (Ballauff et al., 1999). In these women, a threshold leptin level of 1.85 μg/L for minimal LH secretion has been reported (Ballauff et al., 1999) that may also be important for the initiation of the reproductive process in women during puberty (Matkovic et al., 1997). Although the findings in patients with anorexia nervosa indicate that changes in leptin are related to changes in body weight, body composition is also important, since decreased serum levels of leptin have also been found in women with hypothalamic amenorrhea and normal weight (Miller et al., 1998). It is possible that in these subjects inadequate caloric and fat consumption and decreased insulin levels are responsible for the decrease in leptin values (Miller et al., 1998). Further to this, in women athletes of a certain body mass index (BMI), the diurnal 24 hr pattern of leptin levels is abolished if they are amenorrheic, but not if they are normally cycling (Laughlin and Yen, 1997). These changes in leptin values may be induced by low insulin and elevated cortisol levels (Laughlin and Yen, 1997). However, it is also possible that under the strength of body exercise the increased hypothalamic production of CRH, apart from causing hypercortisolemia, stimulates central and then peripheral adrenergic activity resulting in norepinephrine release that inhibits gene expression and leptin secretion (Gettys et al., 1996; Mantzoros et al., 1996; Slieker et al., 1996). Through these mechanisms not only hypoleptinemia, but also inhibition of the activity of GnRH pulse generator occurs (Williams et al., 1990; Laughlin and Yen, 1997).

Although in various animal models leptin under experimental conditions can stimulate the production of GnRH and gonadotropins, and in women hypoleptinemia is accompanied by a reduction in FSH and LH levels, a specific range of leptin levels for maintenance of the reproductive capacity is difficult to define. Evidence has been provided that clinical hyperleptinemia can adversely affect pituitary function. For instance, obese girls with high leptin levels show impaired gonadotropin response to GnRH during late puberty (Bouvattier et al., 1998). In addition, a negative correlation between GnRH-induced LH release and baseline leptin values has been found in a mixed population of lean and obese women during the follicular phase of the cycle (Komorowski et al., 2000). On the other hand, a critical level of leptin is needed to maintain menstruation (Kopp et al., 1997) or for minimal LH secretion in women (Ballauff et al., 1999). It is possible, therefore, that for a normal reproductive function, leptin values should be within a certain limit of variation, since both very low and excessive leptin production can adversely affect neuroendocrine mechanisms in the central reproductive system. The extent to which this information can have an impact on the physiology of the human menstrual cycle is not known. So far, data are limited only to a few studies. In one of them, significant positive correlations between serum leptin and LH concentrations have been found in normal women at midcycle (Teirmaa et al., 1998). Also, synchronicity of LH and leptin pulses has been demonstrated in normal women in the mid- to late follicular phase of the cycle, especially during the night, suggesting thus that leptin may be important for the secretion of LH during the preovulatory period (Licinio et al., 1998). Consistent with this are data in rats indicating that leptin was able to resume LH and prolactin surges that had been abolished after 3-day starvation (Kohsaka et al., 1999).

One possible mechanism through which leptin may affect hypothalamic–pituitary function is the hypothalamic neuropeptide Y (NPY) system. This protein suppressed the increased levels of NPY mRNA in the hypothalamus of fasted rats and at the same time attenuated the suppression of LH secretion that was induced by food deprivation (Kalra et al., 1998). Leptin by inhibiting NPY production stimulates noradrenergic pathways that cause the release of nitric oxide, which then attenuates GnRH release (Yu et al., 1997b). However, intracerebroventricular infusion of leptin into ovariectomized ewes resulted in a reduction in the expression of mRNA for NPY in the arcuate nucleus without affecting the secretion of LH, suggesting first that LH secretion may be independent of the NPY pathways and second that the inhibiting action of leptin on food intake is not mediated through endocrine effects (Henry et al., 1999). Other possible mechanisms via which leptin can influence hypothalamic–pituitary function include changes in the levels of various hormones, such as insulin, as well as changes in metabolic fuels, such as glucose availability (Kamohara et al., 1997; Sivitz et al., 1997). It is still unclear whether leptin affects the secretion of GnRH separately from that of gonadotropins. Receptors of this substance have been found both in the hypothalamus and the pituitary (Zamorano et al., 1997; Hakansson et al., 1998). Since ob receptor is co-localized in estrogen receptor-α containing neurons in the hypothalamus of female rats, it is possible that this steroid controls levels of hypothalamic ob receptor (Diano et al., 1998). In support of this is that ob receptor mRNA in the choroid plexus of cycling rats was inversely correlated with serum estradiol concentrations (Bennett et al., 1999), while in ovariectomized rats treatment with estradiol decreased the expression of *ob* gene in the arcuate and the ventromedial nucleus (Bennett et al., 1998).

4.2. OVARY

4.2.1. Regulation of Leptin Production by Ovarian Steroids

The connection of leptin with the ovary has been supported by several experiments. An interesting question is whether leptin is produced by the ovary. Data available at the moment are conflicting. Expression of leptin at the mRNA and protein level by human granulosa and cumulus cells has been demonstrated with the use of reverse transcriptase–polymerase chain reaction and immunofluorescence (Cioffi et al., 1997). Also, leptin together with other molecules has been localized by immunofluorescence analysis in a highly specialized subpopulation of human granulosa and cumulus oophorus cells (Antczak et al., 1997). These data, however, have not been confirmed in another study in which, with the use of similar techniques, *ob* gene expression was detected in adipose tissue, but not in granulosa, thecal or interstitial cells (Karlsson et al., 1997). In all these cases, granulosa cells were obtained either from women undergoing in vitro fertilization (IVF) treatment (Antczak et al., 1997; Cioffi et al., 1997; Karlsson et al., 1997) or from normally cycling women undergoing laparotomy (Karlsson et al., 1997). A more recent study, using intact ovaries, has contributed further to the uncertainty (Loffler et al., 2001). In particular, leptin-positive cells were localized in the granulosa cells of preantral, but not of antral follicles and in the thecal layer of intact and regressing antral follicles. Also, in the corpus luteum, leptin-positive cells were located in the former thecal layer, but the numbers decreased as the secretory stage progressed (Loffler et al., 2001). Despite these conflicting data, studies in ovariectomized animals and women have provided evidence that the ovary plays a role in the control of leptin production from the adipose tissue.

In rats, the expression of *ob* gene mRNA decreased significantly in subcutaneous and retroperitoneal adipose tissue 8 weeks after ovariectomy (Shimizu et al., 1997). At the same time, concentrations of leptin in serum were significantly lower in the ovariectomized animals than in the controls. These data do not support the assumption that leptin is produced by the ovary. They rather provide evidence that an ovarian mediator stimulates the production of leptin from the adipose tissue, since treatment of the ovariectomized rats with estradiol prevented the decrease of the *ob* gene mRNA expression and of serum leptin concentrations (Shimizu et al., 1997). Subsequent studies in rats have also demonstrated the ability of estradiol to prevent the decrease in the expression of the ob mRNA by fat tissue (Yoneda et al., 1998) and the decline in serum levels of leptin (Chu et al., 1999). It is suggested, therefore, that in rats estradiol exerts a stimulatory effect on leptin production by fat cells. Human studies, however, have not provided supportive evidence of this assumption. In one of them (Messinis et al., 1999), a significant decline in serum leptin levels together with a marked decrease in estradiol and progesterone concentrations was seen within the first 4 days following ovariectomy performed either in the follicular or in the luteal phase of the cycle (Figure 4.2). Since leptin concentrations also declined in a group of women who via laparotomy underwent cholocystectomy without ovariectomy

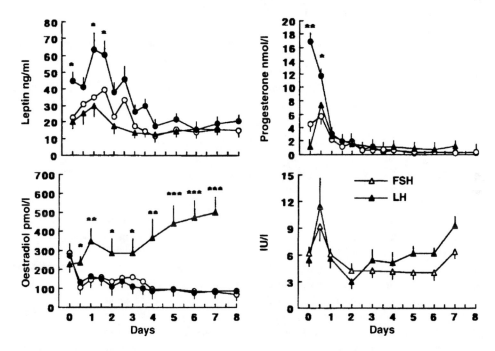

Figure 4.2. Serum leptin levels (mean $\pm$ SEM) in women, who underwent total abdominal hysterectomy plus bilateral ovariectomy (day 0) in the follicular ($\circ$, $n = 7$ women), or in the luteal phase of the cycle ($\bullet$, $n = 7$ women). Comparison with women who underwent cholocystectomy without ovariectomy in the early to midfollicular phase of the cycle ($\blacktriangle$, $n = 6$ women): $\bullet$ * $p < 0.05$, ** $p < 0.01$, and *** $p < 0.001$. Adapted from Messinis et al. (1999) with the permission of Oxford University Press/Human Reproduction.

and, therefore, without a decrease in serum estradiol concentrations, it is suggested that the changes in leptin values seen during the immediate period following ovariectomy are independent of estradiol values (Messinis et al., 1999). There is a difference, however, between animal and human studies in that in animals, the earliest time at which changes in leptin were investigated was 2 weeks after ovariectomy, while in humans only short term changes in leptin values were examined during the first week after the operation.

The role of estradiol in the production of leptin has been further examined in in vivo and in vitro studies. The main source of leptin in women is the subcutaneous fat tissue with little contribution of the visceral fat (Van Harmelen et al., 1998). Normal premenopausal and postmenopausal women have significantly higher serum leptin levels than normal men (Rosenbaum et al., 1996; Shimizu et al., 1997; Castracane et al., 1998). Although this gender difference could be attributed to a stimulating effect of estradiol on leptin production (Shimizu et al., 1997), one can argue that testosterone in men makes the difference because this hormone is able to suppress leptin production when given to transsexuals (Elbers et al., 1997). In favor, however, of a stimulating effect of estradiol on leptin production are also data demonstrating that samples of fat tissue in culture, obtained from omentum in women, produced leptin at a significantly higher rate than samples obtained from omentum in men (Casabiell et al., 1998). In addition, estradiol stimulated leptin secretion significantly in samples taken from women, but did not affect the secretion in the male samples (Casabiell et al., 1998). In vivo data in intact female rats (Brann et al., 1999) have shown that 17β-estradiol given subcutaneously for 2 days stimulates an increase in the levels of leptin mRNA in adipose tissue between 6 and 12 hr after the injection and an increase in plasma leptin values at 12 hr. These data provide direct evidence that estradiol, at least under experimental conditions, is involved in the regulation of leptin secretion in animals and in women and is further supported in rats by the fact, first, that high affinity estrogen binding sites have been found in the cytoplasm of adipose tissue (Wade and Gray, 1978) and, second, that 17β-estradiol stimulates leptin gene expression in isolated adipocytes after 10 hr of incubation (Murakami et al., 1995).

Despite all these data, signifying the importance of estradiol as a stimulator of leptin secretion, this steroid, among several other hormones, is not a principal regulator of leptin production in women. Further to the already mentioned decline in serum leptin concentrations during the first 4 days after ovariectomy (Messinis et al., 1999), when ovariectomized women were treated with either transdermal estradiol or estradiol plus progesterone, starting immediately after the operation, the decline in leptin values was prevented only in the group of women treated with the combined regimen (Messinis et al., 2000) (Figure 4.3). Although progesterone alone was not given to these women, the data suggest that progesterone participates in the control of leptin production by the adipose tissue. In particular, in these studies a temporal increase in leptin values was seen 24 hr after ovariectomy that was preceded by a significant increase in progesterone and cortisol values at a time when estradiol concentrations had decreased markedly (Messinis et al., 1999; 2000). Since both these steroids correlated significantly with leptin increase, it is possible that cortisol (Larsson and Ahren, 1996; Papaspyrou-Rao et al., 1997) and progesterone can stimulate leptin secretion independently of estradiol. The fact, however, that progesterone was unable in vitro to stimulate leptin secretion in samples of adipose tissue obtained from omentum in women (Casabiell et al., 1998) and in vivo to induce an increase in serum leptin and leptin gene expression in white and brown adipose tissue in ovariectomized rats

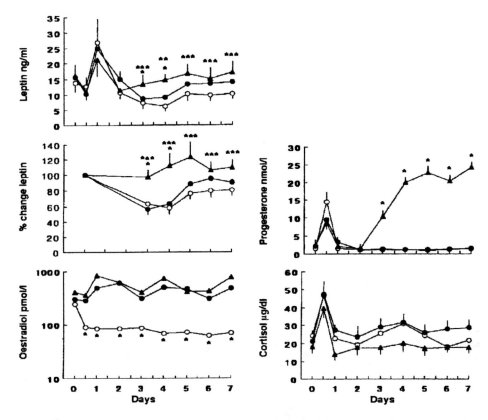

Figure 4.3. Serum leptin levels (mean ± SEM) measured in normally cycling women before and after bilateral ovariectomy plus total abdominal hysterectomy performed in midfollicular phase of the menstrual cycle (day 0). The women received post-operatively oestradiol (days 0, 3, and 6) through skin patches (●, $n = 7$ women) or oestradiol plus progesterone (days 2–7) intravaginally (▲, $n = 7$ women) or no hormonal treatment (O, $n = 7$ women): ● * $p < 0.001$, ** $p < 0.01$, *** $p < 0.05$, and ● † $p < 0.05$. Adapted from Messinis et al. (2000) with the permission of Oxford University Press/Human Reproduction.

(Luukkaa et al., 2001), makes the possibility of the independent action of progesterone less likely. It is rather probable that it is the combining effect of estradiol and progesterone that can stimulate leptin production and that estradiol exerts a priming effect on fat cells sensitizing them to progesterone.

The latter possibility is supported further by a more recent study using a different approach (Messinis et al., 2001). Normally ovulating women were treated during the early follicular phase of the cycle with transdermal estradiol and in a subsequent cycle with estradiol plus progesterone in order to achieve preovulatory concentrations of estradiol and concentrations of progesterone as in the luteal phase. Estradiol alone was unable to induce any changes in serum leptin concentrations, while, during the combined treatment, a significant increase in leptin values occurred (Figure 4.4). These findings confirm the above mentioned data in oophorectomized women demonstrating that estradiol and progesterone when combined in a treatment regimen can stimulate the secretion of leptin (Messinis et al., 2000). That estradiol alone was able to stimulate leptin secretion in animals but not

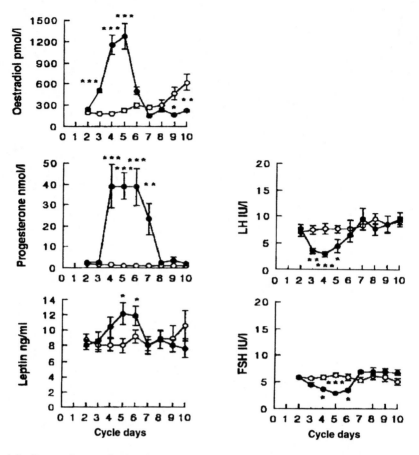

Figure 4.4. Increase in serum leptin values (mean ± SEM) measured in normally cycling women during the exogenous administration of oestradiol (days 2–4) plus progesterone (days 3–5) (●, $n = 6$ women) in the early follicular phase of the cycle. Comparison with untreated spontaneous cycles (○): ● * $p < 0.05$, ** $p < 0.01$, and *** $p < 0.001$. Adapted from Messinis et al. (2001) with the permission of Oxford University Press/Human Reproduction.

in women is difficult to explain. Although species differences may be important, an explanation might be that in animal studies, treatment with estradiol lasted for a longer period than in human studies. Even in rats, in which the stimulatory effect of estrogens on leptin production was seen after at least 4 weeks of treatment following ovariectomy (Shimizu et al., 1997; Yoneda et al., 1998; Chu et al., 1999), when treatment was given for only 18 days serum leptin levels and the expression of leptin gene in white and brown adipose tissue were not affected (Luukkaa et al., 2001). It is likely that a certain period of time is required for estradiol to exert a stimulating effect on leptin secretion, which, however, is shortened significantly when progesterone is added. This phenomenon of a single hormone facilitating the effectiveness of another hormone is well known in endocrinology. An example is the ability of progesterone to induce an endogenous LH surge after pretreatment of normal women with estradiol, even when the estradiol threshold level for the positive feedback effect has not been exceeded (Chang and Jaffe, 1978; Messinis and Templeton, 1990).

Various factors not related to hormones may also be important for the control of leptin secretion during the immediate postoperative period following ovariectomy in women. BMI is one of them, despite the fact that BMI values do not change significantly 1 week after ovariectomy (Messinis et al., 1999). Also, it cannot be excluded that changes in food intake that might have affected body weight particularly during the first 4 days after the operation had an impact. Although in normal or obese subjects dramatic changes in leptin values as a result of changes in food intake are not expected (Korbonits et al., 1997), even a 4 percent reduction in body weight over a few days in women can result in a 61 percent decrease in leptin values (Dubuc et al., 1998). In terms of the factors that may be responsible for the increase in serum leptin values during the first 24 hr following a laparotomy, apart from cortisol and progesterone, total parenteral nutrition may also be important (Messinis et al., 1999; Elimam et al., 2001). Furthermore, the importance of the anesthetic drugs used for these operations should not be underestimated, since a decrease instead of an increase in serum leptin values has been reported during the first 48 hr following gastroplasty in women anesthetized with thiopental or etomidate, a cortisol inhibitor (Montalban et al., 2001).

4.2.2. Effects of Leptin on Ovarian Function

Although it is still unclear if the ovary produces leptin, the receptor of this substance, including the long and the short isoforms, has been identified in the human ovary both in granulosa and thecal cells (Karlsson et al., 1997; Spicer and Francisco, 1997; Agarwal et al., 1999; Loffler et al., 2001). This suggests that leptin exerts a direct action on the ovary. Several experimental studies have shown that this effect is inhibitory to steroidogenesis. In particular, leptin at various concentrations was unable to affect insulin-induced bovine granulosa cell numbers of small or large follicles in culture, but was able to inhibit in vitro insulin-induced increase in progesterone and estradiol production by all kinds of follicles without affecting insulin binding (Spicer and Francisco, 1997). Leptin was also able to inhibit insulin-induced steroidogenesis by bovine granulosa and thecal cells in vitro (Spicer and Francisco, 1998). The inhibitory action of leptin on human chorionic gonadotropin (hCG)-stimulated progesterone production from human luteinized granulosa cells was evident only in the presence of insulin (Brannian et al., 1999).

Further studies have also demonstrated the ability of leptin to modify the effect of various factors on steroidogenesis by granulosa cells in vitro. In one of them, using rat granulosa cells, leptin alone had no effect on basal or FSH-induced estradiol production, but inhibited the synergistic action of insulin like growth factor-I (IGF-I) on FSH-stimulated estradiol production (Zachow and Magoffin, 1997). Also, leptin antagonized the stimulatory effect of transforming growth factor-β (TGF-β) on FSH-induced estrogen production by rat granulosa cells, an effect that was mediated by the attenuation of P450(arom) activity (Zachow et al., 1999). Other factors, such as dexamethazone enhancing FSH-stimulated progesterone production by rat granulosa cells taken from preovulatory follicles, were antagonized by leptin (Barkan et al., 1999). Leptin also antagonized the augmenting effect of IGF-I on FSH-induced estradiol production by human granulosa and thecal cells (Agarwal et al., 1999). However, the stimulatory effect of hCG or IGF-II on estradiol production by human granulosa-lutein cells was not affected by leptin in culture, although basal production of this steroid was reduced (Ghizzoni et al., 2001). In contrast to the above studies, other recent data have shown a stimulatory rather than an inhibitory effect of leptin on steroidogenesis. In particular, leptin was able to stimulate estrogen production by human luteinized granulosa cells in culture possibly through an increase in P450(arom) mRNA and protein expression (Kitawaki et al., 1999). These in vitro data imply that leptin can directly affect ovarian function, but the extent to which leptin may interfere with normal ovarian function in vivo is not known. An assumption can be made that lower circulating concentrations of leptin are without any effect or facilitate the mechanisms that result in normal steroidogenesis and follicle maturation, while high levels exert detrimental effects disrupting steroidogenesis and cyclicity. In rats, an inhibiting effect of leptin on ovulation has been shown in vivo and in vitro (Duggal et al., 2000).

Apart from a role for leptin in steroidogenesis and follicle maturation, this protein seems also to play a regulatory role in oocyte and preembryo development. Human oocytes contain leptin in a polarized manner together with the transcription protein STAT3. Cleaved oocytes by the 8-cell stage demonstrate along the surface a concentration gradient of these proteins with the inner blastomeres being poor and the outer cells being rich in leptin/STAT3, a distribution that persists through the hatched blastocyst stage (Antczak and Van Blerkom, 1997). Furthermore, it has been shown that metaphase II mouse oocytes express mRNA and protein of leptin receptor, while leptin at concentrations found in the follicular fluid cause tyrosine phosphorylation of STAT3 in the oocytes (Matsuoka et al., 1999). These results demonstrate that leptin may exert a direct effect on the oocyte affecting its maturation through activation of STAT transcription factors. The ob/ob mice produce a truncated form of leptin that together with STAT3 is present and polarized in the oocyte (Antczak and Van Blerkom, 1997). The fact that treatment of these animals with leptin restores fertility suggests that this truncated form of leptin may have some functional domains, which may play a role during oocyte maturation and early embryogenesis (Antczak and Van Blerkom, 1997). The importance of STAT3 protein for embryogenesis is demonstrated in STAT3-deficient mice, the embryos of which implant but degenerate after implantation (Takeda et al., 1997). It has been proposed that leptin and STAT3 in mammalian oocytes are derived from intact follicles (Antczak and Van Blerkom, 1997). However, given the fact that expression of the *ob* gene in the granulosa cells of antral follicles has not been verified (Cioffi et al., 1997; Karlsson et al., 1997; Loffler et al., 2001), it is more likely that the oocyte itself is the source of these proteins. It would be interesting to know the maturation status of the oocyte in cases of high serum concentrations of leptin, such as in obese women, since such concentrations at least in in vitro conditions are detrimental for steroidogenesis.

4.2.3. Leptin Levels during the Normal Menstrual Cycle

Whether the described interactions between gonadal steroids and leptin could apply to ovarian physiology is not clear. Several studies have investigated changes in leptin values during the normal menstrual cycle with divergent findings. Most of these, however, have demonstrated a trend for an increase in serum leptin values towards the late follicular phase and higher values in the luteal than in the follicular phase (Hardie et al., 1997;

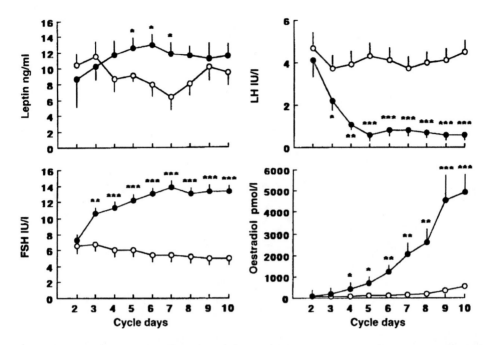

Figure 4.5. Serum leptin values (mean ± SEM) measured in women during the follicular phase of spontaneous (O) and FSH-treated cycles (•): • * $p < 0.05$, ** $p < 0.01$, and *** $p < 0.001$. Adapted from Messinis et al. (1998) with the permission of Oxford University Press/Human Reproduction.

Mannucci et al., 1998; Messinis et al., 1998; Riad-Gabriel et al., 1998; Quinton et al., 1999; Ludwig et al., 2000; Phipps et al., 2001). In the first of these studies (Hardie et al., 1997), blood samples for leptin measurement were taken every 3 days, but no specific pattern of changes in leptin values was found. When, however, blood samples were grouped according to the functional stage of the cycle, leptin concentrations increased significantly from early to late follicular phase and attained peak values during the luteal phase (Hardie et al., 1997). In a subsequent study, in which blood samples were taken every day during the follicular phase, leptin values declined initially from early to midfollicular phase and increased significantly thereafter in the late follicular phase (Messinis et al., 1998) (Figure 4.5). This increase in leptin values correlated significantly with that of estradiol (Messinis et al., 1998). In random blood samples taken during the follicular phase of the cycle, leptin values correlated positively with estradiol concentrations and negatively with those of testosterone and dehydroepiandrosterone sulfate (Paolisso et al., 1998). Moreover, during the mid- to late follicular phase of the cycle, the 24 hr fluctuations in leptin values showed synchronicity with those of LH and estradiol suggesting that the rise in leptin values during the night may control changes in nocturnal LH secretion (Licinio et al., 1998). Although, from these correlations, it is possible that estradiol may be involved in the control of leptin secretion in women during the follicular phase of the cycle, the possibility that estradiol and leptin are also synchronized in the context of a cause and effect relationship needs to be investigated.

Data during the luteal phase of the cycle are also interesting. Riad-Gabriel et al. (1998) have presented a more detailed description of the changes in leptin values during

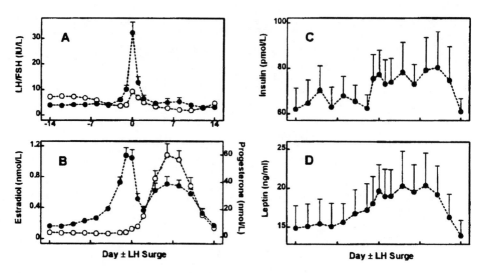

Figure 4.6. Alterations in serum levels (mean ± SEM) during the normal menstrual cycle. Adapted from Riad-Gabriel et al. (1998) with the permission of the Society of the European Journal of Endocrinology.

the luteal phase, the pattern of which resembled that of progesterone (Figure 4.6). Although no significant correlations were found between the absolute values of leptin and those of estradiol and progesterone, net changes in leptin concentrations correlated significantly with the values of the two steroids. In spite of these correlations, progesterone accounted for only 5 percent of the changes in leptin values (Riad-Gabriel et al., 1998). A plausible explanation of the increase in leptin values during the luteal phase is a stimulatory effect of ovarian steroids, since in women exogenous administration of estradiol plus progesterone was able to prevent the decline in serum leptin values induced by ovariectomy (Messinis et al., 2000) and to stimulate a significant increase in leptin values in the early follicular phase of the cycle (Messinis et al., 2001). The mechanism through which ovarian steroids might affect leptin secretion from the adipocytes has not been clarified. At least, estradiol exerts a direct effect on adipose tissue (Wade and Gray, 1978) promoting the replication of fat cells in culture (Roncari and Van, 1978). Since in women estradiol values and BMI show significant relationships (Evans et al., 1983), it may be that gonadal steroids can affect the production of leptin also by influencing the body fat content. Another possible mechanism, however, could be that estradiol may increase the sensitivity of various tissues to leptin, possibly through a change in the balance between the long and the short isoforms of leptin receptor, as has been shown in the brain of rats (Bennett et al., 1998). In terms of progesterone, no specific mechanisms have been proposed as yet.

In contrast to the described changes in leptin values throughout the normal menstrual cycle, other studies have shown only little variation. In one of them (Teirmaa et al., 1998), serum leptin values were similar at the onset and at the end of the cycle as well as at the time of ovulation. However, in that study only three blood samples were collected during the whole cycle. In another study (Stock et al., 1999), in which blood samples were taken three times during the follicular phase and once in the luteal phase, no significant changes were found, although there was a small overall but significant variation in leptin values.

4.2.4. Leptin Values in Superovulated Cycles

The possibility that leptin could be produced by the ovary led several investigators to examine circulating leptin values in women superovulated with FSH. In the first of these studies (Messinis et al., 1998), daily changes of serum leptin values were investigated in normally cycling women superovulated with purified urinary FSH without a GnRH agonist. It was found that leptin concentrations increased gradually from early to midfollicular phase to levels that were significantly higher than in spontaneous cycles of the same women. However, despite the continuation of treatment with FSH, leptin values did not increase further during the late follicular phase, but in the first half of the follicular phase a significant positive correlation was found between leptin and estradiol concentrations (Messinis et al., 1998) (Figure 4.5). Although these findings could be interpreted as indicating that leptin is produced by the ovary, due to the conflicting data regarding the expression of leptin by granulosa and cumulus oophorus cells (Antchak et al., 1997; Cioffi et al., 1997; Karlsson et al., 1997; Loffler et al., 2001), it is suggested that FSH may only indirectly affect the secretion of leptin from the adipocytes. A possible candidate that could mediate the effect of FSH on fat cells is estradiol.

Subsequent studies have confirmed the stimulating effect of FSH on leptin production during superovulation induction, despite the fact that blood samples were collected less frequently (Mannucci et al., 1998; Butzow et al., 1999; Stock et al., 1999; Lindheim et al., 2000; Yamada et al., 2000). Although a significant positive correlation between leptin and estradiol concentrations was found at some stages of the superovulation process (Mannucci et al., 1998; Messinis et al., 1998; Yamada et al., 2000), estradiol increased disproportionately to the leptin levels (Messinis et al., 1998; Butzow et al., 1999), indicating thus that leptin concentrations do not reflect the degree of ovarian hyperstimulation. On the contrary, a negative association between the relative serum leptin increase and the ovarian response to the hyperstimulation regimen has been found (Butzow et al., 1999). This means that high serum leptin concentrations may reduce ovarian response to gonadotropins. Since leptin levels are higher in obese than in lean women (Sinha et al., 1996), it is explained at least in part why obese women require higher doses of gonadotropins to achieve ovulation (Dale et al., 1993; Butzow et al., 1999). That high leptin values can disrupt normal folliculogenesis is also derived from experiments on rats demonstrating that systemic administration of leptin to these animals can inhibit ovulation (Duggal et al., 2000). Women with polycystic appearing ovaries tend to have higher peak values of leptin, estradiol and testosterone during treatment with FSH as compared to women with normal appearing ovaries (Lindheim et al., 2000).

Leptin has been detected in human follicular fluid, obtained from women during a spontaneous cycle or during IVF treatment, at concentrations similar to those in the peripheral circulation (Cioffi et al., 1997; Karlsson et al., 1997; Agarwal et al., 1999; Butzow et al., 1999). Although, initially, an association between follicular fluid leptin concentrations and embryo development was not observed (Cioffi et al., 1997), it was subsequently found that the concentrations of leptin in the follicular fluid represent a marker of the outcome of treatment. In particular, after adjustment for age and BMI, normal women and women with polycystic ovary syndrome, who became pregnant after IVF, had lower concentrations of leptin in the follicular fluid than those who did not become pregnant (Mantzoros et al., 2000). Further to this, leptin concentrations in the follicular fluid of women undergoing superovulation induction and IVF could be used in conjunction with vascular endothelial growth factor and nitric oxide as a marker of follicular hypoxia and poor oocyte quality and embryo development (Barroso et al., 1999). It is possible that high concentrations of leptin in the follicular fluid through suppressing estradiol production adversely affect oocyte maturation.

In terms of serum leptin values, a recent study in women undergoing IVF has shown that after embryo transfer the values tended to be higher in cases resulting in a successful pregnancy than in cases resulting in a miscarriage (Unkila-Kallio et al., 2001). A more recent study, however, has shown data consistent with those in the follicular fluid, that is that high serum leptin values are detrimental for the outcome of IVF (Brannian et al., 2001). In particular, the leptin:BMI ratio was highly predictive of IVF success, since women with a low ratio had embryos of better quality and a higher implantation rate than women with a high leptin:BMI ratio after IVF (Brannian et al., 2001).

4.3. ENDOMETRIUM

Human endometrium is known to undergo changes during the normal menstrual cycle that aim at developing the appropriate environment for the implantation of the blasto-cyst. Polarization of leptin and STAT3 in early embryo is consistent with the possibility that these substances participate in the implantation process. It has been recently shown that endometrium obtained from premenopausal women expresses the long form of leptin receptor mRNA and protein as well as the splice variants of the receptor (Kitawaki et al., 2000). The latter and the long form of the receptor transcripts were expressed at high inci-dences, 90 and 84 percent, respectively (Kitawaki et al., 2000). However, leptin mRNA or protein were not expressed (Alfer et al., 2000; Kitawaki et al., 2000). A fluctuation in abundance of splice variants and the long form receptor transcript expression was found in the human endometrium during the normal menstrual cycle with values increasing gradu-ally from the early to late follicular phase, peaking in the early secretory phase and declin-ing thereafter (Kitawaki et al., 2000). These cyclic changes in leptin receptor indicate that peak values are attained at a time when the endometrium is prepared to accept the preembryo and are probably related to changes in ovarian steroids.

It has been shown that progesterone or medroxyprogesterone acetate in combination with estradiol, but not estradiol alone, were able to suppress the long form of leptin recep-tor mRNA expression by 50 percent in organ-cultured proliferative endometrium speci-mens obtained from premenopausal women (Koshiba et al., 2001). The progestogenic effect of these substances was exerted via the progesterone receptor, because it was blocked with the use of mifepristone. In addition, leptin was able to induce phosphoryla-tion of STAT3 in incubated endometrial specimens (Koshiba et al., 2001). That estradiol did not affect the expression of leptin receptor indicates that this steroid is not the primary factor that participates in the increase in leptin receptor from the early to late follicular phase of the cycle. However, estradiol may play a permissive role facilitating the action of other proliferative factors. In terms of the effect of progesterone, this was shown only in the proliferative and not in the secretory endometrium (Koshiba et al., 2001). This suggests that the expression of the mRNA in the secretory endometrium has been fully suppressed by endogenous progesterone, but even if this is the case, there is still substantial expression of the receptor in the midluteal phase endometrium (Kitawaki et al., 2000). One cannot, however, exclude the possibility that leptin itself participates in the control of its own receptors during the menstrual cycle. These experimental data together with the findings that progesterone stimulates leptin secretion in vivo (Messinis et al., 2000; 2001) indicate that this steroid is the principal determinant that creates the optimal conditions of leptin action during the preimplantation stage. It is interesting that the expression of leptin receptor was found not only in disease-free endometrium, but also in endometrium taken from patients with endometriosis, adenomyosis and leiomyomas (Kitawaki et al., 2000). It would also be interesting to investigate if the expression of leptin receptor in the endometrium changes in patients with infertility.

These findings together with the expression of leptin and leptin receptor in the pre-embryo suggest that leptin may participate in the development of a cross-talk between endometrium and the embryo and, therefore, in the implantation process. This has been recently discussed in a review article by Gonzalez et al. (2000). Original work has demonstrated the role of interleukin-1 system in the cross-talk process during the implantation window (Simon et al., 1993; 1994) and the role of β3 integrine expression in endometrial epithelial cells, induced by interleukin-1 (Simon et al., 1997), as a marker of endometrium receptivity (Lessey et al., 1995; Gonzalez et al., 1999). It has been hypothesized that an inverse association between leptin and interleukin-1 could be the basis of a dialog between the preimplantation embryo and the endometrium (Gonzalez et al., 2000). Leptin may also be important as an autocrine/paracrine regulator during the invasion of trophoblast by stimulating the secretion of metalloproteinases and the expression of integrins in the cytotrophoblast (Gonzalez et al., 2000).

4.4. MENOPAUSE

There is no doubt that gender affects serum leptin concentrations. It has been found that leptin values are higher in women than in men (Rosenbaum et al., 1996; Shimizu et al., 1997; Castracane et al., 1998). Whether this difference is related to the higher concentrations of estradiol in women or to the higher testosterone levels in men has not been clarified (Elbers et al., 1997; Jockenhovel et al., 1997). Another factor that may account for the difference between men and women is the distribution of visceral versus subcutaneous adipose tissue (Hube et al., 1996), although this has not been confirmed (Rosenbaum et al., 2001). After menopause, leptin values are still higher in women than in men, but lower than in premenopausal women (Rosenbaum et al., 1996; Shimizu et al., 1997). This difference, however, between premenopausal and postmenopausal women has not been shown in other studies in which either similar leptin values (Havel et al., 1996; Haffner et al., 1997; Saad et al., 1997; Castracane et al., 1998; Sumner et al., 1998) or even higher concentrations in postmenopausal than in premenopausal women have been found (Di Carlo et al., 2000). Although these discrepancies are difficult to explain, not all studies have used the correction for BMI or for fat mass that is important before a comparison between the different groups is attempted. In the study by Castracane et al. (1998), in which adjusted means were used with the BMI as a covariate, a significant difference existed only between men and women. Similar to premenopausal women, obese postmenopausal women have significantly higher serum leptin concentrations than their non-obese counterparts (Hadji et al., 2000).

Regardless of the total fat mass, fat distribution is altered following menopause in association with estrogen insufficiency. Although the mechanism responsible for this is not clear, at least in ovariectomized rats, excess fat accumulation is associated with insensitivity to central leptin administration and overproduction of NPY in the hypothalamus (Ainslie et al., 2001). Physical activity, however, seems to be an important predictor of abdominal fat distribution in early postmenopausal women, accounting also for a great percentage of the variability of leptin concentrations (Kanaley et al., 2001). Furthermore, the increase in body weight following menopause may be the result of a decreased transport of leptin into the brain (Kastin et al., 2001). A racial difference has been demonstrated, with African American postmenopausal women having 35 percent higher leptin levels than postmenopausal Caucasian women (Berman et al., 2001). This difference seems to be dependent on insulin, as it disappears after adjustment for fasting insulin (Berman et al., 2001). An important parameter that may also affect fluctuation in leptin levels between different individuals as well as regulation of BMI and adiposity are polymorphic variations in leptin receptor and consequently in ligand binding (Quinton et al., 2001).

Several studies have investigated the effect of hormone replacement therapy (HRT) on leptin values in postmenopausal women. Most of them have found no effect. It is surprising that in several of these studies it is not specified which regimen was used (Haffner et al., 1997; Castracane et al., 1998; Di Carlo et al., 2000). There are, however, studies that have shown either an increase or a decrease in serum leptin concentrations after HRT. In a double-blind placebo-controlled randomized study (Elbers et al., 1999), 25 healthy postmenopausal women were treated with unopposed estrogen, 2 mg 17β-estradiol (12 women) or with placebo (13 women). During the treatment period body weight increased significantly in the placebo group; however, the percentage of body fat or the amount of body fat did not change in either group. Serum leptin concentrations increased significantly after 2 months of treatment with estradiol but not with placebo (Elbers et al., 1999). In studies, however, in which postmenopausal women were treated with regimens containing both estrogens and progesterone no increase in leptin values has been found (Havel et al., 1996; Kohrt et al., 1996; Haffner et al., 1997; Baumgartner et al., 1999; Hadji et al., 2000).

In terms of the route of administration, there is no difference in the results. With the transdermal route, both a reduction and no effect on leptin values have been found (Di Carlo et al., 2000; Laivuori et al., 2001). In a recent study, in which oral HRT, consisting of estradiol plus 1 mg norethisterone acetate, was compared with transdermal administration of the same steroids, no change in plasma leptin concentrations was found after 2, 6 and 12 months of either treatment (Laivuori et al., 2001). However, in that study a placebo control group was not used. This is worth mentioning because norethisterone, as an androgenic product, might have suppressed a possible stimulating effect of estradiol on leptin secretion. A similar regimen, however, of 1 mg oral norethisterone combined with transdermal estradiol (80 μg patches) had no effect on leptin levels as compared to placebo after administration for 6 months to postmenopausal women with type 2 diabetes (Perera et al., 2001). Another study in which conjugated estrogens were combined with medrogesterone acetate has shown that changes in serum leptin concentrations during that period were dependent on baseline leptin values (Ongphiphadhanakul et al., 1998). In particular, leptin values decreased significantly together with body weight within 3 months in the women with high baseline leptin and increased significantly in the women with low baseline leptin (Ongphiphadhanakul et al., 1998). Interpretation of these data is difficult and although

HRT in postmenopausal women can induce a slight increase in body weight (Lubbert and Nauert, 1997; Laivuori et al., 2001), it may also prevent the age-induced increase in body weight (Chmouliovsky et al., 1999; Sayegh et al., 1999).

It is evident that there is no consensus regarding the effect of HRT on circulating leptin values in postmenopausal women. Several parameters may play a role. Body activity and exercise may be important, although at least acute treadmill exercise does not cause any changes in diurnal variation of leptin concentrations in postmenopausal women regardless of the HRT status (Kraemer et al., 1999). Other factors that may be important are the percentage fat mass, the age, the body composition, the exact time of blood sampling etc. There is no doubt that prospective controlled studies are required to clarify the effect of HRT on leptin concentrations in postmenopausal women. Finally, intake of isoflavonic phytoestrogens for 3 months even in high doses had no effect on plasma leptin concentrations either in premenopausal or in postmenopausal women (Phipps et al., 2001).

4.5. CONCLUSIONS

Evidence has been provided that leptin produced by the adipose tissue may participate in the functional integrity of the hypothalamic–pituitary–ovarian system (Figure 4.7). Although various interactions have been described, a shared relationship in the context of feedback mechanisms has not been demonstrated. A hypothesis can be developed that during the normal menstrual cycle leptin facilitates the secretion of GnRH and LH especially during the night and promotes estradiol production from the growing follicle. In the other hand, increasing concentrations of estradiol sensitize the adipocytes to progesterone, which then during the periovulatory period and the luteal phase enhances the production of leptin. A regulatory role of leptin for the implantation process has been also hypothesized. Finally, whether leptin concentrations change following menopause or after HRT is a matter of debate.

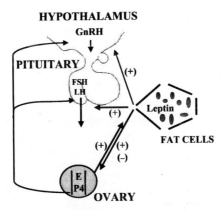

Figure 4.7. Interactions between leptin and the hypothalamic–pituitary–ovarian system. The hypothesis is that ovarian steroids stimulate leptin production from fat cells, while leptin exerts a dual effect on the ovary, that is normal concentrations stimulate and high concentrations inhibit steroidogenesis.

REFERENCES

Agarwal, S. K., Vogel, K., Weitsman, S. R., and Magoffin, D. A. (1999). Leptin antagonizes the insulin-like growth factor-I augmentation of steroidogenesis in granulosa and theca cells of the human ovary. *J. Clin. Endocrinol. Metab.*, *84*, 1072–1076.

Ainslie, D. A., Morris, M. J., Wittert, G., Turnbull, H., Proietto, J., and Thorburn, A. W. (2001). Estrogen deficiency causes central leptin insensitivity and increased hypothalamic neuropeptide Y. *Int. J. Obes. Relat. Metab. Disord.*, *25*, 1680–1688.

Alfer, J., Muller-Schottle, F., Classen-Linke, I., von Rango, U., Happel, L., Beier-Hellwig, K., Rath, W., and Beier, H. M. (2000). The endometrium as a novel target for leptin: differences in fertility and subfertility. *Mol. Hum. Reprod.*, *6*, 595–601.

Antczak, M. and Van Blerkom, J. (1997). Oocyte influences on early development: the regulatory proteins leptin and STAT3 are polarized in mouse and human oocytes and differentially distributed within the cells of the preimplantation stage embryo. *Mol. Hum. Reprod.*, *3*, 1067–1086.

Antczak, M., Van Blerkom, J., and Clark, A. (1997). A novel mechanism of vascular endothelial growth factor, leptin and transforming growth factor-beta2 sequestration in a subpopulation of human ovarian follicle cells. *Hum. Reprod.*, *12*, 2226–2234.

Ballauff, A., Ziegler, A., Emons, G., Sturm, G., Blum, W. F., Remschmidt, H., and Hebebrand, J. (1999). Serum leptin and gonadotropin levels in patients with anorexia nervosa during weight gain. *Mol. Psychiatry*, *4*, 71–75.

Barkan, D., Jia, H., Dantes, A., Vardimon, L., Amsterdam, A., and Rubinstein, M. (1999). Leptin modulates the glucocorticoid-induced ovarian steroidogenesis. *Endocrinology*, *140*, 1731–1738.

Barroso, G., Barrionuevo, M., Rao, P., Graham, L., Danforth, D., Huey, S., Abuhamad, A., and Oehninger, S. (1999). Vascular endothelial growth factor, nitric oxide, and leptin follicular fluid levels correlate negatively with embryo quality in IVF patients. *Fertil. Steril.*, *72*, 1024–1026.

Baumgartner, R. N., Ross, R. R., Waters, D. L., Brooks, W. M., Morley, J. E., Montoya, G. D., and Garry, P. J. (1999). Serum leptin in elderly people: associations with sex hormones, insulin, and adipose tissue volumes. *Obes. Res.*, *7*, 141–149.

Bennett, P. A., Lindell, K., Karlsson, C., Robinson, I. C., Carlsson, L. M., and Carlsson, B. (1998). Differential expression and regulation of leptin receptor isoforms in the rat brain: effects of fasting and oestrogen. *Neuroendocrinology*, *67*, 29–36.

Bennett, P. A., Lindell, K., Wilson, C., Carlsson, L. M., Carlsson, B., and Robinson, I. C. (1999). Cyclical variations in the abundance of leptin receptors, but not in circulating leptin, correlate with NPY expression during the oestrous cycle. *Neuroendocrinology*, *69*, 417–423.

Berman, D. M., Rodrigues, L. M., Nicklas, B. J., Ryan, A. S., Dennis, K. E., and Goldberg, A. P. (2001). Racial disparities in metabolism, central obesity, and sex hormone-binding globulin in postmenopausal women. *J. Clin. Endocrinol. Metab.*, *86*, 97–103.

Bouvattier, C., Lahlou, N., Roger, M., and Bougneres, P. (1998). Hyperleptinaemia is associated with impaired gonadotrophin response to GnRH during late puberty in obese girls, not boys. *Eur. J. Endocrinol.*, *138*, 653–658.

Brann, D. W., De Sevilla, L., Zamorano, P. L., and Mahesh, V. B. (1999). Regulation of leptin gene expression and secretion by steroid hormones. *Steroids*, *64*, 659–663.

Brann, D. W., Wade, M. F., Dhandapani, K. M., Mahesh, V. B., and Buchanan, C. D. (2002). Leptin and reproduction. *Steroids*, *67*, 95–104.

Brannian, J. D., Zhao, Y., and McElroy, M. (1999). Leptin inhibits gonadotrophin-stimulated granulosa cell progesterone production by antagonizing insulin action. *Hum. Reprod.*, *14*, 1445–1448.

Brannian, J. D., Schmidt, S. M., Kreger, D. O., and Hansen, KA. (2001). Baseline non-fasting serum leptin concentration to body mass index ratio is predictive of IVF outcomes. *Hum. Reprod.*, *16*, 1819–1826.

Butzow, T. L., Moilanen, J. M., Lehtovirta, M., Tuomi, T., Hovatta, O., Siegberg, R., Nilsson, C. G., and Apter, D. (1999). Serum and follicular fluid leptin during in vitro fertilization: relationship among leptin increase, body fat mass, and reduced ovarian response. *J. Clin. Endocrinol. Metab.*, *84*, 3135–3139.

Casabiell, X., Pineiro, V., Peino, R., Lage, M., Camina, J., Gallego, R., Vallejo, L. G., Dieguez, C., and Casanueva, F. F. (1998). Gender differences in both spontaneous and stimulated leptin secretion by human omental adipose tissue in vitro: dexamethasone and estradiol stimulate leptin release in women, but not in men. *J. Clin. Endocrinol. Metab.*, *83*, 2149–2155.

Castracane, V. D., Kraemer, R. R., Franken, M. A., Kraemer, G. R., and Gimpel, T. (1998). Serum leptin concentration in women: effect of age, obesity, and estrogen administration. *Fertil. Steril.*, *70*, 472–477.

Chang, R. J. and Jaffe, R. B. (1978). Progesterone effects on gonadotropin release in women pretreated with estradiol. *J. Clin. Endocrinol. Metab.*, *47*, 119–125.

Chehab, F. F., Lim, M. E., and Lu, R. (1996). Correction of the sterility defect in homozygous obese female mice by treatment with the human recombinant leptin. *Nat. Genet.*, *12*, 318–320.

Chmouliovsky, L., Habicht, F., James, R. W., Lehmann, T., Campana, A., and Golay, A. (1999). Beneficial effect of hormone replacement therapy on weight loss in obese menopausal women. *Maturitas.*, *32*, 147–153.

Chu, S. C., Chou, Y. C., Liu, J. Y., Chen, C. H., Shyu, J. C., and Chou, F. P. (1999). Fluctuation of serum leptin level in rats after ovariectomy and the influence of estrogen supplement. *Life Sci.*, *64*, 2299–2306.

Cioffi, J. A., Van Blerkom, J., Antczak, M., Shafer, A., Wittmer, S., and Snodgrass, H. R. (1997). The expression of leptin and its receptors in pre-ovulatory human follicles. *Mol. Hum. Reprod.*, *3*, 467–472.

Dale, O., Tanbo, T., Lunde, O., and Abyholm, T. (1993). Ovulation induction with low-dose follicle-stimulating hormone in women with the polycystic ovary syndrome. *Acta. Obstet. Gynecol. Scand.*, *72*, 43–46.

De Biasi, S. N., Apfelbaum, L. I., and Apfelbaum, M. E. (2001). In vitro effect of leptin on LH release by anterior pituitary glands from female rats at the time of spontaneous and steroid-induced LH surge. *Eur. J. Endocrinol.*, *145*, 659–665.

Di Carlo, C., Tommaselli, G. A., Pisano, G., Nasti, A., Rossi, V., Palomba, S., and Nappi, C. (2000). Serum leptin levels in postmenopausal women: effects of transdermal hormone replacement therapy. *Menopause*, *7*, 36–41.

Diano, S., Kalra, S. P., Sakamoto, H., and Horvath, T. L. (1998). Leptin receptors in estrogen receptor-containing neurons of the female rat hypothalamus. *Brain Res.*, *812*, 256–259.

Dubuc, G. R., Phinney, S. D., Stern, J. S., and Havel, P. J. (1998). Changes of serum leptin and endocrine and metabolic parameters after 7 days of energy restriction in men and women. *Metabolism*, *47*, 429–434.

Duggal, P. S., Van Der Hoek, K. H., Milner, C. R., Ryan, N. K., Armstrong, D. T., Magoffin, D. A., and Norman, R. J. (2000). The in vivo and in vitro effects of exogenous leptin on ovulation in the rat. *Endocrinology*, *141*, 1971–1976.

Elbers, J. M., Asscheman, H., Seidell, J. C., Frolich, M., Meinders, A. E., and Gooren, L. J. (1997). Reversal of the sex difference in serum leptin levels upon cross-sex hormone administration in transsexuals. *J. Clin. Endocrinol. Metab.*, *82*, 3267–3270.

Elbers, J. M., de Roo, G. W., Popp-Snijders, C., Nicolaas-Merkus, A., Westerveen, E., Joenje, B. W., and Netelenbos, J. C. (1999). Effects of administration of 17beta-oestradiol on serum leptin levels in healthy postmenopausal women. *Clin. Endocrinol. (Oxf)*, *51*, 449–454.

Elimam, A., Tjader, I., Norgren, S., Wernerman, J., Essen, P., Ljungqvist, O., and Marcus, C. (2001). Total parenteral nutrition after surgery rapidly increases serum leptin levels. *Eur. J. Endocrinol.*, *144*, 123–128.

Evans, D. J., Hoffmann, R. G., Kalkhoff, R. K., and Kissebah, A. H. (1983). Relationship of androgenic activity to body fat topography, fat cell morphology, and metabolic aberrations in premenopausal women. *J. Clin. Endocrinol. Metab.*, *57*, 304–310.

Finn, P. D., Cunningham, M. J., Pau, K. Y., Spies, H. G., Clifton, D. K., and Steiner, R. A. (1998). The stimulatory effect of leptin on the neuroendocrine reproductive axis of the monkey. *Endocrinology*, *139*, 4652–4662.

Gettys, T. W., Harkness, P. J., and Watson, P. M. (1996). The beta 3-adrenergic receptor inhibits insulin-stimulated leptin secretion from isolated rat adipocytes. *Endocrinology*, *137*, 4054–4057.

Ghizzoni, L., Barreca, A., Mastorakos, G., Furlini, M., Vottero, A., Ferrari, B., Chrousos, G. P., and Bernasconi, S. (2001). Leptin inhibits steroid biosynthesis by human granulosa-lutein cells. *Horm. Metab. Res.*, *33*, 323–328.

Gonzalez, R. R., Palomino, A., Boric, A., Vega, M., and Devoto, L. (1999). A quantitative evaluation of alpha1, alpha4, alphaV and beta3 endometrial integrins of fertile and unexplained infertile women during the menstrual cycle. A flow cytometric appraisal. *Hum. Reprod.*, *14*, 2485–2492.

Gonzalez, R. R., Simon, C., Caballero-Campo, P., Norman, R., Chardonnens, D., Devoto, L., and Bischof, P. (2000). Leptin and reproduction. *Hum. Reprod. Update*, *6*, 290–300.

Hadji, P., Hars, O., Bock, K., Sturm, G., Bauer, T., Emons, G., and Schulz, K. D. (2000). The influence of menopause and body mass index on serum leptin concentrations. *Eur. J. Endocrinol.*, *143*, 55–60.

Haffner, S. M., Mykkanen, L., and Stern, M. P. (1997). Leptin concentrations in women in the San Antonio Heart Study: effect of menopausal status and postmenopausal hormone replacement therapy. *Am. J. Epidemiol.*, *146*, 581–585.

Hakansson, M. L., Brown, H., Ghilardi, N., Skoda, R. C., and Meister, B. (1998). Leptin receptor immunoreactivity in chemically defined target neurons of the hypothalamus. *J. Neurosci.*, *18*, 559–572.

Halaas, J. L., Gajiwala, K. S., Maffei, M., Cohen, S. L., Chait, B. T., Rabinowitz, D., Lallone, R. L., Burley, S. K., and Friedman, J. M. (1995). Weight-reducing effects of the plasma protein encoded by the obese gene. *Science*, *269*, 543–546.

Hardie, L., Trayhurn, P., Abramovich, D., and Fowler, P. (1997). Circulating leptin in women: a longitudinal study in the menstrual cycle and during pregnancy. *Clin. Endocrinol. (Oxf)*, *47*, 101–106.

Havel, P. J., Kasim-Karakas, S., Dubuc, G. R., Mueller, W., and Phinney, S. D. (1996). Gender differences in plasma leptin concentrations. *Nat. Med.*, *2*, 949–950.

Henry, B. A., Goding, J. W., Alexander, W. S., Tilbrook, A. J., Canny, B. J., Dunshea, F., Rao, A., Mansell, A., and Clarke, I. J. (1999). Central administration of leptin to ovariectomized ewes inhibits food intake without affecting the secretion of hormones from the pituitary gland: evidence for a dissociation of effects on appetite and neuroendocrine function. *Endocrinology*, *140*, 1175–1182.

Hube, F., Lietz, U., Igel, M., Jensen, P. B., Tornqvist, H., Joost, H. G., and Hauner, H. (1996). Difference in leptin mRNA levels between omental and subcutaneous abdominal adipose tissue from obese humans. *Horm. Metab. Res.*, *28*, 690–693.

Jockenhovel, F., Blum, W. F., Vogel, E., Englaro, P., Muller-Wieland, D., Reinwein, D., Rascher, W., and Krone, W. (1997). Testosterone substitution normalizes elevated serum leptin levels in hypogonadal men. *J. Clin. Endocrinol. Metab.*, *82*, 2510–2513.

Kalra, S. P., Xu, B., Dube, M. G., Moldawer, L. L., Martin, D., and Kalra, P. S. (1998). Leptin and ciliary neurotropic factor (CNTF) inhibit fasting-induced suppression of luteinizing hormone release in rats: role of neuropeptide Y. *Neurosci. Lett.*, *240*, 45–49.

Kamohara, S., Burcelin, R., Halaas, J. L., Friedman, J. M., and Charron, M. J. (1997). Acute stimulation of glucose metabolism in mice by leptin treatment. *Nature*, *389*, 374–377.

Kanaley, J. A., Sames, C., Swisher, L., Swick, A. G., Ploutz-Snyder, L. L., Steppan, C. M., Sagendorf, K. S., Feiglin, D., Jaynes, E. B., Meyer, R. A., and Weinstock, R. S. (2001). Abdominal fat distribution in pre- and postmenopausal women: The impact of physical activity, age, and menopausal status. *Metabolism*, *50*, 976–982.

Karlsson, C., Lindell, K., Svensson, E., Bergh, C., Lind, P., Billig, H., Carlsson, L. M., and Carlsson, B. (1997). Expression of functional leptin receptors in the human ovary. *J. Clin. Endocrinol. Metab.*, *82*, 4144–4148.

Kastin, A. J., Akerstrom, V., and Pan, W. (2001). Validity of multiple-time regression analysis in measurement of tritiated and iodinated leptin crossing the blood–brain barrier: meaningful controls. *Peptides*, *22*, 2127–2136.

Kitawaki, J., Kusuki, I., Koshiba, H., Tsukamoto, K., and Honjo, H. (1999). Leptin directly stimulates aromatase activity in human luteinized granulosa cells. *Mol. Hum. Reprod.*, *5*, 708–713.

Kitawaki, J., Koshiba, H., Ishihara, H., Kusuki, I., Tsukamoto, K., and Honjo, H. (2000). Expression of leptin receptor in human endometrium and fluctuation during the menstrual cycle. *J. Clin. Endocrinol. Metab.*, *85*, 1946–1950.

Kohrt, W. M., Landt, M., and Birge, S. J. Jr. (1996). Serum leptin levels are reduced in response to exercise training, but not hormone replacement therapy, in older women. *J. Clin. Endocrinol. Metab.*, *81*, 3980–3985.

Kohsaka, A., Watanobe, H., Kakizaki, Y., Habu, S., and Suda, T. (1999). A significant role of leptin in the generation of steroid-induced luteinizing hormone and prolactin surges in female rats. *Biochem. Biophys. Res. Commun.*, *254*, 578–581.

Komorowski, J., Jankiewicz-Wika, J., and Stepien, H. (2000). Effects of Gn-RH, TRH, and CRF administration on plasma leptin levels in lean and obese women. *Neuropeptides*, *34*, 89–97.

Kopp, W., Blum, W. F., von Prittwitz, S., Ziegler, A., Lubbert, H., Emons, G., Herzog, W., Herpertz, S., Deter, H. C., Remschmidt, H., and Hebebrand, J. (1997). Low leptin levels predict amenorrhea in underweight and eating disordered females. *Mol. Psychiatry*, *2*, 335–340.

Korbonits, M., Trainer, P. J., Little, J. A., Edwards, R., Kopelman, P. G., Besser, G. M., Svec, F., and Grossman, A. B. (1997). Leptin levels do not change acutely with food administration in normal or obese subjects, but are negatively correlated with pituitary-adrenal activity. *Clin. Endocrinol. (Oxf)*, *46*, 751–757.

Koshiba, H., Kitawaki, J., Ishihara, H., Kado, N., Kusuki, I., Tsukamoto, K., and Honjo, H. (2001). Progesterone inhibition of functional leptin receptor mRNA expression in human endometrium. *Mol. Hum. Reprod.*, *7*, 567–572.

Kraemer, R. R., Johnson, L. G., Haltom, R., Kraemer, G. R., Hebert, E. P., Gimpel, T., and Castracane, V. D. (1999). Serum leptin concentrations in response to acute exercise in postmenopausal women with and without hormone replacement therapy. *Proc. Soc. Exp. Biol. Med.*, *221*, 171–177.

Lado-Abeal, J., Lukyanenko, Y. O., Swamy, S., Hermida, R. C., Hutson, J. C., and Norman, R. L. (1999). Short-term leptin infusion does not affect circulating levels of LH, testosterone or cortisol in food-restricted pubertal male rhesus macaques. *Clin. Endocrinol. (Oxf)*, *51*, 41–51.

Laivuori, H., Koistinen, H. A., Karonen, S. L., Cacciatore, B., and Ylikorkala, O. (2001). Comparison between 1 year oral and transdermal oestradiol and sequential norethisterone acetate on circulating concentrations of leptin in postmenopausal women. *Hum. Reprod.*, *16*, 1632–1635.

Larsson, H. and Ahren, B. (1996). Short-term dexamethasone treatment increases plasma leptin independently of changes in insulin sensitivity in healthy women. *J. Clin. Endocrinol. Metab.*, *81*, 4428–4432.

Laughlin, G. A. and Yen, S. S. (1997). Hypoleptinemia in women athletes: absence of a diurnal rhythm with amenorrhea. *J. Clin. Endocrinol. Metab.*, *82*, 318–321.

Lessey, B. A., Castelbaum, A. J., Sawin, S. W., and Sun, J. (1995). Integrins as markers of uterine receptivity in women with primary unexplained infertility. *Fertil. Steril.*, *63*, 535–542.

Licinio, J., Negrao, A. B., Mantzoros, C., Kaklamani, V., Wong, M. L., Bongiorno, P. B., Mulla, A., Cearnal, L., Veldhuis, J. D., Flier, J. S., McCann, S. M., and Gold, P. W. (1998). Synchronicity of frequently sampled, 24-h concentrations of circulating leptin, luteinizing hormone, and estradiol in healthy women. *Proc. Natl. Acad. Sci. USA*, *95*, 2541–2546.

Lindheim, S. R., Sauer, M. V., Carmina, E., Chang, P. L., Zimmerman, R., and Lobo, R. A. (2000). Circulating leptin levels during ovulation induction: relation to adiposity and ovarian morphology. *Fertil. Steril.*, *73*, 493–498.

Loffler, S., Aust, G., Kohler, U., and Spanel-Borowski, K. (2001). Evidence of leptin expression in normal and polycystic human ovaries. *Mol. Hum. Reprod.*, *7*, 1143–1149.

Lubbert, H. and Nauert, C. (1997). Continuous versus cyclical transdermal estrogen replacement therapy in post-menopausal women: influence on climacteric symptoms, body weight and bleeding pattern. *Maturitas.*, *28*, 117–125.

Ludwig, M., Klein, H. H., Diedrich, K., and Ortmann, O. (2000). Serum leptin concentrations throughout the menstrual cycle. *Arch. Gynecol. Obstet.*, *263*, 99–101.

Luukkaa, V., Savontaus, E., Rouru, J., Virtanen, K. A., Boss, O., Huhtaniemi, I., Koulu, M., Pesonen, U., and Huupponen, R. (2001). Effects of estrous cycle and steroid replacement on the expression of leptin and uncoupling proteins in adipose tissue in the rat. *Gynecol. Endocrinol.*, *15*, 103–112.

Magni, P., Vettor, R., Pagano, C., Calcagno, A., Beretta, E., Messi, E., Zanisi, M., Martini, L., and Motta, M. (1999). Expression of a leptin receptor in immortalized gonadotropin-releasing hormone-secreting neurons. *Endocrinology*, *140*, 1581–1585.

Mannucci, E., Ognibene, A., Becorpi, A., Cremasco, F., Pellegrini, S., Ottanelli, S., Rizzello, S. M., Massi, G., Messeri, G., and Rotella, C. M. (1998). Relationship between leptin and oestrogens in healthy women. *Eur. J. Endocrinol.*, *139*, 198–201.

Mantzoros, C. S. and Moschos, S. J. (1998). Leptin: in search of role(s) in human physiology and pathophysiology. *Clin. Endocrinol. (Oxf)*, *49*, 551–567.

Mantzoros, C. S., Qu, D., Frederich, R. C., Susulic, V. S., Lowell, B. B., Maratos-Flier, E., and Flier, J. S. (1996). Activation of beta(3) adrenergic receptors suppresses leptin expression and mediates a leptin-independent inhibition of food intake in mice. *Diabetes*, *45*, 909–914.

Mantzoros, C. S., Cramer, D. W., Liberman, R. F., and Barbieri, R. L. (2000). Predictive value of serum and follicular fluid leptin concentrations during assisted reproductive cycles in normal women and in women with the polycystic ovarian syndrome. *Hum. Reprod.*, *15*, 539–544.

Matkovic, V., Ilich, J. Z., Skugor, M., Badenhop, N. E., Goel, P., Clairmont, A., Klisovic, D., Nahhas, R. W., and Landoll, J. D. (1997). Leptin is inversely related to age at menarche in human females. *J. Clin. Endocrinol. Metab.*, *82*, 3239–3245.

Matsuoka, T., Tahara, M., Yokoi, T., Masumoto, N., Takeda, T., Yamaguchi, M., Tasaka, K., Kurachi, H., and Murata, Y. (1999). Tyrosine phosphorylation of STAT3 by leptin through leptin receptor in mouse metaphase 2 stage oocyte. *Biochem. Biophys. Res. Commun.*, *256*, 480–484.

Messinis, I. E. and Templeton, A. A. (1990). Effects of supraphysiological concentrations of progesterone on the characteristics of the oestradiol-induced gonadotrophin surge in women. *J. Reprod. Fert.*, *88*, 513–519.

Messinis, I. E. and Milingos, S. D. (1999). Leptin in human reproduction. *Hum. Reprod. Update*, 5, 52–63.

Messinis, I. E., Milingos, S., Zikopoulos, K., Kollios, G., Seferiadis, K., and Lolis, D. (1998). Leptin concentrations in the follicular phase of spontaneous cycles and cycles superovulated with follicle stimulating hormone. *Hum. Reprod.*, 13, 1152–1156.

Messinis, I. E., Milingos, S. D., Alexandris, E., Kariotis, I., Kollios, G., and Seferiadis, K. (1999). Leptin concentrations in normal women following bilateral ovariectomy. *Hum. Reprod.*, 14, 913–918.

Messinis, I. E., Kariotis, I., Milingos, S., Kollios, G., and Seferiadis, K. (2000). Treatment of normal women with oestradiol plus progesterone prevents the decrease of leptin concentrations induced by ovariectomy. *Hum. Reprod.*, 15, 2383–2387.

Messinis, I. E., Papageorgiou, I., Milingos, S., Asprodini, E., Kollios, G., and Seferiadis, K. (2001). Oestradiol plus progesterone treatment increases serum leptin concentrations in normal women. *Hum. Reprod.*, 16, 1827–1832.

Miller, K. K., Parulekar, M. S., Schoenfeld, E., Anderson, E., Hubbard, J., Klibanski, A., and Grinspoon, S. K. (1998). Decreased leptin levels in normal weight women with hypothalamic amenorrhea: the effects of body composition and nutritional intake. *J. Clin. Endocrinol. Metab.*, 83, 2309–2312.

Montalban, C., Del Moral, I., Garcia-Unzueta, M. T., Villanueva, M. A., and Amado, J. A. (2001). Perioperative response of leptin and the tumor necrosis factor alpha system in morbidly obese patients. Influence of cortisol inhibition by etomidate. *Acta Anaesthesiol. Scand.*, 45, 207–212.

Mounzih, K., Lu, R., and Chehab, F. F. (1997). Leptin treatment rescues the sterility of genetically obese ob/ob males. *Endocrinology*, 138, 1190–1193.

Murakami, T., Iida, M., and Shima, K. (1995). Dexamethasone regulates obese expression in isolated rat adipocytes. *Biochem. Biophys. Res. Commun.*, 214, 1260–1267.

Nagatani, S., Guthikonda, P., Thompson, R. C., Tsukamura, H., Maeda, K. I., and Foster, D. L. (1998). Evidence for GnRH regulation by leptin: leptin administration prevents reduced pulsatile LH secretion during fasting. *Neuroendocrinology*, 67, 370–376.

Ogura, K., Irahara, M., Kiyokawa, M., Tezuka, M., Matsuzaki, T., Yasui, T., Kamada, M., and Aono, T. (2001). Effects of leptin on secretion of LH and FSH from primary cultured female rat pituitary cells. *Eur. J. Endocrinol.*, 144, 653–658.

Ongphiphadhanakul, B., Chanprasertyothin, S., Piaseu, N., Chansirikanjana, S., Puavilai, G., and Rajatanavin, R. (1998). Change in body weight after hormone replacement therapy in postmenopausal women is dependent on basal circulating leptin. *Maturitas*, 30, 283–288.

Paolisso, G., Rizzo, M. R., Mone, C. M., Tagliamonte, M. R., Gambardella, A., Riondino, M., Carella, C., Varricchio, M., and D'Onofrio, F. (1998). Plasma sex hormones are significantly associated with plasma leptin concentration in healthy subjects. *Clin. Endocrinol. (Oxf)*, 48, 291–297.

Papaspyrou-Rao, S., Schneider, S. H., Petersen, R. N., and Fried, S. K. (1997). Dexamethasone increases leptin expression in humans in vivo. *J. Clin. Endocrinol. Metab.*, 82, 1635–1637.

Pelleymounter, M. A., Cullen, M. J., Baker, M. B., Hecht, R., Winters, D., Boone, T., and Collins, F. (1995). Effects of the obese gene product on body weight regulation in ob/ob mice. *Science*, 269, 540–543.

Perera, M., Sattar, N., Petrie, J. R., Hillier, C., Small, M., Connell, J. M., Lowe, G. D., and Lumsden, M. A. (2001). The effects of transdermal estradiol in combination with oral norethisterone on lipoproteins, coagulation, and endothelial markers in postmenopausal women with type 2 diabetes: a randomized, placebo-controlled study. *J. Clin. Endocrinol. Metab.*, 86, 1140–1143.

Phipps, W. R., Wangen, K. E., Duncan, A. M., Merz-Demlow, B. E., Xu, X., and Kurzer, M. S. (2001). Lack of effect of isoflavonic phytoestrogen intake on leptin concentrations in premenopausal and postmenopausal women. *Fertil. Steril.*, 75, 1059–1064.

Quinton, N. D., Laird, S. M., Okon, M. A., Li, T. C., Smith, R. F., Ross, R. J., and Blakemore, A. I. (1999). Serum leptin levels during the menstrual cycle of healthy fertile women. *Br. J. Biomed. Sci.*, 56, 16–19.

Quinton, N. D., Lee, A. J., Ross, R. J., Eastell, R., and Blakemore, A. I. (2001). A single nucleotide polymorphism (SNP) in the leptin receptor is associated with BMI, fat mass and leptin levels in postmenopausal Caucasian women. *Hum. Genet.*, 108, 233–236.

Riad-Gabriel, M. G., Jinagouda, S. D., Sharma, A., Boyadjian, R., and Saad, M. F. (1998). Changes in plasma leptin during the menstrual cycle. *Eur. J. Endocrinol.*, 139, 528–531.

Roncari, D. A. and Van, R. L. (1978). Promotion of human adipocyte precursor replication by 17beta-estradiol in culture. *J. Clin. Invest.*, 62, 503–508.

Rosenbaum, M., Nicolson, M., Hirsch, J., Heymsfield, S. B., Gallagher, D., Chu, F., and Leibel, R. L. (1996). Effects of gender, body composition, and menopause on plasma concentrations of leptin. *J. Clin. Endocrinol. Metab.*, *81*, 3424–3427.

Rosenbaum, M., Pietrobelli, A., Vasselli, J. R., Heymsfield, S. B., and Leibel, R. L. (2001). Sexual dimorphism in circulating leptin concentrations is not accounted for by differences in adipose tissue distribution. *Int. J. Obes. Relat. Metab. Disord.*, *25*, 1365–1371.

Saad, M. F., Damani, S., Gingerich, R. L., Riad-Gabriel, M. G., Khan, A., Boyadjian, R., Jinagouda, S. D., el-Tawil, K., Rude, R. K., and Kamdar, V. (1997). Sexual dimorphism in plasma leptin concentration. *J. Clin. Endocrinol. Metab.*, *82*, 579–584.

Sayegh, R. A., Kelly, L., Wurtman, J., Deitch, A., and Chelmow, D. (1999). Impact of hormone replacement therapy on the body mass and fat compositions of menopausal women: a cross-sectional study. *Menopause.*, *6*, 312–315.

Shimizu, H., Shimomura, Y., Nakanishi, Y., Futawatari, T., Ohtani, K., Sato, N., and Mori, M. (1997). Estrogen increases in vivo leptin production in rats and human subjects. *J. Endocrinol.*, *154*, 285–292.

Simon, C., Piquette, G. N., Frances, A., and Polan, M. L. (1993). Localization of interleukin-1 type I receptor and interleukin-1 beta in human endometrium throughout the menstrual cycle. *J. Clin. Endocrinol. Metab.*, *77*, 549–555.

Simon, C., Frances, A., Piquette, G., Hendrickson, M., Milki, A., and Polan, M. L. (1994). Interleukin-1 system in the materno-trophoblast unit in human implantation: immunohistochemical evidence for autocrine/paracrine function. *J. Clin. Endocrinol. Metab.*, *78*, 847–854.

Simon, C., Gimeno, M. J., Mercader, A., O'Connor, J. E., Remohi, J., Polan, M. L., and Pellicer, A. (1997). Embryonic regulation of integrins beta 3, alpha 4, and alpha 1 in human endometrial epithelial cells in vitro. *J. Clin. Endocrinol. Metab.*, *82*, 2607–2616.

Sinha, M. K., Opentanova, I., Ohannesian, J. P., Kolaczynski, J. W., Heiman, M. L., Hale, J., Becker, G. W., Bowsher, R. R., Stephens, T. W., and Caro, J. F. (1996). Evidence of free and bound leptin in human circulation. Studies in lean and obese subjects and during short-term fasting. *J. Clin. Invest.*, *98*, 1277–1282.

Sivitz, W. I., Walsh, S. A., Morgan, D. A., Thomas, M. J., and Haynes, W. G. (1997). Effects of leptin on insulin sensitivity in normal rats. *Endocrinology*, *138*, 3395–3401.

Slieker, L. J., Sloop, K. W., Surface, P. L., Kriauciunas, A., LaQuier, F., Manetta, J., Bue-Valleskey, J., and Stephens, T. W. (1996). Regulation of expression of ob mRNA and protein by glucocorticoids and cAMP. *J. Biol. Chem.*, *271*, 5301–5304.

Spicer, L. J. and Francisco, C. C. (1997). The adipose obese gene product, leptin: evidence of a direct inhibitory role in ovarian function. *Endocrinology*, *138*, 3374–3379.

Spicer, L. J. and Francisco, C. C. (1998). Adipose obese gene product, leptin, inhibits bovine ovarian thecal cell steroidogenesis. *Biol. Reprod.*, *58*, 207–212.

Stock, S. M., Sande, E. M., and Bremme, K. A. (1999). Leptin levels vary significantly during the menstrual cycle, pregnancy, and in vitro fertilization treatment: possible relation to estradiol. *Fertil. Steril.*, *72*, 657–662.

Sumner, A. E., Falkner, B., Kushner, H., and Considine, R. V. (1998). Relationship of leptin concentration to gender, menopause, age, diabetes, and fat mass in African Americans. *Obes. Res.*, *6*, 128–133.

Takeda, K., Noguchi, K., Shi, W., Tanaka, T., Matsumoto, M., Yoshida, N., Kishimoto, T., and Akira, S. (1997). Targeted disruption of the mouse Stat3 gene leads to early embryonic lethality. *Proc. Natl. Acad. Sci. USA*, *94*, 3801–3804.

Tartaglia, L. A., Dembski, M., Weng, X., Deng, N., Culpepper, J., Devos, R., Richards, G. J., Campfield, L. A., Clark, F. T., Deeds, J. et al. (1995). Identification and expression cloning of a leptin receptor, OB-R. *Cell*, *83*, 1263–1271.

Teirmaa, T., Luukkaa, V., Rouru, J., Koulu, M., and Huupponen, R. (1998). Correlation between circulating leptin and luteinizing hormone during the menstrual cycle in normal-weight women. *Eur. J. Endocrinol.*, *139*, 190–194.

Unkila-Kallio, L., Andersson, S., Koistinen, H. A., Karonen, S. L., Ylikorkala, O., and Tiitinen, A. (2001). Leptin during assisted reproductive cycles: the effect of ovarian stimulation and of very early pregnancy. *Hum. Reprod.*, *16*, 657–662.

Van Harmelen, V., Reynisdottir, S., Eriksson, P., Thorne, A., Hoffstedt, J., Lonnqvist, F., and Arner, P. (1998). Leptin secretion from subcutaneous and visceral adipose tissue in women. *Diabetes*, *47*, 913–917.

Wade, G. N. and Gray, J. M. (1978). Cytoplasmic 17 beta-[3H]estradiol binding in rat adipose tissues. *Endocrinology, 103*, 1695–1701.

Wauters, M., Considine, R. V., and Van Gaal L. F. (2000). Human leptin: from an adipocytes hormone to an endocrine mediator. *Eur. J. Endocrinol., 143*, 293–311.

Williams, C. L., Nishihara, M., Thalabard, J. C., Grosser, P. M., Hotchkiss, J., and Knobil, E. (1990). Corticotropin-releasing factor and gonadotropin-releasing hormone pulse generator activity in the rhesus monkey. Electrophysiological studies. *Neuroendocrinology, 52*, 133–137.

Yamada, M., Irahara, M., Tezuka, M., Murakami, T., Shima, K., and Aono, T. (2000). Serum leptin profiles in the normal menstrual cycles and gonadotropin treatment cycles. *Gynecol. Obstet. Invest., 49*, 119–123.

Yoneda, N., Saito, S., Kimura, M., Yamada, M., Iida, M., Murakami, T., Irahara, M., Shima, K., and Aono, T. (1998). The influence of ovariectomy on ob gene expression in rats. *Horm. Metab. Res., 30*, 263–265.

Yu, W. H., Kimura, M., Walczewska, A., Karanth, S., and McCann, S. M. (1997a). Role of leptin in hypothalamic–pituitary function. *Proc. Natl. Acad. Sci. USA, 94*, 1023–1028.

Yu, W. H., Walczewska, A., Karanth, S., and McCann, S. M. (1997b). Nitric oxide mediates leptin-induced luteinizing hormone-releasing hormone (LHRH) and LHRH and leptin-induced LH release from the pituitary gland. *Endocrinology, 138*, 5055–5058.

Zachow, R. J. and Magoffin, D. A. (1997). Direct intraovarian effects of leptin: impairment of the synergistic action of insulin-like growth factor-I on follicle-stimulating hormone-dependent estradiol-17 beta production by rat ovarian granulosa cells. *Endocrinology, 138*, 847–850.

Zachow, R. J., Weitsman, S. R., and Magoffin, D. A. (1999). Leptin impairs the synergistic stimulation by transforming growth factor-beta of follicle-stimulating hormone-dependent aromatase activity and messenger ribonucleic acid expression in rat ovarian granulosa cells. *Biol. Reprod., 61*, 1104–1109.

Zamorano, P. L., Mahesh, V. B., De Sevilla, L. M., Chorich, L. P., Bhat, G. K., and Brann, D. W. (1997). Expression and localization of the leptin receptor in endocrine and neuroendocrine tissues of the rat. *Neuroendocrinology, 65*, 223–228.

Zhang, Y., Proenca, R., Maffei, M., Barone, M., Leopold, L., and Friedman, J. M. (1994). Positional cloning of the mouse obese gene and its human homologue. *Nature, 372*, 425–432.

5

Impact of Leptin on Ovarian Folliculogenesis and Assisted Reproduction

JOHN D. BRANNIAN AND KEITH A. HANSEN

5.1. INTRODUCTION

With the discovery in the late 1990s that leptin receptors are expressed in high levels within maturing follicles, it became clear that leptin may regulate ovarian function at multiple levels. Not only does leptin participate in the control of gonadotropin secretion via its hypothalamic actions, but circulating or locally produced leptin may also provide direct modulation of thecal, granulosa, and oocyte function. Addition of leptin to the growing list of cytokines and growth factors involved in follicular development and oocyte maturation extends our appreciation of the complex integration of basic cellular and metabolic processes with reproductive function and fertility.

Circulating concentrations of leptin vary widely in humans (<2 ng/ml to >60 ng/ml) as a function of body fat mass (Ruhl and Everhart, 2001). Recognition that a threshold level of leptin is obligatory for maintenance of normal reproductive function has come from collective studies in animals and humans. On the other hand, obesity, which is associated with significantly elevated leptin levels, is known to have a negative impact on fertility. It has been proposed that the hyperleptinemia characteristic of obese individuals is related to leptin resistance. That obesity is a consequence of leptin resistance, as once suggested, seems highly unlikely for most individuals, although obesity may be exacerbated by insensitivity to leptin (Arch et al., 1998). It is much more likely that leptin resistance is a consequence of obesity brought on by lifestyle and diet. The pertinent question then is how does obesity-induced leptin resistance impact human health and fertility. This review

JOHN D. BRANNIAN AND KEITH A. HANSEN • Department of Obstetrics and Gynecology, University of South Dakota School of Medicine.

Leptin and Reproduction. Edited by Henson and Castracane, Kluwer Academic/Plenum Publishers, 2003.

summarizes current knowledge and theories of leptin's role in folliculogenesis with particular emphasis on assisted reproduction. A hypothetical model of how leptin and leptin resistance may impact follicle development, oocyte maturation, and fecundity is presented.

5.2. LEPTIN AND LEPTIN RECEPTORS IN THE OVARY

5.2.1. Leptin Receptors

The presence of functional leptin receptors in the ovary was first reported in 1996 by Cioffi and co-workers (1996), and subsequently confirmed in a variety of species. Leptin receptor mRNA transcripts and protein have been identified in granulosa (GC) and thecal cells (TC) of antral follicles (Cioffi et al., 1997; Karlsson et al., 1997; Zamorano et al., 1997; Agarwal et al., 1999; Ruiz-Cortez et al., 2000; Ryan et al., 2002), corpora lutea (Ruiz-Cortez et al., 2000; Ryan et al., 2002), and oocytes (Matsuoka et al., 1999; Kawamura et al., 2002; Ryan et al., 2002). Moreover, leptin can modulate in vitro mouse preantral follicle development (Kikuchi et al., 2001) suggesting that preantral follicles also express leptin receptor.

5.2.2. Leptin in Follicular Fluid

High expression of leptin receptor in granulosa and thecal compartments of developing follicles implies a functional role for leptin in follicle growth and/or maturation. Although the granulosum of antral follicles has no direct blood supply, follicular fluid (FF) contains leptin protein in concentrations comparable to that in serum or plasma (Karlsson et al., 1997; Zhao et al., 1998; Bützow et al., 1999; Mantzoros et al., 2000). In other words, FF leptin levels correlate with body fat mass as do blood levels. However, interfollicular variation in FF leptin concentrations occurs, and the difference between minimum and maximum levels within a cohort may be as great as 2-fold (Cioffi et al., 1997; Fedorcsák et al., 2000a). This suggests that either transport of leptin from the circulation into the antrum differs among follicles, which could be explained by interfollicular differences in thecal vascularity and blood flow, or that local secretion of leptin contributes to the variability in intrafollicular leptin concentrations.

5.2.3. Evidence for Ovarian Secretion of Leptin

Evidence for leptin production within the ovary is limited and conflicting results have been reported. Cioffi and colleagues (1997) reported that leptin mRNA transcripts and translated protein were present in mural and cumulus GC collected from preovulatory follicles during follicular aspiration for in vitro fertilization (IVF). This was in contrast to another report (Karlsson et al., 1997) that failed to detect leptin gene expression in the ovary. Moreover, although leptin immunoactivity has been demonstrated in follicular cells and oocytes (Antczak and Van Blerkom, 1997; Antczak et al., 1997), leptin immunoactivity was undetectable in spent media from control or gonadotropin-stimulated cultures of luteinized GC recovered from follicular aspirates obtained during IVF oocyte retrieval (Prien and Castracane, 1998). Therefore, the secretory origin of follicular leptin remains in question.

Recent evidence, however, further supports the possibility that leptin may be secreted by the follicle itself. Löffler et al. (2001) immunocytochemically localized leptin in sections of intact human ovaries, and found strong immunostaining in the thecal and granulosa layers of preantral follicles, developing corpora lutea, and the theca of an intact antral follicle. Surprisingly, these investigators did not observe a significant signal in the GC of a healthy antral follicle. But leptin mRNA was detectable after reverse transcriptase-polymerase chain reaction (RT-PCR) in GC aspirated from women undergoing IVF (Löffler et al., 2001) similar to the earlier results of Cioffi et al. (1997). Further studies are needed to confirm leptin secretion within the ovary, and clarify the cellular origins and temporal expression. Nevertheless it appears that intrafollicular concentrations of leptin may be dependent upon both adipose-derived leptin from the circulation and leptin produced locally within the follicle. The possibility of leptin secretion within the developing follicle, together with expression of leptin receptors on TC, GC, and oocytes, strongly suggests that leptin plays an important autocrine or paracrine role in regulation of folliculogenesis, follicle maturation, and/or oocyte maturation.

5.3. LEPTIN AND GONADOTROPIN STIMULATION OF THE OVARY

A great deal of current understanding of leptin dynamics with regard to the ovary has come from studies of exogenous gonadotropin stimulation for ovulation induction and assisted reproductive technologies (ART). Two major findings have arisen. First, leptin concentrations in the bloodstream and FF tend to rise in parallel with estradiol during ovarian stimulation. Second, leptin levels have an impact on the sensitivity and responsiveness of the ovary to gonadotropin stimulation.

5.3.1. Leptin Dynamics during Exogenous Gonadotropin Stimulation

Messinis and co-workers (1998) and Strowitski et al. (1998) first reported that circulating leptin concentrations progressively increased during the first few days of ovarian stimulation with exogenous follicle-stimulating hormone (FSH), and that these levels tended to be higher than those observed in the follicular phases of natural cycles (Messinis et al., 1998). It is unlikely that these changes were a direct action of FSH because FSH decreases as estradiol and leptin rise during natural cycles, and high FSH levels experienced during ovarian stimulation simulate endogenous levels in post-menopausal women, which have lower serum leptin concentrations than pre-menopausal women (Rosenbaum et al., 1996). However, human chorionic gonadotropin (hCG) increased in vitro leptin secretion by an adipocyte cell line (Sivan et al., 1998), and women stimulated with human menopausal gonadotropin (hMG) tended to exhibit a slightly greater rise in serum leptin during ovarian stimulation than those stimulated with pure FSH (Unkila-Kallio et al., 2001), suggesting that LH activity may enhance leptin secretion.

Based on evidence that estrogens can stimulate adipose cell secretion of leptin (Shimuzu et al., 1997; Casabiell et al., 1998; Brann et al., 1999), high estradiol levels experienced during stimulated cycles have been proposed as the principal cause of elevated leptin secretion during ovarian stimulation. However, even though leptin and estradiol rise in parallel, correlation coefficients reported in the literature are surprisingly small, suggesting only a minimal contribution by estrogen to leptin variability. For example, Bützow and colleagues (1999) found no significant correlation between estradiol and leptin at any of four distinct time points during FSH stimulation cycles. Moreover, serum leptin levels after exogenous FSH stimulation were similar to those experienced by the same nine women during the late follicular phase of spontaneous menstrual cycles despite having estradiol levels 10-fold higher in the stimulated cycles (Messinis et al., 1998). Although maturing follicles may produce leptin (Cioffi et al., 1997; Löffler et al., 2001), a follicular contribution would not seem to be enough to significantly alter peripheral concentrations, but this remains to be clarified.

More perplexing is that increases in leptin during gonadotropin stimulation do not occur in all women. In the 37 women we studied (Zhao et al., 2000), mean serum leptin concentrations increased by about 70 percent, on average, in 29 of the women during gonadotropin stimulation. But there was either no change or a slight decrease in leptin in the other eight women. Other investigators have reported similar findings (Bützow et al., 1999; Yamada et al., 2000; Unkila-Kallio et al., 2001). Collectively, data from these studies suggests that about 80 percent of women undergoing FSH stimulation experience an increase in serum leptin concentrations.

Why the other 20 percent of women do not experience a rise in leptin is unclear. A significant positive correlation between the absolute (i.e. ng/ml) change in leptin concentration and percent body fat mass has been reported (Bützow et al., 1999). However, Lindheim and co-workers (2000) compared changes in serum leptin during ovarian stimulation between lean and obese patients, and found that the relative magnitude of the leptin increase was similar (about 55 percent) in both lean and obese women. But interestingly, the maximum leptin level occurred two days sooner on average in the lean women (Lindheim et al., 2000). Unkila-Kallio et al. (2001) recently compared changes in serum leptin during FSH stimulation in a cohort of women grouped on the basis of body mass index (BMI). Whereas serum leptin increased in 10 of 10 overweight (BMI $\geq$ 27) women, only 5 of 9 (56 percent) underweight (BMI $\leq$ 19) women experienced a rise in leptin, and 83 percent of normal weight (BMI 20–26) women had an increase in leptin. Although the mechanism by which leptin increases during ovarian stimulation is of academic interest, the more clinically important question is the impact of leptin and leptin dynamics on follicle development and oocyte maturation in stimulated cycles.

5.3.2. Impact of Leptin on Ovarian Responsiveness

Diminished responsiveness to gonadotropin stimulation is common in obese women. It was first reported more than two decades ago (Shepard et al., 1979) that excess body weight was associated with poor response to clomiphene citrate (CC) stimulation of the ovaries. Imani and co-workers (2000) found that serum leptin levels were significantly higher in normogonadotropic oligo- and amenorrheic women that failed to ovulate after CC therapy than in those that ovulated. Of several parameters measured by these investigators, including insulin, insulin/glucose ratio, inhibin B, IGF-1, IGFBP-1, IGFBP-3, and VEGF, leptin showed the most highly significant differences between CC responders and non-responders (Imani et al., 2000). Moreover, serum leptin was a better predictor of the failure to respond to CC stimulation than was BMI or waist-to-hip ratio (Imani et al., 2000). These data suggest that leptin may be a key component in the mechanism of gonadotropin insensitivity in obese women. However, in those women who ovulated with CC treatment, leptin was not related to pregnancy outcome (Imani et al., 2000).

Bützow and colleagues (1999) related the increase in serum leptin during FSH stimulation for IVF to ovarian response. They found that a larger relative increase in serum leptin concentrations during FSH stimulation was significantly correlated with fewer follicles and fewer retrieved oocytes. Another study (Fedorcsak et al., 2000a) found a significant positive correlation between the ratio of FF leptin to plasma leptin concentrations and the cumulative dose of FSH required for ovarian stimulation in women with polycystic ovary syndrome (PCOS). These reports provide further evidence linking leptin physiology with gonadotropin sensitivity. However, Mantzoros et al. (2000) found no significant correlation between serum or FF leptin concentrations and the amount of gonadotropin used.

5.4. LEPTIN AND ART OUTCOMES

5.4.1. Obesity and Fertility

Although it is generally recognized that obesity impairs fertility (Norman and Clark, 1998), the impact of obesity on pregnancy success from ART has only recently been investigated. In a large retrospective study of 8822 IVF-embryo transfer cycles, the cumulative pregnancy rate progressively decreased as BMI increased from <25 to >35 (Wang et al., 2000). Furthermore, in a population of 383 IVF patients who conceived, overweight patients (BMI > 25) had fewer oocytes retrieved, a higher miscarriage rate, and lower live birth rate than normal weight women (Fedorcsak et al., 2000b). In contrast, others have found no direct relationship between obesity and IVF pregnancy success (Lashen et al., 1999).

A study by Unkila-Kallio et al. (2001) found that the percent increase in serum leptin during FSH stimulation positively correlated with peak estradiol in cycles resulting in successful IVF pregnancy, but no similar correlation was evident in patients who did not become pregnant. Moreover, a rise in leptin was positively correlated with the number of follicles and retrieved oocytes in patients who became pregnant (Unkila-Kallio et al., 2001). This seems inconsistent with previous data (Bützow et al., 1999) indicating that a larger relative increase in serum leptin during the stimulation cycle was associated with fewer follicles and retrieved oocytes, but these authors did not include pregnancy outcome in their analyses.

5.4.2. Leptin and IVF Outcomes

The first direct evidence of a relationship between leptin and ART pregnancy outcome came from a study by Mantzoros et al. (2000). These authors reported that

FF leptin concentrations in women who became pregnant within three cycles of IVF or gamete intra-fallopian transfer (GIFT) were significantly lower than in age- and weight-matched women who did not become pregnant. Surprisingly, serum leptin was not significantly different between pregnant and non-pregnant women even though FF and serum leptin concentrations were strongly correlated (Mantzoros et al., 2000).

Baseline serum leptin levels measured at the beginning of FSH stimulation correlated negatively (likelihood ratio, LR = 5.20, $p = 0.023$) with pregnancy success in 139 women undergoing first attempt IVF cycles (Brannian et al., 2001). But pregnancy success showed no correlation (LR = 0.37, $p > 0.5$) with BMI, even though serum leptin and BMI were strongly correlated (Figure 5.1). However, the serum leptin : BMI ratio was more strongly associated (LR = 7.26, $p = 0.007$) with pregnancy success than was leptin concentration alone (Brannian et al., 2001). An elevated serum leptin : BMI ratio (≥ 0.7; Figure 5.1) was associated with fewer superior quality embryos at the time of embryo transfer, and lower implantation, clinical pregnancy, and live birth rates (Figure 5.2;

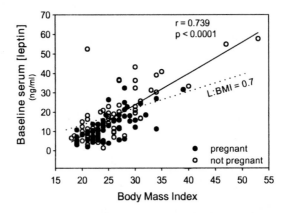

Figure 5.1. Correlation between baseline non-fasting serum leptin concentration and body mass index (BMI) in women ($n = 139$) undergoing first attempt IVF cycles. Solid line = regression line; dotted line = demarcation line of patients with serum leptin : BMI ratio greater or less than 0.7 (adapted from Brannian et al., 2001).

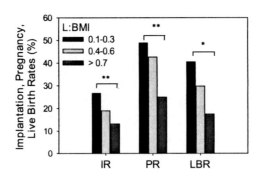

Figure 5.2. Comparison of implantation (IR), clinical pregnancy (PR), and live birth rates (LBR) in IVF cycles ($n = 139$) among three groups of patients defined by serum leptin : BMI ratio (L : BMI). **differ at $p < 0.025$; *$p < 0.05$ (adapted from Brannian et al., 2001).

Brannian et al., 2001). Moreover, the success of subsequent frozen embryo transfers tended to be diminished in women with elevated serum leptin : BMI ratios in their oocyte retrieval cycle (Brannian et al., 2001). In a follow-up study that included 226 IVF cycles, the divergence in embryo quality and pregnancy outcomes between women with low (<0.4) versus high (≥0.7) leptin : BMI ratios was more pronounced in women stimulated with hMG compared with those stimulated with pure FSH (Brannian and Hansen, unpublished data) suggesting that elevated LH activity may have a negative interactive relationship with elevated leptin.

These studies indicate that elevated serum and/or FF leptin may be a key factor in obesity-related sub-fertility. That an elevated leptin : BMI ratio predicted pregnancy success better than leptin alone suggests that leptin negatively impacts fertility independently of body mass. Fewer good embryos and lower implantation rate implies that intrafollicular exposure of maturing oocytes to elevated leptin leads to impaired oocyte developmental competence. Alternatively, the leptin : BMI ratio may be an index of leptin sensitivity, in other words, a high leptin : BMI ratio may indicate leptin resistance. If so, impairment of oocyte developmental competence may instead be a consequence of intrafollicular leptin resistance. This possibility is discussed in more detail later. Since leptin receptors are expressed in the human endometrium (Gonzalez et al., 2000; Kitawaki et al., 2000) a role for leptin in endometrial receptivity is also a possibility.

5.5. LEPTIN, FOLLICULOGENESIS, AND OOCYTE MATURATION

5.5.1. Leptin and Steroidogenesis

Numerous investigators have reported modulatory effects of leptin on in vitro steroid production by TC (Spicer and Francisco, 1998; Agarwal et al., 1999), GC (Spicer and Francisco, 1997; Zachow and Magoffin, 1997; Agarwal et al., 1999), and granulosa-luteal cells (Brannian et al., 1999; Kitawaki et al., 1999; Greisen et al., 2000). Effects of leptin on steroidogenesis are described and discussed in Chapter 6. Whereas it is obvious that regulation of ovarian steroidogenesis by leptin may lead to alterations in follicle growth, maturation, and concomitant oocyte maturation, the present discussion focuses on alternate mechanisms of possible leptin involvement.

5.5.2. Leptin and Follicle Growth

Strong leptin immunoactivity in GC of preantral follicles (Löffler et al., 2001) implies that leptin may affect follicle development at a relatively early stage. Indeed leptin dose-dependently inhibited FSH-stimulated in vitro growth (based on follicle diameter and GC numbers) of cultured preantral mouse follicles (Kikuchi et al., 2001). Leptin suppression of preantral follicle growth was associated with diminished estradiol and inhibin secretion (Kikuchi et al., 2001). In contrast, Spicer and Francisco (1997) could not demonstrate an effect of leptin on basal or insulin-stimulated proliferation of cultured GC isolated from small or large bovine antral follicles. However, these authors found that leptin enhanced insulin-stimulated bovine TC proliferation (Spicer and Francisco, 1998) More research is needed to clarify the effect of leptin on follicular cell proliferation and follicle growth.

5.5.3. Leptin and Ovulation

One recent study suggests that leptin may inhibit ovulation (Duggal et al., 2000). FSH-primed immature rats that received leptin administration for 15 hr during the periovulatory interval ovulated significantly fewer oocytes after hCG injection than controls. There was, however, no effect on the number of preovulatory follicles present in the ovary as determined by histological examination. Moreover, isolated ovaries perfused with luteinizing hormone (LH) and leptin ovulated fewer oocytes demonstrating that this was a direct ovarian effect (Duggal et al., 2001).

5.5.4. Leptin and Perifollicular Blood Flow and Follicular Metabolism

In recent years, it has become increasingly apparent that the vascularity and metabolic state of the developing follicle has a major impact on oocyte developmental competence (Van Blerkom, 1998, 2000). Numerous studies from ART cycles have demonstrated heterogeneity in follicular blood flow and metabolism within a cohort resulting in retrieved oocytes of variable developmental capacity (Nargund et al., 1996; Chui et al., 1997; Van Blerkom et al., 1997; Bhal et al., 1999; Huey et al., 1999). Color Doppler ultrasonography has been used by several investigators to assess perifollicular vascularization during IVF cycle monitoring. Based on these studies, undervascularized follicles have been found to yield oocytes that result in poorer quality embryos (Nargund et al., 1996; Van Blerkom, 1998; Huey et al., 1999) and lower implantation and pregnancy rates (Chui et al., 1997; Bhal et al., 1999; Borini et al., 2001). Moreover, an increased frequency of spindle defects and higher rate of aneuploidy was reported in oocytes from poorly vascularized follicles (Van Blerkom et al., 1997; Van Blerkom, 1998). Van Blerkom (1998, 2000) has proposed that abnormalities in oocyte mitochondrial distribution and reduced ATP generation, which is characteristic of oocytes with impaired developmental competence, may be associated with poor perifollicular vascularity.

Studies of the regulation of perifollicular angiogenesis and vascular permeability have tended to focus on vascular endothelial growth factor (VEGF) (Van Blerkom, 2000). Nevertheless, several lines of evidence implicate leptin as a co-regulator of follicular vascularization and blood flow. Functional leptin receptors are expressed on human endothelial cells, and leptin stimulates in vitro endothelial cell proliferation and angiogenesis (Bouloumie et al., 1998; Sierra-Honigmann et al., 1998; Cao et al., 2001). Leptin's angiogenic action is synergistic with fibroblast growth factor (FGF)-2 and VEGF, and like VEGF, leptin increases vascular permeability when administered to mice (Cao et al., 2001). Conversely, leptin can induce angiopoietin-2 expression, a mediator of vascular regression (Cohen et al., 2001), and leptin dose-dependently promoted ROS formation and oxidative damage in human umbilical vein endothelial cells (Bouloumie et al., 1999). Thus, leptin may play a role in the vascularization and vascular remodeling of the developing and periovulatory follicle.

Barroso and co-investigators (1999) concluded that leptin, along with VEGF and nitric oxide (NO), was a marker of follicular hypoxia. Follicular fluid leptin concentrations correlated negatively with intrafollicular oxygen concentration (pO_2), and positively with intrafollicular VEGF and NO levels (Barroso et al., 1999). Moreover, VEGF levels in FF correlated negatively with embryo quality (Barroso et al., 1999), and mean FF VEGF concentration was lower in women who achieved a clinical pregnancy than in those who did not (Friedman et al., 1998). A similar relationship between embryo development and FF concentrations of NO has been reported. Barrionuevo et al. (2000) found that intrafollicular NO was significantly lower in follicles yielding oocytes that successfully fertilized and developed beyond the 6-cell stage than in follicles containing oocytes that failed to fertilize or develop normally.

As in other tissues, follicular levels of VEGF and NO are inversely related to oxygen concentrations. Human GC production of VEGF is increased several-fold when cells are cultured under hypoxic conditions (Friedman et al., 1997). The relationship among FF leptin, VEGF, NO, and intrafollicular pO_2 (Barroso et al., 1999) suggests that FF leptin levels may also increase in response to a hypoxic environment within the follicle. Leptin secretion by GC (if confirmed), like VEGF secretion, may therefore be stimulated by low oxygen concentrations. This is consistent with the co-sequestration and release of VEGF and leptin by a subpopulation of GC (Antczak et al., 1997).

At least two explanations for the relationship between FF leptin, intrafollicular oxygen, and oocyte quality may be hypothesized. First, elevated circulating leptin, due strictly to extrafollicular factors, could potentially exert a localized negative effect on the perifollicular vasculature resulting in reduced follicular blood flow, hypoxia, and consequently impaired oocyte maturation. Alternatively, intrafollicular leptin may increase by local secretion due to hypoxic conditions within the follicle, and the elevated FF leptin may then exert detrimental actions on the follicle and oocyte.

5.5.5. Leptin Effects on Oocytes and Preimplantation Embryos

Relatively little research so far has focused on direct receptor-mediated effects of leptin on maturing oocytes and pre-implantation embryos. Leptin receptor mRNA and protein is expressed in mouse oocytes (Matsuoka et al., 1999; Kawamura et al., 2002; Ryan et al., 2002), and leptin induced tyrosine phosphorylation of STAT3, a major intracellular leptin signal transduction protein, in mouse metaphase 2 stage oocytes (Matsuoka et al., 1999). Leptin receptor expression in human oocytes has not been confirmed. But leptin and STAT3 protein co-localize in human oocytes and early cleaving embryos (Antczak and Van Blerkom, 1997; Cioffi et al., 1997) suggesting that the leptin receptor pathway is present.

Ryan and co-workers (2002) recently reported that leptin enhanced resumption of meiosis in preovulatory follicle-enclosed oocytes in culture, but did not alter meiotic resumption in denuded or cumulus-enclosed oocytes. However, leptin exposure during in vitro maturation of porcine oocytes and subsequent embryo culture after IVF resulted in the formation of fewer blastocysts relative to controls (Swain et al., 2001). These studies provide the first direct evidence that leptin influences oocyte maturation and developmental competence.

Very recently it was reported that leptin receptor is expressed in mouse 1-cell and 2-cell embryos as well as morulae and blastocysts (Kawamura et al., 2002). These authors demonstrated that the addition of leptin to culture medium promoted blastocyst formation from 2-cell embryos, and that this effect was neutralized by an antibody directed against the extracellular domain of the leptin receptor. Moreover, leptin increased the total cell number of the blastocysts formed, and specifically the number of trophectoderm cells (Kawamura et al., 2002).

Preliminary data from the authors' laboratory confirms that leptin directly influences early embryo development independently of oocyte maturational effects. In thawed 1-cell mouse embryos obtained from a commercial supplier, leptin (10–100 ng/ml) accelerated embryonic cell proliferation and blastocyst formation relative to embryos cultured in serum-free medium alone, and this was similar to the effect of insulin-like growth factors (IGF)-1 (3–10 ng/ml) (Brannian, unpublished data). Leptin induced activation of mitogen-activated protein kinase (MAPK) in a mouse embryonic cell line (Takahashi et al., 1997), suggesting that MAPK may be a mediator of leptin action in promoting early embryonic development. The role of leptin in pre-implantation embryo development awaits further elucidation, including how leptin's effects on oocyte maturation coincide with its effects on post-fertilization events.

5.5.6. Leptin Signaling Interactions

Taken together, research to date suggests that exposure to high leptin levels during follicular growth and maturation alters follicular steroidogenesis and negatively impacts ovarian response to gonadotropin stimulation, follicle growth, oocyte developmental competence, and ovulation. Therefore, it has been proposed that leptin exerts a generally suppressive effect on ovarian function. However, alternative explanations may be hypothesized.

Numerous in vitro studies in non-ovarian tissues have demonstrated that leptin can modulate cellular actions of insulin and IGF by interacting with their intracellular signaling pathways (Cohen et al., 1996; Bjorbaek et al., 1997; Sweeney et al., 2001). It follows logically that leptin may alter insulin/IGF-supported (or other growth factor/cytokine-mediated) events in the developing follicle. But the effect of leptin on insulin and IGF signaling is not uniformly consistent. For example, leptin can attenuate insulin-induced tyrosine phosphorylation of insulin receptor substrate (IRS) proteins, but conversely can enhance IRS activation of phosphoinositol 3'-kinase (PI-3'K) (Cohen et al., 1996). In a hepatic cell line, leptin enhanced insulin-induced tyrosine phosphorylation and PI-3'K binding of IRS-1, but had the opposite effect on IRS-2 (Szanto and Kahn, 2000). Moreover, leptin can activate certain insulin signaling pathways in the absence of insulin (Kim et al., 2000). These studies demonstrate that signaling interactions between leptin and insulin, IGF, and other growth factors is quite complex. This allows for the possibility that leptin may exert mixed actions on the developing follicle dependent upon the specific cellular target, state of differentiation, sensitivity of the target cell, and hormonal milieu of the individual follicle. Elucidating a specific role for leptin in folliculogenesis and oocyte maturation may, therefore, be quite challenging.

Considering leptin's overall physiologic role as a monitor and modulator of nutrient and metabolism balance, it is not clear how leptin would evolve into a negative regulator of ovarian function. On the contrary, from an evolutionary perspective, it seems that leptin's role in regulating reproductive function most probably arose from the need to reserve reproduction for periods of nutrient sufficiency. That this has carried over to humans is supported by the fact that a threshold level of leptin appears to be required for the resumption of menstrual cycles during weight recovery in anorexic women (Audi et al., 1998; Ballauff et al., 1999). Leptin seems to be obligatory for normal reproductive function at the neuroendocrine level, and the high level of leptin receptor expression in the ovary suggests that leptin should necessarily have some significant supportive role in ovarian function. By analogy, insulin and IGF play important roles in normal ovarian function, for example, by mediating gonadotropin action on GC proliferation and steroidogenesis, yet we know that hyperinsulinemia as a common feature of PCOS, can have a negative impact on follicular dynamics (Dunaif, 1997). This is believed to be a consequence of insulin resistance. One can then imagine that the negative effects of high leptin on ovarian function could be related, at least in part, to leptin resistance.

5.6. LEPTIN RESISTANCE

5.6.1. Central and Peripheral Leptin Resistance

Although leptin resistance in humans was first proposed not long after leptin's discovery (Caro et al., 1996; Hamann and Matthaei, 1996; Schwartz et al., 1996), it remains an ill-defined physiological condition. Some have questioned whether leptin resistance exists at all (Arch et al., 1998). Leptin resistance as a major cause of obesity can probably be dismissed. In most cases, hyperleptinemia is almost certainly due primarily to high body fat mass brought on by diet and lifestyle, not to tissue insensitivity to leptin (Arch et al., 1998). Thus the etiology of leptin resistance may not be analogous to insulin resistance. Nevertheless as leptin concentrations in the circulation rise as a consequence of increased fat mass, the emergence of some tissue resistance to leptin may be expected. Indeed, obesity induced in rats by a high-fat diet leads to leptin insensitivity in skeletal muscle tissue (Steinberg and Dyck, 2000).

Leptin resistance could occur as a result of down regulation of leptin receptors (Madiehe et al., 2000; Martin et al., 2000; Scarpace et al., 2001), alterations in leptin signal transduction (Bjorbaek et al., 1999; Emilsson et al., 1999; Wang et al., 2000), decreased leptin transport across the blood–brain barrier (Caro et al., 1996; Schwartz et al., 1996), or differences in soluble leptin binding activity (Lewandowski et al., 1999). Furthermore, considering the complexity of leptin physiology, for example, its intricate signaling interactions with insulin (Kim et al., 2000; Szanto and Kahn, 2000) and other growth factors (Bjorbaek et al., 1997), it follows that considerable variation in leptin sensitivity may occur among individuals of similar body adiposity that is dependent upon numerous variables other than fat mass or baseline leptin levels.

Central leptin resistance was first proposed on the basis of studies in rodents that suggested a saturable transport mechanism for leptin into the brain that was diminished in obese individuals (Caro et al., 1996; Schwartz et al., 1996). Subsequent research has shown that decreased hypothalamic leptin receptor protein expression (Madiehe et al., 2000; Martin et al., 2000) and increased hypothalamic expression of suppressors of cytokine signaling (SOCS-3) protein (Emilsson et al., 1999; Wang et al., 2001) may also contribute to central leptin resistance.

SOCS-3 was first identified as a potential mediator of cellular leptin resistance by Bjorbaek and co-workers (1998) who found that SOCS-3 expression in transfected cell lines blocked post-receptor leptin signal transduction. Moreover, systemic leptin administration induced a rapid increase in SOCS-3 mRNA in leptin receptor-rich regions of the hypothalamus of *ob/ob* mice (Bjorbaek et al., 1998). Subsequent studies have confirmed expression of SOCS-3 in leptin target cells, and its ability to inhibit leptin signaling in transfected cell lines (Bjorbaek et al., 1999) and peripheral tissues (Emilsson et al., 1999; Wang et al., 2000, 2001). SOCS-3 expression in ovarian cells has not been investigated.

It has also been proposed that differential binding of leptin to soluble leptin receptors or other leptin-binding proteins in serum and tissue fluid may be related to peripheral leptin resistance (Sinha et al., 1996; Lewandowski et al., 1999), but the mechanism by which this would occur is not clear. The proportion of bound leptin was found to be much greater in lean subjects compared with obese subjects, suggesting that higher levels of free leptin may be characteristic of leptin resistance (Sinha et al., 1996). Soluble leptin binding activity is present in FF, and was found to be lower than in plasma, at least in women with PCOS (Fedorcsák et al., 2000a).

5.6.2. Evidence of Leptin Resistance in the Ovary

The possibility of leptin resistance impacting ovarian function has interesting implications, and may provide partial explanations for conflicting conclusions regarding leptin's role in ovarian physiology and pathophysiology (e.g., PCOS). No clearly defined markers for leptin resistance have been identified, so the existence of leptin resistance in the ovary can only be inferred. For example, an elevated serum leptin:BMI ratio may be an index of leptin resistance. If so, then the poorer IVF outcomes associated with a high leptin : BMI ratio (Brannian et al., 2001) could be due to leptin resistance in the maturing follicle.

Several studies attempting to show differences in leptin concentrations between PCOS patients and non-PCOS subjects have been inconsistent. In a small population ($n = 11$) of women with PCOS undergoing IVF treatment, the leptin : BMI ratio (1.11 ± 0.11) was significantly greater than in BMI-matched control patients (0.76 ± 1.4, $n = 16$) (Brannian and Hansen, unpublished data). This is consistent with the possibility that leptin resistance plays a role in PCOS as previously suggested (Spritzer et al., 2001; Brannian and Hansen, 2002). Moreover, the ratio of FF leptin to plasma leptin in PCOS patients correlated positively with the cumulative FSH dose implying a relationship with gonadotropin responsiveness (Fedorcsák et al., 2000a). Since FF leptin binding activity was lower than in plasma in these PCOS subjects, the intrafollicular concentration of free leptin may be particularly high (Fedorcsák et al., 2000a). A higher FF:plasma leptin ratio may be a feature of intrafollicular leptin resistance, and opens the possibility that diminished follicular response to gonadotropin is related to leptin resistance.

5.7. CONCLUSIONS

A threshold level of leptin is obligatory for sustained central stimulation of the gonads. Within an evolutionary context, it was critical for animals and early humans to coordinate reproduction with periods of nutrient sufficiency. Leptin's role, therefore, has likely evolved as a signal to communicate the status of the food supply and body fat stores to the hypothalamic–pituitary axis to regulate release of gonadotropic hormones. However, in addition to leptin's neuroendocrine effects, it is apparent that leptin has direct actions on multiple target cells within the ovary. It is much less clear what the physiologic role of leptin is at the gonadal level. But it seems logical to assume that it's ovarian effects should be complementary to its central role, that is to support reproductive function during times of nutrient sufficiency.

Lifestyle-induced obesity prevalent in modern society is associated with sub-fertility. High circulating and intrafollicular leptin concentrations are negatively associated with follicular responsiveness to gonadotropins, follicular metabolism, oocyte developmental competence, and subsequent implantation and pregnancy success. This suggests that elevated leptin levels arising from obesity may exceed a critical ceiling optimal for fertility. Whether the negative relationship between high leptin and fertility is due to overstimulation of leptin-mediated pathways, alteration of insulin and growth factor-mediated pathways, or leptin resistance remains to be determined.

Based on in vitro data, results in non-ovarian tissues, and correlation between leptin and ovarian function derived from clinical studies, a hypothetical model of how leptin may participate in the regulation of follicle development and oocyte maturation is depicted in Figure 5.3. A possible mechanism of how leptin resistance may impact ovarian function is also presented.

First, leptin may modulate thecal cell proliferation and androgen synthesis by interaction with insulin/IGF-mediated pathways (Spicer and Francisco, 1998). In addition, leptin may promote angiogenesis and increase blood flow within the thecal vasculature leading to an enhanced oxygen gradient across the basal lamina facilitating oxygen delivery to the granulosum. Enhanced oxygenation of the FF would support GC and oocyte metabolism essential for optimal oocyte cytoplasmic maturation. Direct modulation of GC estrogen synthesis could further promote an optimal environment for oocyte maturation. Potential receptor-mediated actions on the oocyte may directly facilitate oocyte cytoplasmic maturation. These may include promoting glucose oxidation and ATP production critical for optimizing oocyte developmental competence, as suggested by the positive relationship between mitochondrial numbers and oocyte quality (Van Blerkom, 2000; Reynier et al., 2001).

In contrast, excessively high levels of leptin appear to be associated with suboptimal folliculogenesis and oocyte maturation. Overstimulation of leptin pathways may directly, or indirectly via interaction with insulin, IGF, and other growth factor pathways, perturb the local hormonal milieu and create an environment that is suboptimal. This may be particularly apparent in gonadotropin-stimulated ART cycles in which the normal hormonal balance is already altered.

Alternatively, if it is correct that leptin naturally plays a supportive role in ovarian function, the association between high leptin and poor follicular outcomes may be explained by leptin resistance. Considering the model presented above, a leptin resistant state could have a detrimental impact on follicle development and oocyte maturation (Figure 5.3). For example, leptin insensitivity of theca and thecal endothelial cells may lead to diminished perifollicular vascularization resulting in intrafollicular hypoxia. Granulosal release of VEGF, NO, and leptin in response to the hypoxic environment may

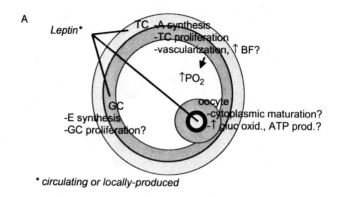

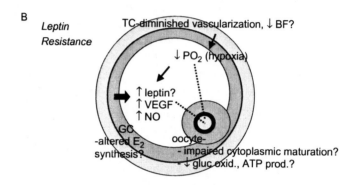

Figure 5.3. (A) Hypothetical model of leptin's role in folliculogenesis and oocyte maturation. (B) Leptin resistance may alternatively result in imparied follicle and oocyte development. See text for description.

explain the correlation among these parameters (Friedman et al., 1997; Barroso et al., 1999). The associated poor oocyte developmental competence may be primarily due to the hypoxic conditions (Van Blerkom et al., 1997; Van Blerkom, 2000). Leptin resistance of the oocyte could contribute to impaired oocyte maturation by further reduction (already diminished by the hypoxia) of glucose oxidation and ATP production, and by disruption of other potential regulatory mechanisms within the oocyte. Steroid and growth factor secretion by TC and GC may also be altered by leptin resistance. Follicular leptin resistance may, therefore, result in an intrafollicular environment characterized by hypoxia and perturbations in autocrine/paracrine signaling.

The use of insulin sensitizing drugs improves fertility in insulin-resistant women with PCOS, either by lowering insulin levels or by direct actions on ovarian cells or both (Nestler et al., 2001). If the relationship between hyperleptinemia and impaired fertility is a consequence of leptin resistance, the possibility exists that a leptin sensitizing drug might be developed that could potentially improve fertility. A leptin sensitizer may also assist in weight loss that might enhance fertility in other ways. For now, women planning to undertake ART cycles may be able to enhance their chances of pregnancy success by lowering leptin concentrations and/or improving leptin sensitivity through weight loss and increased exercise.

REFERENCES

Agarwal, S. K., Vogel, K., Weitsman, S. R., and Magoffin, D. A. (1999). Leptin antagonizes the insulin-like growth factor-I augmentation of steroidogenesis in granulosa and theca cells of the human ovary. *J. Clin. Endocrinol. Metab., 84,* 1072–1076.

Antczak, M. and Van Blerkom, J. (1997). Oocyte influences on early development: the regulatory proteins leptin and STAT3 are polarized in mouse and human oocytes and differentially distributed within the cells of the preimplantation stage embryo. *Mol. Hum. Reprod., 3,* 1067–1086.

Antczak, M., Van Blerkom, J., and Clark, A. (1997). A novel mechanism of vascular endothelial growth factor, leptin, and transforming growth factor-beta2 sequestration in a subpopulation of human ovarian follicle cells. *Hum. Reprod., 12,* 2226–2234.

Arch, J. R., Stock, M. J., and Trayhurn, P. (1998). Leptin resistance in obese humans: does it exist and what does it mean? *Int. J. Obes. Relat. Met. Dis., 22,* 1159–1163.

Audi, L., Mantzoros, C. S., Vidal-Puig, A., Vargas, D., Gussinye, M., and Carrascosa, A. (1998). Leptin in relation to resumption of menses in women with anorexia nervosa. *Mol. Psych., 3,* 477–478.

Ballauff, A., Ziegler, A., Emons, G., Sturm, G., Blum, W. F., Remschmidt, H., and Hebebrand, J. (1999). Serum leptin and gonadotropin levels in patients with anorexia nervosa during weight gain. *Mol. Psych., 4,* 71–75.

Barrionuevo, M. J., Schwandt, R., Rao, P., Graham, L., Maisel, L., and Yeko, T. (2000). Nitric oxide (NO) and interleukin-1beta in follicular fluid and their correlation with fertilization and embryo cleavage. *Am. J. Reprod. Immunol., 44,* 359–364.

Barroso, G., Barrionuevo, M., Rao, P., Graham, L., Danforth, D., Huey, S., Abuhamad, A., and Oehninger, S. (1999). Vascular endothelial growth factor, nitric oxide, and leptin follicular fluid levels correlate negatively with embryo quality in IVF patients. *Fertil. Steril., 72,* 1024–1026.

Bhal, P., Pugh, N., Chui, D., Gregory, L., Walker, S., and Shaw, R. (1999). The use of transvaginal power Doppler ultrasonography to evaluate the relationship between perifollicular vascularity and outcome in in-vitro fertilization treatment cycles. *Hum. Reprod., 14,* 939–945.

Bjorbaek, C., Uotani, S., da Silva, B., and Flier, J. S. (1997). Divergent signaling capacities of the long and short isoforms of the leptin receptor. *J. Biol. Chem., 272,* 32686–32695.

Bjorbaek, C., Elmquist, J. K., Frantz, J. D., Shoelson, S. E., and Flier, J. S. (1998). Identification of SOCS-3 as a potential mediator of central leptin resistance. *Mol. Cells, 1,* 619–625.

Bjorbaek, C., El-Haschimi, K., Frantz, J. D., and Flier, J. S. (1999). The role of SOCS-3 in leptin signaling and leptin resistance. *J. Biol. Chem., 274,* 30059–30065.

Borini, A., Maccolini, A., Tallarini, A., Antonietta, B., Sciajno, R., and Flamigni, C. (2001). Perifollicular vascularity and its relationship with oocyte maturity and IVF outcome. *Anna. N.Y. Acad. Sci., 943,* 64–67.

Bouloumie, A., Drexler, H. C., Lafontan, M., and Busse, R. (1998). Leptin, the product of OB gene, promotes angiogenesis. *Circ. Res., 83,* 1059–1066.

Bouloumie, A., Marumo, T., Lafontan, M., and Busse, R. (1999). Leptin induces oxidative stress in human endothelial cells. *FASEB J., 13,* 1231–1238.

Brann, D. W., DeSevilla, L., Zamorano, P., and Mahesh, V. (1999). Regulation of leptin gene expression and secretion by steroid hormones. *Steroids, 64,* 659–663.

Brannian, J. D. and Hansen, K. A. (2002). Leptin and ovarian folliculogenesis: Implications for ovulation induction and ART outcomes. *Sem. Reprod. Med., 20,* 103–112.

Brannian, J. D., Zhao, Y., and McElroy, M. (1999). Leptin inhibits gonadotrophin-stimulated granulosa cell progesterone production by antagonizing insulin action. *Hum. Reprod., 14,* 1445–1448.

Brannian, J. D., Schmidt, S. M., Kreger, D. O., and Hansen, K. A. (2001). Baseline non-fasting serum leptin concentration to body mass index ratio is predictive of IVF outcomes. *Hum. Reprod., 16,* 1819–1826.

Bützow, T. L., Moilanen, J. M., Lehtovirta, M., Tuomi, T., Hovatta, O., Sieberg, R., Nilsson, C., and Apter, D. (1999). Serum and follicular fluid leptin during in vitro fertilization: relationship among leptin increase, body fat mass, and reduced ovarian response. *J. Clin. Endocrind. Metab., 84,* 3135–3139.

Cao, R., Brakenhielm, E., Wahlestedt, C., Thyberg, J., and Cao, Y. (2001). Leptin induces vascular permeability and synergistically stimulates angiogenesis with FGF-2 and VEGF. *Proc. Natl. Acad. Sci. USA, 98,* 6390–6395.

Caro, J. F., Kolaczynski, J., Nyce, M., Ohannesian, J., Opentanova, I., Goldman, W., Lynn, R., Zhang, P., Sinha, M., and Considine, R. (1996). Decreased cerebrospinal-fluid/serum leptin ratio in obesity: a possible mechanism for leptin resistance. *Lancet, 348,* 159–161.

Casabiell, X., Pineiro, V., Peino, R., Lage, M., Camina, J., Gallego, R., Vallejo, L. G., Dieguez, C., and Casanueva, F. F. (1998). Gender differences in both spontaneous and stimulated leptin secretion by human omental adipose tissue in vitro: dexamethasone and estradiol stimulate leptin release in women, but not in men. *J. Clin. Endocrinol Metab., 83*, 2149–2155.

Chui, D., Pugh, N., Walker, S., Gregory, L., and Shaw, R. (1997). Follicular vascularity—the predictive value of transvaginal power Doppler ultrasonography in an in-vitro fertilization program; a preliminary study. *Hum. Reprod., 12*, 191–196.

Cioffi, J. A., Shafer, A. W., Zupancic, T. J., Smith-Gbur, J., Mikhail, A., Platika, D., and Snodgrass, H. R. (1996). Novel B219/OB receptor isoforms: possible role of leptin in hematopoiesis and reproduction. *Nature Med., 2*, 585–589.

Cioffi, J. A., Van Blerkom, J., Antczak, M., Shafer, A., Wittmer, S., and Snodgrass, H. R. (1997). The expression of leptin and its receptors in pre-ovulatory human follicles. *Mol. Hum. Reprod., 3*, 467–472.

Cohen, B., Novick, D., and Rubinstein, M. (1996). Modulation of insulin activities by leptin. *Science, 274*, 1185–1188.

Cohen, B., Barkan, D., and Levy, Y. (2001). Leptin induces angiopoietin-2 expression in adipose tissues. *J. Biol. Chem., 276*, 7697–7700.

Duggal, P. S., Van Der Hoek, K. H., Milner, C. R., Ryan, N. K., Armstrong, D. T., Magoffin, D. A., and Norman, R. J. (2000). The in vivo and in vitro effects of exogenous leptin on ovulation in the rat. *Endocrinology, 141*, 1971–1976.

Dunaif, A. (1997). Insulin resistance and he polycystic ovary syndrome: mechanism and implications for pathogenesis. *Endocr. Rev., 18*, 774–800.

Emilsson, V., Arch, J. R., de Groot, R. P., Lister, C. A., and Cawthorne, M. A. (1999). Leptin treatment increases suppressors of cytokine signaling in central and peripheral tissues. *FEBS Lett., 455*, 170–174.

Fedorcsák, P., Storeng, R., Dale, P. O., Tanbo, T., Torjesen, P., Urbancsek, T., and Abyholm, T. (2000a). Leptin and leptin binding activity in the preovulatory follicle of polycystic ovary syndrome patients. *Scand. J. Clin. Lab. Invest., 60*, 649–656.

Fedorcsák, P., Storeng, R., Dale, P.O., Tanbo, T., and Abyholm, T. (2000b). Obesity is a risk factor for early pregnancy loss after IVF or ICSI. *Acta Obstet. Gyn. Scand., 79*, 43–48.

Friedman, C., Danforth, D., Herbosa-Encarnacion, C., Arbogast, L., Alak, B., and Seifer, D. (1997). Follicular fluid vascular endothelial growth factor concentrations are elevated in women of advanced reproductive age undergoing ovulation induction. *Fertil. Steril., 68*, 607–612.

Friedman, C., Seifer, D., Kennard, E., Arbogast, L., Alak, B., and Danforth, D. (1998). Elevated level of follicular fluid vascular endothelial growth factor is a marker of diminished pregnancy potential. *Fertil. Steril., 70*, 836–839.

Gonzalez, R. R., Caballero-Campo, P., Jasper, M., Mercader, A., Devoto, L., Pellicer, A., and Simon, C. (2000). Leptin and leptin receptor are expressed in the human endometrium and endometrial leptin secretion is regulated by the human blastocyst. *J. Clin. Endocrinol Metab., 85*, 4883–4888.

Greisen, S., Ledet, T., Moller, N., Jorgensen, J. O., Christiansen, J. S., Petersen, K., and Ovesen, P. (2000). Effects of leptin on basal and FSH stimulated steroidogenesis in human granulosa luteal cells. *Acta Obstet. Gynec. Scand., 79*, 931–935.

Hamann, A. and Matthaei, S. (1996). Regulation of energy balance by leptin. *Exp. Clin. Endocr. Diabet., 104*, 293–300.

Huey, S., Abuhamad, A., Barroso, G., Hsu, M., Kolm, P., Mayer, J., and Oehninger, S. (1999). Perifollicular blood flow Doppler indices, but not follicular pO_2, pCO_2, or pH, predict oocyte developmental competence in in vitro fertilization. *Fertil. Steril., 72*, 707–712.

Imani, B., Eijkemans, M. J., de Jong, F. H., Payne, N. N., Bouchard, P., Giudice, L. C., and Fauser, B. C. J. (2000). Free androgen index and leptin are the most prominent endocrine predictors of ovarian response during clomiphene citrate induction of ovulation in normogonadotropic oligoamenorrheic infertility. *J. Clin. Endocrinol. Metab., 85*, 676–682.

Karlsson, C., Lindell, K., Svensson, E., Bergh, C., Lind, P., Billig, H., Carlsson, L. M., and Carlsson, B. (1997). Expression of functional leptin receptors in the human ovary. *J. Clin. Endocrinol. Metab., 82*, 4144–4148.

Kawamura, K., Sato, N., Fukuda, J., Kodama, H. Kumagai, J., Tanikawa, H., Nakamura, A., and Tanaka, T. (2002). Leptin promotes the development of mouse preimplantation embryos in vitro. *Endocrinology, 143*, 1922–1931.

Kikuchi, N., Andoh, K., Abe, Y., Yamada, K., Mizunuma, H., and Ibuki, Y. (2001). Inhibitory action of leptin on early follicular growth differs in immature and adult female mice. *Biol. Reprod., 65*, 66–71.

Kim, Y.-B., Uotani, S., Pierroz, D., Flier, J. S., and Kahn, B. (2000). In vivo administration of leptin activates signal transduction directly in insulin-sensitive tissues: overlapping but distinct pathways from insulin. *Endocrinology 141*, 2328–2339.

Kitawaki, J., Kusuki, I., Koshiba, H., Tsukamoto, K., and Honjo, H. (1999). Leptin directly stimulates aromatase activity in human luteinized granulosa cells. *Mol. Hum. Reprod., 5*, 708–713.

Kitawaki, J., Koshiba, H., Ishihara, H., Kusuki, I., Tsukamoto, K., and Honjo, H. (2000). Expression of leptin receptor in human endometrium and fluctuation during the menstrual cycle. *J. Clin. Endocrinol. Metab., 85*, 1946–1950.

Lashen, H., Ledger, W., Bernal, A. L., and Barlow, D. (1999). Extremes of body mass do not adversely affect the outcome of superovulation and in-vitro fertilization. *Hum. Reprod., 14*, 712–715.

Lewandowski, K., Horn, R., O'Callaghan, C. J., Dunlop, D., Medley, G. F., O'Hare, P., and Brabant, G. (1999). Free leptin, bound leptin, and soluble leptin receptor in normal and diabetic pregnancies. *J. Clin. Endocrinol. Metab., 84*, 300–306.

Lindheim, S. R., Sauer, M. V., Carmina, E., Chang, P. L., Zimmerman, R., and Lobo, R. A. (2000). Circulating leptin levels during ovulation induction: relation to adiposity and ovarian morphology. *Fertil. Steril., 73*, 493–498.

Löffler, S., Aust, G., Köhler, U., and Spanel-Borowski, K. (2001). Evidence of leptin expression in normal and polycystic human ovaries. *Mol. Hum. Reprod., 7*, 1143–1149.

Madiehe, A. M, Schaffhauser, A. O., Braymer, D. H., Bray, G. A., and York, D. A. (2000). Differential expression of leptin receptor in high- and low-fat-fed Osborne-Mendel and S5B/P1 rats. *Obes. Res., 8*, 467–474.

Mantzoros, C. S., Cramer, D. W., Liberman, R. F., and Barbieri, R. L. (2000). Predictive value of serum and follicular fluid leptin concentrations during assisted reproductive cycles in normal women and in women with the polycystic ovarian syndrome. *Hum. Reprod., 15*, 539–544.

Martin, R. L., Perez, E., He, Y. J., Dawson, R., and Millard, W. J. (2000). Leptin resistance is associated with hypothalamic leptin receptor mRNA and protein downregulation. *Metab. Clin. Exp., 49*, 1479–1484.

Matsuoka, T., Tahara, M., Yokoi, T., Masumoto, N., Takeda, T., Yamaguchi, M., Tasaka, K., Kurachi, H., and Murata, Y. (1999). Tyrosine phosphorylation of STAT3 by leptin through leptin receptor in mouse metaphase 2 stage oocyte. *Biochem. Biophys. Res. Common., 256*, 480–484.

Messinis, I. E., Milingos, S., Zikopoulos, K., Kollios, G., Seferiadis, K., and Lolis, D. (1998). Leptin concentrations in the follicular phase of spontaneous cycles and cycles superovulated with follicle stimulating hormone. *Hum. Reprod., 13*, 1152–1156.

Nargund, G., Bourne, T., Doyle, P., Parsons, J., Cheng, W., Campbell, S., and Collins, W. (1996). Associations between ultrasound indices of follicular blood flow, oocyte recovery, and preimplantation embryo quality. *Hum. Reprod., 11*, 109–113.

Nestler, J. E. (2001). Metformin and the polycystic ovary syndrome. *J. Clin. Endocrinol. Metab., 86*, 1430.

Norman, R. J. and Clark, A. (1998). Obesity and reproductive disorders: a review. *Reprod. Fertil. Dev., 10*, 55–63.

Prien, S. D. and Castracane, D. V. (1998). Leptin and the ovarian follicle. *Proc. Endocr. Soc. Meeting*, 1998 (abstract).

Reynier, P., May-Panloup, P., Chrétien, M.-F., Morgan, C. J., Jean, M., Savagner, F., Barrière, P., and Malthièry, Y. (2001). Mitochondrial DNA content affects the fertilizability of human oocytes. *Mol. Hum. Reprod., 7*, 425–429.

Rosenbaum, M., Nicolson, M., Hirsch, J., Heymsfield, S. B., Gallagher, D., Chu, F., and Leibel, R. L. (1996). Effects of gender, body composition, and menopause on plasma concentrations of leptin. *J. Clin. Endocrinol. Metab., 81*, 3424–3427.

Ruhl, C. E. and Everhart, J. E. (2001). Leptin concentrations in the United States: relations with demographic and anthropometric measures. *Am. J. Clin. Nutr., 74*, 295–301.

Ruiz-Cortez, Z. T., Men, T., Palin, M. F., Downey, B. R., Lacroix, D. A., and Murphy, B. D. (2000). Porcine leptin receptor: molecular structure and expression in the ovary. *Mol. Reprod. Dev., 56*, 465–474.

Ryan, N. K., Woodhouse, C. M., Van Der Hoek, K. H., Gilchrist, R. B., Armstrong, D. T., and Norman, R. J. (2002). Expression of leptin and its receptor in the murine ovary: possible role in the regulation of oocyte maturation. *Biol. Reprod., 66*, 1548–1554.

Scarpace, P. J., Matheny, M., and Tumer, N. (2001). Hypothalamic leptin resistance is associated with impaired leptin signal transduction in aged obese rats. *Neuroscience, 104*, 1111–1117.

Schwartz, M., Peskind, E., Raskind, M., Boyko, E., and Porte, D. (1996). Cerebrospinal fluid leptin levels: relationship to plasma levels and to adiposity in humans. *Nature Med., 2*, 589–593.

Shepard, M. K., Balmaceda, J., and Leija, C. G. (1979). Relationship of weight to successful induction of ovulation with clomiphene citrate. *Fertil. Steril., 32*, 641–645.

Shimizu, H., Shimomura, Y., Nakanishi, Y., Futawatari, T., Ohtani, K., Sato, N., and Mori, M. (1997). Estrogen increases in vivo leptin production in rats and human subjects. *J. Endocrinol., 154*, 285–292.

Sierra-Honigmann, M. R., Nath, A. K., Murakami, C., Garcia-Cardena, G., Papapetropoulos, A., Sessa, W. C., Madge, L. A., Schechner, J. S., Schwabb, M. B., Polverini, P. J., and Flores-Riveros, J. R. (1998). Biological action of leptin as an angiogenic factor. *Science, 281*, 1683–1686.

Sinha, M. K., Opentanova, I., Ohannesian, J. P., Kolaczynski, J. W., Heiman, M. L., Hale, J., Becker, G. W., Bowsher, R. R., Stephens, T. W., and Caro, J. F. (1996). Evidence of free and bound leptin in human circulation: studies in lean and obese subjects and during short-term fasting. *J. Clin. Invest., 98*, 1277–1282.

Sivan, E., Whittaker, P., Sinha, D., Homko, C., Lin, M., Reece, E., and Boden, G. (1998). Leptin in human pregnancy: the relationship with gestational hormones. *Am. J. Obstet. Gynecol., 179*, 1128–1132.

Spicer, L. J. and Francisco, C. C. (1997). The adipose obese gene product, leptin: evidence of a direct inhibitory role in ovarian function. *Endocrinology, 138*, 3374–3379.

Spicer, L. J. and Francisco, C. C. (1998). Adipose obese gene product, leptin, inhibits bovine ovarian thecal cell steroidogenesis. *Biol. Reprod., 58*, 207–212.

Spritzer, P. M., Poy, M., Wiltgen, D., Mylius, L. S., and Capp, E. (2001). Leptin concentrations in hirsute women with polycystic ovary syndrome or idiopathic hirsutism: influence on LH and relationship with hormonal, metabolic, and anthropometric measurements. *Hum. Reprod., 16*, 1340–1346.

Steinberg, G. R. and Dyck, D. J. (2000). Development of leptin resistance in rat soleus muscle in response to high-fat diets. *Am. J. Physiol., 279*, E1374–1382.

Strowitzki, T., Kellerer, M., Capp, E., and Haring, H. U. (1998). Increase in serum leptin concentrations in women undergoing controlled ovarian hyperstimulation for assisted reproduction. *Gynecol. Endocrinol., 12*, 167–169.

Swain, J. E., Bormann, C. L., and Krisher, R. L. (2001). Effects of leptin on porcine oocyte maturation and embryo in vitro. *Biol. Reprod., 64*(Suppl. 1), 212.

Sweeney, G., Keen, J., Somwar, R., Konrad, D., Garg, R., and Klip, A. (2001). High leptin levels acutely inhibit insulin-stimulated glucose uptake without affecting glucose transporter 4 translocation in L6 rat skeletal muscle cells. *Endocrinology, 142*, 4806–4812.

Szanto, I. and Kahn, C. R. (2000). Selective interaction between leptin and insulin signaling pathways in a hepatic cell line. *Proc. Natl. Acad. Sci. USA, 97*, 2355–2360.

Takahashi, Y., Okimura, Y., Mizuno, I., Iida, K., Takahashi, T., Kaji, H., Abe, H., and Chihara, K. (1997). Leptin induces mitogen-activated protein kinase-dependent proliferation of C3H10T1/2 cells. *J. Biol. Chem., 272*, 12897–12900.

Unkila-Kallio, L., Anderson, S., Koistinen, H. A., Karonen, S. L., Ylikorkala, O., and Tiitinen, A. (2001). Leptin during assisted reproductive cycles: the effect of ovarian stimulation and of very early pregnancy. *Hum. Reprod., 16*, 657–662.

Van Blerkom, J. (1998). Epigenetic influences on oocyte developmental competence: Perifollicular vascularity and intrafollicular oxygen. *J. Assist. Reprod. Genet., 15*, 226–234.

Van Blerkom, J. (2000). Intrafollicular influences on human oocyte developmental competence: perifollicular vascularity, oocyte metabolism and mitochondrial function. *Hum. Reprod., 15*(Suppl. 2), 173–188.

Van Blerkom, J., Antczak, M., and Schrader, R. (1997). The developmental potential of the human oocyte is related to the dissolved oxygen content of follicular fluid: association with vascular endothelial growth factor levels and perifollicular blood flow characteristics. *Hum. Reprod., 12*, 1047–1055.

Wang, J. X., Davies, M., and Norman, R. J. (2000). Body mass and probability of pregnancy during assisted reproduction treatment: retrospective study. *Br. Med. J., 321*, 1320–1321.

Wang, Z., Zhou, Y. T., Kakuma, T., Lee, Y., Kalra, S. P., Pan, W., and Unger, R. H. (2000). Leptin resistance of adipocytes in obesity: role of suppressors of cytokine signaling. *Biochem. Biophys. Res. Commun., 277*, 20–26.

Wang, Z. W., Pan, W. T., Lee, Y., Kakuma, T., Zhou, Y. T., and Unger, R. H. (2001). The role of leptin resistance in the lipid abnormalities of aging. *FASEB J., 15*, 108–114.

Yamada, M., Irahara, M., Tezuka, M., Murakami, T., Shima, K., and Aono, T. (2000). Serum leptin profiles in the normal menstrual cycles and gonadotropin treatment cycles. *Gynecol. Obstet. Invest., 49*, 119–123.

Zachow, R. J. and Magoffin, D. A. (1997). Direct intraovarian effects of leptin: impairment of the synergistic action of insulin-like growth factor-I on follicle-stimulating hormone-dependent estradiol-17 beta production by rat ovarian granulosa cells. *Endocrinology, 138*, 847–850.

Zamorano, P. L., Mahesh, V. B., De Sevilla, L. M., Chorich, L. P., Bhat, G. K., and Brann, D. W. (1997). Expression and localization of the leptin receptor in endocrine and neuroendocrine tissues of the rat. *Neuroendocrinology, 65*, 223–228.

Zhao, Y., Kreger, D. O., and Brannian, J. D. (1998). Leptin concentrations in serum and follicular fluid during gonadotropin stimulation cycles. *J. Soc. Gynecol. Invest., 5*(Suppl.), 50A.

Zhao, Y., Kreger, D. O., and Brannian, J. D. (2000). Serum leptin concentrations in women during gonadotropin stimulation cycles. *J. Reprod. Med., 45*, 121–125.

6

The Effect of Leptin on Ovarian Steroidogenesis

LEON J. SPICER

6.1. INTRODUCTION

Since the determination of the amino acid sequence of leptin (Zhang et al., 1994), which is highly (i.e., 84–97 percent) conserved across species (Zhang et al., 1994; Masuzaki et al., 1995; Ogawa et al., 1995; Blache et al., 2000), leptin's role as a metabolic hormone has captured the attention of many (for reviews see Hossner, 1998; Barb, 1999; Spicer 2001), creating "leptinomania" within the scientific literature (Spicer, 2001). Subsequently, leptin has emerged as a potential regulator of reproduction (for reviews see other chapters of this book).

Genetically obese (ob/ob) mice, which lack endogenous leptin (Zhang et al., 1994), are sterile and when given exogenous leptin (50 µg twice per day) the sterility defect is eliminated (Barash et al., 1996; Chehab et al., 1996). Associated with the return to reproductive competence in female ob/ob mice were increased ovarian and uterine weight, increased concentrations of serum luteinizing hormone (LH) (but not follicle-stimulating hormone [FSH]), and increased numbers of primary and graafian follicles (Barash et al., 1996; Chehab et al., 1996). Also in ob/ob mice, exogenous leptin treatment for 2 weeks increased whole ovarian levels of mRNA for side-chain cleavage (scc) and 17α-hydroxylase (Zamorano et al., 1997). Although these studies did not determine whether leptin had direct or indirect effects on the reproductive system, they goaded researchers to conduct studies designed to evaluate which of the specific reproductive organs or tissues respond to leptin. Results of many of these studies are summarized in this textbook. Those that have focused on the effects of leptin in ovarian steroidogenesis will be summarized in this chapter.

LEON J. SPICER • Department of Animal Science, Oklahoma State University.

Leptin and Reproduction. Edited by Henson and Castracane, Kluwer Academic/Plenum Publishers, 2003.

6.2. LEPTIN LEVELS DURING FEMALE REPRODUCTIVE CYCLES

In order to better understand how leptin may be involved in ovarian steroidogenesis, a brief synopsis of the changes in leptin during female reproductive cycles is needed. As detailed in Chapter 4, mid-luteal levels of plasma leptin are significantly greater than during the follicular phase in women (Hardie et al., 1997; Shimizu et al., 1997; Riad-Gabriel et al., 1998; Teirmaa et al., 1998; Cella et al., 2000). How these increased levels of leptin may impact luteal function directly is described in the next section of this review. Decreases in leptin concentrations have also been reported after menopause (Rosenbaum et al., 1996; Shimizu et al., 1997). These changes do not appear to be due to variations in circulating estrogen and(or) progesterone concentrations since sex steroids have no effect on circulating leptin in young women (Teirmaa et al., 1998) or post menopausal women (Haffner et al., 1997; Castracane et al., 1998; Gower et al., 2000; Salbach et al., 2000). However, estrogens may increase leptin secretion in rats (Chu et al., 1999). At present, only a few studies have described changes in systemic leptin concentrations that occur during the estrous cycle of domestic animals (Ehrhardt et al., 2000; Maciel et al., 2001). As specific leptin radioimmunoassays are developed for domestic animals (Kauter et al., 2000; Richards et al., 2000; Henry et al., 2001; see Chapter 7), in vivo associations can then be made between circulating leptin concentrations and ovarian function in these species.

6.3. ROLE OF LEPTIN IN OVARIAN FUNCTION

Several lines of evidence indicate that leptin has direct action at the level of the ovary. First, ovarian granulosa (Spicer and Francisco, 1997) and thecal (Spicer and Francisco, 1998) cells have high affinity receptors for leptin. Second, leptin receptor mRNA has been identified in the adult human ovary (Cioffi et al., 1996, 1997), rat ovary (Zamorano et al., 1997), human granulosa (GCs) and thecal cells (TCs) (Karlson et al., 1997; Agarwal et al., 1999), human corpora lutea (CL) (Loffler et al., 2001), mouse oocytes (Matsuoka et al., 1999), and porcine ovary/CL (Lin et al., 2000; Ruiz-Cortes et al., 2000). In fact, the human (Cioffi et al., 1996) and porcine (Lin et al., 2000) ovary had one of the highest levels of leptin receptor mRNA when compared to other organs. The chronology of the discovery of leptin receptors and its mRNA in the mammalian ovary is summarized in Table 6.1. The RNA sequence that encodes for the leptin receptor, like leptin itself, is highly (that is, >70 percent) conserved across species in which it has been sequenced (Ruiz-Cortes et al., 2000). Only partial sequences of the ovine and bovine leptin receptor have been reported (Dyer et al., 1997; Pfister-Genskow et al., 1997). Long and short isoforms of the leptin receptor exist, which exhibit divergent signaling capacities (Bjorbaek et al., 1997; Uotani et al., 1999). Third, leptin has direct effects on cultured GCs and TCs in vitro, the details of which are described in the next sections. Fourth, leptin (1 μg/ml) treatment concomitantly with LH decreased ovulation rate 77 percent compared with LH treatment alone in FSH-primed whole rat ovaries perifused in vitro (Duggal et al., 2000). Exogenous leptin caused a 67 percent decrease in ovulation rate in

Table 6.1. Chronology of evidence for ovarian leptin (OB) receptors (Rc)

Reference	Species/cell type	Measurement
Cioffi et al. (1996)	Human ovary	OB-Rc mRNA
Cioffi et al. (1997)	Human GC	OB-Rc mRNA
Spicer and Francisco (1997)	Bovine GC	OB-Rc binding
Zamorano et al. (1997)	Rat ovary	OB-Rc mRNA
Karlsson et al. (1997)	Human GC & TC	OB-Rc mRNA
Spicer and Francisco (1998)	Bovine TC	OB-Rc binding
Zachow et al. (1999)	Rat TC & GC	OB-Rc mRNA
Green et al. (2000)	Baboon CL	OB-Rc mRNA
Lin et al. (2000)	Porcine ovary	OB-Rc mRNA
Ruiz-Cortes et al. (2000)	Porcine CL, GC, TC	OB-Rc mRNA/protein
Loffler et al. (2001)	Human CL & GC	OB-Rc mRNA

Note: GC = granulosa cells; TC = thecal cells; CL = corpora lutea.

immature gonadotropin-primed rats (Duggal et al., 2000). Collectively, these studies indicate that the ovary is a likely target organ for leptin. In the following sections, the specific effects of leptin on GC and TC steroidogenesis will be described.

6.3.1. Leptin and GC Steroidogenesis

Leptin (10–300 ng/ml) directly antagonizes insulin's stimulatory effect on bovine GC estradiol and progesterone production (Spicer and Francisco, 1997), but in the absence of insulin, leptin has little or no effect on bovine GC steroidogenesis (Spicer and Francisco, 1997; Spicer et al., 2000). Similarly, Brannian et al. (1999) found that the inhibitory effect of leptin on hCG-stimulated progesterone production by human luteinized GC was only manifested in the presence of insulin. However, Ghizzoni et al. (2001) observed that leptin (10^{-9} M) decreased estradiol secretion but had no effect on progesterone production by human granulosa luteal cells grown in serum-free medium. Because leptin inhibited both progesterone and estradiol productin in two of three studies, the locus of the inhibitory effect of leptin is likely not focused on a single steroidogenic enzyme. Furthermore, leptin is a more potent inhibitor of insulin-induced aromatase activity of undifferentiated (IC_{50}=30 ng/ml) than differentiated (IC_{50}=90 ng/ml) bovine GCs (Figure 6.1; Spicer and Francisco, 1997). This latter observation implies that numbers of leptin receptors in GCs may decrease as follicles grow and develop, rendering mature graafian follicles less sensitive to the negative effects of leptin. In support of this suggestion, a recent study (Tena-Sempere et al., 2001b) in rats found that testicular leptin receptor gene expression is developmentally regulated and sensitive to regulation by LH and FSH. Because concentrations of leptin in blood of lean and obese women range from 2 to 10 ng/ml and 10 to 100 ng/ml, respectively (Caro et al., 1996; Mantzoros et al., 2000) these in vitro results indicate that leptin at physiological concentrations may have a direct impact on ovarian GC function. The maximal inhibitory effect of leptin on bovine GC aromatase activity is similar to or greater than that of other inhibitory hormones that have been tested in vitro (Table 6.2).

Interestingly, leptin had no effect on basal or insulin-induced increases in bovine GC numbers (Spicer and Francisco, 1997; Spicer et al., 2000). These latter observations

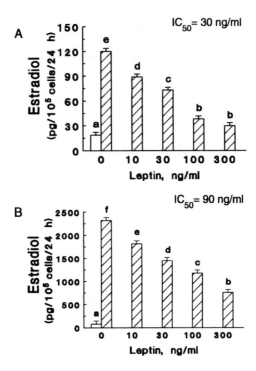

Figure 6.1. Summary of the inhibitory effect of leptin on insulin-induced aromatase activity in bovine granulosa cells. Panel A: Small (1–5 mm) follicle granulosa cells were cultured for 24 hr with FSH (open bar) or FSH plus insulin (hatched bars) and various doses of leptin. Panel B: Large ($\geq$8 mm) follicle granulosa cells were cultured for 24 hr with FSH (open bar) or FSH plus insulin (hatched bars) and various doses of leptin. IC_{50}=concentration of leptin required to inhibit estradiol production by 50%. Within a panel, means without a common letter differ ($P < 0.05$). Data are redrawn from Spicer and Francisco (1997), with permission: The adipose gene product, leptin: evidence of a direct inhibitory role in ovarian function; *Endocrinology, 138*, 3374–3379, 1998; The Endocrine Society.

Table 6.2. Summary of the in vitro effects of various hormones that affect bovine ovarian cell steroidogenesis

Hormone	Reference	Granulosa cell aromatase activity[a]	Thecal cell androstenedione production[b]
Leptin	Spicer and Francisco (1997, 1998)	90%	80%
bFGF[c]	Spicer and Stewart (1996a) Vernon and Spicer (1994)	90%	80%
TNFα[c]	Spicer and Alpizar (1994) Spicer (1998)	75%	40%
IL-6[c]	Alpizar and Spicer (1994)	75%	Not tested
EGF[c]	Spicer and Stewart (1996a)	Not tested	75%
IL-2[c]	Raja et al. (1995)	50%	No effect
Cortisol	Spicer and Chamberlain (1998)	No effect	Stimulation ($<$2-fold)
GH[c]	Spicer and Stewart (1996b)	$<$50%	Stimulation ($<$2-fold)
T_3 or T_4[c]	Spicer et al. (2001)	Stimulation ($<$2-fold)	Stimulation ($>$2-fold)

[a]Values are the maximum percentage decrease from control cultures treated with insulin and FSH.
[b]Values are the maximum percentage decrease from control cultures treated with insulin and LH.
[c]bFGF = basic fibroblast growth factor; TNFα = tumor necrosis factor α; IL-6 = interleukin-6; EGF = epidermal growth factor; IL-2 = interleukin-2; GH = growth hormone; T_3 = triiodothryonine; T_4 = thyroxine.

indicate that leptin does not influence the mitotic action of insulin inspite of the fact that leptin reduces insulin-induced GC steroidogenesis. In addition, follicular fluid levels of leptin correlate negatively with embryo quality in IVF patients (Barroso et al., 1999; Butzow et al., 1999). Whether this intrafollicular leptin is produced within the follicle, as suggested by others (Cioffi et al., 1997; Loffler et al., 2001), remains to be determined. Butzow et al. (1999) concluded that high concentrations of leptin may reduce ovarian responsiveness to gonadotropins in in vitro fertilization (IVF) patients, and thus, explain why obese individuals exhibiting high concentrations of leptin require higher amounts of gonadotropins than lean subjects to achieve ovarian hyperstimulation.

In the absence of insulin but presence of FSH, leptin inhibits IGF-I-induced (Zachow and Magoffin, 1997) and transforming growth factor-α-induced (Zachow et al., 1999) estradiol production by cultured rat GCs. In contrast, leptin has no effect on IGF-I plus FSH-induced estradiol production by bovine GCs (Spicer et al., 2000), FSH-induced progesterone production by rat GCs (Zachow and Magoffin, 1997; Barkan et al., 1999), or LH-induced estradiol and progesterone production by whole rat ovaries (Duggal et al., 2000). Zachow and Magoffin (1997) found that leptin inhibited FSH plus IGF-I-induced estradiol production by <30 percent at optimal doses of FSH and by >80 percent at suboptimal doses of FSH. Thus, dose of FSH dramatically influences the magnitude of leptin's inhibition of IGF-I-induced steroidogenesis. Leptin also inhibits CRH-induced progesterone release from cultured rat GCs (Roguski et al., 2000), and inhibits dexamethasone-induced but not FSH-induced progesterone production by rat GCs in serum-free medium (Barkan et al., 1999). In cultured human GCs, leptin inhibits FSH plus IGF-I-induced estradiol production (Agarwal et al., 1999) and LH-induced estradiol production in the presence of 1 percent fetal bovine serum (Karlsson et al., 1997). In another study using serum-free medium, leptin (20 ng/ml) reduced basal and FSH-stimulated estradiol and progesterone secretion by human granulosa luteal cells (Greisen et al., 2000). To date, only one study has reported stimulatory effects of leptin on aromatase activity (Kitawaki et al., 1999). As summarized from the studies that evaluated insulin action, the locus of the inhibitory effect of leptin is likely not focused on a single steroidogenic enzyme because leptin inhibited both progesterone and estradiol production in the majority of the published studies. In addition, these studies indicate that high concentrations of leptin (i.e., 100 ng/ml), characteristic of obese women (Caro et al., 1996; Mantzoros et al., 2000), mainly inhibit insulin-, IGF-I-, and(or) gonadotropin-stimulated G C steroidogenesis. Further studies will be required to elucidate the hormonal specificity of leptin's inhibitory action within the ovary of these various species.

The mechanism of the inhibitory action of leptin on GC steroidogenesis has not been studied extensively. Zachow et al. (1999) reported that the inhibitory effect of leptin on FSH-induced estradiol production by rat GCs was due in part to leptin inhibition of P450 aromatase mRNA levels. Another study reported that the inhibitory effects of leptin on estradiol and progesterone production in human granulosa–luteal cells could not be explained by a change in mRNA levels for steroidogenic acute regulatory protein (StAR), aromatase or CYP17 (Ghizzoni et al., 2001). This latter observation is in sharp contrast to studies with adrenal cells where leptin was found to suppress ACTH-induced P450-17α, P450-C21 and P450-scc mRNA levels in bovine cortical cells (Bornstein et al., 1997; Kruse et al., 1998). Spicer and Francisco (1997) concluded from studies with bovine GCs that the inhibitory effect of leptin on insulin-induced estradiol and progesterone

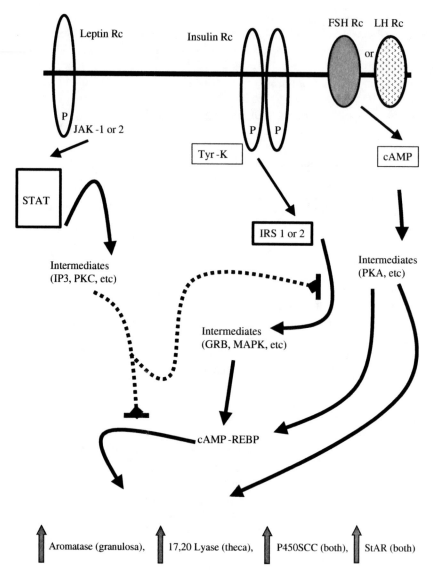

Figure 6.2. Summary of the possible inhibitory effects of leptin on ovarian granulosa and thecal cell steroidogenesis. As suggested from studies using non-ovarian cells, leptin likely induces phosphorylation (P) of janus kinase (JAK)-1 or -2 and subsequently STAT and other intracellular insulin intermediates (e.g., insulin receptor substrate-1, IRS-1) that act on components of the intracellular insulin cascade and part of the intracellular gonadotropin cascade (e.g., cAMP-REBP), ultimately leading to decreases in steroidogenic enzyme activity (e.g., aromatase, 17,20-lyase, and side-chain cleavage enzyme [P450scc]).

production was not due to leptin inhibiting insulin binding to its receptor as has been suggested for rat adipocytes (Waldner et al., 1997; Nowak et al., 1998). Based on mechanistic studies in non-ovarian cells, leptin may inhibit activation of cAMP-response element binding proteins (cAMP-REBP; Shimizu-Albergine et al., 2001; Sweeney et al., 2001) which are known to be present in ovarian cells (see Figure 6.2). In support of this latter

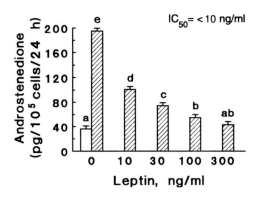

Figure 6.3. Summary of the inhibitory effect of leptin on insulin-induced androstenedione production by bovine thecal cells. Thecal cells from large ($\geq$8 mm) follicles were cultured for 24 hr with LH (open bar) or LH plus insulin (hatched bars) and various doses of leptin. IC_{50} = concentration of leptin required to inhibit androstenedione production by 50%. Means without a common letter differ ($p < 0.05$). Data are redrawn from Spicer and Francisco (1998), with permission: Adipose obese gene product, leptin, inhibits bovine ovarian thecal cell steroidogenesis; *Biol. Reprod., 58*, 207–212, 1998; The Society for the Study of Reproduction, Inc.

suggestion, leptin inhibited 8 br-cAMP-induced murine follicle growth in vitro (Kikuchi et al., 2001). Leptin also attenuates tyrosine phosphorylation of the insulin receptor substrate-1 (IRS-1) in cultured HepG2 cells (Cohen et al., 1996) and activates phosphatidylinositol-3 (PI-3) kinase (Kellerer et al., 1997), but whether these or some other point(s) of regulation is(are) the mechanism of action of leptin on GC steroidogenesis will require further study. These possible intracellular sites of action of leptin on ovarian cell steroidogenesis are summarized in Figure 6.2.

6.3.2. TC Steroidogenesis

Leptin also directly antagonizes insulin's stimulatory effect on bovine TC androstenedione and progesterone production (Spicer and Francisco, 1998). The IC_{50} for leptin on thecal androstenedione production is less than 10 ng/ml (Figure 6.3; Spicer and Francisco, 1998). However, leptin had no effect on LH-induced or LH plus TGFβ-induced androsterone production by rat theca-intersitial cells (Zachow et al., 1999). Leptin also inhibits IGF-I plus LH-induced androstenedione production by human TCs (Agarwal et al., 1999) but not bovine TCs (Spicer et al., 2000). Whether these discrepancies in results are due to species differences remains to be determined. The maximal inhibitory effect of leptin on bovine TC androstenedione production is similar to or greater than other inhbitory hormones that we have tested in vitro (Table 6.2).

Although the locus of leptin's action on thecal steroidogenesis will require additional studies, in vitro effects of leptin observed in bovine TCs are not due to a change in cell viability since insulin, leptin, and their combined treatment had no effect on TC viability (Spicer and Francisco, 1998). In addition, as observed for bovine GCs (Spicer and Francisco, 1997), the inhibitory effect of leptin on insulin action is not mediated by direct inhibition of insulin binding to its receptor in bovine TCs (Spicer and Francisco, 1998). Similar to TCs, leptin inhibits hCG-induced testosterone secretion from adult rat testicular slices (Tena-Sempere et al., 1999, 2000) and Leydig cells (Caprio et al., 1999) with an IC_{50}

of about 20 ng/ml. Because leptin had no effect on hCG-induced cAMP levels but inhibited 8 br-cAMP-induced testosterone production (Caprio et al., 1999), the mechanism of leptin inhibition of Leydig cell steroidogenesis is likely at steps distal to cAMP generation. Indeed, whether the inhibitory effect leptin has on thecal steroidogenesis is due to down-regulation of the steroid producing enzyme cascade (e.g., StAR and P450-scc) as it does in Leydig cells (Caprio et al., 1999; Tena-Sempere et al., 2001a) and adrenal cells (Bornstein et al., 1997; Kruse et al., 1998; Cherradi et al., 2001) and(or) involves various intracellular signal transducers such as IRS-1, PI-3 or cAMP-REBP as proposed for GCs remains to be elucidated (see Figure 6.2).

Because women with polycystic ovarian disease (PCOD) are generally obese (Caro, 1997; Krassas et al., 1998; Jacobs and Conway, 1999), recent research has focused on whether elevated leptin was associated with PCOD in women. As described in Chapter 21, it appears that systemic levels of leptin in women with PCOD do not differ from age and BMI-matched control women (Caro, 1997; Laughlin et al., 1997; Mantzoros et al., 1997; Rouru et al., 1997), and that PCOD is not commonly a consequence of mutations of the leptin or leptin receptor genes (Oksanen et al., 2000).

6.3.3. Luteal Cell Steroidogenesis

As mentioned earlier and summarized in Table 6.1, leptin receptor mRNA is present in CL of several species including baboons (Green et al., 2000), pigs (Ruiz-Cortes et al., 2000), and humans (Loffler et al., 2002). Interestingly, the levels of leptin receptor protein and its mRNA increase during development of CL in pigs (Ruiz-Cortes et al., 2000). Also, serum concentrations of leptin and progesterone are positively correlated during the menstrual cycle (Cella et al., 2000). However, the effects of leptin on luteal cell steroidogenesis has not been directly investigated. As summarized in the "Leptin and Granulosa Cell Steroidogenesis" section, a few studies have reported inhibitory effects of leptin on basal, FSH- and hCG- plus insulin-induced progesterone production by human luteinized GCs (Brannian et al., 1999; Greisen et al., 2000). But whether these effects are due to leptin inhibition of "granulosa" or "luteal" cell function remains to be determined.

6.4. SUMMARY AND CONCLUSIONS

Collectively, in vitro data indicate that leptin may exert a direct inhibitory effect on ovarian function by inhibiting both GC and TC steroidogenesis, ultimately leading to a decrease in estradiol secretion (Figure 6.2). Why leptin has positive in vivo effects on reproduction and ovarian function in leptin-deficient ob/ob mice (Barash et al., 1996; Chehab et al., 1996; Zamarano et al., 1997) but negative in vitro effects on ovarian cell steroidogenesis is unclear. This discrepancy may be explained by the possible multiple sites of leptin action on the hypothalamic–pituitary–ovarian axis in vivo. We hypothesize that increasing but moderate concentrations of leptin (i.e., <10 ng/ml) that occur during puberty or during refeeding from a nutritionally deprived state act to trigger the reproductive axis at the level of the hypothalamus and pituitary, whereas in a low leptin environment (i.e., poor nutrition), ovarian function is dictated primarily by gonadotropins and insulin/IGF-I (Spicer, 2001). In a moderate to high leptin environment (e.g., obesity), the ovary is kept from "over" producing estradiol via leptin inhibition of insulin-induced androstenedione production by the TCs and hormone-induced aromatase activity by GCs (Figure 6.2).

In conclusion, studies conducted over the past few years indicate that leptin has a predominant inhibitory effect on ovarian steroidogenesis in mammals. The ovarian tissues that have leptin receptors and are influenced by leptin include the GC, TC, and luteal cell. The precise role and mechanism of action of leptin in regulating hormone-induced ovarian steroidogenesis of these cell types may be species dependent and will require further elucidation.

ACKNOWLEDGMENTS

The authors thank Paula Cinnamon for preparation of this manuscript. This research was approved for publication by the Director, Oklahoma Agricultural Experiment Station, and supported under project H-2329 and HR4-032 from the Oklahoma Center for Advancement of Science and Technology.

REFERENCES

Agarwal, S. K., Vogel, K., Weitsman, S. R., and Magoffin, D. A. (1999). Leptin antagonizes the insulin-like growth factor-I augmentation of steroidogenesis in granulosa and theca cells of the human ovary. *J. Clin. Endocrinol. Metab., 84*, 1072–1076.

Alpizar, E. and Spicer, L. J. (1994). Effects of interleukin-6 on proliferation and follicle-stimulating hormone-induced estradiol production by bovine granulosa cells in vitro: dependence on size of follicle. *Biol. Reprod., 50*, 38–43.

Barash, I. A., Cheung, C. C., Weigle, D. S., Ren, H., Kabigting, E. B., Kuijper, J. L., Clifton, D. K., and Steiner, R. A. (1996). Leptin is a metabolic signal to the reproductive system. *Endocrinology, 137*, 3144–3147.

Barb, C. R. (1999). The brain–pituitary–adipose axis: role of leptin in modulating neuroendocrine function. *J. Anim. Sci., 77*(Suppl. 3), 1249–1257.

Barkan, D., Jia, H., Dantes, A., Vardimon, L., Amsterdam, A., and Rubinstein, M. (1999). Leptin modulates the glucocorticoid-induced ovarian steroidogenesis. *Endocrinology, 140*, 1731–1738.

Barroso, G., Barrionuevo, M., Rao, P., Graham, L., Danforth, D., Huey, S., Abuhamad, A., and Oehninger, S. (1999). Vascular endothelial growth factor, nitric oxide, and leptin follicular fluid levels correlate negatively with embryo quality in IVF patients. *Fertil. Steril., 72*, 1024–1026.

Bjorbaek, C., Uotani, S., da Silva, B., and Flier, J. S. (1997). Divergent signaling capacities of the long and short isoforms of the leptin receptor. *J. Biol. Chem., 272*, 32686–32695.

Blache, D., Tellam, R. L., Chagas, L. M., Blackberry, M. A., Vercoe, P. E., and Martin, G. B. (2000). Level of nutrition affects leptin concentrations in plasma and cerebrospinal fluid in sheep. *J. Endocrinol., 165*, 625–637.

Bornstein, S. R., Uhlmann, K., Haidan, A., Ehrhart-Bornstein, M., and Scherbaum, W. A. (1997). Evidence for a novel peripheral action of leptin as a metabolic signal to the adrenal gland: leptin inhibits cortisol release directly. *Diabetes, 46*, 1235–1238.

Brannian, J. D., Zhao, Y., and McElroy, M. (1999). Leptin inhibits gonadotrophin-stimulated granulosa cell progesterone production by antagonizing insulin action. *Hum. Reprod., 14*, 1445–1448.

Butzow, T. L., Moilanen, J. M., Lehtovirta, M., Tuomi, T., Hovatta, O., Siegberg, R., Nilsson, C. G., and Apter, D. (1999). Serum and follicular fluid leptin during in vitro fertilization: relationship among leptin increase, body fat mass, and reduced ovarian response. *J. Clin. Endocrinol. Metab., 84*, 3135–3139.

Caprio, M., Isidori, A. M., Carta, A. R., Moretti, C., Dufau, M. L., and Fabbri, A. (1999). Expression of functional leptin receptors in rodent Leydig cells. *Endocrinology, 140*, 4939–4947.

Caro, J. F., Sinha, M. K., Kolaczynski, J. W., Zhang, P. L., and Considine, R. V. (1996). Leptin: the tale of an obesity gene. *Diabetes, 45*, 1455–1462.

Caro, J. F. (1997). Leptin is normal in PCOS, an editorial about three "negative" papers. *J. Clin. Endocrinol. Metab., 82*, 1685–1686.

Castracane, V. D., Kraemer, R. R., Franken, M. A., Kraemer, G. R., and Gimpel, T. (1998). Serum leptin concentration in women: effect of age, obesity, and estrogen administration. *Fertil. Steril., 70*, 472–477.

Cella, F., Giordano, G., and Cordera R. (2000). Serum leptin concentrations during the menstrual cycle in normal-weight women: effects of an oral triphasic estrogen-progestin medication. *Eur. J. Endocrinol., 142*, 174–178.

Chehab, F. F., Lim, M. E., and Lu, R. (1996). Correction of the sterility defect in homozygous obese female mice by treatment with the human recombinant leptin. *Nat. Genet., 12*, 318–320.

Cherradi, N., Capponi, A. M., Gaillard, R. C., and Pralong, F. P. (2001). Decreased expression of steroidogenic acute regulatory protein: a novel mechanism participating in the leptin-induced inhibition of glucocorticoid biosynthesis. *Endocrinology, 142*, 3302–3308.

Chu, S. C., Chou, Y. C., Liu, J. Y., Chen, C. H., Shyu, J. C., and Chou, F. P. (1999). Fluctuation of serum leptin level in rats after ovariectomy and the influence of estrogen supplement. *Life Sci., 64*, 2299–2306.

Cioffi, J. A., Shafer, A. W., Zupancic, T. A., Smith-Gbur, J., Mikhail, A., Platika, D., and Snodgrass, H. R. (1996). Novel B219/OB receptor isoforms: possible role of leptin in hematopoiesis and reproduction. *Nat. Med., 2*, 585–289.

Cioffi, J. A., Van Blerkom, J., Antczak, M., Shafer, A., Wittmer, S., and Snodgrass, H. R. (1997). The expression of leptin and its receptors in pre-ovulatory human follicles. *Mol. Hum. Reprod., 3*, 467–472.

Cohen, B., Novick, D., and Rubinstein, M. (1996). Modulation of insulin activities by leptin. *Science, 274*, 1185–1188.

Duggal, P. S., Van Der Hoek, K. H., Milner, C. R., Ryan, N. K., Armstrong, D. T., Magoffin, D. A., and Norman, R. J. (2000). The in vivo and in vitro effects of exogenous leptin on ovulation in the rat. *Endocrinology, 141*, 1971–1976.

Dyer, C. J., Simmons, J. M., Matteri, R. L., and Keisler, D. H. (1997). Leptin receptor mRNA is expressed in ewe anterior pituitary and adipose tissues and is differentially expressed in hypothalamic regions of well-fed and feed-restricted ewes. *Domest. Anim. Endocrinol., 14*, 119–128.

Ehrhardt, R. A., Slepetis, R. M., Siegal-Willott, J., Van Amburgh, M. E., Bell, A. W., and Biosclair, Y. R. (2000). Development of a specific radioimmunoassay to measure physiological changes in circulating leptin in cattle and sheep. *J. Endocrinol., 166*, 519–528.

Ghizzoni, L., Barreca, A., Mastorakos, G., Furlini, M., Vottero, A., Ferrari, B., Chrousos, G. P., and Gernasconi, S. (2001). Leptin inhibits steroid biosynthesis by human granulosa-lutein cells. *Horm. Metab. Res., 33*, 323–328.

Gower, B. A., Nagy, T. R., Goran, M. I., Smith, A., and Kent, E. (2000). Leptin in postmenopausal women: influence of hormone therapy, insulin, and fat distribution. *J. Clin. Endocrinol. Metab., 85*, 1770–1775.

Green, A. E., O'Neil, J. S., Swan, K. F., Bohm, R. P. Jr., Ratterree, M. S., and Henson, M. C. (2000). Leptin receptor transcripts are constitutively expressed in placenta and adipose tissue with advancing baboon pregnancy. *Proc. Soc. Exp. Biol. Med., 223*, 362–366.

Greisen, S., Ledet, T., Moller, N., Jorgensen, J. O., Christiansen, J. S., Petersen, K., and Ovesen, P. (2000). Effects of leptin on basal and FSH stimulated steroidogenesis in human granulosa luteal cells. *Acta Obstet. Gynecol. Scand., 79*, 931–935.

Haffner, S. M., Mykkanen, L., and Stern, M. P. (1997). Leptin concentrations in women in the San Antonio Heart Study: effect of menopausal status and postmenopausal hormone replacement therapy. *Am. J. Epidemiol., 146*, 581–585.

Hardi, L., Trayhurn, P., Abramovich, D., and Fowler, P. (1997). Circulating leptin in women: a longitudinal study in the menstrual cycle and during pregnancy. *Clin. Endocrinol., 47*, 101–106.

Henry, B. A., Goding, J. W., Tilbrook, A. J., Dunshea, F. R., and Clarke, I. J. (2001). Intracerebroventricular infusion of leptin elevates the secretion of luteinising hormone without affecting food intake in long-term food-restricted sheep, but increases growth hormone irrespective of body weight. *J. Endocrinol., 168*, 67–77.

Hossner, K. L. (1998). Cellular, molecular and physiological aspects of leptin: potential application in animal production. *Can. J. Anim. Sci., 78*, 463–472.

Jacobs, H. S. and Conway, G. S. (1999). Leptin, polycystic ovaries and polycystic ovary syndrome. *Hum. Reprod. Update, 5*, 166–171.

Karlsson, C., Lindell, K., Svensson, E., Bergh, C., Lind, P., Billig, H., Carlsson, L. M., and Carlsson, B. (1997). Expression of functional leptin receptors in the human ovary. *J. Clin. Endocrinol. Metab., 82*, 4144–4148.

Kauter, K., Ball, M., Kearney, P., Tellam, R., and McFarland, J. R. (2000). Adrenaline, insulin and glucagon do not have acute effects on plasma leptin levels in sheep: development and characterisation of an ovine leptin ELISA. *J. Endocrinol., 166*, 127–135.

Kellerer, M., Koch, M., Metzinger, E., Mushack, J., Capp, E., and Haring, H. U. (1997). Leptin activates PI-3 kinase in C_2C_{12} myotubes via janus kinase-2 (JAK-2) and insulin receptor substrate-2 (IRS-2) dependent pathways. *Diabetologia, 40*, 1358–1362.

Kikuchi, N., Andoh, K., Abe, Y., Yamada, K., Mizunuma, H., and Ibuki, Y. (2001). Inhibitory action of leptin on early follicular growth differs in immature and adult female mice. *Biol. Reprod., 65*, 66–71.

Kitawaki, J., Kusuki, I., Koshiba, H., Tsukamoto, K., and Honjo, H. (1999). Leptin directly stimulates aromatase activity in human luteinized granulosa cells. *Mol. Hum. Reprod., 5*, 708–713.

Krassas, G. E., Kaltsas, T. T., Pontikides, N., Jacobs, H., Blum, W., and Messinis, I. (1998). Leptin levels in women with polycystic ovary syndrome before and after treatment with diazoxide. *Eur. J. Endocrinol., 139*, 184–189.

Kruse, M., Bornstein, S. R., Uhlmann, K., Paeth, G., and Scherbaum, W. A. (1998). Leptin down-regulates the steroid producing system in the adrenal. *Endocr. Res., 24*, 587–590.

Laughlin, G. A., Morales, A. J., and Yen, S. S. (1997). Serum leptin levels in women with polycystic ovary syndrome: the role of insulin resistance/hyperinsulinemia. *J. Clin. Endocrinol. Metab., 82*, 1692–1696.

Lin, J., Barb, C. R., Matteri, R. L., Kraeling, R. R., Chen, X., Meinersmann, R. J., and Rampacek, G. B. (2000). Long form leptin receptor mRNA expression in the brain, pituitary, and other tissues in the pig. *Domest. Anim. Endocrinol., 19*, 53–61.

Loffler, S., Aust, G., Kohler U., and Spanel-Borowski, K. (2001). Evidence of leptin expression in normal and polycystic human ovaries. *Mol. Hum. Reprod., 7*, 1143–1149.

Maciel, S. M., Chamberlain, C. S., Wettemann, R. P., and Spicer, L. J. (2001). Dexamethasone influences endocrine and ovarian function in cattle. *J. Dairy Sci., 84*, 1998–2009.

Mantzoros, C. S., Cramer, D. W., Liberman, R. F., and Barbieri, R. L. (2000). Predictive value of serum and follicular fluid leptin concentrations during assisted reproductive cycles in normal women and in women with the polycystic ovarian syndrome. *Hum. Reprod., 15*, 539–544.

Mantzoros, C. S., Dunaif, A., and Flier, J. S. (1997). Leptin concentrations in the polycystic ovary syndrome. *J. Clin. Endocrinol. Metab., 82*, 1687–1691.

Masuzaki, H., Ogawa, Y., Isse, N., Satoh, N., Okazaki, T., Shigemoto, M., Mori, K., Tamura, N., Hosoda, K., and Yoshimasa, Y. (1995). Human obese gene expression. Adipocyte-specific expression and regional differences in the adipose tissue. *Diabetes, 44*, 855–858.

Matsuoka, T., Tahara, M., Yokoi, T., Masumoto, N., Takeda, T., Yamaguchi, M., Tasaka, K., Kurachi, H., and Murata, Y. (1999). Tyrosine phosphorylation of STAT3 by leptin through leptin receptor in mouse metaphase 2 stage oocyte. *Biochem. Biophys. Res. Commun., 256*, 480–484.

Nowak, K. W., Mackowiak, P., Nogowski, L., Szkudelski, T., and Malendowicz, L. K. (1998). Acute leptin action on insulin blood level and liver insulin receptor in the rat. *Life Sci., 63*, 1347–1352.

Ogawa, Y., Masuzaki, H., Isse, N., Okazaki, T., Mori, K., Shigemoto, M., Satoh, N., Tamura, N., Hosoda, K., Yoshimasa, Y., Jingami, H., Kawada, T., and Nakao, K. (1995). Molecular-cloning of rat obese cDNA and augmented gene-expression in genetically-obese Zucker fatty (FA/FA) rats. *J. Clin. Invest., 96*, 1647–1652.

Oksanen, L., Tiitinen, A., Kaprio, J., Koistinen, H. A., Karonen, S., and Kontula, K. (2000). No evidence for mutations of the leptin receptor genes in women with polycystic ovary syndrome. *Mol. Hum. Reprod., 6*, 8873–8876.

Pfister-Genskow, M., Hayes, H., Eggen, A., and Bishop, M. D. (1997). The leptin receptor (LEPR) gene maps to bovine chromosome 3q33. *Mamm. Genome., 8*, 227.

Raja, C. A. R., Spicer, L. J., and Stewart, R. E. (1995). Interleukin-2 affects steroidogenesis and numbers of bovine ovarian granulosa cells but not thecal cells in vitro. *Endocrine, 3*, 899–905.

Riad-Gabriel, M. G., Jinagouda, S. D., Sharma, A., Boyadjian, R., and Saad, M. F. (1998). Changes in plasma leptin during the menstrual cycle. *Eur. J. Endocrinol., 139*, 528–531.

Richards, M. P., Caperna, T. J., Elsasser, T. H., Ashwell, C. M., and McMurtry, J. P. (2000). Design and application of a polyclonal peptide antiserum for the universal detection of leptin protein. *J. Biochem. Biophys. Methods, 45*, 147–156.

Roguski, K., Baranowska, B., Chmielowska, M., and Borowiec, M. (2000). Leptin modulates the corticotropin-releasing hormone (CRH) action on progesterone release from cultured rat granulosa cells. *Neuroendocrinol. Lett., 21*, 383–389.

Rosenbaum, M., Nicolson, M., Hirsch, J., Heymsfield, S. B., Gallagher, D., Chu, F., and Leibel, R. L. (1996). Effects of gender, body composition, and menopause on plasma concentrations of leptin. *J. Clin. Endocrinol. Metab., 81*, 3424–3427.

Rouru, J., Anttila, L., Koskinen, P., Penttila, T. A., Irjala, K., Huupponen, R., and Koulu, M. (1997). Serum leptin concentrations in women with polycystic ovary syndrome. *J. Clin. Endocrinol. Metab., 82*, 1697–1700.

Ruiz-Cortes, Z. T., Men, T., Palin, M. F., Downey, B. R., Lacroix, D. A., and Murphy, B. D. (2000). Porcine leptin receptor: Molecular structure and expression in the ovary. *Mol. Reprod. Dev., 56*, 465–474.

Salbach, B., Nawroth, P. P., Kubler, W., von Holst, T. H., and Salbach, P. B. (2000). Serum leptin levels and body weight in postmenopausal women under transdermal hormone replacement therapy. *Eur. J. Med. Res., 28*, 63–66.

Shimizu, H., Shimomura, Y., Nakanishi, Y., Futawatari, T., Ohtani, K., Sato, N., and Mori, M. (1997). Estrogen increases in vivo leptin production in rats and human subjects. *J. Endocrinol., 154*, 285–292.

Shimizu-Albergine, M., Ippolito, D. L., and Beavo, J. A. (2001). Downregulation of fasting-induced cAMP response element-mediated gene induction by leptin in neuropeptide Y neurons of the arcuate nucleus. *J. Neurosci., 21*, 1238–1246.

Spicer, L. J. (1998). Tumor necrosis factor-α (TNF-α) inhibits steroidogenesis of bovine ovarian granulosa and thecal cells in vitro: Involvement of TNF-α receptors. *Endocrine, 8*, 109–115.

Spicer, L. J. (2001). Leptin: a possible metabolic signal affecting reproduction. *Domest. Anim. Endocrinol., 21*, 251–270.

Spicer, L. J., Alonzo, J., and Chamberlain, C. S. (2001). Effects of thyroid hormones on bovine granulosa and thecal cell function in vitro: dependence on insulin and gonadotropins. *J. Dairy Sci., 84*, 1069–1076.

Spicer, L. J. and Alpizar, E. (1994). Effects of cytokines on FSH-induced estradiol production by bovine granulosa cells in vitro: dependence on size of follicle. *Domest. Anim. Endocrinol., 11*, 25–34.

Spicer, L. J. and Chamberlain, C. S. (1998). Influence of cortisol on insulin- and insulin-like growth factor I (IGF-I)-induced steroid production and on IGF-I receptors in cultured bovine granulosa cells and thecal cells. *Endocrine, 9*, 151–161.

Spicer, L. J., Chamberlain, C. S., and Francisco, C. C. (2000). Ovarian action of leptin: effects on insulin-like growth factor-I-stimulated function of granulosa and thecal cells. *Endocrine, 12*, 53–59.

Spicer, L. J. and Francisco, C. C. (1997). The adipose obese gene product, leptin: evidence of a direct inhibitory role in ovarian function. *Endocrinology, 138*, 3374–3379.

Spicer, L. J. and Francisco, C. C. (1998). Adipose obese gene product, leptin, inhibits bovine ovarian thecal cell steroidogenesis. *Biol. Reprod., 58*, 207–212.

Spicer, L. J. and Stewart, R. E. (1996a). Interactions among basic fibroblast growth factor, epidermal growth factor, insulin, and insulin-like growth factor-I (IGF-I) on cell numbers and steroidogenesis of bovine thecal cells: role of IGF-I receptors. *Biol. Reprod., 54*, 255–263.

Spicer, L. J. and Stewart, R. E. (1996b). Interaction among bovine somatotropin, insulin, and gonadotropins on steroid production by bovine granulosa and thecal cells. *J. Dairy Sci., 79*, 813–821.

Sweeney, G., Keen, J., Somwar, R., Konrad, D., Garg, R., and Klip, A. (2001). High leptin acutely inhibit insulin-stimulated glucose uptake without affecting glucose transporter 4 translocation of 16 rat skeletal muscle cells. *Endocrinology, 142*, 4806–4812.

Teirmaa, T., Luukkaa, V., Rouru, J., Koulu, M., and Huupponen, R. (1998). Correlation between circulating leptin and luteinizing hormone during the menstrual cycle in normal-weight women. *Eur. J. Endocrinol., 139*, 190–194.

Tena-Sempere, M., Manna, P. R., Zhang, F. P., Pinilla, L., Gonzalez, L. C., Dieguez, C., Huhtaniemi, I., and Aguilar, E. (2001a). Molecular mechanisms of leptin action in adult rat testis: potential targets for leptin-induced inhibition of steroidogenesis and pattern of leptin receptor messenger ribonucleic acid expression. *J. Endocrinol., 170*, 413–423.

Tena-Sempere, M., Pinilla, L., Gonzalez, L. C., Dieguez, C., Casanueva, F. F., and Aguilar, E. (1999). Leptin inhibits testosterone secretion from adult rat testis in vitro. *J. Endocrinol., 161*, 211–218.

Tena-Sempere, M., Pinilla, L., Gonzalez, L. C., Navarro, J., Dieguez, C., Casanueva, F. F., and Aguilar, E. (2000). In vitro pituitary and testicular effects of the leptin-related synthetic peptide leptin (116–130) amide involve actions both similar to and distinct from those of the native leptin molecule in the adult rat. *Eur. J. Endocrinol., 142,* 406–410.

Tena-Sempere, M., Pinilla, L., Zhang, F.-P., Gonzales, L. C., Huhtaniemi, I., Casanueva, F. F., Dieguez, C., and Aguilar, E. (2001b). Developmental and hormonal regulation of leptin receptor (Ob-R) messenger ribonucleic acid expression in rat testis. *Biol. Reprod., 64,* 634–643.

Uotani, S., Bjorbaek, C., Tornoe, J., and Flier, J. S. (1999). Functional properties of leptin receptor isoforms: internalization and degradation of leptin and ligand-induced receptor downregulation. *Diabetes, 48,* 279–286.

Vernon, R. K. and Spicer, L. J. (1994). Effects of basic fibroblast growth factor and heparin on follicle-stimulating hormone-induced steroidogenesis by bovine granulosa cells. *J. Anim. Sci., 72,* 2696–2702.

Waldner, K., Filippis, A., Clark, S., Zimmet, P., and Collier, G. R. (1997). Leptin inhibits insulin binding in isolated rat adipocytes. *J. Endocrinol., 155,* R5–R7.

Zachow, R. J. and Magoffin, D. A. (1997). Direct intraovarian effects of leptin: impairment of the synergistic action of insulin-like growth factor-I on follicle-stimulating hormone-dependent estradiol-17β production by rat ovarian granulosa cells. *Endocrinology, 138,* 847–850.

Zachow, R. J., Weitsman, S. R., and Magoffin, D. A. (1999). Leptin impairs the synergistic stimulation by transforming growth factor-β of follicle-stimulating hormone-dependent aromatase activity and messenger ribonucleic acid expression in rat ovarian granulosa cells. *Biol. Reprod., 61,* 1104–1109.

Zamorano, P. L., Mahesh, V. B., De Sevilla, L. M., Chorich, L. P., Bhat, G. K., and Brann, D. W. (1997). Expression and localization of the leptin receptor in endocrine and neuroendocrine tissues of the rat. *Neuroendocrinology, 65,* 223–228.

Zhang, Y., Proenca, R., Maffei, M., Barone, M., Leopold, L., and Friedman, J. M. (1994). Positional cloning of the mouse obese gene and its human homologue. *Nature, 372,* 425–432.

Regulation of Leptin and Leptin Receptor in the Human Uterus: Possible Roles in Implantation and Uterine Pathology

V. Daniel Castracane and Michael C. Henson

Early studies following the identification of leptin in (1994) by Zhang and colleagues indicated a number of roles for leptin in the reproductive system. These would include the regulation of gonadotropin secretion, a contribution to the onset of puberty, steroidgenesis and ovarian function, and of course a role in pregnancy, which was presaged by the early identification of the expression of the leptin message in the placenta. All of these aspects are currently reviewed in other chapters in this volume, however the question of uterine physiology has not been studied as extensively as the other areas. The purpose of this chapter is to introduce this potentially important area of leptin physiology and the relationship of the leptin system to normal and pathological uterine function. We are clearly in the beginning stages of understanding any roles that leptin may play in the regulation of uterine physiology and these will be introduced with appropriate citations that promise to be the subject of important research in the near future.

V. Daniel Castracane • Department of Obstetrics and Gynecology and the Women's Health Institute Center of Amarillo, Texas Tech University Health Sciences Center—Amarillo; Diagnostic Systems Laboratories, Inc., Webster, Texas.

Michael C. Henson • Departments of Obstetrics and Gynecology, Physiology, and Structural and Cellular Biology, the Tulane National Primate Research Center, and Interdisciplinary Program in Molecular and Cellular Biology, Tulane University Health Sciences Center.

Leptin and Reproduction. Edited by Henson and Castracane, Kluwer Academic/Plenum Publishers, 2003.

7.1. LEPTIN AND LEPTIN RECEPTOR IN THE ENDOMETRIUM

The first study to investigate leptin physiology in the human endometrium was that of Gonzales et al. (2000), who demonstrated the expression of the long form of the leptin receptor. These authors also reported that leptin could be detected in endometrial tissue with reverse transcriptase-polymerase chain reaction (RT-PCR). Subsequent studies from other laboratories have been able to demonstrate the message for the leptin receptor, but generally have not been able to demonstrate the expression of the leptin message and this area remains controversial. Whether this demonstrates a methodological difference or differences in sensitivity between laboratories has not been established. In addition, Gonzales and colleagues (2000) were able to demonstrate not only expression of the leptin receptor, but also a distinct signal for leptin was demonstrated in the secretory endometrium using RT-PCR, immunohistochemistry, and western blotting. These studies were also able to demonstrate that the production of leptin could be confirmed by the appearance of leptin in conditioned medium from cultured endometrial epithelial cells. These authors were the first to suggest that the presence of leptin receptor denoted a role for leptin in endometrial physiology and implantation.

Kitawaki et al. (2000) were also able to demonstrate that the leptin receptor was expressed in the human endometrium. They were able to identify the long form of the leptin receptor, but were not able to demonstrate leptin mRNA. All the splice variants of the leptin receptor were identified in this tissue. Changes during the cycle were observed with the expression of the leptin receptor mRNA peaking in the early secretory phase. Decidual tissue in early gestation also expressed several variants of the leptin receptor, including the long extracellular domain form. The abundance of leptin receptor in this tissue was not related to the degree of adiposity in individuals within the normal range of body weight.

The factors that regulate endometrial leptin receptor probably involve ovarian steroid hormones, which are so important in changing the nature of endometrial histology and function. Koshiba and co-workers (2001) demonstrated that the expression of the leptin receptor mRNA changes during the course of the cycle with the peak in the early secretory phase. Proliferative endometrium specimens were grown in organ culture and estradiol had no acute effect on the expression of this message. When estradiol was combined with either progesterone or medroxyprogesterone acetate the expression of the leptin message declined to one half. This progesterone induced suppression was inhibited by the use of the progesterone antagonist, mifepristone. These studies suggest that there is steroid hormone regulation of leptin within the human endometrium and that progesterone, acting through its receptor, is able to suppress the expression of the leptin receptor. These factors may contribute to the sensitivity of the endometrium to leptin and may be important in establishing the window of implantation.

7.2. LEPTIN AND IMPLANTATION

The localization of the leptin receptor in the endometrium and the discovery that leptin is produced by the blastocyst suggest that some relationship exists between the implanting blastocyst and the receptive endometrium. When blastocyst development was arrested, the production of leptin was significantly reduced. When blastocysts were co-cultured with endometrial epithelial cells, results suggested that the human endometrium was a site for local production, as well as a target tissue for circulating leptin that originated in the blastocyst (Gonzales et al., 2000). These were perhaps the first studies to suggest that the leptin system may be implicated in the implantation process.

Further indications that leptin may facilitate implantation include the polypeptide's ability, similar to that of interleukins, to affect a marked change in the endometrium, thereby facilitating trophoblastic invasion. In these studies, cytotrophoblast cultures were used to demonstrate an effect of leptin and cytokines on the expression on $\alpha2$, $\alpha5$, and $\alpha6$ integrin subunits in the endometrium, as well as an effect on MMP2 and MMP9. The $\alpha6$ subunit was greatly upregulated by leptin, with other subunits enhanced to a lesser degree. MMP9 was upregulated by both leptin and IL1α. These studies demonstrate an effect of leptin from the cytotrophoblast and demonstrate mechanisms in the endometrium that may facilitate a more invasive role for the blastocyst. These authors suggested, therefore, that leptin may play an autocrine or paracrine role in cytotrophoblast invasion during implantation (Gonzales and Leavis, 2001). In addition, leptin also upregulates $\beta3$ integrin expression in human endometrial epithelial cell cultures (Gonzales et al., 2001). These investigators were also able to demonstrate that interleukin 1β upregulates leptin and leptin receptor expression in the same cell cultures. These studies further support the observation that an autocrine–paracrine role for leptin exists in respect to endometrial receptivity. In this capacity, leptin may serve as a necessary component in establishing a window of receptivity and initiating the appearance of integrin in the endometrial cells where blastocyst attachment first takes place.

In a recent preliminary report (Baumgartner et al., 2002) the leptin receptor was identified in the cervical stroma, although not in the epithelium, with long form of the leptin receptor being upregulated in late pregnancy. Administration of leptin and relaxin increased both cervical weight and distensibility. These results suggest a role for leptin in cervical ripening and further studies, particularly in primate species, seem warranted.

7.3. LEPTIN AND ENDOMETRIAL PATHOLOGY

Endometriosis, frequently associated with infertility and/or pelvic pain, remains an important clinical condition for women of reproductive age. A better appreciation of normal physiology may help to better understand endometriosis as well as other endometrial pathologies. A role for leptin in the progression of endometriosis has been suggested in preliminary studies.

Alfer and colleagues (2002) have also demonstrated the expression of mRNA for the long form of the leptin receptor in the human endometrium, which was expressed in both glandular and luminal tissue. Expression changes throughout the menstrual cycle with major production in the follicular and midluteal phase. These investigators were not able to detect leptin expression using RT-PCR but suggested that the human endometrium may be a novel target tissue for leptin action. They were able to demonstrate leptin and leptin receptor in glands and luminal epithelium of the endometrium using immunohistochemical techniques with weaker staining in the stroma. They concluded that the leptin identified via immunohistochemistry was derived from adipose tissue. They investigated a group of subfertile women, taking two biopsies from each throughout the menstrual cycle. These patients had an endometrial maturation defect, despite normal serum hormone levels and steroid hormone receptor expression in the endometrium. Because patients were deficient in functional leptin receptors, the authors concluded that the lack of leptin receptor may contribute to their subfertility.

Interestingly, serum leptin levels are higher in both serum and peritoneal fluid in patients with pelvic endometriosis. This is not simply the result of relative adiposity, as when corrected for adiposity, levels of leptin were highest in endometriosis patients when compared to normal controls. These changes were also independent of the menstrual cycle phase. The sources of this increased leptin remain to be determined, but an intra-abdominal increase either from endometriotic islands or increased omental fat production of leptin was suggested. Since angiogenesis is thought to play an important role in the development of endometriotic implants and leptin has been related to angiogenesis, the possibility that leptin may stimulate angiogenesis in endometriotic islands might be considered (Bouloumie et al., 1998; Sierra-Honigmann et al., 1998). Further studies are warranted to better understand the role of the leptin system in the pathophysiology of endometriosis (Matarese et al., 2000).

Wu and co-workers (2002) studied the roles of leptin and leptin receptor expression in eutopic and ectopic endometria in order to determine any potential roles for leptin in the development of endometriosis. Leptin mNRA was undetectable in 7 of 17 eutopic endometria with minimal amounts detected in the remaining samples. In contrast, there was a marked increase in leptin mRNA and protein expression in endometriotic lesions. Conversely, the levels of mRNA transcripts for the leptin receptor were suppressed in association with the severity of endometriosis. In addition, leptin significantly enhanced both eutopic and ectopic endometrial stromal cell proliferation in vitro. The authors concluded, therefore, that leptin and its receptor may have important autocrine and paracrine roles in human endometriosis. The study of other uterine pathologies, to date, has been neglected. Whether or not fibroids express leptin or its receptor is not clear. The only study with regard to leptin and uterine myomas and the use of LHRH analogs was not associated with serum changes in leptin. These results suggest that there are no changes with the decrease in reproductive endocrinology nor does uterine myoma contribute to circulating levels. The design of this study precluded the examination of whether paracrine actions of leptin on fibroid growth may occur. These studies did not examine the production of leptin or changes in leptin receptor concentration in uterine fibroids. Based on changes in endometriotic tissue, such studies seem worthy of further investigation (Nowicki et al., 2002). In a single report, serum leptin levels were increased in women with endometrial cancer, but when corrected for BMI were not different from controls (Petridou et al., 2002).

REFERENCES

Alfer, J., Müller-Schöttle, F., Classen-Linke, I., von Rango, U., Happel, L., Beier-Hellwig, K., Rath, W., and Beier, H. M. (2000). The endometrium as a novel target for leptin: differences in fertility and subfertility. *Mol. Hum. Reprod., 6*, 595–601.

Baumgartner, W. W., Rivera, A., and Bahr, J. M. (2002). A possible role for leptin in the rat cervix during gestation and parturition. *Biol. Reprod., 66*(Suppl. 1), Abstract 222, 187–188.

Bouloumie, A., Drexler, H. C. A., Lafontan, M., and Busse, R. (1998). Leptin, the product of ob gene, promotes angiogenesis. *Circ. Res., 83*, 1059–1066.

Gonzales, R. R. and Leavis, P. (2001). Leptin upregulates $\beta3$-integrin expression and interleukin-1β upregulates leptin and leptin receptor expression in human endometrial epithelial cell cultures. *Endocrine, 16*, 21–28.

Gonzales, R. R., Caballero-Campo, P., Jasper, M., Mercader, A., Devoto, L., Pellicer, A., and Simon, C. (2000). Leptin and leptin receptors are expressed in the human endometrium and endometrial leptin secretion is regulated by the human blastocyst. *J. Clin. Endocrinol. Metab., 85*, 4883–4888.

Gonzales, R. R., Devoto, L., Campana, A., and Bischof, P. (2001). Effects of leptin, interleukin-1α, interleukin-6, and transforming growth factor-β on markers of trophoblast invasive phenotype. *Endocrine, 15*, 157–164.

Kitawake, J., Koshiba, H., Ishihara, H., Kusuki, I., Tsukamoto, K., and Honjo, H. (2000). Expression of leptin receptor in human endometrium and fluctuation during the menstrual cycle. *J. Clin. Endocrinol. Metab., 85*, 1946–1950.

Koshiba, H., Kitawaki, J., Ishihara, H., Kado, N., Kusuki, I., Tsukamoto, K., and Honjo, H. (2001). Progesterone inhibition of functional leptin receptor mRNA expression in human endometrium. *Mol. Hum. Reprod., 7*, 567–572.

Matarese, G., Alviggi, C., Sanna, V., Howard, J. K., Lord, G. M., Carravetta, C., Fontana, S., Lechler, R. I., Bloom, S. R., and De Placido, G. (2000). Increased leptin levels in serum and peritoneal fluid of patients with pelvic endometriosis. *J. Clin. Endocrinol. Metab., 85*, 2483–2487.

Nowicki, M., Adamkiewicz, G., Bryc, W., and Kokot, F. (2002). The influence of luetinizing hormone-releasing hormone analog on serum leptin and body composition in women with solitary uterine myoma. *Am. J. Obstet. Gynecol., 186*, 340–344.

Petridou, E., Belechri, M., Dessypris, N., Koukoulomatis, P., Diakomanolis, E., Spanos, E., and Trichopoulos, D. (2002). Leptin and body mass index in relation to endometrial cancer risk. *Ann. Nutr. Metab., 46*, 147–151.

Sierra-Honigmann, M. R., Nath, A. K., Murakami, C., Garcia-Cardena, G., Papapetropoulos, A., Sessa, W. C., Madge, L. A., Schechner, J., Schwabb, M. B., Polverini, P. J., and Flores-Riveros, J. R. (1998). Biological action of leptin as an angiogenic factor. *Science, 281*, 1683–1686.

Wu, M. H., Chuang, P. C., Chen, H. M., Lin, C. C., and Tsai, S. J. (2002). Increased leptin expression in endometriosis cells is associated with endometrial stromal cell proliferation and leptin gene upregulation. *Mol. Hum. Reprod., 5*, 456–464.

Zhang, Y., Proenca, R., Maffei, M., Barone, M., Leopold, L., and Friedman, J. M. (1994). Positional cloning of the mouse obese gene and its human homologue. *Nature, 372*, 425–432.

8

Leptin and Reproduction in the Male

JOAQUIN LADO-ABEAL AND REID L. NORMAN

8.1. INTRODUCTION

The association of fertility with fatness is an ancient concept. Artifacts from France and Austria thought to be at least twenty thousand years old and considered to be representations of fertility goddesses (Schneider and Wade, 2000) depict women who would be considered obese by today standards. Although generous stores of body fat in women is associated with fertility, obese mice (Zhang et al., 1994) and humans (Montague et al., 1997; Strobel et al., 1998) who suffer from a mutation in the gene that codes for either the adipocyte hormone leptin, or for its receptor, are infertile. This suggests that a circulating signal from fat might be important for both appetite control and for fertility.

The concept of a circulating factor produced by fat that was sensed by the hypothalamus was suggested (Kennedy, 1953) and its existence demonstrated (Hervey, 1958; Hausberger, 1959) long before it was identified and given the name leptin (Zhang et al., 1994). As elegantly shown by parabiosis experiments, the metabolic consequences of genetic mutations that rendered this circulating factor inactive (Hausberger, 1959) were similar to those in animals with mutations that made animals insensitive to the factor (Coleman and Hummel, 1969). These conclusions derived from physiological experiments have been confirmed in recent years in studies showing that there are genetic defects in both the gene for leptin (Zhang et al., 1994) and in the gene for the leptin receptor (Tartaglia et al., 1995; Chen et al., 1996; Chua et al., 1996; Lee et al., 1996).

Soon after the discovery of leptin by the Friedman group (Zhang et al., 1994), it was shown that the deficit of biologically active leptin resulting from a mutation in the *ob* gene

JOAQUIN LADO-ABEAL • Department of Medicine, The University of Chicago Medical Center.
REID L. NORMAN • Department of Pharmacology, Texas Tech University Health Sciences Center—Lubbock.

Leptin and Reproduction. Edited by Henson and Castracane, Kluwer Academic/Plenum Publishers, 2003.

117

was the cause of not only the obesity, but also of the infertility of ob/ob mice (Chehab et al., 1996). Homozygous ob/ob mice remain prepubertal due to a functional hypothalamic defect (Swerdloff et al., 1976; 1978) and leptin administration corrects the infertility in both female and male ob/ob mice (Chehab et al., 1996; Mounzih et al., 1997). These initial observations were followed by a myriad of papers searching for a role of leptin in mammalian reproduction.

An interesting recent finding was that the genetic background determines not only the metabolic phenotype (Ewart-Toland et al., 1999) but also the reproductive phenotype of ob/ob mice. Inbred ob/ob mice with the C57BL/6J background crossed with mice with the BALB/cJ background generated fertile BALB/cJ ob/ob mice (Ewart-Toland et al., 1999; Qui et al., 2001), indicating that modifier genes control leptin action and suggesting that the role of leptin in reproduction could be reduced in highly heterogeneous species.

8.2. FATNESS AND FERTILITY

8.2.1. Fat and Fertile—Native Leptin

Twenty years before the discovery of leptin, the association between energy stored as fat and the attainment of fertility was articulated into a hypothesis by Frisch and McArthur (1974) who reported that there was a strong correlation between body fat and menarche. There is little doubt that much of the interest in the role of leptin in reproduction is due to the popularity of this fatness hypothesis (Frisch and McArthur, 1974). According to this hypothesis, women need a critical body weight and certain percentage of body fat to attain menarche and subsequently, to avoid the secondary amenorrhea associated with low body weight. Because serum leptin levels are directly proportional to body fat mass in individuals on a normal diet (Considine et al., 1996), leptin has been viewed as an adipocytic signal to the brain relaying information regarding nutritional reserves stored as fat.

8.2.2. Fat and Infertile—Inactive Leptin

Ob/ob mice that have a defect in the leptin molecule are obese and infertile and have several metabolic abnormalities that could be related to the decreased fertility. The combination of insulin resistance, uncontrolled diabetes, impaired hypothalamic glucose utilization and high plasma corticosterone levels (Herberg and Coleman, 1977) could all contribute to the infertility of the ob/ob mice. Insulin has an important role in the neuroendocrine control of reproduction, and mice with a neuron-specific disruption of the insulin receptor gene (Bruning et al., 2000) and insulin receptor substrate-2 (Burks et al., 2000) have impaired fertility caused in part, by hypothalamic dysregulation. Ob/ob mice have moderate diabetes (Westman, 1968; Coleman and Hummel, 1973) but they can develop severe hyperglycemia (Halaas et al., 1995; Pelleymounter et al., 1995) that subsequently produces hyperosmolarity, a cause of central hypogonadism (Lado-Abeal and Norman, unpublished observations). Ob/ob mice also show a decrease in glucose utilization in the ventromedial nucleus of hypothalamus (Herberg and Coleman, 1977) and neuroglucopenia is a known inhibitor of GnRH release (Chen et al., 1992).

Both leptin administration and food restriction resulted in weight loss in genetically obese (ob/ob) mice, but only leptin rescued fertility. This is a strong argument that it is the leptin deficiency and not other metabolic abnormalities that cause infertility in ob/ob mice. However, these two experiments have different physiological effects and conclusions should be drawn cautiously. Leptin decreases appetite (Halaas et al., 1995) but food restriction does not, and although weight loss induced by both treatments improves hyperglycemia and insulin resistance, food restricted animals are under considerable stress as shown by activation of the adrenal axis in this situation (Schwartz and Seely, 1997). Increased activity in the hypothalamic–pituitary–adrenal (HPA) axis as evidenced by high circulating cortisol levels can also result in decreased fertility. It has been shown that both intracerebroventricular administration of corticotropin-releasing hormone (CRH; Ferin, 1999) and chronically increased serum glucocorticoid levels (Lado-Abeal et al., 1998) inhibit GnRH release.

8.3. CENTRAL ACTIONS OF LEPTIN

Leptin receptors are expressed in the hypothalamus (Elmquist et al.,1998) but not in GnRH neurons in vivo (Hakansson et al., 1998), suggesting that leptin actions on the neuroendocrine control of reproduction are not directly on GnRH neurons. At the hypothalamic level, leptin interacts with many peptidergic systems involved in energy balance. Neuropeptide Y (NPY) is one of the key peptidergic signals driving the hypothalamic control of reproductive and eating behavior (Kalra and Kalra,1996). NPY neurons express leptin receptors and ob/ob mice with NPY deficiency ($NPY^{-/-}$ ob/ob mice) show an improvement in the metabolic and reproductive abnormalities found in ob/ob mice (Erickson et al., 1996) indicating that NPY is a central effector of leptin deficiency in ob/ob mice. However, mice deficient for NPY ($NPY^{-/-}$ mice) have normal or even increased sensitivity to leptin (Erickson et al., 1997) indicating that NPY is not essential for certain leptin actions.

8.4. LEPTIN AND FERTILITY

8.4.1. Human Leptin Mutations

Two human families with congenital leptin deficiency (Montague et al., 1997; Strobel et al., 1998) and one family with a mutation in the leptin receptor (Clement et al., 1998) have been identified. Homozygous affected members have morbid obesity but no major abnormalities in glucose metabolism or adrenal axis function. The older affected members have central hypogonadism and the only adult male reported has low total and free testosterone, normal LH and FSH, and normal response to GnRH (Ozata et al., 1999). It is interesting that one affected female entered spontaneously into puberty in her mid-thirties maintaining regular menstrual cycles (Ozata et al., 1999). A 9-year-old prepubertal girl treated with recombinant leptin for 12 months presented a gradual increase in basal and stimulated serum LH and FSH and nocturnal pulsatile gonadotropin secretion after 12 months of treatment (Farooqi et al., 1999).

8.4.2. Lessons from Lipodystrophies (Life without Fat)

In contrast to patients with a defect in the leptin signaling pathway, patients with generalized lipoatrophy, a disorder characterized by nearly complete absence of body fat and very low serum leptin levels (Garg, 2000), entered puberty at the expected age or with minimal delay (Pardini et al., 1998). Women with this condition have regular menstrual cycles (Andreelli et al., 2000) or oligomenorrhea due to polycystic ovaries (Garg, 2000), and are able to conceive (Andreelli et al., 2000), whereas men with this disorder have normal reproductive potential (Garg, 2000). The fact that leptin is very low in both types of patients but only patients with *ob* gene mutations develop central hypogonadism suggests that low leptin levels alone are not sufficient to explain the infertility of the patients with *ob* gene mutation.

8.4.3. Non-Primate Studies

In regard to the effects of leptin on gonadotropin secretion, leptin administration totally or partially reverses the inhibitory effects of starvation on gonadotropin and steroid levels in mice (Ahima et al., 1996), and the inhibition of pulsatile LH secretion in both food restricted rats (Nagatami et al., 1998) and castrated male sheep on estrogen replacement (Nagatani et al., 2000). Other effects of leptin administration on reproductive function include reversal of the decrease in ovulation in starved rats (Schneider et al., 1998; Otukonyong et al., 2000) and alleviation of the fasting-induced anestrus in hamsters (Schneider et al., 1998), and the pubertal delay in food restricted mice and rats (Cheung et al., 1997; Gruaz et al., 1998; Cheung et al., 2001).

8.4.4. Non-Human Primate Studies

8.4.4.1. Leptin and Puberty. In rhesus monkeys, a primate model for human reproductive physiology, the role of leptin in reproductive function is less clear. In young male macaques, the GnRH "pulse generator" activity is very sensitive to energy availability and pulsatile LH secretion decreases dramatically after a short period of fasting (Cameron and Nobisch, 1991). Similar results are found in pubertal male macaques where the reactivation of the GnRH "pulse generator" is first observed by reappearance of pulsatile gonadotropin secretion during the evening and nighttime hours (Germak and Knobil, 1990). Within a few hours after missing the afternoon meal, pubertal male macaques show a decrease in serum leptin levels and inhibition of the nocturnal secretion of LH (Figure 8.1; Lado-Abeal et al., 1999). If leptin were an obligatory signal required for the sustained activity of the GnRH "pulse generator," leptin infusion should restore LH pulsatility in food-restricted macaques. However, as shown in Figure 8.1, a supraphysiological dose of leptin administered to a fasted pubertal male macaque was unable to overcome the suppression of the nocturnal pulsatile secretion of LH (Lado-Abeal et al., 1999).

8.4.4.2. Leptin and Adult Reproductive Function. In young adult male macaques, inhibition of LH pulsatility is normally apparent within 48 hr of food restriction (Cameron and Nobisch, 1991; Lado-Abeal et al., 2000). The continuous infusion of recombinant

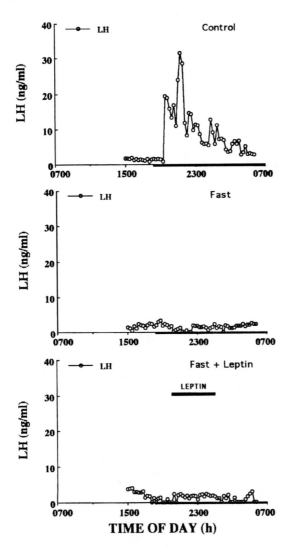

Figure 8.1. Serum LH levels during the afternoon and nighttime hours in a pubertal male macaque. The top panel shows a fed control, the middle panel a fasted animal and the bottom panel a fasted animal that received leptin (100 μg/hr) for 4 hr. The dark bar on the abscissa indicated when the lights were out.

rhesus monkey leptin during a 48 hr period of fasting in adult macaques did not restore the regular pulsatile pattern of LH and testosterone observed in fed animals (Figure 8.2). Likewise, neither the 2-fold increase in mean serum cortisol level nor the increase in frequency of GH pulses caused by fasting was relieved with leptin infusions (Figure 8.2; Lado-Abeal et al., 2000). The results of these studies suggest that changes in circulating leptin levels are not responsible for the adjustments in secretion of reproductive (LH) and metabolic (GH and cortisol) hormones that accompany acute fluctuations in energy intake. Finn et al. (1998) reported different results in peripubertal male macaques where they were

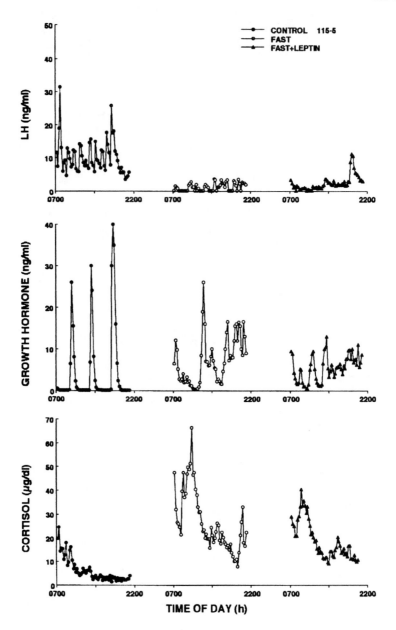

Figure 8.2. Serum LH (top), GH (middle), and cortisol (bottom) concentrations in an adult male rhesus macaque that was fed his normal diet (control), fasted (48 hr), or fasted and infused with recombinant rhesus leptin (100 μg/hr) beginning 11 hr after the first missed meal. Blood samples were collected through an indwelling catheter from a remote site at 15 min intervals for the last 15 hr of the fast. The individual studies were separated by at least a 2-week interval.

able to reverse the inhibitory effect of 48 hr of fasting on nocturnal LH secretion with recombinant human leptin. In Finn's study (1998), serum leptin levels after 2 days of fasting were 4–5 times higher than those reported in non-fasted pubertal and adult rhesus monkeys (Plant and Durrant, 1997; Urbanski and Pau, 1998; Lado-Abeal et al., 1999; Lado-Abeal et al., 2000), suggesting that the state of energy balance in the peripubertal male macaque is different from the adult male macaque (Lado Abeal et al., 2000).

8.5. ENERGY AVAILABILITY, LEPTIN, AND FERTILITY

8.5.1. Energy Availability and the HPA

In both human (Kolaczynski et al., 1996) and non-human primates (Lado-Abeal et al., 1999) and in rodents (Ahima et al., 1996), serum leptin levels change with energy intake and expenditure, reflecting short-term energy balance (Kolaczynski et al., 1996). Since starved rodents have low leptin levels and neuroendocrine abnormalities similar to ob/ob mice, it was proposed that the main physiological role of leptin is to serve as a signal whose absence indicates starvation (Ahima et al., 1996). Therefore, low leptin levels reflect reduced energy intake and during periods of low energy availability there are metabolic adjustments, such as increased cortisol and GH secretion to mobilize stored energy. As discussed above, it is well documented that in many species, GnRH release can be blocked in conditions that activate the HPA axis such as psychological stress. Thus, the combination of both low energy availability as a result of food restriction and the psychological stress of hunger may exert a more powerful inhibitory effect on the GnRH pulse generator than either situation alone.

8.5.2. Energy Availability and Gonadotropin Secretion

The metabolic hypothesis developed to explain the effects of food restriction on reproductive function states that reproduction is influenced by the availability of metabolic fuels for cellular oxidation (Schneider and Wade, 2000). Food deprivation inhibits estrous cyclicity in hamsters, but only after glycogen and fat stores have been depleted (Schneider and Wade, 2000). Pubertal male macaques are more sensitive than are adult males to the inhibitory effect of fasting on pulsatile LH secretion (Finn et al., 1998; Lado-Abeal et al., 2000) most likely because pubertal macaques have less energy reserves to mobilize when energy intake is reduced. We tested the hypothesis that it was the lack of available energy rather than decreased circulating leptin that resulted in suppressed gonadotropin levels in fasted male macaques (Lado-Abeal et al., 2000). Dextrose was infused (20 kCal/hr) into fasted male macaques beginning at the time of the first missed meal. In agreement with the metabolic hypothesis, we found that a continuous intravenous supply of glucose given to male rhesus monkeys fasted for 48 hr prevented the activation of adrenal axis, normalized GH secretion, and partially prevented the inhibition of LH release (Figure 8.3; Lado-Abeal et al., 2002). Previous studies have shown that energy sources that do not raise

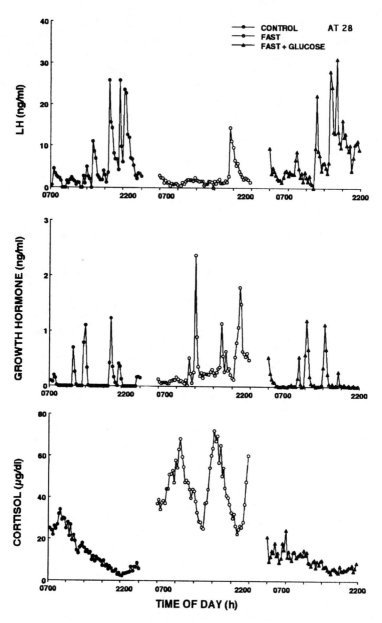

Figure 8.3. Serum LH, GH, and cortisol concentrations in an adult male rhesus macaque that was fed his normal diet (control), fasted (48 hr) or fasted and infused with glucose (20 kCal/hr) beginning at the time of the first missed meal. Blood samples were collected through an indwelling catheter from a remote site at 15 min intervals for the last 15 hr of the fast. The individual studies were separated by at least a 2-week interval.

blood glucose can also reverse the inhibition of pulsatile LH secretion caused by fasting (Schreihofer et al., 1996). Therefore, it appears that the availability of energy is more important for ongoing reproductive function than is the circulating level of leptin that reflects the level of stored energy. Studies in rodents also support the notion that available energy is more important than leptin in regulating reproductive function. Leptin reversed fasting-induced anestrus and anovulation in hamsters only when glucose and free fatty acid (FFA) mobilization were not inhibited (Schneider and Wade, 2000). In severely food-restricted prepubertal rats, leptin administration cannot overcome the hormonal consequences of caloric deprivation (Cheung et al., 2001). Leptin increases the availability and oxidation of glucose in different tissues (Kamohara et al., 1997), and leptin effects on the neuroendocrine control of reproduction could be mediated by this unique mechanism while metabolic fuels are available.

8.6. LEPTIN AND GONADAL STEROIDS

8.6.1. Does Leptin Affect Testosterone Secretion?

Severely obese men have a decrease in total and free plasma testosterone, impaired testicular response to human chorionic gonadotropin (hCG), and low serum LH with lower LH pulse amplitude suggesting a functional testicular defect and impairment of LH secretion (Bray, 1997; Isidori et al., 1999; Lima et al., 2000). Since serum leptin levels are elevated in morbid obesity and leptin receptors are expressed in the testes including Sertoli and Leydig cells (Zamorano et al., 1997; Caprio et al., 1999), it was hypothesized that leptin could decrease testosterone release. In vitro testosterone secretion stimulated by hCG in Leydig cells (Caprio et al., 1999) or in testicular slices (Tena-Sempere et al., 1999; 2001) was inhibited by leptin in experiments with incubation periods shorter than 180 min but not when Leydig cells were exposed to leptin for 24 hr (Lado-Abeal et al., 1999). Leptin was able to partially suppress the hCG-stimulated expression of SF-1, StAR, and $P450_{scc}$ enzyme transcript levels, which could explain the inhibition of Leydig cell steroidogenesis (Tena-Sempere et al., 2001). The question of whether leptin has any physiological effect on in vivo androgen secretion at the level of the testis has not been adequately addressed, but existing data suggest that the primary effects of leptin on reproductive potential appear to be at the hypothalamic level.

8.6.2. Does Testosterone Affect Leptin Levels?

Circulating leptin levels are gender dependent with women having higher leptin concentrations than men (Hickey et al., 1996). Females have a higher percentage of body fat and a higher proportion of subcutaneous fat versus visceral fat than males, both of which influence serum leptin levels. However, after correction for these variables, sex differences in leptin levels still persist (Rosenbaum and Leibel, 1999), suggesting that they are not solely dependent on the anatomic distribution of fat. Several studies indicate that testosterone may be responsible for gender differences in leptin levels. Cross-sectional (Garcia-Mayor et al., 1997) and longitudinal (Mantzoros et al., 1997) studies suggest that testosterone inhibits circulating leptin levels. Leptin levels decrease during male pubertal development and suppression of testosterone secretion in children with central precocious puberty was associated with increased leptin levels whereas resumption of puberty resulted in decreased leptin levels (Palmert et al., 1998). Furthermore, androgen replacement therapy and anabolic androgenic steroid use, both, reduced leptin levels (Wauters et al., 2000). However, changes in body composition caused by the circumstances described above could also explain the changes in leptin secretion. Unfortunately, there are also conflicting results from in vitro studies; testosterone and dihydrotestosterone decreased leptin secretion from subcutaneous adipocytes in females (Wabitsch et al., 1997) but had no effect on leptin secretion from omental adipocytes in males (Pineiro et al., 1999).

8.7. SUMMARY

Although it is possible to demonstrate that leptin can influence gonadotropin secretion and fertility in rodents and primates, it has been difficult to separate the effects of low leptin from some of the other consequences of low energy availability. Except where there is a mutation in the gene that codes for leptin that renders the protein inactive, or a mutation in the leptin receptor that results in the inability to detect leptin, there is little evidence that leptin influences fertility in humans. It is more likely that among the physiological adjustments that occur in response to reduced energy intake, such as increased GH secretion to mobilize fat and elevated cortisol secretion to maintain circulating glucose levels, gonadotropin secretion is suppressed to conserve energy.

REFERENCES

Ahima, R. S., Prabakaran, D., Mantzoros, C., Qu, D., Lowell, B., Marotos-Flier, E., and Flier, J. S. (1996). Role of leptin in the neuroendocrine response to fasting. *Nature, 382*, 250–252.

Andreelli, F., Hanaire-Broutin, H., Laville, M., Tauber, J. P., Riou, J. P., and Thivolet, C. (2000). Normal reproductive function in leptin-deficient patients with lipoatropic diabetes. *J. Clin. Endocrinol. Metab., 85*, 715–719.

Bray, G. A. (1997). Obesity and reproduction. *Hum. Repro., 12*, 26–32.

Bruning, J. C., Gautan, D., Burks, D. J., Gillette, J., Schubert, M., Orban, P. C., Klein, R., Drone, W., Muller-Weiland, D., and Kahn, C. R. (2000). Role of brain insulin receptor in control of body weight and reproduction. *Science, 289*, 2122–2125.

Burks, D. J., Font de Mora, J., Schubert, M., Withers, D. J., Myers, M. G., Towery, H. H., Altamuro, S. L., Flint, C. L., and White, M. F. (2000). IRS-2 pathways integrate female reproduction and energy homeostasis. *Nature, 407*, 377–382.

Cameron, J. L. and Nosbisch, C. (1991). Suppression of pulsatile luteinizing hormone and testosterone secretion during short-term food restriction in adult male rhesus monkey (Macaca mulatta). *Endocrinology, 128*, 1532–1540.

Caprio, M., Isidori, A. M., Carta, A. R., Moretti, C., Dufau, M. L., and Fabbri, A. (1999). Expression of functional leptin receptors in rodent Leydig cells. *Endocrinology, 140*, 4939–4947.

Chehab, F. F., Lim, M. E., and Lu, R. (1996). Correction of the sterility defect in homozygous obese female mice by treatment with the human recombinant leptin. *Nat. Gen., 12*, 318–320.

Chen, M. D., O'Byrne, K. T., Chiappini, S. E., Hotchkiss, J., and Knobil, E. (1992). Hypoglycemic stress and gonadotropin-releasing hormone pulse generator activity in the rhesus monkey: role of the ovary. *Neuroendocrinology, 56*, 666–673.

Chen, H., Charlat, O., Tartaglia, L. A., Woolf, E. A., Weng, X., Ellis, S., Lakey, N. D., Culpepper, J., Moore, K. J., Brietbart, R. E., Duyk, G. M., Tepper, R. I., and Morgenstern, J. P. (1996). Evidence that the diabetes gene encodes the leptin receptor: identification of a mutation in the leptin receptor gene in db/db mice. *Cell, 84*, 491–495.

Cheung, C. C., Thornton, J. E., Kuijper, J. L., Weigle, D. S., Clifton, D. K., and Steiner, R. A. (1997). Leptin is a metabolic gate for the onset of puberty in the female rat. *Endocrinology, 138*, 855–858.

Cheung, C. C., Thornton, J. E., Nurani, S. D., Clifton, D. K., and Steiner, R. A. (2001). A reassessment of leptin's role in triggering the onset of puberty in the rat and mouse. *Neuroendocrinology, 74*, 12–21.

Chua, S. C., Chung, W. K., Wu-Peng, X. S., Zhang, Y., Lui, S. M., Tartaglia, L., and Leibel, R. L. (1996). Phenotypes of mouse diabetes and rat fatty due to mutation in the Ob (leptin) receptor. *Science, 271*, 994–996.

Clement, K., Vaisse, C., Lahlous, N., Cabro, S., Pelloux, V., Cassuto, D., Gourmelen, M., Dina, C., Chambaz, J., Lacorte, J. M., Basdevent, A., Bougneres, P., Lebouc, Y., Froguel, P., and Guy-Grand, B. (1998). A mutation in the human leptin receptor gene causes obesity and pituitary dysfunction. *Nature, 392*, 398–401.

Coleman, D. L. and Hummel, K. P. (1969). Effects of parabiosis of normal with genetically diabetic mice. *Am. J. Physiol., 217*, 1298–1304.

Coleman, D. L. and Hummel, K. P. (1973). The influence of genetic background on the expression of the obese (ob) gene in the mouse. *Diabetologia, 9*, 287–293.

Considine, R. V., Sinha, M. K., Heiman, M. L., Kriauciunas, A., Stephens, T. W., Nyce, M. R., Ohannesian, J. P., Marco, C. C., McKee, L. J., Bauer, T. L., and Caro, F. J. (1996). Serum-immunoreactive leptin concentrations in normal-weight and obese humans. *N. Engl. J. Med., 334*, 292–295.

Elmquist, J. K., Bjorbaek, C. J., Ahima, R. S., Flier, J. S., and Saper, C. B. (1998). Distributions of leptin receptor mRNA isoforms in the rat brain. *J. Comp. Neurol., 395*, 535–547.

Erickson, J. C., Hollopeter, G., and Palmiter, R. D. (1996). Attenuation of the obesity syndrome of ob/ob mice by loss of neuropeptide Y. *Science, 274*, 1704–1707.

Erickson, J. C., Ahima, R. S., Hollopeter, G., Flier, J. S., and Palmiter, R. D. (1997). Endocrine function of neuropeptide Y knockout mice. *Regul. Pept., 70*, 199–202.

Ewart-Toland, A., Mounzih, K., Qiu, J., and Chehab, F. F. (1999). Effect of the genetic background on the reproduction of leptin-deficient obese mice. *Endocrinology, 140*, 732–735.

Farooqi, I. S., Jebb, S. A., Langmack, G., Lawrence, E., Cheetham, C. H., Prentice, A. M., Hughes, I. A., McCamish, M. A., and O'Rahilly, S. (1999). Effects of recombinant leptin therapy in a child with congenital leptin deficiency. *N. Engl. J. Med., 341*, 879–884.

Ferin, M. (1999). Stress and the reproductive cycle. *J. Clin. Endocrinol. Metab., 84*, 1768–1774.

Finn, P. D., Cunningham, M. J., Pau, K.-Y. F., Spies, H. G., Clifton, D. K., and Steiner, R. A. (1998). The stimulatory effect of leptin on the neuroendocrine reproductive axis of the monkey. *Endocrinology, 139*, 4652–4662.

Frisch, R. E. and McArthur, J. W. (1974). Menstrual cycles: fatness as a determinant of minimum weight for height necessary for their maintenance or onset. *Science, 185*, 949–951.

Garcia Mayor, R. V., Andrade, M. A., Rios, M., Lage, M., Dieguez, C., and Casanueva, F. F. (1997). Serum leptin levels in normal children: relationship to age, gender, body mass index, pituitary–gonadal hormones, and pubertal stage. *J. Clin. Endocrinol. Metab., 82*, 2849–2855.

Garg, A. (2000). Lipodystrophies. *Am. J. Med., 108*, 143–152.

Germak, J. A. and Knobil, E. (1990). Control of puberty in the rhesus monkey. In M. M. Grumbach, P. G. Sizonenko, and M. L. Aubert (Eds.). *Control of the Onset of Puberty*, Williams & Wilkins, Baltimore, pp. 69–81.

Gruaz, M. M., Lalaoui, M., Pierroz, D. D., Englaro, P., Sizonenko, P. C., Blum, W. F., and Aubert, M. L. (1998). Chronic administration of leptin into the lateral ventricle induces sexual maturation in severely food-restricted rats. *J. Neuroendocrinol., 10*, 627–633.

Hakansson, M.-L., Brown, H., Ghilardi, N., Skoda, R. C., and Meister, B. (1998). Leptin receptor immunoreactivity in chemically defined target neurons of the hypothalamus. *J. Neurosci., 18*, 559–572.

Halaas, J. L., Gajiwala, K. S., Maffei, M., Cohen, S. L., Chait, T. T., Rabinowitz, D., Lallone, R. L., Burley, S. K., and Friedman, J. M. (1995). Weight-reducing effects of the plasma protein encoded by the obese gene. *Science, 269*, 543–546.

Hausberger, F. X. (1959). Parabiosis and transplantation experiments in hereditarily obese mice. *Anat. Rec., 130*, 313.

Herberg, L. and Coleman, D. L. (1977). Laboratory animals exhibiting obesity and diabetes syndromes. *Metabolism, 26*, 59–99.

Hervey, G. R. (1958). The effects of lesions in the hypothalamus in parabiotic rats. *J. Physiol., 145*, 336–352.

Hickey, M. S., Israel, R. G., Gardiner, S. N., Considine, R. V., McCammon, M. R., Tyndall, G. L., Houmard, J. A., Marks, R. H., and Caro, J. F. (1996). Gender differences in serum leptin levels in humans. *Biochem. Mol. Med., 59*, 1–6.

Isidori, A. M., Caprio, M., Strollo, F., Moretti, C., Frajese, G., Isidori, A., and Fabbri, A. (1999). Leptin and androgens in male obesity: evidence for leptin contribution to reduced androgen levels. *J. Clin. Endocrinol. Metab., 84*, 3673–3680.

Kalra, S. P. and Kalra, P. S. (1996). Nutritional infertility: the role of the interconnected hypothalamic neuropeptide Y-galanin-opioid network. *Fron. Neuroendo., 17*, 371–401.

Kamohara, S., Burcelin, R., Halaas, J. L., Friedman, J. M., and Charron, M. J. (1997). Acute stimulation of glucose metabolism in mice by leptin treatment. *Nature, 389*, 374–377.

Kennedy, G. C. (1953). The role of depot fat in the hypothalamic control of food intake in the rat. *Proc. Roy. Soc. B, 140*, 578–592.

Kolaczynski, J. W., Considine, R. V., Ohannesian, J., Marco, C., Opentanova, I., Nyce, M. R., Myint, M., and Caro, J. F. (1996). Responses of leptin to short-term fasting and refeeding in humans: a link with ketogenesis but not ketones themselves. *Diabetes, 45*, 1511–1515.

Lado-Abeal, J., Rodriguez Arnao, J., Newell-Price, J. D., Perry, L. A., Grossman, A. B., and Besser, G. M. (1998). Menstrual abnormalities in women with Cushing's disease are correlated with hypercortisolemia rather than raised circulating androgen levels. *J. Clin. Endocrinol. Metab., 83*, 3083–3088.

Lado-Abeal, J., Lukyanenko, Y. O., Swamy, S., Hermida, R. C., Hutson, J. C., and Norman, R. L. (1999). Short-term leptin infusion does not affect circulating levels of LH, testosterone or cortisol in food-restricted pubertal male rhesus macaques. *Clin. Endocrinol., 51*, 41–51.

Lado-Abeal, J., Hickox, J. R., Cheung, T. L., Veldhuis, J. D., Hardy, D. M., and Norman, R. L. (2000). Neuroendocrine consequences of fasting in adult male macaques: effects of recombinant rhesus macaque leptin infusion. *Neuroendocrinology, 71*, 196–208.

Lado-Abeal, J., Veldhuis, J. D., and Norman, R. L. (2002). Glucose relays information regarding nutritional status to the neural circuits that control the somatotropic, corticotropic, and gonadotropic axes in adult male rhesus macaques. *Endocrinology, 143*, 403–410.

Lee, G. H., Proenca, R., Montez, J. M., Carroll, K. M., Darvishzadeh, J. G., Lee, J. I., and Friedman, J. M. (1996). Abnormal splicing of the leptin receptor in diabetic mice. *Nature, 379*, 632–635.

Lima, N., Cavaliere, H., Knobel, M., Halpern, A., and Medeiros-Neto, G. (2000). Decreased androgen levels in massively obese men may be associated with impaired function of the gonadostat. *Int. J. Obes. Relat. Metab. Disord., 24*, 1433–1437.

Mantzoros, C. S., Flier, J. S., and Rogol, A. D. (1997). A longitudinal assessment of hormonal and physical alterations during normal puberty in boys. V. Rising leptin levels may signal the onset of puberty. *J. Clin. Endocrinol. Metab., 82*, 1066–1070.

Montague, C. T., Farooqi, I. S., Whitehead, J. P., Soos, M. A., Rau, H., Wareham, N. J., Sewter, C. P., Digby, J. E., Mohanned, S. N., Hurst, J. A., Cheetham, C. H., Early, A. R., Barnett, A. H., Prins, J. B., and O'Rahilly, S. (1997). Congenital leptin deficiency is associated with severe early-onset obesity in humans. *Nature, 387*, 903–908.

Mounzih, K., Lu, R., and Chehab, F. F. (1997). Leptin treatment rescues the sterility of genetically obese ob/ob males. *Endocrinology, 138*, 1190–1193.

Nagatami, S., Guthikonda, P., Thompson, R. C., Tsukamura, H., Maeda, K-I., and Foster, D. L. (1998). Evidence for GnRH regulation by leptin: leptin administration prevents reduced pulsatile LH secretion during fasting. *Neuroendocrinology, 67*, 370–376.

Nagatani, S., Zeng, Y., Keisler, D. H., Foster, D. L., and Jaffe, G. A. (2000). Leptin regulates pulsatile luteinizing hormone and growth hormone secretion in the sheep. *Endocrinology, 141*, 3965–3975.

Otukonyong, E. E., Okutani, F., Takahashi, S., Murata, T., Morioka, N., Kaba, H., and Higuchi, T. (2000). Effect of food deprivation and leptin repletion on plasma levels of estrogen (E_2) and NADPH-d reactivity in the ventromedial and arcuate nuclei of the hypothalamus in the female rats. *Brain Res., 887*, 70–79.

Ozata, M., Ozdemir, C., and Licinio, J. (1999). Human leptin deficiency caused by a missense mutation: multiple endocrine defects, decreased sympathetic tone, and immune system dysfunction indicate new targets for leptin action, greater central than peripheral resistance to the effects of leptin, and spontaneous correction of leptin-mediated defects. *J. Clin. Endocrinol. Metab., 84*, 3686–3695.

Palmert, M. R., Radovick, S., and Boepple, P. A. (1998). The impact of reversible gonadal sex steroid suppression on serum leptin concentrations in children with central precocious puberty. *J. Clin. Endocrinol. Metab., 83*, 1091–1096.

Pardini, V. C., Victoria, I. M. N., Rocha, S. M. V., Andrade, D. G., Rocha, A. M., Pierone, F. B., Milagres, G., Purisch, S., and Velho, G. (1998). Leptin levels, β-cell function, and insulin sensitivity in families with congenital and acquired generalized lipoatropic diabetes. *J. Clin. Endocrinol. Metab., 83*, 503–508.

Pelleymounter, M. A., Cullen, M. J., Baker, M. B., Hecht, R., Winters, D., Boon, T., and Collins, F. (1995). Effects of the obese gene product on body weight regulation on ob/ob mice. *Science, 269*, 540–543.

Pineiro, V., Casabiell, X., Peino, R., Lage, M., Camina, J. P., Menendez, C., Baltar, J., Dieguez, C., and Casanueva, F. (1999). Dihydrotestosterone, stanazol, androstenedione and dehydroepiandrosterone sulphate inhibit leptin secretion in female but not in male samples of omental adipose tissue in vitro: lack of effect of testosterone. *J. Endocrinol., 160*, 425–432.

Plant, T. M. and Durrant, A. R. (1997). Circulating leptin does not appear to provide a signal for triggering the initiation of puberty in the male rhesus monkey (Macaca mulatta). *Endocrinology, 138*, 4505–4508.

Qiu, J., Ogus, K., Mounzih, K., Ewart-Toland, A., and Chehab, F. F. (2001). Leptin-deficient mice backcrossed to the BALB/cj genetic background have reduced adiposity, enhanced fertility, normal body temperature, and severe diabetes. *Endocrinology, 14*, 3421–3425.

Rosenbaum, M. and Leibel, R. L. (1999). Role of gonadal steroids in the sexual dimorphisms in body composition and circulating concentrations of leptin. *J. Clin. Endocrinol. Metab., 84*, 1784–1789.

Schneider, J. E. and Wade, G. N. (2000). Inhibition of reproduction in service of energy balance. In K. Wallen, and J. E. Schneider (Eds.). *Reproduction in Context: Social and Environmental Influences on Reproduction*, MIT Press, Cambridge, pp. 35–82.

Schneider, J. E., Goldman, M. D., Tang, S., Bean, B., Ji, H., and Friedman, M. I. (1998). Leptin indirectly affects estrous cycles by increasing metabolic fuel oxidation. *Horm. Behav., 33*, 217–228.

Schreihofer, D. A., Renda, F., and Cameron, J. L. (1996). Feeding-induced stimulation of luteinizing hormone secretion in male rhesus monkeys is not dependent on a rise in blood glucose concentration. *Endocrinology, 137*, 3770–3776.

Schwartz, M. W. and Seely, R. J. (1997). Neuroendocrine responses to starvation and weight loss. *N. Engl. J. Med., 336*, 1802–1811.

Strobel, A., Issad, T., Camoin, L., Ozata, M., and Strosberg, A. D. (1998). A leptin missense mutation associated with severe early-onset obesity in humans. *Nat. Gen., 18*, 903–908.

Swerdloff, R. S., Batt, R., and Bray, G. (1976). Reproductive hormonal function in the genetically obese (ob/ob) mouse. *Endocrinology, 98*, 1359–64.

Swerdloff, R. S., Peterson, M., Vera, A., Batt, R. A., Heber, D., and Bray, G. (1978). The hypothalamic–pituitary axis in genetically obese (ob/ob) mice: response to luteinizing hormone-releasing hormone. *Endocrinology, 103*, 542–547.

Tartaglia, L. A., Dembski, M., Weng, X., Deng, N., Culpepper, J., Devos, R., Richards, G. J., Campfile, L. A., Clark, F. T., Deeds, J., Muri, C., Sanker, S., Moriarty, A., Moore, K. J., Smutko, J. S., Mays, G. G., Woolf, E. A., Monroe, C. A., and Tepper, R. I. (1995). Identification and expression cloning of a leptin receptor, OB-R. *Cell, 83*, 1263–1271.

Tena-Sempere, M., Pinilla, L., Gonzalez, L. C., Dieguez, C., Casaneuva, F. F., and Aguilar, E. (1999). Leptin inhibits testosterone secretion from adult rat testis in vitro. *J. Endocrinol., 161*, 211–218.

Tena-Sempere, M., Manna, P. R., Zhang, F.-P., Pinilla, L., Gonzalez, L. C., Dieguez, C., Huhtaniemi, I., and Aguilar, E. (2001). Molecular mechanism of leptin action in adult rat testis: potential targets for leptin-induced inhibition of steroidogenesis and pattern of leptin receptor messenger ribonucleic acid expression. *J. Endocrinol., 170*, 413–423.

Urbanski, H. F. and Pau, K.-Y. F. (1998). A biphasic developmental pattern of circulating leptin in the male rhesus macaque (*Macaca mulatta*). *Endocrinology, 139*, 2284–2286.

Wabitsch, M., Blum, W. F., Muche, R., Braun, M., Hube, F., Rascher, W., Heinze, E., Teller, W., and Hauner, H. (1997). Contribution of androgens to the gender difference in leptin production in obese children and adolescents. *J. Clin. Invest., 100*, 808–813.

Wauters, M., Considine, R. V., and Van Gaal, L. (2000). Human leptin: from an adipocyte hormone to an endocrine mediator. *Eur. J. Endocrinol., 143*, 293–311.

Westman, S. (1968). Development of the obese-hyperglycaemic syndrome in mice. *Diabetologia, 4*, 141–149.

Zamorano, P. L., Mahesh, V. B., De Sevilla, L., Chorich, L., Bhat, G. K., and Brann, D. W. (1997). Expression and localization of the leptin receptor in endocrine and neuroendocrine tissues of the rat. *Neuroendocrinology, 65*, 223–228.

Zhang, Y., Proenca, R., Maffei, M., Barone, M., Leopold, L., and Friedman, J. M. (1994). Positional cloning of the mouse obese gene and its human homologue. *Nature, 372*, 425–432.

III

Puberty

Leptin and Pubertal Development in Higher Primates

DAVID R. MANN AND TONY M. PLANT

9.1. INTRODUCTION

According to Romer's classification (Romer, 1959), higher primates are comprised of three families; namely Cercopithecidae (Old World monkeys), Simiidae (man-like apes), and Hominidae (human). Although puberty in these mammalian families, as in non-primate species, is defined as the period of becoming first capable of reproducing sexually, the mechanism that times the onset of this developmental milestone in monkeys, apes, and man appears to be unique to higher primates (Plant, 1994). The distinguishing feature of puberty in higher primates, as exemplified in our own species, is that despite a competent hypothalamic–pituitary–gonadal axis during the greater part of prepubertal development, the initiation of robust sex steroid production in association with gametogenesis is delayed for years. The delay of puberty in monkeys and man is occasioned at the hypothalamic level by a prolonged reduction in the activity of the neuronal network governing gonadotropin-releasing hormone (GnRH) release that extends from infancy until termination of the juvenile phase of development. Puberty in primates thus comprises the subsequent re-augmentation of pulsatile GnRH release and the profound sequelae set in motion by this neurobiological event (Plant, 1994).

Cardinal questions limiting our understanding of primate puberty are: (1) what are the signals that lead, postnatally, first to a decrease and then to a subsequent re-augmentation in

DAVID R. MANN • Department of Physiology and the Cooperative Reproductive Science Research Center, Morehouse School of Medicine.

TONY M. PLANT • Department of Cell Biology and Physiology and the Specialized Cooperative Center Program of Reproduction Research, University of Pittsburgh Medical School.

Leptin and Reproduction. Edited by Henson and Castracane, Kluwer Academic/Plenum Publishers, 2003.

the activity of the GnRH neuronal network at the time of infancy and puberty, respectively; and (2) what is the nature of the physiological control system that dictates the timing of the arrival of these signals at the level of the hypothalamus. In this regard, it is generally recognized that growth preceding sexual maturity is required for reproductive competence (Bullough, 1951) and the notion that attainment of a particular body size or composition may provide the trigger that initiates the onset of puberty has attracted considerable attention. In such a model, the hypothalamus would track growth by monitoring a circulating metabolic or hormonal indicator of this developmental process. Accordingly, the attainment of an adult soma would be registered by the growth-tracking device, the so-called somatometer (Plant et al., 1989a), which in turn would relay this information to the hypothalamic GnRH pulse generator. If a circulating factor of somatic origin is to serve as a positive trigger to time the pubertal re-augmentation of pulsatile GnRH release, two scenarios must be satisfied. First, an increase in the concentration of the candidate substance must be temporally associated with the developmental change in hypothalamic function. Second, an experimental or pathophysiological imposition of a premature increment in the circulating factor should be associated with precocious re-augmentation of pulsatile GnRH release. It is important to note that while a circulating factor may not serve as a cue timing the pubertal increase in GnRH release, this scenario does not necessarily preclude the possibility that the same factor may be nevertheless obligatory, in a permissive sense, for this developmental event to become manifest once the timing mechanism has been activated. One set of such permissive factors, for example, is that which mediates nutritional influences on GnRH release. Weight loss and undernutrition in the adult primate leads to an arrest of the GnRH pulse generator, while at the end of juvenile development, this perturbation prevents or compromises initiation of the pubertal re-augmentation of pulsatile GnRH release (Warren, 1990; Cameron, 1996). In both instances, although the influence of nutrition on GnRH secretion is obligatory, it is nonetheless permissive. Thus, conceptually, the notion of a permissive role of a circulating factor in regulating GnRH release must be separated from a discussion of the events that actively time the pubertal re-augmentation of GnRH release and therefore the onset of primate puberty.

With the discovery of the *ob/ob* gene in 1994, leptin, the subject of this book, has emerged as a candidate for the circulating signal responsible for timing the onset of puberty, and, in this chapter, we will examine the role of leptin in the regulation of this critical developmental event in higher primates.

9.2. DEVELOPMENT OF THE HYPOTHALAMIC–PITUITARY–GONADAL AXIS

The pituitary–gonadal axis in both male and female primates is relatively quiescent during juvenile development and this underlies the characteristic sexually immature phenotype of this stage of development (Plant, 1994). Nevertheless, shortly after the transition into the juvenile phase of primate development, the reproductive axis acquires the potential to support adult patterns of gametogenesis and steroidogenesis. This is manifest in our own species by the pathophysiology of central precocious puberty in both boys and girls, and may be demonstrated experimentally in nonhuman primates by providing the pituitary–gonadal axis of juvenile monkeys with a premature intermittent GnRH stimulation of either endogenous or exogenous origin (Plant et al., 1989b; Grumbach and Styne, 1992; Plant, 1994). Thus, the pituitary–gonadal axis of the juvenile and adult primate must be viewed as a slave to the GnRH pulse generator.

The hypothalamic GnRH pulse generator in primates is comprised of several hundred diffusely distributed GnRH neurons and associated neural and glial inputs (Plant, 2000). The neurobiological mechanisms that synchronize the activity of these widely dispersed neurons are incompletely understood, but the increasing application of contemporary molecular biological techniques in conjunction with electrophysiological approaches (Herbison et al., 2001) should help to resolve this fascinating problem. In primates, as in rats, GnRH neurons are born in the olfactory placode and migrate to the fetal hypothalamus early in embryonic life (Ronnekleiv and Resko, 1990). The fundamental organization of the GnRH pulse generator appears to be completed by mid-gestation as reflected by the ability of fetal GnRH neurons to restore ovarian cyclicity in adult monkeys with hypothalamic lesions (Saitoh et al., 1995), by an intermittent pattern of gonadotropin release from the fetal pituitary (Jaffe, 1989), and by an established negative feedback loop between the fetal pituitary and gonad (Resko and Ellinwood, 1985). During later fetal life, pulsatile GnRH release is probably suppressed by rising levels of feto-placental steroids, but following the withdrawal of this steroid inhibition concomitantly with parturition, the GnRH pulse generator enters a phase of robust activity (Plant, 1994). In the infantile male this leads to a similarly robust pattern of luteinizing hormone (LH) and follicle-stimulating hormone (FSH) release, and to an adult-like pattern of testosterone secretion from the testis at this early stage of development. The frequency of pulsatile GnRH release in the infantile female monkey is less than that in the male, and this sex difference in GnRH pulse generator activity is responsible for the relatively high FSH:LH ratio observed during this phase of female development. At the end of the infantile period, the GnRH pulse generator is brought into check and the protracted juvenile phase of relative inactivity of the pituitary–gonadal axis is established (Plant, 1994).

Studies of the male rhesus monkey have provided evidence for the view that neuropeptide Y (NPY) is an integral component of the neurobiological brake which holds pulsatile GnRH in check for the greater part of prepuberal development (El Majdoubi et al., 2000). The biphasic pattern of pulsatile GnRH release from birth to puberty is inversely correlated with the expression of the gene encoding NPY in the mediobasal hypothalamus (El Majdoubi et al., 2000). Moreover, injection of NPY into the cerebroventricle system of the postpubertal monkey inhibits GnRH release (Kaynard et al., 1990, Pau et al., 1995; Plant, 2001a). Although gamma aminobutyric acid and transforming growth factor alpha have also been implicated in the timing of the onset of primate puberty (Ojeda et al., 2000; Terasawa and Fernandez, 2001), the role of NPY is of particular interest in the present context because NPY neurons in the primate hypothalamus are established targets for leptin action (Finn et al., 1998).

9.3. LEPTIN RECEPTORS AND CIRCULATING FORMS OF LEPTIN

Leptin receptors are not only present in the brain, but have been identified in a wide variety of tissues including liver, kidney, heart, adipocytes, lung, spleen, intestine, and gonads (Lee et al., 1996). The leptin receptor is a member of the class I cytokine family of receptors and apparently exists in a number of alternatively spliced forms (OB-Ra, OB-Rb, OB-Rc,

OB-Rd, and OB-Re) (Lee et al., 1996). Only the OB-Rb isoform (the full length form of the receptor) is apparently involved in signal transduction (Fruhbeck et al., 1998). The truncated OB-Re isoform of the receptor that corresponds to the receptor's extracellular domain is secreted into the blood in sufficient amounts so that it likely functions as a leptin buffer (Lollmann et al., 1997). Since leptin-binding activity in human blood co-elutes during size exclusion chromatography with this soluble form of the leptin receptor, it appears that OB-Re isoform is the major circulating binding protein for leptin in humans (Quinton et al., 1999; Lammert et al., 2001). It is likely that the relationship between free and bound leptin is not static but dynamic, and that bioavailability of leptin is regulated by changes in the balance of free to bound hormone as it is for other hormones. Related to this issue, recent data suggest that growth hormone/insulin-like growth factor-1 (GH/IGF1)-induced changes in body composition in GH deficient patients may be effected by changes in the balance of free/bound leptin in the circulation (Randeva et al., 2002). In a preliminary study from our laboratory, circulating leptin and leptin receptor concentrations were inversely correlated in girls and boys (Mann and Castracane, 2001; Figures 9.1 and 9.2), and leptin binding activity

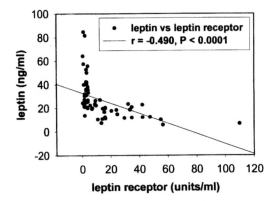

Figure 9.1. Correlation between serum leptin and leptin receptor concentrations in girls between 0 and 20 years of age. r and P values are shown.

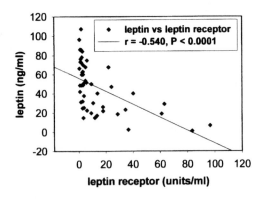

Figure 9.2. Correlation between serum leptin and leptin receptor concentrations in boys between 0 and 20 years of age. r and P values are shown.

and circulating leptin receptor levels decline with age and pubertal stage in children (Quinton et al., 1999; Kratzch et al., 2001; Mann and Castracane, 2001). Accordingly, bioavailable leptin in children may be increasing more rapidly during development than changes in total concentrations of the hormone in the circulation might suggest (see Section 9.5).

9.4. POTENTIAL SITES OF ACTION OF LEPTIN WITHIN THE HYPOTHALAMIC–PITUITARY–GONADAL AXIS

Although leptin receptors and their mRNA are found within the hypothalamus, pituitary, and gonad of rodents, monkeys, and humans (see Cunningham et al., 1999), in the context of the question of pubertal triggers those expressed in the hypothalamus are of greatest interest. A saturable system exists for the transport of leptin across the blood–brain barrier (Banks et al., 1996) and, as described above, the primate GnRH pulse generator is located within the MBH. Leptin receptor expression in this hypothalamic area has been demonstrated within the arcuate and ventromedial nuclei (Huang et al., 1996; Schwartz et al., 1996; Finn et al., 1998). To date, however, evidence for an action of leptin on GnRH release has been obtained largely from studies of rodents. Treatment of hypothalamic explants from adult female, but not male rats, with leptin increased GnRH pulse amplitude without influencing pulse frequency (Parent et al., 2000), and administration of a bolus of leptin (0.01–1.0 µg) directly into 3rd cerebroventricle of peripubertal female rats stimulated LH secretion (Dearth et al., 2000). It appears that any effects of leptin on GnRH neurons are indirect because these neurons do not appear to express leptin receptors (Finn et al., 1998; Hakansson et al., 1998). Leptin receptors within the hypothalamus of both rodents and primates, however, are co-localized with pro-opiomelanocortin and NPY neurons (Cheung et al., 1997; Finn et al., 1998; Hakansson et al., 1998): two neuronal systems that have been shown to be involved in the regulation of GnRH secretion. The latter observation is of particular interest because, as described above, developmental changes in NPY gene expression in the MBH have been argued to play a primary role in pubertal development in the male monkey. While the role of NPY in pubertal development of the female monkey has not been investigated, data from the male strongly suggest that one of the likely potential sites for leptin effects on pubertal development (whether it be as a permissive modulator or causal trigger) is the NPY neuronal system.

Leptin receptor expression also has been documented within the gonads of rodents (Caprio et al., 1999) and primates (Karlsson et al., 1997), and expression of isoforms of the leptin receptor in the rat testis has been shown to change during development (Tena-Sempere et al., 2001). Furthermore, receptor expression was regulated by leptin and gonadotropin. In the human ovary both granulosa and theca cells express the leptin receptor (Karlsson et al., 1997). Culturing of human granulosa cells with leptin (100 ng/ml) attenuated estradiol production in response to LH (Karlsson et al., 1997). It should be noted that in contrast to the brain that has a saturable transport system for leptin, such a system does not appear to exist for transporting leptin across the blood–testis barrier (Banks et al., 1999). Only small amounts of blood leptin were able to enter the testis in this study. In a number of these studies that have demonstrated a direct effect of leptin on gonadal function (usually inhibitory) the concentrations of leptin utilized were quite high and may have exceeded those achieved under normal physiological conditions. The direct effects of leptin on the gonad may be more important in conditions such as severe obesity in which reproductive function is often impaired. For example, leptin levels were very high (38 ng/ml) in adult male humans with severe obesity (mean body mass index [BMI] = 46 kg/m^2) compared to control values of 6 ng/ml for males with a mean BMI of 24.5 kg/m^2 stimulation (Isidori et al., 1999). There was a negative correlation between leptin levels in the circulation and the total and free testosterone response to LH. The testicular steroidogenic hormone pattern from these men with severe obesity was consistent with a defect of 17,20-lyase activity. Isidori et al. (1999) proposed that during severe obesity, excess leptin might be a major contributor to subnormal androgen concentrations in these patients.

Leptin receptors are also present in the anterior pituitary of the rat (Zamorano et al., 1997; Sone et al., 2001) and monkey (Finn et al., 1998). Leptin has been reported to increase both basal and GnRH-stimulated gonadotropin secretion from cultured rat pituitary cells (Yu et al., 1997; Ogura et al., 2001). The significance, if any, of a direct pituitary effect of leptin on gonadotropin secretion during pubertal development in primates remains to be determined.

9.5. PATTERNS OF CIRCULATING LEPTIN DURING PUBERTAL DEVELOPMENT

9.5.1. Nonhuman Primate

Studies of the role of leptin on pubertal development in nonhuman primates are limited and confined to the male of the species. The first study that examined developmental changes in circulating leptin concentrations from birth to adulthood in a nonhuman primate was a cross-sectional investigation of rhesus monkeys ranging from 0 to 20 years of age (Urbanski and Pau, 1998). The developmental pattern of leptin was triphasic and paralleled that of testicular testosterone secretion. Levels of leptin were elevated in infancy, declined in the juvenile and through the onset of puberty, and then increased again in adults. Throughout development, leptin levels were positively correlated with testosterone concentrations. Conversely, leptin concentrations were negatively correlated with body weight in prepubertal animals, but after puberty that relationship changed and body weight and leptin increased in tandem. The peripubertal transition in the relationship of leptin to body weight likely resulted from a shift in body composition and/or changes in the dynamics of body energy expenditure and storage and the control of appetite at this developmental stage. No significant peripubertal rise in leptin was detected that might serve as a trigger for the initiation of puberty. However, only a few blood samples ($N = 5$–7) were drawn from what could be considered peripubertal animals (3–5 years of age), and therefore, a transient peak of leptin could have gone undetected. In any case, serum leptin concentrations in these peripubertal animals were substantially less than those from infants and juvenile animals.

Plant and Durrant (1997) published the first detailed study that examined peripubertal changes in leptin concentrations in a nonhuman primate. Longitudinal blood samples were drawn from intact male rhesus macaques on a weekly or biweekly basis during peripubertal development. A rise in nocturnal testosterone (week 0) was used to mark the

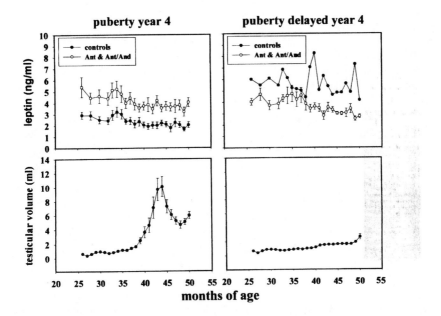

Figure 9.4. Panels on the left illustrate data from male monkeys that became sexually mature during their fourth year of life, those on the right illustrate data from animals in which puberty was delayed until a year later. (Top) Changes in mean serum leptin concentrations ($\pm$SEM) during the peripubertal period in control and Ant- (treated as infants with a GnRH antagonist) and Ant/And- (treated as infants with GnRH antagonist and androgen replacement) treated male monkeys. The data for the Ant and Ant/And-treated animals are combined in this panel for simplification. (Bottom) Changes in mean testicular volume ($\pm$SEM) during the peripubertal period in male monkeys that reached puberty in the fourth year and those with delayed puberty. Data from all treatment groups are combined in the lower two panels. Data in this figure is redrawn from Mann et al. (2002) with permission of the copyright holder, the *Society of the European Journal of Endocrinology.*

a signal of adequate rather than excessive fat stores and that leptin signaling occurs at low concentrations.

The relationship between circulating leptin levels and the pubertal re-awakening of the GnRH pulse generator has also been examined in agonadal male rhesus monkeys (Plant and Durrant, 1997; Suter et al., 2000). In the study by Plant and Durrant (1997), leptin values in the blood were similar to those in intact male monkeys, and did not change in association with the pubertal rise in LH and FSH secretion. In contrast, Suter et al. (2000) reported that an increase in circulating leptin preceded the onset of puberty. However, a reassessment of the latter data by one of us failed to substantiate this view (Plant, 2001b). Thus, the current balance of evidence does not support the hypothesis that leptin acts as a trigger to initiate the onset of puberty in the male nonhuman primate.

9.5.2. Human

A number of studies (both cross-sectional and longitudinal) have examined developmental changes in circulating levels of leptin in children. Among these are two longitudinal studies published by Mantzoros et al. (1997) and Ahmed et al. (1999). The former study was confined to boys; the latter included both boys and girls. Mantzoros et al. (1997) examined changes in circulating leptin secretion in eight boys at 4 month intervals beginning at 9.8–11.9 years of age and continuing for 2.5–5.1 years. The study by Ahmed et al. (1999) included 20 boys and 20 girls ranging from 8 to 9 years at the beginning of the study and continuing for approximately 8 years. Subjects in this study were observed at 6 month intervals. In the latter study, as has been reported in cross-sectional analyses (see below), leptin levels increased gradually with age in prepubertal children of both sexes, but more so in girls than boys. With the onset of puberty, a sexual dimorphism was evident with levels of leptin continuing to increase with advancing development in girls but declining in boys.

A unique finding of the Mantzoros et al. (1997) was that boys exhibited a transient rise in circulating leptin concentrations that appeared to precede the initiation of the pubertal rise in testosterone secretion. Levels of leptin were at or near peak concentrations prior to the initial increase in testosterone. These observations have been largely responsible for the premise that a rise in leptin serves as a trigger to initiate human puberty. However, as we pointed out in a recent review of this area (Mann and Plant, 2002), basal leptin values in these boys varied widely (as much as 15-fold), and the magnitude of the rise in circulating leptin in several of the boys was very limited. Furthermore, these data need to be interpreted with some caution because others studies of boys, like those of male monkeys (Plant and Durrant, 1997; Urbanski and Pau, 1998; Mann et al., 2000, 2002), have failed to demonstrate this late prepubertal transient increase in circulating leptin (Carlsson et al., 1997; Garcia-Mayor et al., 1997; Ahmed et al., 1999; reviewed below).

Garcia-Mayor et al. (1997) published a comprehensive cross-sectional study of leptin levels in 789 children (446 boys and 343 girls) between the ages of 5 and 15 years. Leptin concentrations increased in both boys and girls from 5 through 10 years of age and then the pattern diverged, declining in boys through 13 years of age but continuing to increase in girls. The changes in leptin with age occurred in parallel to an increase in body weight through 10 years of age in boys, but then with the decline in leptin values, the relationship was inverted. Leptin values were lower in boys than in girls even in the youngest group (5–6 years of age) despite there being no gender differences at that age in body weight, height, or adiposity. Others have reported a similar sexual dimorphism in serum leptin during childhood (Nakanishi et al., 2001). In boys, leptin levels at 10 years of age were elevated above values at 5–6 years of age; a 5 year period when neither circulating FSH, LH, or testosterone changed significantly (Garcia-Mayor et al., 1997). The progressive age-related rise in FSH, LH, and testosterone observed in this study in boys did not become significant (relative to 5–6 years of age) until 11, 12, and 12 years of age, respectively. Thus, the increase in leptin concentrations appeared to precede pubertal activation of the reproductive axis. From 10 through 15 years of age, leptin was negatively correlated with testosterone levels. When data from children were divided into stages of pubertal development (prepubertal, early puberty, and overt puberty) according to circulating gonadal hormone levels, levels of leptin increased between the prepubertal and early pubertal periods in boys, but then declined in boys with established pubertal development stages. The results from this study do not provide any definitive data as to whether there is a transient rise in leptin that precedes timing of re-augmentation of GnRH pulse generator activity and data for leptin concentrations were normalized to this index. The animal's age at the time of the pubertal rise in testosterone levels ranged from 27 to 31 months, and there were no significant fluctuations in circulating leptin concentrations during the period ranging from week 26 before until week 9 after the nocturnal rise in testosterone secretion. Again, the data do not support the contention that a transient rise in leptin serves as a trigger for the onset of puberty in the male nonhuman primate. Two additional developmental studies of the male rhesus macaque have been published that also fail to support the idea that leptin serves as a trigger to initiate puberty in the male of this species (Mann et al., 2000, 2002). In the first of these, developmental changes in leptin were monitored longitudinally in eight male rhesus macaques from 10 until 80 months of age. Animals were maintained throughout the study in a large social group under natural environmental lighting. Under these conditions, rhesus monkeys exhibit seasonal sexual activity and young animals usually reach puberty during the breeding season of their fourth year (41–46 months of age), or if sexual maturity is not achieved at this time, it is delayed a year until the subsequent breeding season. Apparently, in male rhesus monkeys housed under natural lighting conditions, reemergence of the activity of the GnRH pulse generator, the driving force behind pubertal development, can only occur during the fall breeding season. Morning blood samples were drawn bimonthly before 24 months of age and monthly thereafter. As reported in the earlier cross-sectional study (Urbanski and Pau, 1998), leptin concentrations declined during the prepubertal period, and were negatively correlated with body weight. Furthermore, in animals that reached sexual maturity during the breeding season of their fourth year (seven of eight monkeys), no transient increase in leptin concentrations was detected preceding or in association with the initiation of the pubertal rise in testosterone (Figure 9.3). Over the period from 12 months before to 12 months after the onset of puberty, leptin levels declined from approximately 4 ng/ml to 2 ng/ml and over the immediate period (4 months) before the onset of puberty, and no transient increase in leptin concentrations was detected. Interestingly, the one animal in which puberty was delayed a year had the highest leptin levels throughout prepubertal development. The data suggest that within animals of a social group, factors other than fat mass and circulating leptin

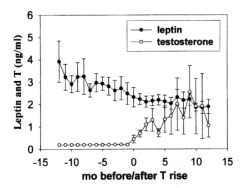

Figure 9.3. Peripubertal changes in serum leptin and testosterone concentrations in seven male monkeys. Leptin concentrations are normalized to the beginning of a sustained rise in serum testosterone levels. Data in this figure is redrawn from Mann et al. (2000) with permission of the Society for the Study of Reproduction, Inc.

levels (e.g., social rank and photoperiod) play a more critical role in determining the timing of the onset of puberty. In young adult animals (4.5 years of age), leptin levels were positively correlated with body fat mass and were inversely proportional to lean body mass, but there was no association between serum testosterone and leptin. After this age, there was no significant association between any of these parameters and serum leptin concentrations. These data confirm the earlier work of Plant and Durrant (1997) showing that peripubertal changes in circulating leptin concentrations in the male rhesus monkey are unremarkable and not consistent with the idea of leptin serving as a trigger for the onset of puberty in the male primate.

Another line of evidence, generated in a study performed in our laboratory to investigate the potential involvement of the neonatal testosterone elevation in sexual development in the male monkey (Mann et al., 1998), also argues against the premise that leptin plays a critical role in the initiation of puberty in the male primate. In this experiment, infant male rhesus monkeys were treated continuously with a GnRH antagonist for the first four postnatal months to prevent the neonatal elevation of circulating testosterone and the effects on pubertal development were examined. A control group of animals received androgen replacement therapy in addition to the GnRH antagonist. Because several of the animals in this study showed a 1-year delay in the onset of puberty, it provided an opportunity to compare, retrospectively, the developmental changes in leptin in animals with a normal ($N = 17$) and delayed ($N = 7$) puberty (Mann et al., 2002). In animals that reached puberty during their fourth year of life (regardless of treatment), serum leptin levels declined gradually over the 14 months prior to the pubertal increase in testicular size (Figure 9.4, top left and bottom left panels). There were no significant fluctuations in leptin either before or in association with pubertal testicular growth, although animals treated with the GnRH antagonist as neonates had higher levels of leptin throughout this developmental period. In addition, the pattern of leptin concentrations in these animals over this period did not differ from those in which puberty was to be delayed until the following year (Figure 9.4, right panels). The one control animal in which puberty was delayed until Year 5 actually had higher levels of leptin during Year 4 than those animals that were experiencing puberty at that time. It appears then that while leptin may be an important metabolic signal to the CNS that nutritional status is sufficient to initiate puberty, it is not the proverbial trigger for this process since the patterns and concentrations of circulating leptin were essentially the same between pubertal animals and those in which puberty was to be delayed for another year.

The decline in circulating leptin concentrations despite increasing body weight during prepubertal development in male rhesus monkeys (Urbanski and Pau, 1998; Mann et al., 2000, 2002) may be related to decreasing body fat mass in the face of increasing lean body mass. In a recent study, major differences in serum leptin levels were reported between captive and wild baboons (Banks et al., 2001). Leptin levels were 3-fold higher in captive than in wild animals and differences were most pronounced in younger animals and leptin levels declined with increasing age as they do in rhesus monkeys. The decline in leptin with age and the lower levels of leptin in wild animals likely reflect differences in adiposity. The level of adiposity has been reported to be three times greater in captive (6.1 percent) (Altmann et al., 1993) versus wild (1.9 percent) (Rutenberg et al., 1987) baboons (as per Banks et al., 2001). The authors (Banks et al., 2001) propose that the low levels of leptin in wild baboons is consistent with the idea that leptin evolved originally as pubertal activation of the hypothalamic–pituitary–testicular axis as has been suggested for boys in the Mantzoros et al. (1997) longitudinal study. The decline in circulating leptin levels during pubertal development in boys likely results from the well-documented inhibitory effect of testosterone on leptin secretion (Elbers et al., 1997; Jockenhovel et al., 1997; Wabitsch et al., 1997). Testosterone is thought to be partially responsible for the lower levels of leptin in men versus women (Ostlund et al., 1996; Rosenbaum et al., 1996).

Similar cross-sectional studies exploring developmental changes in circulating leptin concentrations in boys and girls, but with substantially different results were published by Blum et al. (1997) and Carlsson et al. (1997). In the former study, leptin levels did not change significantly with age in boys although levels did rise in girls between 5.8 and 19.9 years of age. In the latter study, levels of leptin increased with age in girls as reported in the other two studies, but declined in boys. Thus, while the developmental pattern for leptin in girls was similar in the three studies, the results reported for boys were an increase (Garcia-Mayor et al., 1997), no change (Carlsson et al., 1997), or a decrease (Blum et al., 1997) with age. There was, however, general agreement that leptin concentrations are lower in boys than girls even before puberty, and that the pattern of leptin was divergent between boys and girls with successive pubertal stages (increasing from Tanner's stage 1 through 5 in girls, but declining from stage 2 through 5 in boys). It should be noted that the gender differences in circulating leptin concentrations in adolescents and during late puberty were evident even after the data were adjusted for BMI or percentage body fat (Blum et al., 1997). Some of the gender differences in leptin in these subjects are likely related to higher testosterone levels in boys (especially in late puberty) since circulating leptin levels, when normalized for fat mass, fall in boys, but rise in girls during late puberty (Horlick et al., 2000). Other factors that may be involved in establishing this sexual dimorphism include a more rapid developmental increase in lean body mass to fat mass in boys (Ahmed et al., 1999) or some as yet unknown gender-related factor.

Leptin concentrations in the circulation exhibit a 24 hr rhythm in both children and adults with peak levels in the early morning hours and a nadir in the late morning or early afternoon (Pombo et al., 1997; Licinio et al., 1998; Palmert et al., 1998; Saad et al., 1998; Stoving et al., 1998; Ankarberg-Lindgren et al., 2001). However, there do not appear to be any substantial differences in the diurnal rhythm of leptin between the prepubertal and pubertal child, or in the pattern or magnitude of the leptin rhythm at successive stages of pubertal development. While a sexual dimorphism in the leptin rhythm existed before puberty in children (e.g., higher amplitude in girls versus boys), there were no significant differences in this rhythm between prepubertal and pubertal boys and girls, respectively (Pombo et al., 1997). In another study, mean levels of leptin changed during pubertal development (increasing in boys before puberty and then declining, or in girls rising throughout pubertal development), but there appeared to be little change in the characteristics of the rhythm during this process (Ankarberg-Lindgren et al., 2001). Therefore, existence of the diurnal leptin rhythm was not dependent on reproductive status. It does not appear from the extant data that the development of the diurnal rhythm of leptin or changes in its characteristics serve as a trigger to initiate the onset of puberty.

9.6. EXPERIMENTAL EVIDENCE (INCLUDING THOSE OF NATURE) THAT LEPTIN IS INVOLVED IN PUBERTAL DEVELOPMENT IN HIGHER PRIMATES

9.6.1. Nonhuman Primate

The only direct test of the role of leptin in triggering the onset of puberty in nonhuman primates is a study originating from Plant's laboratory (Barker-Gibb et al., 2002). Here, agonadal juvenile male rhesus monkeys were infused intravenously with recombinant human leptin (5 μg/kg/h) for up to 20 days to produce a square wave increment (approximately 20 ng/ml) in circulating leptin. Nocturnal GnRH release was assessed indirectly by measuring LH secretion from the in situ pituitary that had been sensitized to GnRH stimulation before the start of the study. Despite the sustained elevation in circulating leptin, precocious GnRH release was not observed although an elevation of GH secretion indicated that the human leptin was bioactive in the monkey. These data, which fail to support the premise that leptin acts as a trigger to initiate the onset of puberty in primates, are entirely consistent with reports that endogenous leptin levels in blood do not increase before the onset of spontaneous puberty in the nonhuman primate (see above). It must be emphasized that comparable studies have not been performed in the female monkey. Such studies would appear to be of crucial importance for two reasons. First, gender differences in the neurobiological regulatory systems controlling puberty appear to exist in the nonhuman primate (Plant, 2001a); second, the developmental pattern of circulating leptin concentrations in peripubertal boys and girls are markedly different.

9.6.2. Human

Autosomal homozygous mutations of the genes coding for either the leptin (ob) or the leptin receptor (rb) are associated with early onset morbid obesity and retarded sexual development in humans (Montague et al., 1997; Clement et al., 1998; Strobel et al., 1998; Farooqi et al., 1999). Heterozygous parents of these individuals do not exhibit morbid obesity or abnormal sexual maturation (Farooqi and O'Rahilly, 2000). In a Turkish family, three offspring were homozygous for a mutation of the *ob* gene leading to a situation similar to that of the *ob/ob* condition in mice (Strobel et al., 1998). The children (two girls and a boy) had low circulating leptin levels and exhibited hyperphagia leading to overt obesity. The 22-year old male and 32-year old female remained sexually immature and exhibited signs of hypogonadism. A similar clinical situation was apparent in three female offsprings of a family with a homozygous mutation of the gene coding for the leptin receptor (Clement et al., 1998). All exhibited morbid obesity beginning early in life and none had experienced the onset of puberty by 19 years of age. Secondary sexual characteristics were undeveloped in these patients. While both mutations of the *ob* and *rb* gene result in morbid obesity and retardation of sexual development, growth hormone production is normal in the former situation, but growth retardation is evident in patients with mutational defects of the leptin receptor (Farooqi et al., 1999). These

neuroendocrine axes, this remains to be conclusively shown by interventional studies in humans.

10.2.2. Neuroendocrine Aspects

The onset of puberty is characterized by a significant change in the pulsatile characteristics of the hypothalamic "gonadostat." The amplitude of pulses of hypothalamic GnRH neurons is increased, first leading to a substantial rise in nocturnal FSH, and then to a rise in LH pulsatile release by the pituitary (Jakacki et al., 1982). This results in a remarkably higher output of sex steroids by the gonads, which has also been linked to an increased production of GH and IGF-1. The combined increase in GH and IGF-1 affects linear growth, muscle bulk, and mineralization of the skeleton (Mauras et al., 1996). Although the secretion of GH, its effector peptide IGF-1, the GH dependent IGF-1 carrier protein IGF-BP3, and the sex steroids, peak through mid-puberty, leptin levels already increase during pre-puberty. This suggests an interaction between leptin and the gonadotropins in late childhood (Clayton et al., 1997) and a role of leptin in the initiation of puberty (see Section 10.3).

10.3. LEPTIN AND NORMAL PUBERTAL DEVELOPMENT

10.3.1. Leptin in Childhood

10.3.1.1. Circulating Leptin Levels in Neonates. During pregnancy, high leptin levels are present in the amniotic fluid and in the arterial and venous cord blood. (Mantzoros et al., 1997b; Matsuda et al., 1997; Schubring et al., 1997; Sivan et al., 1997; Varvarigou et al., 1999; Christou et al., 2001). Umbilical cord blood leptin levels are derived from both the placenta (Masuzaki et al., 1997) and the fetal tissue (Schubring et al., 1997) and correlate positively with the absolute body weight and fat mass of the neonate. These levels are lower in pre-term and small-for-gestational-age neonates, and are higher in large-for-gestational-age neonates than in normal neonates (Mantzoros et al., 1997b).

High leptin concentrations have also been reported in arterial and venous cord blood at birth as well as in capillary and venous blood shortly after birth. Interestingly, neonate girls have been shown to have higher leptin levels than boys probably because of their higher estrogen levels immediately after birth. Circulating leptin levels decrease rapidly and dramatically after birth (Kiess et al., 1998; Schubring et al., 1999).

10.3.1.2. Circulating Leptin Levels during Childhood. A strong positive correlation between leptin levels and body fat mass, and to a lesser extent between leptin and body mass index (BMI) as well as weight, has been reported in all age groups during childhood. The well-established pulsatile and nyctohemeral rhythm of leptin secretion in adults is also thought to be present in children (Hassink et al., 1996; Havel et al., 1996; Kiess et al., 1996; Blum et al., 1997; Clayton et al., 1997; Ellis and Nicolson, 1997; Lahlou et al., 1997; Matkovic et al., 1997a). These findings are unaffected by ethnicity (Ellis and Nicolson, 1997). The major gender-related differences in leptin secretion seem to be due to pulse

Andreelli, F., Hanaire-Broutin, H., Laville, M., Tauber, J. P., Riou, J. P., and Thivolet, C. (2000). Normal reproductive function in leptin-deficient patients with lipoatropic diabetes. *J. Clin. Endocrinol. Metab.*, 85, 715–719.

Ankarberg-Lindgren, C., Dahlgren, J., Carlsson, B., Rosberg, S., Carlsson, L., Wikland, K. A., and Norjavaara, E. (2001). Leptin levels show diurnal variation throughout puberty in healthy children, and follow a gender-specific pattern. *Eur. J. Endocrinol.*, 145, 43–51.

Banks, W. A., Kastin, A. J., Huang, W., Jaspan, J. B., and Maness, L. M. (1996). Leptin enters the brain by a saturable system independent of insulin. *Peptides*, 17, 305–311.

Banks, W. A., McLay, R. N., Kastin, A. J., Sarmiento, U., and Scully, S. (1999). Passage of leptin across the blood–testis barrier. *Am. J. Physiol.*, 276, E1099–E1104.

Banks, W. A., Phillips-Conroy, J. E., Jolly, C. J., and Morley, J. E. (2001). Serum leptin levels in wild and captive populations of baboons (Papio): implications for the role of leptin. *J. Clin. Endocrinol. Metab.*, 86, 4315–4320.

Barker-Gibb, M. L., Sahu, A., Pohl, C. R., and Plant, T. M. (2002). Elevating circulating leptin in prepubertal male rhesus monkeys (macca mulatta) does not elicit precocious gonadotropin-releasing hormone release, assessed indirectly. *J. Clin. Endocrinol. Metab.*, 87, 4976–4983.

Blum, W. F., Englaro, P., Hanitsch, S., Juul, A., Hertel, N. T., Muller, J., Skakkebaek, N. E., Heiman, M. L., Birkett, M., Attanasio, A. M., Kiess, W., and Rascher, W. (1997). Plasma leptin levels in healthy children and adolescents: dependence on body mass index, body fat mass, gender, pubertal stage, and testosterone. *J. Clin. Endocrinol. Metab.*, 82, 2904–2910.

Bullough, W. S. (1951). *Vertebrate Sexual Cycles*, Methuen, London.

Cameron, J. L. (1996). Regulation of reproductive hormone secretion in primates by short-term changes in nutrition. *Rev. Reprod.*, 1, 117–126.

Caprio, M., Isidori, A. M., Carta, A. R., Moretti, C., Dufau, M. L., and Fabbri, A. (1999). Expression of functional leptin receptors in rodent Leydig cells. *Endocrinology*, 140, 4929–4947.

Carlsson, B., Ankarberg, C., Rosberg, S., Norjavaara, E., Albertsson-Wikland, K., and Carlsson, L. M. S. (1997). Serum leptin concentrations in relation to pubertal development. *Arch. Dis. Child.*, 77, 396–400.

Cheung, C. C., Clifton, D. K., and Steiner, R. A. (1997). Proopiomelanocortin neurons are direct targets for leptin in the hypothalamus. *Endocrinology*, 138, 4489–4492.

Clement, K., Vaisse, C., and Lahlou, N. (1998). A mutation in the human leptin receptor gene causes obesity and pituitary dysfunction. *Nature*, 392, 398–401.

Cunningham, M. J., Clifton, D. K., and Steiner, R. A. (1999). Leptin's actions on the reproductive axis: perspectives and mechanisms. *Biol. Reprod.*, 60, 216–222.

Dearth, R. K., Hiney, J. K., and Dees, W. L. (2000). Leptin acts centrally to induce the prepubertal secretion of luteinizing hormone in the female rat. *Peptides*, 21, 387–392.

Elbers, J. M. H., Asscheman, H., Seidell, J. C., Frolich, M., Meinders, A. E., and Gooren, L. J. G. (1997). Reversal of the sex hormone difference in serum leptin levels upon cross-sex hormone administration in transsexuals. *J. Clin. Endocrinol. Metab.*, 82, 3267–3270.

El Majdoubi, M., Sahu, A., Ramaswamy, S., and Plant, T. M. (2000). Neuropeptide Y: a hypothalamic brake restraining the onset of puberty in primates. *Proc. Natl. Acad. Sci.*, 97, 6179–6184.

Farooqi, I. S. and O'Rahilly, S. (2000). Recent advances in the genetics of severe childhood obesity. *Arch. Dis. Child.*, 83, 31–34.

Farooqi, I. S., Jebb, S. A., and Langmack, G. (1999). Effects of recombinant leptin therapy in a child with congenital leptin deficiency. *N. Engl. J. Med.*, 341, 879–884.

Finn, P. D., Cunningham, M. J., Francis Pau, K.-Y., Spies, H. G., Clifton, D. K., and Steiner, R. A. (1998). The stimulatory effect of leptin on the neuroendocrine reproductive axis of the monkey. *Endocrinology*, 139, 4652–4662.

Fruhbeck, G., Jebb, S. A., and Prentice, A. M. (1998). Leptin: physiology and pathophysiology. *Clin. Physiol.*, 18, 399–419.

Galler, A., Schuster, V., and Kiess, W. (2001). Pubertal adipose tissue: is it really necessary for normal sexual maturation? *Eur. J. Endocrinol.*, 145, 807–808.

Garcia-Mayor, R. V., Andrade, M. A., Rios, M., Lage, M., Dieguez, C., and Casanueva, F. F. (1997). Serum leptin levels in normal children: relationship to age, gender, body mass index, pituitary-gonadal hormones, and pubertal stage. *J. Clin. Endocrinol. Metab.*, 82, 2849–2855.

Gill, M. S., Hall, C. M., Tillman, V., and Clayton, P. E. (1999). Constitutional delay in growth and puberty (CDGP) is associated with hypoleptinemia. *Clin. Endocrinol.*, 50, 721–726.

Mantzoros, C. S., Flier, J. S., and Rogol, A. D. (1997). A longitudinal assessment of hormonal and physical alterations during normal puberty in boys. V. rising leptin levels may signal the onset of puberty. *J. Clin. Endocrinol. Metab., 82*, 1066–1070.

Montague, C. T., Farooqi, I. S., and Whitehead, J. P. (1997). Congenital leptin deficiency is associated with severe early-onset obesity in humans. *Nature, 387*, 903–908.

Nakanishi, T., Li, R., Liu, Z., Yi, M., Nakagawa, Y., and Ohzeki, T. (2001). Sexual dimorphism in relationship of serum leptin and relative weight for the standard in normal-weight, but not in overweight, children as well as adolescents. *Eur. J. Clin. Nutr., 55*, 989–993.

Ogura, K., Irahara, M., Kiyokawa, M., Tezuka, M., Matsuzaki, T., Yasui, T., Kamada, M., and Aono, T. (2001). Effects of leptin on secretion of LH and FSH from primary cultured female rat pituitary cells. *Eur. J. Endocrinol., 144*, 653–658.

Ojeda, S. R., Ma, Y. J., Dziedzic, B., and Prevot, V. (2000). Astrocyte-neuron signaling and the onset of female puberty. In J.-P. Bourguignon and T. M. Plant (Eds.). *The Onset of Puberty in Perspective*, Elsevier Science Publishers B. V., Amsterdam, pp. 41–83.

Ostlund, R. E., Yang, J. W., Klein, S., and Gingerich, R. (1996). Relation between plasma leptin concentration and body fat, gender, diet, age, and metabolic covariates. *J. Clin. Endocrinol. Metab., 81*, 3909–3913.

Palmert, M. R., Radovick, S., and Boepple, P. A. (1998). The impact of reversible gonadal sex steroid suppression on serum leptin concentrations in children with central precocious puberty. *J. Clin. Endocrinol. Metab., 83*, 1091–1096.

Parent, A. S., Lebrethon, M. C., Gerard, A., Vandersmissen, E., and Bourguignon, J. P. (2000). Leptin effects on pulsatile gonadotropin releasing hormone secretion from the adult rat hypothalamus and interaction with cocaine and amphetamine regulated transcript peptide and neuropeptide Y. *Regul. Pept., 92*, 17–24.

Pau, K.-Y. F., Berria, M., Hess, D. L., and Spies, H. G. (1995). Hypothalamic site-dependent effects of neuropeptide Y on gonadotropin-releasing hormone secretion in rhesus macaques. *J. Neuroendocrinol., 7*, 63–67.

Plant, T. M. (1994). Puberty in primates. In E. Knobil and J. D. Neill (Eds.). *The Physiology of Reproduction*, Raven Press, New York, pp. 453–485.

Plant, T. M. (2000). Ontogeny of GnRH gene expression and secretion in primates. In J.-P. Bourguignon and T. M. Plant (Eds.). *The Onset of Puberty in Perspective*. Elsevier Science Publisher B. V., Amsterdam, pp. 3–13.

Plant, T. M. (2001a). Neurobiological bases underlying the control of the onset of puberty in the rhesus monkey: a representative higher primate. *Frontiers Neurobiol., 22*, 107–139.

Plant, T. M. (2001b). Leptin, growth hormone, and the onset of primate puberty. *J. Clin. Endocrinol. Metab., 86*, 458–460.

Plant, T. M. and Durrant, A. R. (1997). Circulating leptin does not appear to provide a signal for triggering the initiation of puberty in the male rhesus monkey (*Macaca mulatta*). *Endocrinology, 138*, 4505–4508.

Plant, T. M., Fraser, M. O., Medhamurthy, R., and Gay, V. L. (1989a). Somatogenic control of GnRH neuronal synchronization during development in primates: a speculation. In H. A. Delemarre van de Waal, T. M. Plant, G. P. van Rees, and J. Schoemaker (Eds.). *Control of the Onset of Puberty, III*, Elsevier Science Publishers B. V., Amsterdam, pp. 111–121.

Plant, T. M., Gay, V. L., Marshall, G. R., and Arslan, M. (1989b). Puberty in monkeys is triggered by chemical stimulation of the hypothalamus. *Proc. Natl. Acad. Sci. USA, 86*, 2506–2510.

Pombo, M., Herrera-Justiniano, E., Considine, R. V., Hermida, R. C., Galvez, M. J., Martin, T., Barreiro, J., Casanueva, F. F., and Dieguez, C. (1997). Nocturnal rise of leptin in normal prepubertal and pubertal children and in patients with perinatal stalk-transection syndrome. *J. Clin. Endocrinol. Metab., 82*, 2751–2754.

Quinton, N. D., Smith, R. F., and Clayton, P. E. (1999). Leptin binding activity changes with age: the link between leptin and puberty. *J. Clin. Endocrinol. Metab., 84*, 2336–2341.

Randeva, H. S., Murray, R. D., Lewandowski, K. C., O'Callaghan, C. J., Horn, R., O'Hare, P., Brabant, G., Hillhouse, E. W., and Shalet, S. M. (2002). Differential effects of GH replacement on the components of the leptin system in GH-deficient individuals. *Endocrine Soc., 87*, 798–804.

Resko, J. A. and Ellinwood, W. E. (1985). Negative feedback regulation of gonadotropin secretion by androgens in fetal rhesus macaques. *Biol. Reprod., 33*, 346–352.

Romer, A. S. (1959). *The Vertebrate Story*, University of Chicago Press, Chicago.

Ronnekleiv, O. K. and Resko, J. A. (1990). Ontogeny of gonadotropin-releasing hormone-containing neurons in early fetal development of rhesus macaques. *Endocrinology, 126*, 498–511.

pubertal activation of the hypothalamic–pituitary–testicular axis as has been suggested for boys in the Mantzoros et al. (1997) longitudinal study. The decline in circulating leptin levels during pubertal development in boys likely results from the well-documented inhibitory effect of testosterone on leptin secretion (Elbers et al., 1997; Jockenhovel et al., 1997; Wabitsch et al., 1997). Testosterone is thought to be partially responsible for the lower levels of leptin in men versus women (Ostlund et al., 1996; Rosenbaum et al., 1996).

Similar cross-sectional studies exploring developmental changes in circulating leptin concentrations in boys and girls, but with substantially different results were published by Blum et al. (1997) and Carlsson et al. (1997). In the former study, leptin levels did not change significantly with age in boys although levels did rise in girls between 5.8 and 19.9 years of age. In the latter study, levels of leptin increased with age in girls as reported in the other two studies, but declined in boys. Thus, while the developmental pattern for leptin in girls was similar in the three studies, the results reported for boys were an increase (Garcia-Mayor et al., 1997), no change (Carlsson et al., 1997), or a decrease (Blum et al., 1997) with age. There was, however, general agreement that leptin concentrations are lower in boys than girls even before puberty, and that the pattern of leptin was divergent between boys and girls with successive pubertal stages (increasing from Tanner's stage 1 through 5 in girls, but declining from stage 2 through 5 in boys). It should be noted that the gender differences in circulating leptin concentrations in adolescents and during late puberty were evident even after the data were adjusted for BMI or percentage body fat (Blum et al., 1997). Some of the gender differences in leptin in these subjects are likely related to higher testosterone levels in boys (especially in late puberty) since circulating leptin levels, when normalized for fat mass, fall in boys, but rise in girls during late puberty (Horlick et al., 2000). Other factors that may be involved in establishing this sexual dimorphism include a more rapid developmental increase in lean body mass to fat mass in boys (Ahmed et al., 1999) or some as yet unknown gender-related factor.

Leptin concentrations in the circulation exhibit a 24 hr rhythm in both children and adults with peak levels in the early morning hours and a nadir in the late morning or early afternoon (Pombo et al., 1997; Licinio et al., 1998; Palmert et al., 1998; Saad et al., 1998; Stoving et al., 1998; Ankarberg-Lindgren et al., 2001). However, there do not appear to be any substantial differences in the diurnal rhythm of leptin between the prepubertal and pubertal child, or in the pattern or magnitude of the leptin rhythm at successive stages of pubertal development. While a sexual dimorphism in the leptin rhythm existed before puberty in children (e.g., higher amplitude in girls versus boys), there were no significant differences in this rhythm between prepubertal and pubertal boys and girls, respectively (Pombo et al., 1997). In another study, mean levels of leptin changed during pubertal development (increasing in boys before puberty and then declining, or in girls rising throughout pubertal development), but there appeared to be little change in the characteristics of the rhythm during this process (Ankarberg-Lindgren et al., 2001). Therefore, existence of the diurnal leptin rhythm was not dependent on reproductive status. It does not appear from the extant data that the development of the diurnal rhythm of leptin or changes in its characteristics serve as a trigger to initiate the onset of puberty.

9.6. EXPERIMENTAL EVIDENCE (INCLUDING THOSE OF NATURE) THAT LEPTIN IS INVOLVED IN PUBERTAL DEVELOPMENT IN HIGHER PRIMATES

9.6.1. Nonhuman Primate

The only direct test of the role of leptin in triggering the onset of puberty in nonhuman primates is a study originating from Plant's laboratory (Barker-Gibb et al., 2002). Here, agonadal juvenile male rhesus monkeys were infused intravenously with recombinant human leptin (5 μg/kg/h) for up to 20 days to produce a square wave increment (approximately 20 ng/ml) in circulating leptin. Nocturnal GnRH release was assessed indirectly by measuring LH secretion from the in situ pituitary that had been sensitized to GnRH stimulation before the start of the study. Despite the sustained elevation in circulating leptin, precocious GnRH release was not observed although an elevation of GH secretion indicated that the human leptin was bioactive in the monkey. These data, which fail to support the premise that leptin acts as a trigger to initiate the onset of puberty in primates, are entirely consistent with reports that endogenous leptin levels in blood do not increase before the onset of spontaneous puberty in the nonhuman primate (see above). It must be emphasized that comparable studies have not been performed in the female monkey. Such studies would appear to be of crucial importance for two reasons. First, gender differences in the neurobiological regulatory systems controlling puberty appear to exist in the nonhuman primate (Plant, 2001a); second, the developmental pattern of circulating leptin concentrations in peripubertal boys and girls are markedly different.

9.6.2. Human

Autosomal homozygous mutations of the genes coding for either the leptin (ob) or the leptin receptor (rb) are associated with early onset morbid obesity and retarded sexual development in humans (Montague et al., 1997; Clement et al., 1998; Strobel et al., 1998; Farooqi et al., 1999). Heterozygous parents of these individuals do not exhibit morbid obesity or abnormal sexual maturation (Farooqi and O'Rahilly, 2000). In a Turkish family, three offspring were homozygous for a mutation of the *ob* gene leading to a situation similar to that of the *ob/ob* condition in mice (Strobel et al., 1998). The children (two girls and a boy) had low circulating leptin levels and exhibited hyperphagia leading to overt obesity. The 22-year old male and 32-year old female remained sexually immature and exhibited signs of hypogonadism. A similar clinical situation was apparent in three female offsprings of a family with a homozygous mutation of the gene coding for the leptin receptor (Clement et al., 1998). All exhibited morbid obesity beginning early in life and none had experienced the onset of puberty by 19 years of age. Secondary sexual characteristics were undeveloped in these patients. While both mutations of the *ob* and *rb* gene result in morbid obesity and retardation of sexual development, growth hormone production is normal in the former situation, but growth retardation is evident in patients with mutational defects of the leptin receptor (Farooqi et al., 1999). These experiments of nature suggest the importance of leptin for pubertal development, but if leptin plays a critical role in this process, then treatment of individuals with a homozygous mutation of the *ob* gene with leptin should result in the initiation of a pubertal pattern of GnRH and gonadotropin secretion and subsequent gonadal development, the appearance of secondary sexual characteristics and the achievement of reproductive competence.

A nine-year old female patient with a homozygous mutation of the *ob* gene was treated with recombinant human leptin for a 12-month period (Farooqi et al., 1999). At the time of initiation of treatment, the patient showed no signs of pubertal development. Although bone age was advanced by 3.5 years, gonadotropin and estradiol levels were in the prepubertal range. At the end of leptin treatment, a pulsatile pattern of nocturnal gonadotropin secretion characteristic of early puberty was observed. These fascinating data suggest that a threshold level of circulating leptin may be necessary in order to permit pubertal development to proceed, but they do not address whether leptin's role in this developmental process is permissive or causal. This is because the patient had both a pubertal chronological and bone age at the time the pubertal pattern in pulsatile gonadotropin secretion was observed.

Other recent case studies suggest that, if circulating leptin is indeed necessary for pubertal development in girls, then the threshold is probable very low. A female patient with Beals syndrome (congenital contractural arachnodactyly) had a body weight 10 kg below the 3rd percentile for her age, a BMI at the 3rd percentile and a blood leptin level of 1.6 ng/ml (Galler et al., 2001). Despite the low BMI, low body weight, and low leptin, menarche occurred at 15 years of age and menstrual cycles were reported to be regular since that time. Two patients with lipoatropic diabetes (atrophy of visceral and subcutaneous adipose tissue), that developed the disease early in life, had even lower circulating leptin concentrations (<1 ng/ml) than the patient with Beals syndrome but a normal age of menarche and appearance of secondary sexual characteristics (Andreelli et al., 2000). The diabetic patients both indicated that they had regular menstrual cycles and one of the subjects had been pregnant on three occasions. The findings from these patients suggest that if leptin is obligatory for pubertal events to occur then the actual leptin threshold must be quite low. In any event, it should be apparent from these data and from those in nonhuman primates that puberty in primates can occur over a wide range of leptin concentrations.

There are some additional studies that do not support a dynamic role for leptin in triggering the onset of puberty in higher primates. Leptin concentrations were reported to be higher in patients with delayed puberty than in normal prepubertal or pubertal children (Giusti et al., 1999). In normal pubertal boys, leptin concentrations were higher than normal prepubertal boys, but leptin concentrations did not differ between pubertal boys with delayed puberty and prepubertal boys with delayed puberty (Gill et al., 1999). The 24 hr pattern of leptin secretion did not vary between prepubertal and pubertal boys and girls (Pombo et al., 1997). The data once again suggest that puberty occurs over a wide range of leptin values and that an elevation of blood leptin concentrations is not a prerequisite for puberty to proceed. These data in addition to our own results that peripubertal changes in leptin in the male monkey are unremarkable (no evidence of a transient rise in leptin levels prior to the onset of puberty), and that the developmental pattern of leptin did not differ between peripubertal male monkeys and age-matched monkeys in which puberty was being delayed (data reviewed earlier), suggest that leptin is not the primary factor for timing puberty in higher primates. The data are more supportive of a permissive rather than a dynamic role in primate pubertal development.

9.7. CONCLUSIONS

A role of leptin in pubertal development in many non-primate species (particularly in rodents) has been clearly established. Thus, from a comparative perspective, it is reasonable to propose that this adipocyte hormone may subserve an analogous role in monkeys and man. Studies of this relationship in primates, however, have been largely correlative and the only nonhuman primate investigated to date is the male rhesus monkey. Moreover, since gender differences exist in the developmental pattern of leptin in humans, one may anticipate this to be the case in nonhuman primates, and, therefore, there is a need for studies that investigate the role of leptin in pubertal development in female monkeys. Direct evidence for the view that leptin is necessary for pubertal development in primates is limited. Human subjects with mutations of the *ob/ob* gene fail to progress through puberty, and, in one such individual of pubertal age, a pulsatile pattern of gonadotropin secretion characteristic of early puberty was detected at the time of leptin replacement. Primate puberty occurs over a wide range of circulating leptin concentrations, and only very low levels seems to be required (threshold) because in subjects with lipoatropic dystrophy and Beals syndrome (two other pathophysiological conditions of low circulating leptin) pubertal development was not interrupted. The finding that low circulating leptin concentrations permit pubertal development in man is consistent with the suggestion that leptin evolved as a signal of adequate rather than excessive fat stores, and that this endocrine signal functions at low concentrations. Whether leptin serves a purely permissive role in allowing puberty to progress once the signal that initiates the re-augmentation of pulsatile GnRH at this stage of development has been delivered to the hypothalamus, or whether leptin is itself the timing signal has not been examined in a comprehensive manner. The only study directly addressing this issue is a report from our laboratories demonstrating that imposition of a premature rise in circulating leptin concentrations in the prepubertal male monkey did not lead to precocious activation of pulsatile GnRH release. Taking this finding together with the inconsistent results on the temporal relationship between changes in circulating leptin concentrations immediately prior to the onset of puberty, we conclude that this adipocyte hormone is not the signal that times hypothalamic activation of the pituitary–gonadal axis at the initiation of puberty.

ACKNOWLEDGMENTS

The authors' work was supported in part by NIH grants HD26423, RR03034, and HD41749 to Morehouse School of Medicine and HD13254 and HD08610 to the University of Pittsburgh School of Medicine.

REFERENCES

Ahmed, M. L., Ong, K. K. L., Morrell, D. J., Cox, L., Drayer, N., Perry, L., Preece, M. A., and Dunger, D. B. (1999). Longitudinal study of leptin concentrations during puberty: sex differences and relationship to changes in body composition. *J. Clin. Endocrinol. Metab., 84,* 899–905.

Altmann, J., Schoeller, D., Altmann, S. A., Muruthi, P., and Sapolsky, R. M. (1993). Body size and fatness of free-living baboons reflect food availability and activity levels. *Am J. Physiol., 30,* 149–161.

Andreelli, F., Hanaire-Broutin, H., Laville, M., Tauber, J. P., Riou, J. P., and Thivolet, C. (2000). Normal reproductive function in leptin-deficient patients with lipoatropic diabetes. *J. Clin. Endocrinol. Metab., 85*, 715–719.

Ankarberg-Lindgren, C., Dahlgren, J., Carlsson, B., Rosberg, S., Carlsson, L., Wikland, K. A., and Norjavaara, E. (2001). Leptin levels show diurnal variation throughout puberty in healthy children, and follow a gender-specific pattern. *Eur. J. Endocrinol., 145*, 43–51.

Banks, W. A., Kastin, A. J., Huang, W., Jaspan, J. B., and Maness, L. M. (1996). Leptin enters the brain by a saturable system independent of insulin. *Peptides, 17*, 305–311.

Banks, W. A., McLay, R. N., Kastin, A. J., Sarmiento, U., and Scully, S. (1999). Passage of leptin across the blood–testis barrier. *Am. J. Physiol., 276*, E1099–E1104.

Banks, W. A., Phillips-Conroy, J. E., Jolly, C. J., and Morley, J. E. (2001). Serum leptin levels in wild and captive populations of baboons (Papio): implications for the role of leptin. *J. Clin. Endocrinol. Metab., 86*, 4315–4320.

Barker-Gibb, M. L., Sahu, A., Pohl, C. R., and Plant, T. M. (2002). Elevating circulating leptin in prepubertal male rhesus monkeys (macca mulatta) does not elicit precocious gonadotropin-releasing hormone release, assessed indirectly. *J. Clin. Endocrinol. Metab., 87*, 4976–4983.

Blum, W. F., Englaro, P., Hanitsch, S., Juul, A., Hertel, N. T., Muller, J., Skakkebaek, N. E., Heiman, M. L., Birkett, M., Attanasio, A. M., Kiess, W., and Rascher, W. (1997). Plasma leptin levels in healthy children and adolescents: dependence on body mass index, body fat mass, gender, pubertal stage, and testosterone. *J. Clin. Endocrinol. Metab., 82*, 2904–2910.

Bullough, W. S. (1951). *Vertebrate Sexual Cycles*, Methuen, London.

Cameron, J. L. (1996). Regulation of reproductive hormone secretion in primates by short-term changes in nutrition. *Rev. Reprod., 1*, 117–126.

Caprio, M., Isidori, A. M., Carta, A. R., Moretti, C., Dufau, M. L., and Fabbri, A. (1999). Expression of functional leptin receptors in rodent Leydig cells. *Endocrinology, 140*, 4929–4947.

Carlsson, B., Ankarberg, C., Rosberg, S., Norjavaara, E., Albertsson-Wikland, K., and Carlsson, L. M. S. (1997). Serum leptin concentrations in relation to pubertal development. *Arch. Dis. Child., 77*, 396–400.

Cheung, C. C., Clifton, D. K., and Steiner, R. A. (1997). Proopiomelanocortin neurons are direct targets for leptin in the hypothalamus. *Endocrinology, 138*, 4489–4492.

Clement, K., Vaisse, C., and Lahlou, N. (1998). A mutation in the human leptin receptor gene causes obesity and pituitary dysfunction. *Nature, 392*, 398–401.

Cunningham, M. J., Clifton, D. K., and Steiner, R. A. (1999). Leptin's actions on the reproductive axis: perspectives and mechanisms. *Biol. Reprod., 60*, 216–222.

Dearth, R. K., Hiney, J. K., and Dees, W. L. (2000). Leptin acts centrally to induce the prepubertal secretion of luteinizing hormone in the female rat. *Peptides, 21*, 387–392.

Elbers, J. M. H., Asscheman, H., Seidell, J. C., Frolich, M., Meinders, A. E., and Gooren, L. J. G. (1997). Reversal of the sex hormone difference in serum leptin levels upon cross-sex hormone administration in transsexuals. *J. Clin. Endocrinol. Metab., 82*, 3267–3270.

El Majdoubi, M., Sahu, A., Ramaswamy, S., and Plant, T. M. (2000). Neuropeptide Y: a hypothalamic brake restraining the onset of puberty in primates. *Proc. Natl. Acad. Sci., 97*, 6179–6184.

Farooqi, I. S. and O'Rahilly, S. (2000). Recent advances in the genetics of severe childhood obesity. *Arch. Dis. Child., 83*, 31–34.

Farooqi, I. S., Jebb, S. A., and Langmack, G. (1999). Effects of recombinant leptin therapy in a child with congenital leptin deficiency. *N. Engl. J. Med., 341*, 879–884.

Finn, P. D., Cunningham, M. J., Francis Pau, K.-Y., Spies, H. G., Clifton, D. K., and Steiner, R. A. (1998). The stimulatory effect of leptin on the neuroendocrine reproductive axis of the monkey. *Endocrinology, 139*, 4652–4662.

Fruhbeck, G., Jebb, S. A., and Prentice, A. M. (1998). Leptin: physiology and pathophysiology. *Clin. Physiol., 18*, 399–419.

Galler, A., Schuster, V., and Kiess, W. (2001). Pubertal adipose tissue: is it really necessary for normal sexual maturation? *Eur. J. Endocrinol., 145*, 807–808.

Garcia-Mayor, R. V., Andrade, M. A., Rios, M., Lage, M., Dieguez, C., and Casanueva, F. F. (1997). Serum leptin levels in normal children: relationship to age, gender, body mass index, pituitary-gonadal hormones, and pubertal stage. *J. Clin. Endocrinol. Metab., 82*, 2849–2855.

Gill, M. S., Hall, C. M., Tillman, V., and Clayton, P. E. (1999). Constitutional delay in growth and puberty (CDGP) is associated with hypoleptinemia. *Clin. Endocrinol., 50*, 721–726.

Giusti, M., Guido, R., Valenti, S., and Giordano, G. (1999). Serum leptin levels in males with delayed puberty during short-term pulsatile GnRH administration. *J. Endocrinol. Invest., 22*, 6–11.

Grumbach, M. M. and Styne, D. M. (1992). Puberty: ontogeny, neuroendocrinology, physiology, and disorders. In J. D. Wilson and D. W. Foster (Eds.). *Williams Textbook of Endocrinology*, (8th ed.), W. B. Saunders and Company, Philadelphia pp. 1139–1221.

Hakansson, M.-L., Brown, H., Ghilardi, N., Skoda, R. C., and Meister, B. (1998). Leptin receptor immunoreactivity in chemically defined target neurons of the hypothalamus. *J. Neurosci., 18*, 559–572.

Herbison, A. E., Pape, J.-R., Simonian, S. X., Skynner, M. J., and Sim, J. A. (2001). Molecular and cellular properties of GnRH neurons revealed through transgenics in the mouse. *Mol. Cell. Endocrinol., 185*, 185–194.

Horlick, M. B., Rosenbaum, M., Nicolson, M., Levine, L. S., Fedun, B., Wang, J., Pierson, R. N., Jr., and Leibel, R. (2000). Effect of puberty on the relationship between circulating leptin and body composition. *J. Clin. Endocrinol. Metab., 85*, 2509–2518.

Huang, X. F., Koutcherov, I., Lin, S., Wang, H. Q., and Storlien, L. (1996). Localization of leptin receptor mRNA expression in mouse brain. *Neuroreport, 7*, 2635–2638.

Isidori, A. M., Caprio, M., Strollo, F., Moretti, C., Frajese, G., Isidori, A., and Fabbri, A. (1999). Leptin and androgens in male obesity: evidence for leptin contribution to reduced androgen levels. *J. Clin. Endocrinol. Metab., 84*, 3673–3680.

Jaffe, R. B. (1989). Fetal neuroendocrinology. In S. Mancuso (Ed.). *Achievements in Gynecology*, Karger, Basel, pp. 104–110.

Jockenhovel, F., Blum, W. F., Englaro, P., Muller-Wieland, D., Reinwein, D., Rascher, W., and Krone, W. (1997). Testosterone substitution normalizes elevated serum leptin levels in hypogonadal men. *J. Clin. Endocrinol. Metab., 82*, 2510–2513.

Karlsson, C., Lindell, K., Svensson, E., Bergh, C., Lind, P., Billig, H., Carlsson, L. M. S., and Carlsson, B. (1997). Expression of functional leptin receptors in the human ovary. *J. Clin. Endocrinol. Metab., 82*, 4144–4148.

Kaynard, A. H., Pau, K.-Y. F., Hess, D. L., and Spies, H. G. (1990). Third-ventricular infusion of neuropeptide Y suppresses luteinizing hormone secretion in ovariectomized rhesus macaques. *Endocrinology, 127*, 2437–2444.

Kratzch, J., Lammert, A., Mueller, G., Reich, A., De Paly, H., Seidel, B., and Kiess, W. (2001). Serum levels of the soluble leptin receptor decrease with age in healthy children and adolescents. *Annual Meeting of the Endocrine Society*, 267 (abstract PI-557).

Lammert, A., Kiess, W., Bottner, A., Glasow, A., and Kratzch, J. (2001). Soluble leptin receptor represents the main leptin binding activity in human blood. *Biochem. Biophys. Res. Comm., 283*, 982–988.

Lee, G.-H., Proenca, R., Montez, J. M., Carroll, K. M., Darvishzadeh, J. G., Lee, J. L., and Friedman, J. M. (1996). Abnormal splicing of the leptin receptor in Diabetic mice. *Nature, 379*, 632–635.

Licinio, J., Negrao, A. B., Mantzoros, C., Kaklamani, V., Wong, M., Bongiorno, P. B., Mulla, A., Cearnal, L., Veldhuis, J. D., Flier, J. S., McCann, S. M., and Gold, P. W. (1998). Synchronicity of frequently sampled, 24-h concentrations of circulating leptin, luteinizing hormone, and estradiol in healthy women. *Proc. Natl. Acad. Sci., 95*, 2541–2546.

Lollmann, B. D., Gruninger, S., Stricker-Krongrad, A., and Chiesi, M. (1997). Detection and quantification of the leptin receptor splice variants Ob-Ra, b, and e in different mouse tissue. *Biochem. Biophys. Res. Comm., 238*, 648–652.

Mann, D. R. and Castracane, V. D. (2001). A cross-sectional study of circulating leptin, and testosterone in normal male development in humans. *Annual Meeting of the Society of Study of Reproduction* 190 (abstract 214).

Mann, D. R. and Plant, T. M. (2002). Leptin and pubertal development. *Seminars in Reproductive Medicine, 20*, 93–102.

Mann, D. R., Akinbami, M. A., Gould, K. G., and Castracane, V. D. (2000). A longitudinal study of leptin during development in the male rhesus monkey: the effect of body composition and season on circulating leptin levels. *Biol. Reprod., 62*, 285–291.

Mann, D. R., Akinbami, M. A., Gould, K. G., and Castracane, V. D. (2002). Leptin and thyroxine during sexual development in male monkeys: effect of neonatal GnRH antagonist treatment and delayed puberty on the developmental pattern of leptin and thyroxine secretion. *Eur. J. Endocrinol., 46*, 891–898.

Mann, D. R., Akinbami, M. A., Gould, K. G., Paul, K., and Wallen, K. (1998). Sexual maturation in male rhesus monkeys: importance of neonatal testosterone exposure and social rank. *J. Endocrinol., 156*, 493–501.

Mantzoros, C. S., Flier, J. S., and Rogol, A. D. (1997). A longitudinal assessment of hormonal and physical alterations during normal puberty in boys. V. rising leptin levels may signal the onset of puberty. *J. Clin. Endocrinol. Metab., 82,* 1066–1070.

Montague, C. T., Farooqi, I. S., and Whitehead, J. P. (1997). Congenital leptin deficiency is associated with severe early-onset obesity in humans. *Nature, 387,* 903–908.

Nakanishi, T., Li, R., Liu, Z., Yi, M., Nakagawa, Y., and Ohzeki, T. (2001). Sexual dimorphism in relationship of serum leptin and relative weight for the standard in normal-weight, but not in overweight, children as well as adolescents. *Eur. J. Clin. Nutr., 55,* 989–993.

Ogura, K., Irahara, M., Kiyokawa, M., Tezuka, M., Matsuzaki, T., Yasui, T., Kamada, M., and Aono, T. (2001). Effects of leptin on secretion of LH and FSH from primary cultured female rat pituitary cells. *Eur. J. Endocrinol., 144,* 653–658.

Ojeda, S. R., Ma, Y. J., Dziedzic, B., and Prevot, V. (2000). Astrocyte-neuron signaling and the onset of female puberty. In J.-P. Bourguignon and T. M. Plant (Eds.). *The Onset of Puberty in Perspective,* Elsevier Science Publishers B. V., Amsterdam, pp. 41–83.

Ostlund, R. E., Yang, J. W., Klein, S., and Gingerich, R. (1996). Relation between plasma leptin concentration and body fat, gender, diet, age, and metabolic covariates. *J. Clin. Endocrinol. Metab., 81,* 3909–3913.

Palmert, M. R., Radovick, S., and Boepple, P. A. (1998). The impact of reversible gonadal sex steroid suppression on serum leptin concentrations in children with central precocious puberty. *J. Clin. Endocrinol. Metab., 83,* 1091–1096.

Parent, A. S., Lebrethon, M. C., Gerard, A., Vandersmissen, E., and Bourguignon, J. P. (2000). Leptin effects on pulsatile gonadotropin releasing hormone secretion from the adult rat hypothalamus and interaction with cocaine and amphetamine regulated transcript peptide and neuropeptide Y. *Regul. Pept., 92,* 17–24.

Pau, K.-Y. F., Berria, M., Hess, D. L., and Spies, H. G. (1995). Hypothalamic site-dependent effects of neuropeptide Y on gonadotropin-releasing hormone secretion in rhesus macaques. *J. Neuroendocrinol., 7,* 63–67.

Plant, T. M. (1994). Puberty in primates. In E. Knobil and J. D. Neill (Eds.). *The Physiology of Reproduction,* Raven Press, New York, pp. 453–485.

Plant, T. M. (2000). Ontogeny of GnRH gene expression and secretion in primates. In J.-P. Bourguignon and T. M. Plant (Eds.). *The Onset of Puberty in Perspective.* Elsevier Science Publisher B. V., Amsterdam, pp. 3–13.

Plant, T. M. (2001a). Neurobiological bases underlying the control of the onset of puberty in the rhesus monkey: a representative higher primate. *Frontiers Neurobiol., 22,* 107–139.

Plant, T. M. (2001b). Leptin, growth hormone, and the onset of primate puberty. *J. Clin. Endocrinol. Metab., 86,* 458–460.

Plant, T. M. and Durrant, A. R. (1997). Circulating leptin does not appear to provide a signal for triggering the initiation of puberty in the male rhesus monkey (*Macaca mulatta*). *Endocrinology, 138,* 4505–4508.

Plant, T. M., Fraser, M. O., Medhamurthy, R., and Gay, V. L. (1989a). Somatogenic control of GnRH neuronal synchronization during development in primates: a speculation. In H. A. Delemarre van de Waal, T. M. Plant, G. P. van Rees, and J. Schoemaker (Eds.). *Control of the Onset of Puberty, III,* Elsevier Science Publishers B. V., Amsterdam, pp. 111–121.

Plant, T. M., Gay, V. L., Marshall, G. R., and Arslan, M. (1989b). Puberty in monkeys is triggered by chemical stimulation of the hypothalamus. *Proc. Natl. Acad. Sci. USA, 86,* 2506–2510.

Pombo, M., Herrera-Justiniano, E., Considine, R. V., Hermida, R. C., Galvez, M. J., Martin, T., Barreiro, J., Casanueva, F. F., and Dieguez, C. (1997). Nocturnal rise of leptin in normal prepubertal and pubertal children and in patients with perinatal stalk-transection syndrome. *J. Clin. Endocrinol. Metab., 82,* 2751–2754.

Quinton, N. D., Smith, R. F., and Clayton, P. E. (1999). Leptin binding activity changes with age: the link between leptin and puberty. *J. Clin. Endocrinol. Metab., 84,* 2336–2341.

Randeva, H. S., Murray, R. D., Lewandowski, K. C., O'Callaghan, C. J., Horn, R., O'Hare, P., Brabant, G., Hillhouse, E. W., and Shalet, S. M. (2002). Differential effects of GH replacement on the components of the leptin system in GH-deficient individuals. *Endocrine Soc., 87,* 798–804.

Resko, J. A. and Ellinwood, W. E. (1985). Negative feedback regulation of gonadotropin secretion by androgens in fetal rhesus macaques. *Biol. Reprod., 33,* 346–352.

Romer, A. S. (1959). *The Vertebrate Story,* University of Chicago Press, Chicago.

Ronnekleiv, O. K. and Resko, J. A. (1990). Ontogeny of gonadotropin-releasing hormone-containing neurons in early fetal development of rhesus macaques. *Endocrinology, 126,* 498–511.

Rosenbaum, M., Nicoloson, M., Hirsch, J., Heymsfield, S. B., Gallagher, D., Chu, F., and Leibel, R. L. (1996). Effects of gender, body composition, and menopause on plasma concentrations of leptin. *J. Clin. Endocrinol. Metab., 81*, 3424–3427.

Rutenberg, G. W., Coelho, A. M., Jr., Lewis, D. S., Carey, K. D., and McGill, H. C., Jr. (1987). Body composition in baboons: evaluating a morphometric method. *Am J. Primat., 12*, 275–285.

Saad, M. F., Riad-Gabriel, M. G., Khan, A., Sharma, A., Michael, R., Jinagouda, S. D., Boyadjian, R., and Steil, G. M. (1998). Diurnal and ultradian rhythmicity of plasma leptin: effects of gender and adiposity. *J. Clin. Endocrinol. Metab., 83*, 453–459.

Saitoh, Y., Luchansky, L. L., Claude, P., and Terasawa, E. (1995). Transplantation of the fetal olfactory placode restores reproductive cycles in female rhesus monkeys (*Macaca mulatta*) bearing lesions in the medial basal hypothalamus. *Endocrinology, 136*, 2760–2769.

Schwartz, M. W., Seeley, R. J., Campfield, L. A., Burn, P., and Baskin, D. G. (1996). Identification of targets of leptin action in rat hypothalamus. *J. Clin. Invest., 98*, 1101–1106.

Sone, M., Nagata, H., Takekoshi, S., and Osamura, R. Y. (2001). Expression and localization of leptin receptor in the normal rat pituitary gland. *Cell Tissue Res., 305*, 351–356.

Stoving, R. K., Vinten, J., Handberg, A., Ebbesen, E. N., Hangaard, J., Hansen-Nord, M., Kristiansen, J., and Hagen, C. (1998). Diurnal variation of the serum leptin concentration in patients with anorexia nervosa. *Clin. Endocrinol., 48*, 761–768.

Strobel, A., Issad, I., Camoin, L., Ozata, M., and Strosberg, A. D. (1998). A leptin missense mutation associated with hypogonadism and morbid obesity. *Nat. Genet., 18*, 213–215.

Suter, K. J., Pohl, C. R., and Wilson, M. E. (2000). Circulating concentrations of nocturnal leptin, growth hormone, and insulin-like growth factor-I increase before the onset of puberty in agonadal male monkeys: potential signals for the initiation of puberty. *J. Clin. Endocrinol. Metab., 85*, 808–814.

Tena-Sempere, M., Pinilla, L., Zhang, F., Gonzalez, L. C., Huhtaniemi, I., Casanueva, F. F., Dieguez, C., and Aguilar, E. (2001). Developmental and hormonal regulation of leptin receptor (Ob-R) messenger ribonucleic acid expression in rat testis. *Biol. Reprod., 64*, 634–643.

Terasawa, E. and Fernandez, D. L. (2001) Neurobiological mechanisms of the onset of puberty in primates. *Endo. Rev., 22*, 111–151.

Urbanski, F. H. and Pau, F. K. Y. (1998). A biphasic developmental pattern of circulating leptin in the male rhesus macaque (*Macaca mulatta*). *Endocrinology, 139*, 2284–2286.

Wabitsch, M., Blum, W. F., Muche, R., Braun, M., Hube, F., Rascher, W., Heinze, E., Teller, W., and Hauner, H. (1997). Contribution of androgens to the gender difference in leptin production in obese children and adolescents. *J. Clin. Invest., 100*, 808–813.

Warren, M. P. (1990). Metabolic factors and the onset of puberty. In M. M. Grumbach, P. C. Sizonenko, and M. L. Aubert (Eds.). *Control of the Onset of Puberty*, Williams & Wilkins, Baltimore, pp. 553–573.

Yu, W. H., Kimura, M., Walczewska, A., Karanth, S., and McCann, S. M. (1997). Role of leptin in hypothalamic-pituitary function. *Proc. Natl. Acad. Sci. USA, 94*, 1023–1028.

Zamorano, P. L., Mahesh, V. B., De Sevilla, L. M., Chorich, L. P., Bhat, G. K., and Brann, D. W. (1997). Expression and localization of the leptin receptor in endocrine and neuroendocrine tissues of the rat. *Neuroendocrinology, 65*, 223–228.

Leptin and Pubertal Development in Humans

SUSANN BLÜHER AND CHRISTOS S. MANTZOROS

10.1. ROLE OF LEPTIN IN THE REGULATION OF THE HYPOTHALAMIC–PITUITARY–GONADAL AXIS

10.1.1. Production of Leptin and Expression of the Leptin Receptor

Since the discovery of leptin through positional cloning of the *ob* gene (Zhang et al., 1994) more than 8 years ago, extensive research efforts have focused on the biology and physiology of this hormone. Leptin is a 167-amino acid protein that is structurally related to members of the long-chain helical cytokine family (Madej et al., 1995; Zhang et al., 1997). It is expressed primarily in white adipose tissue (Maffei et al., 1995; Klein et al., 1996; Leroy et al., 1996). Leptin expression has also been detected in the stomach (Bado et al., 1998), hypothalamus (Morash et al., 1999), pituitary (Jin et al., 2000), skeletal muscle (Wang et al., 1998) placenta (Masuzaki et al., 1997), mammary gland (Smith-Kirwin et al., 1998; Aoki et al., 1999), and possibly in brown fat.

Leptin exerts its effects by activating the leptin receptor (OB-R), which belongs to the cytokine class I receptor family. Several isoforms of this receptor, resulting from alternative splicing, are involved in mediating leptin's actions in the hypothalamus and a variety of peripheral organs as well as the transport and clearance of this hormone (Tartaglia et al., 1995; Tartaglia, 1997). The long isoform (OB-Rb) is expressed abundantly in the hypothalamic

SUSANN BLÜHER AND CHRISTOS S. MANTZOROS • Division of Endocrinology, Department of Medicine, Beth Israel Deaconess Medical Center, Harvard Medical School.

Leptin and Reproduction. Edited by Henson and Castracane, Kluwer Academic/Plenum Publishers, 2003.

arcuate, ventromedial, and dorsomedial nuclei (Heretier et al., 1997; Huang et al., 1997). The short isoforms of the leptin receptor (OB-Ra, c, d, e, d) are present in a wide variety of peripheral tissues, including the ovary, prostate and testis, suggesting a direct effect of leptin on the gonads (Cioffi et al., 1996; Mantzoros, 1999). This wide distribution of the expression of both leptin and its receptor suggests that leptin is not only an "antiobesity hormone," but also a crucial endocrine factor in the regulation of physiological processes including immune function, inflammation, angiogenesis, hematopoiesis, and most importantly, reproductive function.

10.1.2. Leptin is Secreted in a Pulsatile Fashion, and Circulating Leptin Levels are Associated with those of Other Hormones

Plasma leptin is secreted in a pulsatile, circadian fashion with a peak in the early morning and a nadir in the afternoon. This circadian pattern has been found in lean and obese subjects, healthy controls and patients with type 2 diabetes mellitus as well as adults and teenagers (Sinha et al., 1996a,b; Licinio et al., 1997). Leptin's pulsatility characteristics are similar in lean and obese subjects with the only exception being pulse amplitude, which is higher in obese subjects. Diurnal and circadian oscillations are also physiological characteristics of several other hormones. Interestingly, thyrotropin (TSH) rhythms have been observed to be similar to that of leptin and more importantly, appear to be synchronous to those of leptin (Mantzoros et al., 2001). In addition, the circadian rhythm of leptin is similar to that of prolactin, free fatty acids, and melatonin (Matkovic et al., 1997a), but inversely related to that of ACTH and cortisol (Licinio et al., 1997). More importantly, leptin pulsatility is synchronous to the pulsatility of serum luteinizing hormone (LH) and estradiol levels in normal women, especially during the night when leptin levels are relatively high. This implies a role for leptin in the regulation of physiologic levels and rhythmicity of reproductive hormones (Licinio et al., 1998a,b).

10.1.3. Central Role of Leptin in Regulating Reproductive Function—Leptin and the Hypothalamic–Pituitary Axis

The long isoform of the leptin receptor, OB-Rb, is highly expressed in the arcuate and ventromedial hypothalamic nuclei, regions that control both food intake and sexual behavior. Thus, many studies have attempted to elucidate a possible role for leptin not only in regulating feeding and energy homeostasis, but also in regulating the onset of puberty/reproduction and the regulation of the hypothalamic–pituitary–gonadal (HPG) axis in the post-pubertal life. Basic research studies have demonstrated that leptin receptors are expressed on GnRH-secreting neurons, and that leptin accelerates GnRH-pulsatility (but not the pulse amplitude) in arcuate hypothalamic neurons thus regulating the release of gonadotropins (Yu et al., 1997a; Cunningham et al., 1999; Lebrethon et al., 2000a). It has therefore been suggested that leptin serves as a signal to convey information to the brain that the metabolic resources as well as body fat stores are adequate and acts as a permissive signal to activate the reproductive axis and to trigger the onset of puberty. It is believed that leptin may facilitate GnRH secretion predominantly via indirect mechanisms, acting through altering the secretion of neuropeptides (Terasawa, 1998). Although the downstream effectors of leptin important in mediating leptin's effect on GnRH release remain to be fully elucidated, neuropeptide Y (NPY) does not appear to be the only or main mediator of leptin's regulation on GnRH release, since NPY knockout mice show normal reproductive function (Erickson et al., 1997). NPY, however, has been shown to have both stimulatory and inhibitory effects on GnRH secretion depending on several factors such as gender, acute versus chronic action, and the state of sexual maturation (i.e., pre-pubertal versus adult). Furthermore, it has been shown that leptin and NPY use different signaling pathways to stimulate GnRH secretion (Lebrethon et al., 2000a). In addition, melanocortins have not been shown to play a significant role in reproduction, as recent experimental evidence suggests that activation of melanocortin receptors does not mediate the effect of leptin on reproduction (Argiolas et al., 2000; Hohmann et al., 2000). In concordance with these data, melanocortin-4-receptor [MC4-R] knockout mice have been shown to be obese but reproductively competent (Huszar et al., 1997). Other neuropeptides important in the regulation of the reproductive axis include melanocortin-concentrating hormone (MCH), cocaine-and-amphetamine regulated transcript (CART) and galanin-like peptide (GALP) (Jureus et al., 2000; Lebrethon et al., 2000b; Murray et al., 2000; Parent et al., 2000). Leptin also may increase, probably by acting on its own receptor at the hypothalamic level, the release of nitric oxide (NO) from adrenergic interneurons, which then induces GnRH release from GnRH neurons by activating both guanylate cyclase and cyclooxigenase. (Yu et al., 1997a, b).

In the pituitary, the leptin receptor is expressed in almost 90 percent of the gonadotropes in the pars tuberalis and in 30 percent of the gonadotropes in the pars distalis (Iqbal et al., 2000) suggesting that leptin may have direct effects at the level of the pituitary. Moreover, almost 25 percent of anterior pituitary cells, predominantly folliculostellate cells and corticotropes, express leptin. This suggests a potential regulatory function of leptin on growth and differentiation of pituitary cells (Jin et al., 2000) in addition to its effect to directly stimulate LH and follicle-stimulating hormone (FSH) release by the pituitary gland, in part via NO synthase activation in gonadotropes (Yu et al., 1997b).

10.1.4. Peripheral Role of Leptin in Regulating Reproductive Function: Leptin and the Gonads

Leptin has also been found to have important effects on the gonads. Ovarian follicular cells, including granulosa, theca, and interstitial cells, as well as Leydig cells express a functional leptin receptor (Cioffi et al., 1996; Karlsson et al., 1997; Caprio et al., 1999). These findings, and the fact that leptin-mRNA is also expressed in granulosa and cumulus cells of pre-ovulatory human follicles (Cioffi et al., 1997), suggest endocrine and/or direct paracrine effects of leptin on the gonads. In-vivo studies have demonstrated that relatively high concentrations of leptin exert an inhibitory effect on ovulation although no changes in steroid secretion were noted in the absence of growth factor augmentation (Duggal et al., 2000). Other studies have shown that leptin, in medium to high physiological doses, antagonizes the stimulating effects of several growth factors and hormones such as IGF-1, insulin, and glucocorticoids, on gonadotropin stimulated steroidogenesis in ovarian cells (Spicer and Francisco, 1998; Agarwal et al., 1999; Barkan et al., 1999; Brannian et al., 1999; Kitawaki et al., 1999). Similarly, leptin in supraphysiologic doses, has been shown to inhibit dose dependently testosterone production in the Leydig cells of the testis (Caprio et al., 1999). It is thus tempting to speculate that leptin exerts a bimodal action on the HPG axis depending on serum leptin levels. Specifically, leptin deficiency results in down-regulation of the HPG axis (Farooqi et al., 1999), and leptin may also have an inhibitory effect on the gonads at the high serum leptin levels seen in extreme obesity. This may explain the reproductive abnormalities seen in states of both leptin deficiency and excess.

10.2. NORMAL PUBERTAL MATURATION

10.2.1. Physiological Aspects

Puberty is associated with accelerated linear growth, achievement of reproductive maturity, rapid increase of muscle mass, and mineralization of the skeleton. Pubertal development is divided into five stages (Tanner Stages 1–5). Each stage represents the extent of pubic hair growth and breast development (girls) respectively pubic hair growth and genital development (boys). Tanner Stage 1 is regarded to be pre-pubertal. In Tanner Stage 2, breast bud and papilla are elevated in girls, and scrotum and testes have enlarged in boys. Pubic hair starts growing sparsely. Tanner Stage 3 is characterized by further enlargement of breast mound and increased palpable glandular tissue in girls, whereas growth of the penis has occurred in boys. Pubic hair growth continues and spreads to the pubic junction. Tanner Stage 4 includes elevation of the female areola and papilla as well as development of the male glans penis and further enlargement of the testes and scrotum. The pubic hair is now adult in type, and the underarm hair growth is noticeable. In Tanner Stage 5, genitalia as well as pubic and underarm hair are adult in size and shape. The first menstrual period in girls usually occurs during Tanner Stage 4 or 5. Normal progression through the Tanner stages varies widely depending on the individuality and ethnicity of the adolescent (Marshall and Tanner, 1969, 1970). The onset of puberty is closely related to hypothalamic activation which subsequently leads to stimulation of the gonads. More specifically, up-regulation or activation of several hormonal axes, including leptin and both the HPG and growth hormone/insulin-like growth factor-I (GH/IGF-I) axis seem to play a role in the initiation and completion of the complex process of puberty (Moschos et al., 2002). It has been hypothesized that a critical body weight, a critical fat mass, and/or crit-ical levels of metabolites linked to fat mass have to be achieved before puberty can occur (Kennedy and Mitra, 1963; Frisch and Revelle, 1970; Frisch, 1980). More recent clinical observations (Jaruratanasirikul et al., 1997) have led to the conclusion that obese girls reach menarche at an earlier age than normal weight subjects. Moreover, observational studies in both genders indicate that leptin levels rise prior to the initiation of puberty and may trigger the onset of puberty in humans. These studies are in accordance with the hypothesis that increasing leptin levels contribute to the initiation of puberty. However, although there is evidence from observational studies in humans and interven-tional studies in animals that leptin might be the fat derived signal that activates the other neuroendocrine axes, this remains to be conclusively shown by interventional studies in humans.

10.2.2. Neuroendocrine Aspects

The onset of puberty is characterized by a significant change in the pulsatile characteristics of the hypothalamic "gonadostat." The amplitude of pulses of hypothalamic GnRH neurons is increased, first leading to a substantial rise in nocturnal FSH, and then to a rise in LH pulsatile release by the pituitary (Jakacki et al., 1982). This results in a remarkably higher output of sex steroids by the gonads, which has also been linked to an increased production of GH and IGF-1. The combined increase in GH and IGF-1 affects linear growth, muscle bulk, and mineralization of the skeleton (Mauras et al., 1996). Although the secretion of GH, its effector peptide IGF-1, the GH dependent IGF-1 carrier protein IGF-BP3, and the sex steroids, peak through mid-puberty, leptin levels already increase during pre-puberty. This suggests an interaction between leptin and the gonadotropins in late childhood (Clayton et al., 1997) and a role of leptin in the initiation of puberty (see Section 10.3).

10.3. LEPTIN AND NORMAL PUBERTAL DEVELOPMENT

10.3.1. Leptin in Childhood

10.3.1.1. Circulating Leptin Levels in Neonates. During pregnancy, high leptin levels are present in the amniotic fluid and in the arterial and venous cord blood. (Mantzoros et al., 1997b; Matsuda et al., 1997; Schubring et al., 1997; Sivan et al., 1997; Varvarigou et al., 1999; Christou et al., 2001). Umbilical cord blood leptin levels are derived from both the placenta (Masuzaki et al., 1997) and the fetal tissue (Schubring et al., 1997) and correlate positively with the absolute body weight and fat mass of the neonate. These levels are lower in pre-term and small-for-gestational-age neonates, and are higher in large-for-gestational-age neonates than in normal neonates (Mantzoros et al., 1997b).

High leptin concentrations have also been reported in arterial and venous cord blood at birth as well as in capillary and venous blood shortly after birth. Interestingly, neonate girls have been shown to have higher leptin levels than boys probably because of their higher estrogen levels immediately after birth. Circulating leptin levels decrease rapidly and dramatically after birth (Kiess et al., 1998; Schubring et al., 1999).

10.3.1.2. Circulating Leptin Levels during Childhood. A strong positive correlation between leptin levels and body fat mass, and to a lesser extent between leptin and body mass index (BMI) as well as weight, has been reported in all age groups during childhood. The well-established pulsatile and nyctohemeral rhythm of leptin secretion in adults is also thought to be present in children (Hassink et al., 1996; Havel et al., 1996; Kiess et al., 1996; Blum et al., 1997; Clayton et al., 1997; Ellis and Nicolson, 1997; Lahlou et al., 1997; Matkovic et al., 1997a). These findings are unaffected by ethnicity (Ellis and Nicolson, 1997). The major gender-related differences in leptin secretion seem to be due to pulse amplitude rather than frequency, as the leptin release from adipose tissue into blood is 2–3-fold higher in females than in males (Licinio et al., 1998b; Saad et al., 1998). During the entire physical and sexual maturation period and in adulthood, female subjects have higher serum leptin levels than males, even after normalization for BMI. This is probably due to the positive effects of estrogen on leptin production as well as the higher percentage of subcutaneous versus intraabdominal fat mass in women (Roemmich et al., 1998).

10.3.2. Leptin in Puberty

10.3.2.1. Circulating Leptin Levels and the Onset of Puberty. It has been proposed that since the timing of puberty is predicted more precisely by body weight than by chronological age, a "critical body weight" or a "critical fat mass" must be achieved before puberty occurs (Kennedy and Mitra, 1963; Frisch and Revelle, 1970; Frisch, 1980). Several factors have been suggested as potential signals that might link growth, adiposity, or metabolic status with pubertal development (Barash et al., 1996; Cameron, 1997). Recent evidence implicates that leptin treatment is capable of normalizing the HPG axis in mouse models lacking leptin (Barash et al., 1996; Chebab et al., 1996a) as well as in normal mice with decreased leptin levels in response to starvation (Ahima et al., 1996). In addition, animal experiments have shown that leptin administration accelerates the onset of puberty in normal mice (Chebab et al., 1996b; Ahima et al., 1997). However, many issues related to this hypothesis still remain unanswered, such as whether animal experiments can be directly extrapolated to humans and/or whether leptin is merely a permissive factor or provides a direct signal for triggering pubertal onset (Hileman et al., 2000).

Leptin circulates in the blood and is present in both a free and protein-bound form. The leptin binding activity (LBA) has been discussed as a potential link between leptin and puberty, as it has been reported to be low at birth, to peak in the pre-pubertal years, and to fall through puberty. This leads to an increase in free leptin levels and therefore an increase in leptin activity. Furthermore, LBA is correlated with pubertal status in both boys and girls. It has been postulated that the fall in LBA, which is associated with increasing age and stage of puberty, makes leptin available to the full-length receptors. This action could potentially transmit the biological signals for leptin (Quinton et al., 1999).

10.3.2.2. Leptin in Male Puberty. Several longitudinal and cross-sectional studies aimed to investigate the physiology of circulating leptin during Tanner stages in both genders. After the initiation of puberty and during pubertal maturation, leptin levels are not only dependent on body fat stores, but also on the Tanner stage of sexual development (Hassink et al., 1996). As in childhood and in adults, circulating leptin levels are also significantly higher in female than in male subjects. This sexual dimorphism seems to increase with proceeding Tanner stages of pubertal development. Roemmich et al. (1998) reported that the gender differences in leptin levels during puberty are related to both the subcutaneous fat depot and the androgens. In boys, leptin levels rise by approximately 50 percent just before the onset of puberty and reach their peak immediately after the initiation of puberty, independently of increasing BMI (Clayton et al., 1997; Mantzoros et al., 1997a). As testosterone increases with the evolution of puberty through the different Tanner stages, leptin declines progressively to baseline levels (Blum et al., 1997; Horlick et al., 2000). Thus, there is a significant negative correlation between circulating testosterone levels and leptin levels in male, but not female subjects, as well as between testicular volume and leptin levels in male adolescents (Clayton et al., 1997; Roemmich et al., 1998). The sexual dimorphism evident in leptin concentrations in later puberty is therefore likely to be due to the suppressive effect of testosterone on both leptin mRNA expression and leptin secretion (Blum et al., 1997; Hislop et al., 1999; Horlick et al., 2000). This is consistent with observations in post-pubertal boys: the leptin concentration per kilogram body fat is less than before puberty (Lahlou et al., 1997; Wabitsch et al., 1997).

In contrast to testosterone, estradiol seems not to be a major determinant of leptin secretion in men, as short-term reduction in serum estradiol by an aromatase inhibitor does not affect serum leptin levels in male adolescents (Luukkaa et al., 2000).

10.3.2.3. Leptin in Female Puberty. Several clinical studies have focused on leptin's secretion pattern during puberty in female subjects. Similar to boys, girls show a progressive increase in leptin levels from pre-puberty into early puberty. This suggests a role of leptin as a permissive factor or a factor triggering the onset of puberty. However, unlike boys, leptin levels in girls continue rising in late puberty (Blum et al., 1997; Clayton et al., 1997), a fact which leads to significantly higher leptin levels, normalized for body fat mass, in girls during late puberty. This persists into adulthood (Castracane et al., 1998; Horlick et al., 2000).

Early studies reported an inverse relation between body weight and fat mass and between body weight and age at menarche (Frisch, 1980). Recent findings suggest that the link between age at menarche and fat mass is leptin since there is an inverse relationship between leptin and the age at menarche, with an increase of 1 ng/ml in serum leptin levels associated with an earlier onset of menarche by 1 month (Matkovic et al., 1997b). Finally, although most observational studies have focused on the association between the relative total body fat and the average daytime serum leptin level, it has also been reported that the percentage increase in the nocturnal leptin level which is inversely related to the percentage gain in total body weight may also be of importance (Matkovic et al., 1997a). However, it remains unclear, what role circadian fluctuations of leptin concentrations might play in the triggering of sexual maturation (Kiess et al., 2000).

10.3.3. Leptin's Role in Activating the HPG Axis during Human Puberty

As mentioned above, leptin has been shown to have direct stimulatory effects on the HPG axis on both the hypothalamic and the pituitary level. To summarize the data discussed in Section 10.1.3., GnRH-secreting neurons express leptin receptors (Cunningham et al., 1999), and leptin is capable of increasing the pulsatility of GnRH in the arcuate hypothalamus (Lebrethon et al., 2000a). On the pituitary, leptin directly stimulates LH, and to a lesser extend, FSH release, mainly via NO synthetase activation in gonadotropes (Yu et al., 1997b). Finally, leptin has been found to exert endocrine and/or direct paracrine effects on the gonadal organs. The strong interaction between leptin and the gonadotropins led to the hypothesis that leptin may be one of the factors sending information to the brain about the body's fat stores and thus conveying information on when the body is "ready" for pubertal maturation (Weiss et al. 1999). In conclusion, available evidence indicates that leptin signals

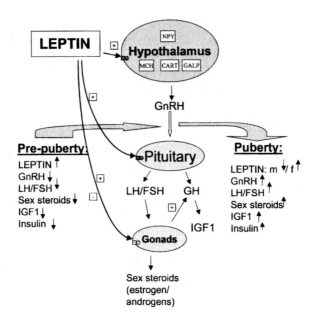

Figure 10.1. Hypothalamic–pituitary–gonadal axis and leptin in pre-puberty and puberty: leptin levels increase during pre-puberty. At the onset of puberty, the pulse amplitude of hypothalamic GnRH increases leading to a substantial increase in FSH and LH release by the pituitary. This results in a higher output of sex steroids by the gonads, thus increasing the production of GH and IGF-1. Sexual dimorphism of leptin levels increases with advancing Tanner stages of pubertal development.

to the brain information about metabolic fuel availability (Roemmich and Rogol, 1999) and thus triggers increase in GnRH pulses at the onset of puberty (Figure 10.1).

10.3.4. Leptin and Sexual Maturation in Nonhuman Primates

To help elucidate the physiological role of leptin during pubertal development and sexual maturation in the primate, several studies addressing this issue have been performed in macaques. However, a potential role of leptin in triggering the onset of puberty in nonhuman primates remains controversial: In a longitudinal study employing male rhesus monkeys, leptin levels declined through the juvenile period until the onset of puberty. Furthermore, leptin levels were negatively correlated with body weight and did not exhibit any association with peripubertal changes in serum concentrations of LH, testosterone, or testicular volume. (Mann et al., 2000). This study confirms earlier investigations by Plant and Durrant (1997) as well as Urbanski and Francis Pau (1998) suggesting that leptin may not act as a metabolic trigger toward the initiation of puberty in male rhesus monkeys. Supporting this data, short-term leptin administration to food-restricted male rhesus macaques did not effect circulating LH, testosterone, or cortisol levels either during or after the infusion. Finally, leptin had no direct effect on basal or LH-stimulated testosterone production in Leydig cells (Lado-Abeal et al., 1999) indicating that the respective role of leptin may be different in rodents than nonhuman primates. Whether leptin plays a role in triggering puberty in humans remains to be shown by interventional studies.

However, several studies strongly suggest that leptin may serve as an important factor regulating gestation and pregnancy in both nonhuman and human primates. In baboon pregnancy, maternal leptin levels increase with advancing gestational age and may derive from both adipose tissue and the placenta (O'Neil et al., 2001). In addition, maternal peripheral leptin concentrations arc positively and placental leptin mRNA levels are negatively correlated with gestational age (Henson et al., 1999). Leptin might thus regulate growth and development of the conceptus, fetal, and placental angiogenesis, embryonic hematopoiesis, and hormone biosynthesis within the maternal-fetoplacental unit during pregnancy (Henson and Castracane, 2000).

10.4. LEPTIN AND ABERRANT PUBERTAL DEVELOPMENT

10.4.1. Obesity and Puberty

Animal studies as well as rare clinical observations in leptin or leptin receptor-deficient humans have shown that obesity, caused by either mutations in the leptin gene or in the leptin receptor gene, is associated with reproductive dysfunction/hypogonadism. These clinical observations included a leptin-deficient 34-year-old woman with primary amenorrhea (Strobel et al., 1998), a 22-year-old male with leptin deficiency who had failed to undergo puberty and presented with hypothalamic hypogonadism (Strobel et al., 1998), as well as three leptin-resistant girls (age 13–19) who failed to enter puberty and presented with primary amenorrhea (Clement et al., 1998). There are several reports about children with leptin deficiency, including a 9-year-old leptin deficient girl who was treated with daily injections of leptin and subsequently developed a pulsatile, nocturnal, secretory pattern of gonadotropins consistent with early puberty (Montague et al., 1997; Farooqi et al., 1999).

With the exception of the rare cases of gene mutations, human obesity is normally associated with high leptin levels due to leptin resistance rather than leptin deficiency (Matsuoka et al., 1997). Both obese children and obese adults have elevated leptin levels compared to age-matched controls (Hassink et al., 1996). Interestingly, Bray (1997) has described a positive association between increasing adiposity and rising numbers of anovulatory cycles. High leptin levels directly inhibit ovarian steroidogenesis leading to ineffective follicular maturation (Duggal et al., 2000). In perimenarcheal and young adult girls, LH and FSH responses to GnRH are negatively correlated with BMI and circulating leptin levels. Decreased LH and FSH responses to GnRH were associated with increased adiposity and hyperleptinemia suggesting a direct neuroendocrine effect of excess leptin on the central reproductive system of female adolescents, but not in boys of comparable adiposity (Bouvattier et al., 1998).

10.4.2. Eating Disorders and Puberty

Not only extremely obese adolescents, but also subjects with low body fat mass and subjects who are starving have delayed puberty and impaired development of the reproductive system. Eating disorders, including anorexia nervosa and bulimia nervosa, are associated with significantly lower serum leptin levels in both females and males compared to healthy, normal weight controls. However, leptin levels still correlate with body weight and percentage of body fat (Grinspoon et al., 1996; Mantzoros et al., 1997c; Jimerson et al., 2000; Wabitsch et al., 2001).

Anorexia nervosa, caused by a drastic decrease in food intake, results in progressive weight loss and reproductive disturbances such as hypothalamic amenorrhea. When weight falls below the threshold of about 70 percent of the mean weight of a matched population, multiple endocrine axes apart from the reproductive axis are disrupted (Cantopher et al., 1988). In female patients with anorexia nervosa, dietary treatment leading to weight gain interestingly causes an increase in leptin levels and serum LH/FSH, implying that the rise in leptin levels may activate the HPG axis (Ballauff et al., 1999). However, amenorrheic and eumenorrheic female subjects with anorexia nervosa frequently show no difference in serum leptin concentrations, implying either that there are significant interindividual differences with respect to responsiveness to leptin or that other endocrine axes, such as the GH/IGF1-axis, also play an important role in normalizing reproductive function in this disease (Audi et al., 1998). Similarly in male patients with anorexia nervosa, changes in leptin levels during weight gain are significantly correlated with changes in gonadotropins, testosterone, and free androgen index. This suggests that leptin may regulate the HPG axis in both genders with this disease.

In bulimia nervosa, characterized by binge eating, the body weight tends to be normal. Similar to patients with anorexia nervosa, patients with bulimia nervosa have significantly lower serum leptin levels compared to weight- and age-matched healthy subjects. These levels remain low even during remission periods but are not as low as in patients with anorexia nervosa (Jimerson et al., 2000). Disturbances in the regular onset of puberty are still poorly understood in bulimia nervosa, but menstrual irregularities seem to occur less frequently than in patients with anorexia nervosa (Cantopher et al., 1988).

10.4.3. Strenuous Exercise and Puberty

Extreme exercise, as commonly seen in elite athletes or ballet dancers, also results in very low leptin levels and disrupted reproductive system. Elite gymnasts, both male and female, frequently present with low body fat mass and retarded pubertal development and growth. They have significantly lower serum leptin levels than age- and weight-matched controls (Weimann et al., 1999). Female elite gymnasts have delayed menarche and amenorrhea resulting from the suppression of GnRH pulsatility leading to low estrogen levels. Female athletes not only present with hypoleptinemia, but also with absence of diurnal rhythm in leptin pulsatility. It has therefore been suggested that the amenorrhea in these women is not only related to a negative energy balance, but also due to a low body fat mass resulting in low leptin levels (Laughlin and Yen, 1996, 1997).

Similar to females, high impact training in male athletes appears to suppress the HPG axis thus resulting in decreased testosterone levels and impaired reproductive function (MacConnie et al., 1986).

10.4.4. Precocious and Delayed Puberty

Observational studies focussing on the role of leptin in precocious and delayed puberty have shown that leptin levels in girls with precocious puberty are significantly elevated compared to age-matched girls, but are appropriate for BMI, fat mass, gender, and developmental stage in these subjects. Thus, precocious puberty is not a determining factor of serum leptin concentrations (Heger et al., 1999).

Constitutional delay in growth and pubertal development, which is a disorder characterized by delayed timing and progression of physical maturation, represents the other end of the spectrum. As expected, constitutional delay in puberty and growth in boys is associated with hypoleptinemia, even when corrected for BMI (Gill et al., 1999).

Elite gymnasts represent another group of subjects which frequently present with retarded pubertal development and growth as well as low body fat mass. Both female and male elite gymnasts have significantly lower serum leptin levels than age- and weight-matched controls (Weimann et al., 1999). Female elite gymnasts frequently present with delayed menarche and amenorrhea supporting the "critical body weight theory" (Frisch and Revelle, 1970). Ongoing clinical trials involving leptin administration will provide definitive evidence on whether their low leptin levels are responsible for the reproductive abnormalities seen in these subjects.

10.5. CONCLUSIONS AND FUTURE DIRECTIONS

Leptin has been suggested to be a permissive factor conveying information to the brain on body's fat stores and metabolic fuel availability, thus triggering the onset of pubertal development and maintaining normal function of the HPG axis in post-pubertal life. Several conditions including eating disorders, extreme cases of obesity, and strenuous exercise, are associated with altered serum leptin levels and abnormalities in both pubertal development and reproductive function. However, because most of the data available so far has been derived from animal studies or observational studies in humans, there is no conclusive evidence for the role of leptin in normal pubertal maturation and reproduction. Interventional studies that are currently underway in our institution are expected to provide conclusive evidence that may lead to the development of new therapeutic strategies for these pathophysiological conditions.

REFERENCES

Agarwal, S. K., Vogel, K., Weitsman, S. R., and Magoffin, D. A. (1999). Leptin antagonizes the insulin-like growth factor-I augmentation of steroidogenesis in granulosa and theca cells of the human ovary. *J. Clin. Endocrinol. Metab., 84*, 1072–1076.

Ahima, R. S., Prabakaran, D., Mantzoros, C., Qu, D., Lowell, B., Maratos-Flier, E., and Flier, J. S. (1996). Role of leptin in the neuroendocrine response to fasting. *Nature, 382*, 250–252.

Ahima, R. S., Dushay, J., Flier, S. N., Prabakaran, D., and Flier, J. S. (1997). Leptin accelerates the timing of puberty in normal female mice. *J. Clin. Invest., 99*, 391–395.

Aoki, N., Kawamura, M., and Matsuda, T. (1999). Lactation-dependent down regulation of leptin production in mouse mammary gland. *Biochim. Biophys. Acta, 1427*, 298–306.

Argiolas, A., Melis, M. R., Murgia, S., and Schioth, H. B. (2000). ACTH- and alpha-MSH-induced grooming, stretching, yawning, and penile erection in male rats: sites of action in the brain and role of melanocortin receptors. *Brain Res. Bull., 51*, 425–431.

Audi, L., Mantzoros, C. S., Vidal-Puig, A., Vargas, D., Gussinye, M., and Carrascosa, A. (1998). Leptin in relation to resumption of menses in women with anorexia nervosa. *Mol. Psychiatry, 3*, 544–547.

Bado, A., Levasseur, S., Attoub, S., Kermorgant, S., Laigneau, J. P., Bortoluzzi, M. N., Moizo, L., Lehy, Th., Guerre-Millo, M., Le Marchand-Brustel, Y., and Lewin, M. J. M. (1998). The stomach is a source of leptin. *Nature, 394*, 790–793.

Ballauff, A., Ziegler, A., Emons, G., Sturm, G., Blum, W. F., Remschidt, H., and Hebebrand, J. (1999). Serum leptin and gonadotropin levels in patients with anorexia nervosa during weight gain. *Mol. Psychiatry, 4*, 71–75.

Barash, I. A., Cheung, C. C., Weigle, D. S., Ren, H., Kabigting, E. B., Kuijper, J. L., Clifton, D. K., and Steiner, R. A. (1996). Leptin is a metabolic signal to the reproductive system. *Endocrinology, 137*, 3144–3147.

Barkan, D., Jia, H., Dantes, A., Vardimon, L., Amsterdam, A., and Rubinstein, M. (1999). Leptin modulates the glucocorticoid-induced ovarian steroidogenesis. *Endocrinology, 140*, 1731–1738.

Blum, W. F., Englaro, P., Hanitsch, S., Juul, A., Hertel, N. T., Kiess, W., Skakkebaek, N. E., Birkett, M., Heiman, M., Attanasio, A., and Rascher, W. (1997). Plasma leptin levels in healthy children and adolescents: dependence on body mass index, body fat mass, gender, pubertal stage, and testosterone. *J. Clin. Endocrinol. Metab., 82*, 2904–2910.

Bouvattier, C., Lahlou, N., Roger, M., and Bougneres, P. (1998). Hyperleptinemia is associated with impaired gonadotrophin response to GnRH during late puberty in obese girls, not boys. *Eur. J. Endocrinol., 138*, 653–658.

Brannian, J. D., Zhao, Y., and McElroy, M. (1999). Leptin inhibits gonadotrophin-stimulated granulosa cell progesterone production by antagonizing insulin action. *Hum. Reprod., 14*, 1445–1448.

Bray, G. A. (1997). Obesity and reproduction. *Hum. Reprod., 12*, 26–32.

Cameron, J. L. (1997). Search for the signal that conveys metabolic status to the reproductive axis. *Curr. Opin. Endocrinol. Diabetes, 4*, 158–163.

Cantopher, T., Evans, C., Lacey, J. H., and Pearce, J. M. (1988). Menstrual and ovulatory disturbance in bulimia. *BMJ, 297*, 836–837.

Castracane, V. D., Kraemer, R. R., Franken, M. A., Kraemer, G. R., and Gimpel, T. (1998). Serum leptin concentration in women: effect of age, obesity, and estrogen administration. *Fertil. Steril., 70*(3), 472–477.

Caprio, M., Isidori, A. M., Carta, A. R., Moretti, C., Dufau, M. L., and Fabbri, A. (1999). Expression of functional leptin receptors in rodent Leydig cells. *Endocrinology, 140*, 4939–4947.

Chebab, F., Lim, M., and Lu, R. (1996a). Correction of the sterility defect in homozygous obese female mice by treatment with human recombinant leptin. *Nat. Genet., 12*, 318–320.

Chebab, F. F., Mounzih, K., Lu, R., and Lim, M. E. (1996b). Early onset of reproductive function in normal mice treated with leptin. *Science, 275*, 88–90.

Christou, H., Connors, J. M., Ziotopoulou, M., Hatzidakis, V., Papathanassoglou, E., Ringer, S. A., and Mantzoros, C. S. (2001). Cord blood leptin and insulin-like growth factor levels are independent predictors of fetal growth. *J. Clin. Endocrinol. Metab., 86*, 935–938.

Cioffi, J. A., Shafer, A. W., Zupanicic, T. J., Smih-Gbur, J., Mikhail, A., Platika, D., and Snodgrass, H. R. (1996). Novel B219/OB receptor isoforms: possible role of leptin in hematopoiesis and reproduction. *Nat. Med., 2*, 585–589

Cioffi, J. A., Van Blerkom, J., Antczak, M., Shafer, A., Wittmer, S., and Snodgrass, H. R. (1997). The expression of leptin and its receptors in pre-ovulatory human follicles. *Mol. Hum. Reprod., 3*, 467–472.

Clayton, P. E., Gill, M. S., Hall, C. M., Tillmann, V., Whatmore, A. J., and Price, D. A. (1997). Serum leptin through childhood and adolescence. *Clin. Endocrinol., 46*(6), 727–733.

Clement, K., Vaisse, C., Lahlou, N., Cabrol, S., Pelloux, V., Cassuto, D., Gourmelen, M., Dina, C., Chambaz, J., Lacorte, J. M., Basdevant, A., Bougneres, P., Lebouc, Y., Froguel, P., and Guy-Grand, B. (1998). A mutation in the human leptin receptor gene causes obesity and pituitary dysfunction. *Nature, 392*, 398–401.

Cunningham, M. J., Clifton, D. K., and Steiner, R. A. (1999). Leptin's actions on the reproductive axis: perspectives and mechanisms. *Biol. Reprod., 60*, 216–222.

Duggal, P. S., Van Der Hoek, K. H., Milner, C. R., Ryan, N. K., Armstrong, D. T., Magoffin, D. A., and Norman, R. J. (2000). The in vivo and in vitro effects of exogenous leptin on ovulation in the rat. *Endocrinology, 141*, 1971–1976.

Erickson, J. C., Ahima, R. S., Hollopeter, G., Flier, J. S., and Palmiter, R. D. (1997). Endocrine function of neuropeptide Y knockout mice. *Regul. Pept., 70*, 199–202.

Ellis, K. J. and Nicolson, M. (1997). Leptin levels and body fatness in children: effects of gender, ethnicity, and sexual development. *Pediatr. Res., 42*, 484–488.

Farooqi, I. S., Jebb, S. A., Langmack, G., Lawrence, E., Cheetham, C. H., Prentice, A. M., Hughes, I. A., McCamish, M. A., and O'Rahilly, S. (1999). Effects of recombinant leptin therapy in a child with congenital leptin deficiency. *N. Engl. J. Med., 341*, 879–884.

Frisch, R. E. and Revelle, R. (1970). Height and weight at menarche and a hypothesis of critical body weights and adolescent events. *Science, 169*, 397–399.

Frisch, R. E. (1980). Pubertal adipose tissue: it is necessary for normal sexual maturation? Evidence from the rat and human female. *Fed. Proc., 39*, 2395–2400.

Gill, M. S., Hall, C. M., Tillmann, V., and Clayton, P. E. (1999). Constitutional delay in growth and puberty (CDGP) is associated with hypoleptinemia. *J. Clin. Endocrinol. Metab., 50*, 721–736.

Grinspoon, S., Gulick, T., Askari, H., Landt, M., Lee, K., Anderson, E., Ma, Z., Vignati, L., Bowsher, R., Herzog, D., and Klibanski, A. (1996). Serum leptin levels in women with anorexia nervosa. *J. Clin. Endocrinol. Metab., 81*, 3861–3863.

Hassink, S. G., Sheslow, D. V., de Lancey, E., Opentanova, I., Considine, R. V., and Caro, J. F. (1996). Serum leptin in children with obesity: relationship to gender and development. *Pediatrics, 98*, 201–203.

Havel, P. J., Kasim-Karakas, S., Dubuc, G. R., Mueller, W., and Phinney, S. D. (1996). Gender differences in plasma leptin concentrations. *Nat. Med., 2*, 949–950.

Heger, S., Partsch, C. J., Peter, M., Blum, W. F., Kiess, W., and Sippell, W. (1999). Serum leptin levels in patients with progressive central precocious puberty. *Pediatr. Res., 46*, 71–75.

Henson, M. C. and Castracane, V. D. (2000). Leptin in pregnancy. *Biol. Reprod., 63*(5), 1219–1228.

Henson, M. C., Castracane, V. D., O'Neil, J. S., Gimpel, T., Swan, K. F., Green, A. E., and Shi, W. (1999). Serum leptin concentrations and expression of leptin transcripts in placental trophoblast with advancing baboon pregnancy. *J. Clin. Endocrinol. Metab., 84*, 2543–2549.

Heretier, A., Charnay, Y., and Aubert, M. L. (1997). Regional distribution of mRNA encoding the long form of receptor in mouse brain. *Neurosci. Res. Commun., 21*, 113–118.

Hileman, S. M., Pierroz, D. D., and Flier, J. S. (2000). Leptin, nutrition, and reproduction: timing is everything. *J. Clin. Endocrinol. Metab., 85*, 804–807.

Hislop, M. S., Ratanjee, B. D., Soule, S. G., and Marais, A. D. (1999). Effects of anabolic-androgenic steriod use or gonadal testosterone suppression on serum leptin concentration in men. *Eur. J. Endocrinol., 141*, 40–46.

Hohmann, J. G., Teal, T. H., Clifton, D. K., Davis, J., Hruby, V. J., Han. G., and Steiner, R. (2000). Differential role of melanocortins in mediating leptin's central effects on feeding and reproduction. *Am. J. Physiol. Regul. Integr. Comp. Physiol., 278*, R50–59.

Horlick, M. B., Rosenbaum, R. V., Nicolson, M., Levine, L. S., Fedun, B., Wang, J., Pierson, R. N., Jr., and Leibel, R. L. (2000). Effect of puberty on the relationship between circulating leptin and body composition. *J. Clin. Endocrinol. Metab., 85*, 2509–2518.

Huang, X. F., Lin, S., and Zhang, R. (1997). Upregulation of leptin receptor mRNA expression in obese mouse brain. *Neuroreport, 8*, 1035–1038.

Huszar, D., Lynch, C. A., Fairchild-Huntress, V., Dunmore, J. H., Fang, Q., Berkemeir, L. R., Gu, W., Kesterson, R. A., Boston, B. A., Cone, R. D., Smith, F. J., Campfield, L. A., Burn, P., and Lee, F. (1997). Targeted disruption of the melanocortin-4 receptor results in obesity in mice. *Cell, 88*, 131–141.

Iqbal, J., Pompolo, S., Considine, R. V., and Clarke, I. J. (2000). Localization of leptin receptor-like immunoreactivity in the corticotropes, somatotropes, and gonadotropes in the ovine anterior pituitary. *Endocrinology, 141*, 1515–1520.

Jakacki, R. I., Kelch, R. D., Saunder, S. E., Lloyd, J. S., Hopwood, J. N., and Marshall, J. C. (1982). Pulsatile secretion of luteinizing hormone in children. *J. Clin. Endocrinol. Metab., 55*, 453–458.

Jaruratanasirikul, S., Mosuwan, L., and Lebel, L. (1997). Growth pattern and age at menarche of obese girls in a transitional society. *J. Pediatr. Endocrinol. Metab., 10*, 487–490.

Jimerson, D. C., Mantzoros, C., Wolfe, B. E., and Metzger, E. D. (2000). Decreased serum leptin in bulimia nervosa. *J. Clin. Endocrinol. Metabol., 85*, 4511–4514.

Jin, L., Zhang, S., Burguera, B. G., Couce, M. E., Osamura, R. Y., Kulig, E., and Lloyd, R. V. (2000). Leptin and leptin receptor expression in rat and mouse pituitary cells. *Endocrinology, 141*, 333–339.

Jureus, A., Cunningham, M. J., McClain, M. E., Clifton, D. K., and Steiner, R. A. (2000). Galanin-like peptide (GALP) is a target for regulation by leptin in the hypothalamus of the rat. *Endocrinology, 141*, 2703–2706.

Karlsson, C., Lindell, K., Svensson, E., Bergh, C., Lind, P., Billig, H., Carlsson, L. M., and Carlsson, B. (1997). Expression of functional leptin receptors in the human ovary. *J. Clin. Endocrinol. Metab., 82*, 4144–4148.

Kennedy, G. C. and Mitra, J. (1963). Body-weight and food intake as initiating factors for puberty in the rat. *J. Physiol., 166*, 375–381.

Kiess, W., Englaro, P., Hanitsch, S., Rascher, W., Attanasio, A., and Blum, W. F. (1996). High leptin concentrations in serum of very obese children are further stimulated by dexamethasone. *Horm. Metabol. Res., 28*, 708–710.

Kiess, W., Siebler, T., Englaro, P., Kratzsch, J., Deutsch, J., Meyer, K., Gallaher, B., and Blum, W. F. (1998). Leptin as a metabolic regulator during fetal and neonatal life and in childhood and adolescence. *J. Pediatr. Endocrinol. Metab., 11*, 1–13.

Kiess, W., Reich, A., Meyer, K., Glasow, A., Deutsch, J., Klammt, J., Yang, Y., Müller, G., and Kratzsch, J. (1999). A role for leptin in sexual maturation and puberty? *Horm. Res., 51*, 55–63.

Kiess, W., Müller, G., Galler, A., Reich, A., Deutscher, J., Klammt, J., and Kratzsch, J. (2000). Body fat mass, leptin and puberty. *J. Pediatr. Endocrinol. Metab., 13*, 717–722.

Kitawaki, J., Kusuk, I., Koshiba, H., Tsukamoto, K., and Honjo, H. (1999). Leptin directly stimulates aromatase activity in human luteinized granulosa cells. *Mol. Hum. Reprod., 5*, 708–713.

Klein, S., Coppack, S. W., Mohamed Ali, V., and Landt, M. (1996). Adipose tissue leptin production and plasma leptin kinetics in humans. *Diabetes, 45*, 984–987.

Lado-Abeal, J., Lukyuanenko, Y. O., Swamy, S., Hermida, R. C., Hutson, J. C., and Norman, R. L. (1999). Short-term leptin infusion does not affect circulating levels of LH, testosterone or cortisol in food-restricted pubertal males rhesus macaques. *Clin. Endocrinol., 51*(1), 41–51.

Lahlou, N., Landais, P., De Boissieu, D., and Bougneres, P. F. (1997). Circulating leptin in normal children and during the dynamic phase of juvenile obesity: relation to body fatness, energy metabolism, caloric intake and sexual dimorphism. *Diabetes, 46*, 989–993.

Laughlin, G. A. and Yen, S. S. (1996). Nutritional and endocrine-metabolic aberrations in amenorrheic athletes. *J. Clin. Endocrinol. Metab., 81*, 4301–4309.

Laughlin, G. A. and Yen, S. S. (1997). Hypoleptinemia in women athletes: absence of a diurnal rhythm with amenorrhea. *J. Clin. Endocrinol. Metab., 82*, 318–321.

Lebrethon, M. C., Vandersmissen, E., Gerard, A., Parent, A. S., Junien, J. L., and Bourguignon, J. P. (2000a). In vitro stimulation of the pre-pubertal rat gonadotropin-releasing hormone pulse generator by leptin and neuropeptide Y through distinct mechanisms. *Endocrinology, 141*, 1464–1469.

Lebrethon, M. C., Vandersmissen, E., Gerard, A., Parent, A. S., and Bourguignon, J. P. (2000b). Cocaine and amphetamine-regulated-transcript peptide mediation of leptin stimulatory effect on the rat gonadotropin-releasing hormone pulse generator in vitro. *J. Neuroendocrinol., 12*, 383–385.

Leroy, P., Dessolin, S., Villageois, P., Moon, B. C., Friedman, J. M., Ailhaud, G., and Dani, C. (1996). Expression of ob gene in adipose cells. Regulation by insulin. *J. Biol. Chem., 271*, 2365–2368.

Licinio, J., Mantzoros, C., Negrao, A. B., Cizza, G., Wong, M. L., Bongiorno, P. B., Chrousos, G. P., Karp, B., Allen, C., Flier, J. S., and Gold, P. W. (1997). Human leptin levels are pulsatile and inversely related to pituitary-adrenal function. *Nat. Med., 3*, 575–579.

Licinio, J., Negrao, A. B., Mantzoros, C., Kaklamani, V., Wong, M. L., Bongiorno, P. B., Mulla, A., Cearnal, L., Veldhuis, J. D., Flier, J. S., McCann, S. M., and Gold, P. W. (1998a). Synchronicity of frequently sampled, 24-h concentrations of circulating leptin, luteinizing hormone, and estradiol in healthy women. *Proc. Natl. Acad. Sci. USA, 95*(5), 2541–2546.

Licinio, J., Negrao, A. B., Mantzoros, C., Kaklamani, V., Wong, M. L., Bongiorno, P. B., Negro, P. P., Mulla, A., Veldhuis, J. D., Cearnal, L., Flier, J. S., and Gold, P. W. (1998b). Sex differences in circulating human leptin pulse amplitude: Clinical implications. *J. Clin. Endocrinol. Metab., 83*, 4140–4147.

Luukkaa, V., Rouru, J., Ahokoski, O., Scheinin, H., Irjala, K., and Huupponen, R. (2000). Acute inhibition of oestrogen biosynthesis does not affect serum leptin levels in young men. *Eur. J. Endocrinol., 142*, 164–169.

MacConnie, S. E., Barkan, A., and Lampman, R. M. (1986). Decreased hypothalamic gonadotropin-releasing hormone secretion in male marathon runners. *N. Engl. J. Med., 315*, 411–417.

Madej, T., Boguski, M. S., and Bryant, S. H. (1995). Threading analysis suggests that the obese gene product may be a helical cytokine. *FEBS Lett., 373*, 13–18.

Maffei, M., Halaas, J., Ravussin, E., Pratley, R. E., Lee, G. H., Zhang, Y., Fei, H., Kim, S., Lallone, R., Ranganathan, S., Kern, P. A., and Friedman, J. M. (1995). Leptin levels in human and rodent: measurement of plasma leptin and ob RNA in obese and weight-reduced subjects. *Nat. Med., 1*, 1155–1161.

Mann, D. R., Akinbami, M. A., Gould, K. G., and Castracane, V. D. (2000). A longitudinal study of leptin during development in the male rhesus monkey: the effect of body composition and season on circulating leptin levels. *Biol. Reprod., 62,* 285–291.

Mantzoros, C. S., Flier, J. S., and Rogol, A. D. (1997a). A longitudinal assessment of hormonal and physical alterations during normal puberty in boys. Rising leptin levels may signal the onset of puberty. *J. Clin. Endocrinol. Metab., 82,* 1066–1070.

Mantzoros, C. S., Varvarigou, A., Kaklamani, V. G., Beratis, N. G., and Flier, J. S. (1997b). Effect of birth weight and maternal smoking on cord blood leptin concentrations of full-term and preterm newborns. *J. Clin. Endocrinol. Metab., 82,* 2856–2861.

Mantzoros, C., Flier, J. S., Lesem, M. D., Brewerton, T. D., and Jimerson, D. C. (1997c). Cerebrospinal fluid leptin in anorexia nervosa: correlation with nutritional status and potential role in resistance to weight gain. *J. Clin. Endocrinol. Metab., 82,* 1845–1851.

Mantzoros, C. S. (1999). The role of leptin in human obesity and disease: a review of current evidence. *Ann. Intern. Med., 130,* 671–680.

Mantzoros, C. S., Ozata, M., Negrao, A. B., Suchard, M. A., Ziotopoulou, M., Caglayan, S., Elashoff, R. N., Cogswell, R. J., Negro, P., Liberty, V., Wong, M., Veldhuis, J., Ozdemir, I. C., Gold, P. W., Flier, J. S., and Licinio, J. (2001). Synchronicity of frequently sampled thyrotropin (TSH) and leptin concentrations in healthy adults and leptin-deficient subjects: evidence for possible partial TSH regulation by leptin in humans. *J. Clin. Endocrinol. Metab., 86,* 3284–3291.

Marshall, W. A. and Tanner, J. M. (1969). Variations in the pattern of pubertal changes in girls. *Arch. Dis. Child., 44*(235), 291–303.

Marshall, W. A. and Tanner, J. M. (1970). Variations in the pattern of pubertal changes in boys. *Arch. Dis. Child., 45*(239), 13–23.

Masuzaki, H., Ogawa, Y., Sagawa, N., Hosoda, K., Matsumoto, T., Mise, H., Nishimura, H., Yoshimasa, Y., Tanaka, I., Mori, T., and Nakao, K. (1997). Nonadipose tissue production of leptin: leptin as a novel placenta-derived hormone in humans. *Nat. Med., 3,* 1029–1033.

Matkovic, V., Ilich, J. Z., Badenhop, N. E., Skugor, M., Clairmont, A., Klisovic, D., and Landoll, J. D. (1997a). Gain in body fat is inversely related to the nocturnal rise in serum leptin levels in young females. *J. Clin. Endocrinol. Metab., 82,* 1368–1372.

Matkovic, V., Ilich, J. Z., Skugor, M., Badenhop, N. E., Goel, P., Clairmont, A., Klisovic, D., Nahhas, R. W., and Landoll, J. D. (1997b). Leptin is inversely related to age at menarche in human females. *J. Clin. Endocrinol. Metab., 82,* 3239–3245.

Matsuda, J., Yokota, I., Iida, M., Murakami, T., Naito, E., Ito, M., Shima, K., and Kuroda, Y. (1997). Serum leptin concentration in cord blood: relationship to birth weight and gender. *J. Clin. Endocrinol. Metab., 82,* 1642–1644.

Matsuoka, N., Ogawa, Y., Hosoda, K., Matsuda, J., Masuzaki, H., Miyawaki, T., Azuma, N., Natsui, K., Nishimura, H., Yoshimasa, Y., Nishi, S., Thompson, D. B., and Nakao, K. (1997). Human leptin receptor gene in obese Japanese subjects: evidence against either obesity-causing mutations or association of sequence variants with obesity. *Diabetologia, 40,* 1204–1210.

Mauras, N., Rogol, A. D., Haymond, M. W., and Veldhuis, J. (1996). Sex steroids, growth hormone, insulin-like growth factor-1: neuroendocrine and metabolic regulation in puberty. *Horm. Res., 45,* 74–80.

Montague, C. T., Farooqi, I. S., Whitehead, J. P., Soos, M. A., Rau, H., Wareham, N. J., Sewter, C. P., Digby, J. E., Mohammed, S. N., Hurst, J. A., Cheetham, C. H., Earley, A. R., Barnett, A. H., Prins, J. B., and O'Rahilly, S. (1997). Congenital leptin deficiency is associated with severe early-onset obesity in humans. *Nature, 387,* 903–908.

Morash, B., Li, A., Murphy, P. R., Wilkinson, M., and Ur, E. (1999). Leptin gene expression in the brain and pituitary gland. *Endocrinology, 140,* 5995–5998.

Moschos, S., Chan, J. L., and Mantzoros, C. S. (2002). Leptin and reproduction: a review. *Fertil. Steril., 77*(3), 433–444.

Murray, J. F., Mercer, J. G., Adan, R. A., Datta, J. J., Aldairy, C., Moar, K. M., Baker, B. I., Stock, M. J., and Wilson, C. A. (2000). The effect of leptin on luteinizing hormone release is exerted in the zona incerta and mediated by melanin-concentrating hormone. *J. Neuroendocrinol., 12,* 1133–1139.

O'Neil, J. S., Green, A. E., Edwards, D. E., Swan, K. F., Gimpel, T., Castracane, V. D., and Henson, M. C. (2001). Regulation of leptin and leptin receptor in baboon pregnancy: effects of advancing gestation and fetectomy. *J. Clin. Endocrinol. Metab., 86,* 2518–2524.

Parent, A. S., Lebrethon, M. C., Gerard, A., Vandersmissen, E., and Bourguignon, J. P. (2000). Leptin effects on pulsatile gonadotropin releasing hormone secretion from the adult rat hypothalamus and interaction with cocaine and amphetamine regulated transcript peptide and neuropeptide Y. *Regul. Pept.*, *92*, 17–24.

Plant, T. M. and Durant, A. R. (1997). Circulating leptin does not appear to provide a signal for triggering the initiation of puberty in the male rhesus monkey (Macaca mulatta). *Endocrinology, 138*, 4505–4508.

Quinton, N. D., Smith, R. F., Clayton, P. E., Gill, M. S., Shalet, S., Justice, S. K., Simon, S. A., Walters, S., Postel-Vinay, M.-C., Blakemore, A. I. F., and Ross, R. J. M. (1999). Leptin binding activity changes with age: the link between leptin and puberty. *J. Clin. Endocrinol. Metab., 84*, 2336–2341.

Roemmich, J. N., Clark, P. A., Berr, S. S., Mai, V., Mantzoros, C. S., Flier, J. S., Weltman, A., and Rogol, A. D. (1998). Gender differences in leptin levels during puberty are related to subcutaneous fat depot and sex steroids. *Am. J. Physiol., 275*, E543–E 551.

Roemmich, J. N. and Rogol, A. D. (1999). Role of leptin during childhood growth and development. *Endocrinol. Metab. Clin. North Am., 28*(4), 749–758.

Saad, M. F., Riad-Gabriel, M. G., Khan, A., Sharma, A., Michael, R., Jinagouda, S. D., Boyadjian, R., and Steil, G. M. (1998). Diurnal and ultradian rhythmicity of plasma leptin: effects of gender and adiposity. *J. Clin. Endocrinol. Metab., 83*, 453–459.

Schubring, C., Kiess, W., Englaro, P., Rascher, W., Dötsch, J., Hanitsch, S., Attanasio, A., and Blum, W. (1997). Levels of leptin in maternal serum, amniotic fluid, and arterial and venous cord blood: relation to neonatal and placental weight. *J. Clin. Endocrinol. Metab., 82*, 1480–1483.

Schubring, C., Siebler, T., Englaro, P., Blum, W. F., Kratzsch, J., Triep, K., and Kiess, W. (1999). Rapid decline of leptin serum levels in healthy neonates after birth. *Clin. Endocrinol., 51*, 199–204.

Sinha, M. K., Ohannesian, J. P., Heiman, M. L., Kriauciunas, A., Stephens, T. W., Magosin, S., Marco, C., and Caro, J. F. (1996a). Nocturnal rise of leptin in lean, obese, and non-insulin-dependent diabetes mellitus subjects. *J. Clin. Invest., 97*, 1344–1347.

Sinha, M. K., Sturis, J., Ohannesian, J., Magosin, S., Stephens, T., Heiman, M. L., Polonsky, K. S., and Caro, J. F. (1996b). Ultradian oscillations of leptin secretion in humans. *Biochem. Biophys. Res. Commun., 228*, 733–738.

Sivan, E., Lin, W. M., Homko, C. J., Reece, E. A., and Boden, G. (1997). Leptin is present in human cord blood. *Diabetes, 46*, 917–919.

Smith-Kirwin, S. M., O'Connor, D. M., De Johnston, J., Lancey, E. D., Hassink, S. G., and Funanage, V. L. (1998). Leptin expression in human mammary epithelial cells and breast milk. *J. Clin. Endocrinol. Metab., 83*, 1810–1813.

Spicer, L. J. and Francisco, C. C. (1998). Adipose obese gene product, leptin, inhibits bovine ovarian thecal cell steroidogenesis. *Biol. Reprod., 58*, 207–212.

Strobel, A., Issad, T., Camoin, L., Ozata, M., and Strosberg, A. D. (1998). A leptin missense mutation associated with hypogonadism and morbid obesity. *Nat. Genet., 18*, 213–215.

Tartaglia, L. A., Dembski, M., Weng, X., Deng, N., Culpepper, J., and Devos, R. (1995). Identification and cloning of a leptin receptor, OB-R. *Cell, 83*, 1263–1271.

Tartaglia, L. A. (1997). The leptin receptor. *J. Biol. Chem., 272*, 6093–6096.

Terasawa, E. (1998). Cellular mechanism of pulsatile LHRH release. *Gen. Comp. Endocrinol., 112*, 283–295.

Urbanski, H. F. and Francis Pau, K.-Y. (1998). A biphasic developmental pattern of circulating leptin in the male rhesus macaque (*Macaca mulatta*). *Endocrinology, 139*(5), 2284–2286.

Varvarigou, A., Mantzoros, C. S., and Beratis, N. G. (1999). Cord blood leptin concentrations in relation to intrauterine growth. *Clin. Endocrinol., 50*, 177–183.

Wabitsch, M., Blum, W. F., Muche, R., Braun, M., Hube, F., Rascher, W., Heinze, E., Teller, W., and Hauner, H., (1997). Contribution of androgens to the gender difference in leptin production in obese children and adolescents. *J. Clin. Invest., 100*, 808–813.

Wabitsch, M., Ballauff, A., Holl, R., Blum, W. F., Heinze, E., Remschidt, H., and Hebebrand, J. (2001). Serum leptin, gonadotropin, and testosterone concentrations in male patients with anorexia nervosa during weight gain. *J. Clin. Endocrinol. Metab., 86*, 2982–2988.

Wang, J., Liu, R., Hawkins, M., Barzilai, N., and Rossetti, L. (1998). A nutrient-sensing pathway regulates leptin gene expression in muscle and fat. *Nature, 393*, 684–688.

Weimann, E., Blum, W. F., Witzel, C., Schwidergall, S., and Bohles, H. J. (1999). Hypoleptinemia in female and male elite gymnasts. *Eur. J. Clin. Invest., 29*, 853–860.

Yu, W. H., Kimura, M., Walczewska, A., Karanth, S., and McCann, S. M. (1997a). Role of leptin in hypothalamic-pituitary function. *Proc. Natl. Acad. Sci. USA, 94*, 1023–1028.

Yu, W. H., Walczewska, A., Karanth, S., and McCann, S. M. (1997b). Nitric oxide mediates leptin-induced luteinizing hormone-releasing hormone (LHRH) and LHRH and leptin-induced LH release from the pituitary gland. *Endocrinology, 138*, 5055–5058.

Zhang, Y., Proenca, M., Maffei, M., Barone, M., Leopold, L., and Friedman, J. M. (1994). Positional cloning of the obese gene and its human homologue. *Nature, 372*, 425–432.

Zhang, Y., Basinski, M. B., Beals, J. M., Briggs, S. L., Churgay, L. M., Clawson, D. K., Di Marchi, R. D., Furman, T. C., Hale, J. E., Hsiung, H. M., Schoner, B. E., Smith, D. P., Zhang, X. Y., Wery, J. P., and Schevitz, R. W. (1997). Crystal structure of the obese protein leptin E-100. *Nature, 387*, 206–209.

11

Integration of Leptin with Other Signals Regulating the Timing of Puberty

Lessons Learned from the Sheep Model

DOUGLAS L. FOSTER AND LESLIE M. JACKSON

11.1. OVERVIEW

How puberty is timed remains a major unanswered question in reproductive biology. Perhaps there is no single answer because there are many factors that time puberty. Leptin may be one of these answers, but the challenge is to fit this metabolic cue into our overall understanding of how sexual maturity is timed during the life of the individual. There are a variety of considerations, among which are the definition of puberty, level of inquiry, internal versus external influences, and differences between sexes and species. This chapter attempts to place leptin into the constellation of common and unique mechanisms that have been proposed to time the irreversible transition into adulthood. Studies using sheep as an experimental model have strongly influenced the formulation of our overall view of how puberty is timed in a large mammal living in a natural environment. As will become evident, in this species, the immediate cue timing puberty is different in males and females. Therefore, while we have been historically interested in how puberty is timed in the female, the mechanisms are more complex than for the male. Thus, our investigations into the importance of leptin as a pubertal signal in the sheep have been in the male. As a

DOUGLAS L. FOSTER AND LESLIE M. JACKSON • Department of Obstetrics and Gynecology and Department of Ecology and Evolutionary Biology, University of Michigan.

Leptin and Reproduction. Edited by Henson and Castracane, Kluwer Academic/Plenum Publishers, 2003.

prelude to discussing the role of leptin in puberty, it is essential to lay the groundwork for what constitutes puberty. A backdrop to this discussion is a consideration of the level at which the investigations are being conducted. Although it is obvious that there is no gene for puberty, the level of investigation can range from integrative biology to molecular biology. Finally, the development of animal models and approaches even within a single species, is fraught with pitfalls, and the interpretation of data depends on the model from which they are obtained.

11.2. THE MANY FACETS OF PUBERTY

11.2.1. Event versus Process

As noted by Donovan and van der Werff ten Bosch (1965), "Whilst it is easy to distinguish a child from an adult, the transitional stage—puberty or adolescence—is ill-defined and vaguely delimited." Puberty can be viewed as either an event or a process depending upon the orientation of the investigator or observer. A common view is essential to any discussion, and a historical perspective is useful to understand the evolution of the modern notion of "puberty" and its root causes. The classical definition was inspired by the changing outward appearance of the human body, and was derived from the Latin, "pubes" (hair) and French, "puber" (grown up). In addition to hair distribution, there are other outward, external, or visible signs that maturity is impending in young women. These include breast development and first menses, which reflect, respectively, an increase in the amount of estrogen being secreted by the ovary and its cyclical effects on the uterus. Because the original anthrocentric definition is not particularly useful in most mammals, which are already covered with hair at birth or shortly thereafter, other markers note activation of the reproductive system in various species. One common thread is that all the markers of puberty reflect changes in steroid action on somatic and/or neural tissues. However, the degree to which the reproductive system is activated at the time when the sign(s) of sexual maturation first appears differs among species. For example, in the female rat, vaginal opening is a response to rising concentrations of circulating estrogens, but ovulation does not occur until several days later (Ojeda and Urbanski, 1994). In subhuman primates, ovulations do not occur until several weeks or even months after perineal swelling, menarche and breast development begin (Plant, 1994). By contrast, the only obvious sign of puberty in the sheep, first estrus, is preceded by one or more ovulations; in this species, progesterone priming (from a corpus luteum) is necessary for estrogen to induce sexual behavior (Foster, 1994). Health care professionals have long considered puberty as a *process*, rather than an *event*, because multiple signs of puberty are easily recognizable and appear gradually over an extended period in children. The Tanner designation of stages of puberty, when combined with indices of somatic growth, has provided a classic example of the clinical usefulness of this idea to gauge the normal progression of the pubertal process in any single individual (Tanner, 1962). This allows for the detection, diagnosis, and potential treatment of individuals in which this process is either too rapid or too slow. Now puberty is being more widely considered a process in most species because of our ability to characterize changes in hormonal patterns over time.

11.2.2. Level of Inquiry

Our notion of what times puberty is strongly influenced by the level of inquiry and the associated experimental approach. It is well-established that it is the brain, not the gonad that plays the pivotal role as the timer of the pubertal process. The gonads develop the potential to function at a relatively young age, as does the pituitary with respect to its secretion of gonadotropic hormones. Because the neural peptide, gonadotropin-releasing hormone (GnRH) regulates gonadotropin secretion, the essential question is what causes the increase in secretion of GnRH that activates the gonads during puberty. Multiple approaches and levels of inquiry have been used to attack this question. Some investigators have studied the development of neurosecretion. Morphologic investigations have focused on whether puberty is the result of a profound change in the number, type, or distribution of GnRH neurons. The answer may depend on the species examined. In the rat and Djungarian hamster, alterations occur in the structure (Wray and Hoffman, 1986) and number (Yellon and Newman, 1991) of GnRH neurons at puberty, but in the sheep, no major postnatal reorganization occurs, and the GnRH neuronal system becomes established before birth (Wood et al., 1992). This would be expected because the sheep, a precocial species, is born at a relatively advanced stage of neuronal development compared to the rat and hamster (altricial species). The ability of the GnRH neurons to secrete GnRH early in life has been studied in the context of the properties of the GnRH neural complex and how the dispersed GnRH neurons, which are also relatively few in number, communicate to fire synchronously as the "pulse generator" (Terasawa, 2001 for review). Again, the general understanding is that the ability of GnRH neurons to fire synchronously is an early property of this neurosecretory system and that to comprehend fully how the pubertal process is controlled neurobiologically, we must more completely understand what type of changes in incoming information is supplied to GnRH neurons during development. This includes unraveling the phenotypes of the presynaptic neurons, and how they communicate with GnRH neurons. Clearly, this regulation is multifaceted and involves both stimulatory and inhibitory neurons and their regulation by nonneuronal elements (glia, astrocytes) that physically and chemically interact with GnRH neurons to result in their increased activity during puberty (Bourguignon and Plant, 2000 for review). Whatever the detailed neurobiological mechanism, an interesting related debate is whether GnRH neurosecretion is under active inhibition by steroid hormones secreted by the developing gonads during the prepubertal period. Some believe that it is, and that the pubertal GnRH rise reflects a decrease in sensitivity to their negative feedback (Foster, 1994); others believe that there is an increased steroid-independent drive to GnRH secretion during puberty (Plant, 1994). Most likely, both occur (Foster et al., 1989). All of these foregoing reductionist approaches seek explanations for how neuronal regulation of GnRH neurons changes during puberty, and the question that must be raised is whether these changes are unique to the initiation of fertility. It is most likely that they are not and that such changes occur whenever an individual makes the transition from a reproductively quiescent state to a fertile one. Thus, the elusive "trigger" for puberty may not arise from within the brain and other explanations must be sought. In an attempt to learn why the pubertal increase in GnRH secretion occurs when it does, integrative approaches have been used to study how growth times puberty during the lifespan. The logic here is that growth is unique to development.

11.3. GROWTH CUES

11.3.1. Energetics of Puberty

The conventional wisdom that decreased food availability retards puberty leads to the unassailable tenet that growth, not simply age or developmental time plays a dominant role in timing puberty. Energy partitioning is the underpinning of our understanding of how developmental changes occur with growth to initiate puberty. With advanced growth, less energy is expended for maintenance of basal metabolism per unit of body mass. This energy surfeit is sensed by the brain through metabolites or metabolic hormones and serves to signal for an increase in GnRH secretion. These considerations are based on the work of Kennedy and Mitra (1963) nearly 40 years ago that yielded the yet contemporary hypothesis that the timing of puberty is ultimately based on energetics. Their thesis was that the *first* transition between an infertile and fertile state (puberty) is tightly coupled to a change in somatic metabolism. According to their view, once growth was sufficient as reflected by appropriate changes in energy balance, the reproductive system would become active. Their hypothesis considered that the brain might somehow detect the decrease in basal metabolic rate that is necessary to maintain a stable core temperature because growth produces a disproportionate increase in heat production by the body mass relative to the ability to dissipate heat through the increase in surface area. This overall concept still drives the search for growth-related signals timing the transition into adulthood, but the metabolic signals which could serve as the molecular links between growth and reproduction remained elusive until the discovery of leptin.

11.3.2. Criteria for a Blood-Borne Metabolic Signal Timing Puberty

To qualify a substance as being an important *blood-borne* metabolic signal timing the pubertal increase in GnRH secretion, we believe that at least two major criteria must be satisfied (Foster and Nagatani, 1999): (1) the circulating substance must be quantitatively different between the sexually immature and sexually mature individual and must increase/decrease during growth; (2) when administered prepubertally, the putative metabolic signal must lead to an increase in GnRH secretion. While this last criterion relates to a substance being agonistic to GnRH secretion, in theory, a substance could be an inhibitor of GnRH secretion and a decrease in its concentration could serve to derepress GnRH secretion, and thus, lead to its pubertal rise. Whether the candidate substance increases or decreases is not an important issue as long as it clearly changes to provide a signal. It could be argued that the sensitivity or responsiveness to the substance could change during growth and development, rather than concentrations of the substance itself changing. This would argue against the candidate metabolic signal being a pubertal signal, although it would be considered permissive. One must then determine what is causing the change in sensitivity to the substance.

11.3.3. Studies in the Sheep Model

11.3.3.1. Exogenous Leptin Prevents Fasting-Induced Hypogonadotropism. We have used a sheep model to determine if leptin is an important signal timing puberty. The sheep is a long-lived, precocial species that becomes sexually mature within the first year after birth given adequate nutrition. We have taken care to optimize our model. The young male is used in all of our studies of metabolic signals timing puberty to avoid the confounding photoperiod. As is discussed in a later section, the female is highly photoperiodic with respect to the initiation of puberty, but the developing male is not. In our model, the testes are removed and the males are treated chronically with estrogen (subcutaneous implant) to stabilize the sex steroid environment that regulates GnRH secretion. As part of our approach, we tested the efficacy of leptin to modulate GnRH secretion in the adult sheep before studies of puberty were launched. This was done for two reasons. The first was the practical one of determining the dose, frequency, and route of administration. The second reason was conceptual. We believe that the relative importance of a candidate metabolic signal in controlling the timing of puberty can be assessed by determining its importance in the maintenance of adult reproductive function. In both our adult investigations and puberty studies in the sheep, leptin was administered peripherally, rather than centrally, to simulate the peripheral production of the hormone by adipocytes. We also believe that the first physiological question to address is whether leptin has the ability to increase GnRH secretion, yet recognize that identifying the site of leptin action is essential. Central administration of the hormone is an important approach to use and the results of others in the sheep are discussed in the next section. Our previous experience with leptin in the rat formed the basis for our experimental design. In the castrated adult female rat, either in the presence or absence of steroids, peripheral administration of leptin (recombinant mouse, 3 μg/g BW, ip. each 8 hr) was found to prevent the decrease in luteinizing hormone (LH) pulse frequency that occurred during a 48-hr fast (Nagatani et al., 1998). This study was replicated in the adult sheep model (castrated, estrogen-treated male), using recombinant human leptin at a dose (0.050 μg/g BW, sc each 8 hr) calculated to approximate concentrations of circulating leptin reported in obese men and women; these were verified by measurement of the circulating leptin during treatment (15–20 ng/ml) (Nagatani et al., 2000b). The dose used in the sheep model was only 1 percent of that used in the rat model, but comparisons of the biological effectiveness of the two treatments are difficult. Importantly, the results of leptin treatment during an acute fast were the same in the sheep (Figure 11.1) as in the rat. During the 78-hr fast, all sheep exhibited the classical signs of fasting (hypoinsulinemia, hypoglycemia, hypercortisolemia) and hypogonadotropism. Despite no differences in the metabolic fasting parameters, peripheral treatment with leptin was able to maintain high LH pulse frequency. However, in contrast to the female rat model (Nagatani et al., 1998), there were no apparent effects of fasting on LH secretion in the absence of estrogen. Whether this is a sex difference or a species difference remains to be determined.

11.3.3.2. Exogenous Leptin Stimulates LH during Fasting-Induced Hypogonadotropism. Having determined a dose of leptin in the sheep that would prevent hypogonadotropism, our adult model was further optimized to align it with one that could be used for studies of puberty. By definition, a prepubertal individual is hypogonadotropic, and the experimental question relates to whether leptin can *increase* GnRH pulse frequency. The adult model for any such comparison would be an individual that is hypogonadotropic before leptin treatment. This condition was produced by first fasting the agonadal, steroid-clamped adult male for 48 h. During an additional 48 h of fasting, leptin was administered peripherally, and from the depths of this fast, exogenous leptin

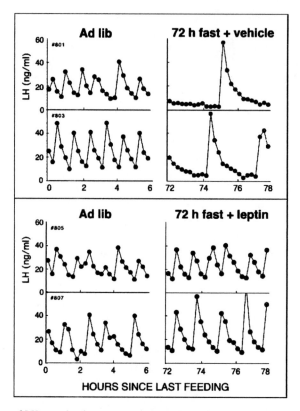

Figure 11.1. Patterns of LH secretion in representative gonadectomized male sheep treated chronically with low dose estradiol (sc implant, <1 pg/ml) during ad libitum feeding (0–6 hr) and fasting (72–78 hr after last feeding). Sheep in the upper panel received saline vehicle and those in the lower panel were treated sc each 8 hr for 72 hr with recombinant human leptin (50 μg/kg). Note that leptin maintains LH pulse frequency during fasting. (Redrawn from Nagatani et al., 2000b.)

(0.013 μg/g BW, sc each 4 hr) was capable of increasing LH pulse frequency (Jackson et al., 2002) (Figure 11.2). In addition, the response was the same as that found during refeeding when the adults were returned to the standard diet after a 48-h fast (data not shown). It is noteworthy that the *amount* of leptin used to drive LH pulses to a high-frequency during fasting was half that used in the previous experiment (Figure 11.1). We reduced the dose of leptin administered and increased the frequency of treatment to minimize the amount of exogenous hormone required to treat a large animal peripherally, while maintaining steady-state concentrations of leptin equivalent to those measured in response to less frequent treatment with a higher dose.

11.3.3.3. Exogenous Leptin Stimulates LH in the Hypogonadotropic Sexually Immature Male. With the model and dose that had been successful in stimulating LH secretion in the adult, we began our studies in the agonadal, steroid-replaced male lamb. For reference, in the presence of steroid negative feedback, developing male lambs gradually increase the frequency of LH pulses, and in this model, the production of hourly pulses

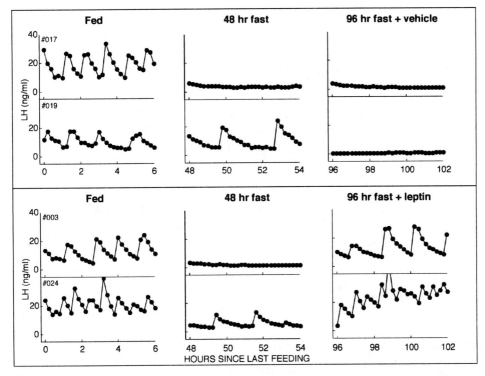

Figure 11.2. Patterns of LH secretion in representative gonadectomized male sheep treated chronically with low dose estradiol (sc implant, <1 pg/ml) during feeding (0–6 h) and during fasting (48–54 hr and 96–102 hr after last feeding). After 54 hours of fasting, sheep in the top panel received saline vehicle, and those in the bottom panel were treated sc with recombinant human leptin (12.5 µg/kg) each 4 hr between hours 54 and 96 of the fast. Note that leptin increases LH pulse frequency in the fasted adult sheep. (Data from Jackson et al., 2002.)

(neuroendocrine puberty) is typically achieved between 9 and 13 weeks of age (Olster and Foster, 1986). Thus, we chose to administer leptin at 6 and 10 weeks of age at the same dose as that used in the adult for 48 hr to drive LH secretion during fasting; the length of treatment in the lamb was extended to 72 hr. The response of the young male to exogenous leptin was variable, and it appeared to be related to the frequency of endogenous LH release that ranged from nonpulsatile to hourly pulses. In all lambs producing low frequency LH pulses before treatment, leptin increased pulse frequency. In those lambs that were already producing high-frequency (hourly) pulses, exogenous leptin did not increase LH pulse frequency further. Interestingly, leptin did not consistently increase LH pulse frequency in those lambs that were not producing LH pulses before treatment as some responded while others did not. These different responses to leptin were evident at both six and 10 weeks of age. Thus, in this model, leptin cannot increase GnRH pulse frequency beyond the normal pubertal frequency, but it can increase pulse frequency when the lamb has achieved the ability to produce low frequency pulses. It is not immediately clear why leptin could stimulate GnRH secretion in some nonpulsatile lambs, but not in others (Figure 11.3). This finding is in contrast to the ability of leptin to drive LH pulses in fasted, nonpulsatile, adult males (see Figure 11.2). Endogenous LH pulse frequency in prepubertal

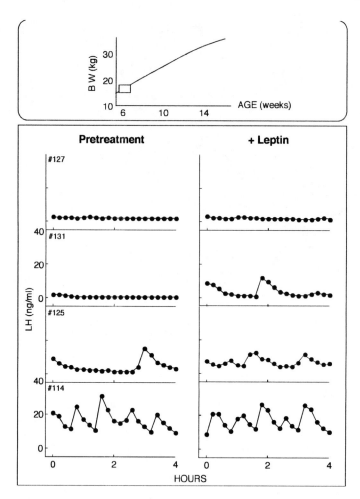

Figure 11.3. Patterns of LH secretion in representative prepubertal, male sheep at 6 weeks of age. The males had been neonatally gonadectomized, treated chronically with low dose estradiol (sc implant, <1 pg/ml), and fed to maintain normal growth as depicted in the top panel. Patterns are presented before treatment and after 72 h of treatment with recombinant human leptin (12.5 μg/kg, sc each 4 hr). Various responses were evident (see text for discussion). Prepubertal males receiving saline vehicle did not increase LH pulse frequency (data not presented). (Data from Jackson et al., 2002.)

lambs correlates positively with body weight, but the smallest lambs that do not produce LH pulses are not invariably those that do not respond to leptin treatment. Similarly, those males who had already achieved puberty, as evidenced by pretreatment production of hourly LH pulses, were not always the largest individuals. Based on our experience with the sheep model, leptin appears to be a permissive pubertal signal that interacts with at least one other developmental or growth cue regulating the early onset of GnRH secretion.

11.3.3.4. Exogenous Leptin Stimulates LH in the Hypogonadotropic Growth-Retarded Male Lamb. Growth (and its attendant internal changes) is influenced by quantity and quality of nutrition. However, most experiments in the laboratory are conducted in genetically homogenous animals raised on a standardized diet designed to produce optimum growth, so quantity and quality of nutrition are not variables. In the real world they are. Many investigators shun dietary manipulations suggesting they are unphysiological while others believe that without this approach, a full understanding of how growth times puberty may be limited. Because the mechanisms controlling species-specific rates of development and sexual maturation evolved to optimize reproductive fitness in response to environmental constraints, it is reasonable to study how they operate under suboptimal conditions characteristic of the real world. We recently began a study of growth-restricted lambs to determine how other developmental signals control leptin's ability to stimulate GnRH secretion. In this ongoing study, lambs are both developmentally and nutritionally hypogonadotropic because they are maintained at weaning body weight (20–24 kg) until past the normal age of puberty. Our preliminary results reveal that leptin can increase LH secretion in growth-restricted lambs (Figure 11.4), and also reinforce the notion that body weight per se is not a steadfast predictor of either neuroendocrine function or responsiveness to leptin. Within the group of growth-restricted lambs we observed a range of LH pulse patterns from nonpulsatile to hourly suggesting individuals differ in the extent to which the same level of chronic undernutrition, even during development, suppresses reproductive function. As in our study of normal prepubertal males (Figure 11.3), some growth-retarded individuals that were not producing LH pulses before treatment responded to leptin treatment while others did not. Those with a low pretreatment pulse frequency all respond to leptin. Our current focus is on potential morphologic, metabolic, or growth-related differences between leptin-responsive and leptin-unresponsive individuals with the intent to determine how leptin interacts with other signals to regulate the onset of puberty.

Overall, our findings lead to the conclusion that leptin provides information on energy reserves and plays a role in the onset of neuroendocrine puberty, once the individual has reached the developmental stage when allocating energy for reproduction becomes important. We must now determine if leptin is the unique peripheral energetic signal timing the pubertal GnRH rise, is an important signal, or is but one of a constellation of such signals. Leptin satisfies one of the two criteria as a signal for puberty as evidenced by its ability to increase GnRH secretion in the immature sheep (male). However, because leptin can augment GnRH secretion in the adult, it cannot be considered as a unique energetic signal timing puberty. It may be an important one, but physiologically whether it works in concert with other metabolic signals (orexin, IGF-1, insulin, etc.) is not clear. The second criterion has not yet been satisfied in the sheep model. Peripheral concentrations of leptin should increase during puberty, but whether this occurs during the pubertal increase in GnRH pulse frequency has not been determined. Currently, the assays for sheep leptin must be optimized before such measurements can be made. Nonetheless, careful measurements will be needed from multiple ages and at several times of day. In this respect, we have noted the appearance of a nocturnal increase in circulating leptin in the female rat during puberty (Nagatani et al., 2000a).

11.3.3.5. Central versus Peripheral Action of Leptin in the Sheep. The results of our studies with peripheral administration of leptin in the lamb and in the adult sheep differ from those of others in which leptin is administered centrally. Our rationale for administering leptin peripherally is that the hormone is secreted by fat cells and its action to increase GnRH secretion could be direct in the brain, indirect by one or more other

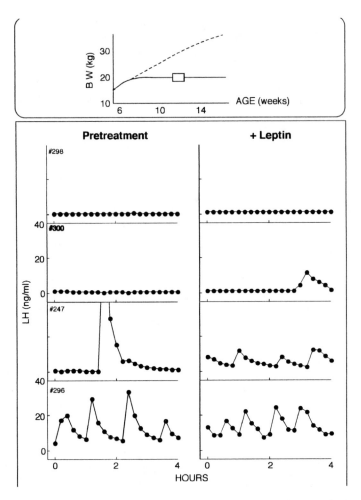

Figure 11.4. Patterns of LH secretion in representative growth-retarded prepubertal, male sheep at 12 weeks of age. The males had been neonatally gonadectomized, treated chronically with low dose estradiol (sc implant, <1 pg/ml), and fed a restricted diet to maintain body weight at ~20 kg as depicted in the top panel. Patterns are presented before treatment and after 72 h of treatment with recombinant human leptin (12.5 µg/kg, sc each 4 hr). Various responses were evident (see text for discussion). Prepubertal males receiving saline vehicle did not increase LH pulse frequency (data not presented). (L. M. Jackson, T. Ichimura, C. A. Jaffe, and D. L. Foster, unpublished.)

mechanisms, or a combination of direct and indirect effects. Certainly leptin may act in the brain as evidenced by receptors located there (Mercer et al., 1996; Elmquist et al., 1998), although GnRH neurons contain few if any leptin receptors (Håkansson et al., 1998; Finn et al., 1998). Arguments have also been made that leptin could work more indirectly through regulation of glucose availability (Foster and Nagatani, 1999), but whether this is a central or peripheral action is not clear. Administration of leptin intracerebroventricularly is used as an approach to assess the central action of leptin and the results depend on the model used (well-fed versus limit-fed). In this regard, studies in adult sheep have demonstrated that central leptin suppresses appetite without affecting LH secretion under well-fed conditions (Henry et al., 1999, 2001; Blache et al., 2000); in food-restricted individuals central leptin does not decrease voluntary food intake but does stimulate LH secretion (Henry et al., 2001). In pubertal ovariectomized lambs, central administration of leptin decreases appetite in well-fed females, but not in nutritionally restricted females (Morrison et al., 2001); in either well-fed or limit-fed females, 8 days of treatment does not alter LH pulse frequency. However, the fact that the dramatic decrease in food consumption in these well-fed lambs was *not* accompanied by a decrease in LH pulse frequency suggests that leptin was, in fact, stimulating GnRH secretion. Similar to the findings of central leptin administration in nutritionally restricted immature or adult sheep discussed above, we have no evidence for appetite suppression during peripheral leptin treatment in our studies. This could be due to the unimportance of leptin as an anorexigenic agent in a rapidly growing or fasted individual in which nutritional demands are very high. Assessing the effect of either central or peripheral leptin treatment on GnRH secretion requires careful consideration of the interactions between leptin as an appetite suppressant, the resulting changes in energy balance, and neuroendocrine function. Given the differences in type and dose of leptin used, preexisting nutritional conditions, seasonal variables, sex, and steroid milieu that exist in studies to date, it is currently difficult to compare the effects of central versus peripheral leptin treatment.

11.4. INTEGRATION OF EXTERNAL AND INTERNAL INFORMATION TO TIME PUBERTY

11.4.1. Multiple Cues for Puberty

As discussed above, the amount of energy intake depends on the quantity and quality of food, and information about energy balance is thought to be relayed to mechanisms controlling GnRH secretion through metabolic signals such as leptin. Because of the great species differences in size, lifespan, and environment, there may be multiple cues timing puberty. In long-lived species, when energy availability is low and growth is retarded, puberty is delayed for months or even years until there is greater energy intake. In small, short-lived species with high metabolism the consequences of reduced nutrition and reduced growth are much more severe as they may not attain puberty before they die. Thus, it is not surprising that in addition to quality and quantity of nutrition, other external information is used to govern the tempo of sexual maturation. In most species, seasonal and social influences play a prominent role. Seasonal information is necessary for species living in their natural environment to time the onset of their reproductive activity such that the young develop when energy is abundant. This is especially critical during the post-weaning period. While photoperiod serves as a primary seasonal cue, specific seasonal nutritional signals have been described as well (e.g., MBOA, a plant factor is used by some microtine rodents). Pheromonal or behavioral social cues are often important as a means of regulating population size and quality; in the developing female (in many species), pheromones from adult females typically delay puberty while those from adult males advance puberty. The necessity for using such complex information to begin reproductive activity reinforces the contention that a simple developmental clock mechanism to time puberty would not meet the needs of the developing individual to adjust to its changing environment and maximize its reproductive success.

11.4.2. Growth and Seasonal Cues Timing Initiation of Ovulations in Female Sheep

The integration of multiple cues to time puberty is readily illustrated by a simple experiment in developing female sheep (Suffolk breed). In this breed, photoperiod is important for the timing of reproductive cycles, and adult females require a decreasing day length to begin their annual reproductive cycles. Young female lambs need similar photoperiodic conditions to initiate their first reproductive cycles, but they have growth requirements as well. In the experiment illustrated in Figure 11.5, groups of young females were all born at the same time of year (spring), and the growth rates were altered by changing the availability of food. Normally growing female lambs exhibited first ovulations

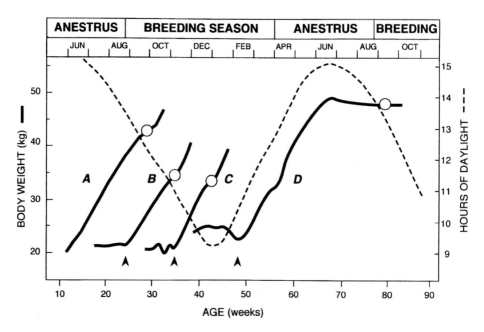

Figure 11.5. Season and growth influence timing of onset of reproductive cycles in female lambs. Females were born in spring (March) and were raised outdoors (broken line, photoperiod). Female lambs in Group A, fed ad libitum after weaning, grew rapidly and first ovulations (puberty) occurred at the usual age (~30 weeks) during the decreasing day lengths of autumn. Other lambs (Groups B, C, D) were fed a restricted diet after weaning and then placed on ad libitum feeding at various ages (arrows). Solid lines represent mean growth curves for each group and open circles denote mean age and body weight at the first luteal phase. Note that puberty is only possible when both growth and photoperiod criteria are satisfied. Sufficient growth must occur such that an adequate metabolic state is attained during the breeding season. (Redrawn from Foster and Ryan, 1990.)

(puberty) at approximately 30 weeks of age during the decreasing day lengths of autumn. The onset of reproductive cycles was delayed in female lambs on a restricted diet despite their having experienced the appropriate photoperiod for initiation of ovulations. Puberty did not occur when lambs were growth-retarded because metabolic cues did not provide the appropriate message to the brain to permit high GnRH secretion. This underscores the importance of information about metabolism and somatic growth relative to photoperiod. When growth-retarded females were provided additional food to induce growth during the autumn and winter breeding season (Groups B and C), cycles began at an earlier stage of growth (i.e., a smaller body size, 35 kg) than for controls (45 kg). In retrospect, this raises the possibility that the normally growing females (Group A) had achieved the appropriate metabolic state and stage of growth for puberty much earlier in the year (August), but day lengths were too long at those younger ages (25 weeks) to permit high GnRH secretion. Later in the year (October at 30 weeks of age), when day length decreased, sensitivity to estradiol negative feedback could be reduced to initiate reproductive cycles. Finally, when the phase of rapid growth was induced during the spring anestrous season (Group D), the lambs remained anovulatory (due to long days of summer) despite their growth well beyond the normal size required to initiate reproductive cycles. It was not until the decreasing day lengths of autumn that reproductive cycles began. Thus, in this example in the sheep, both season and size serve as important codeterminants timing the initiation of reproductive cycles through photoperiod and growth-related cues timing the decrease in sensitivity to steroid feedback and expression of the high-frequency GnRH pulses. Although not discussed here, some of the mechanisms by which photoperiod controls GnRH secretion in the sheep are known, and these include the regulation of sensitivity to estradiol negative feedback inhibition by duration, not amplitude, of the nightly melatonin rise (Foster, 1994; Malpaux et al., 2002). However, sex differences exist in the sheep concerning the relative importance of photoperiod and growth-related cues that regulate the pubertal GnRH rise. As discussed below, these differences are programmed before birth.

11.4.3. Prenatal Programming of the Type of Cues Used for Puberty in the Sheep

Experience with the sheep as an experimental animal has revealed that there is a marked sex difference in the timing of puberty, and this has theoretic significance in terms of which cues are involved. Importantly, this sex difference has practical implications in terms of which growth cues time puberty and how one might develop models to study them. As illustrated in Figure 11.5, photoperiod and growth serve as codeterminants to time first ovulation. Regardless of how rapid the growth of the developing female sheep, the decrease in sensitivity to estradiol negative feedback that times the pubertal GnRH rise will not occur during increasing or long daylengths. The strict requirement for short days has been determined using artificial photoperiods (for review, Foster, 1994). By contrast, the developing male will begin its pubertal LH rise at 7–10 weeks of age, which is very early relative to that of the female. This will occur at the same age in the male regardless of photoperiod be it long, short, increasing, or decreasing daylengths. Thus, by functionally ignoring photoperiod cues the first few months of life the male sheep has developed a very interesting strategy to develop its testes during a period when the adults are reproductively quiescent. This is necessary because of the extended time needed for complete development of spermatogenesis ($\sim$60 days). Although GnRH secretion is not under photoperiod control in the developing male, it is yet highly photoperiodic in other respects much like the female, for example, melatonin secretion and prolactin secretion (Claypool, et al., 1989; Ebling et al., 1989). Of equal interest, is the well-established fact that control of gonadotropin secretion in the adult male sheep is regulated by photoperiod (Lincoln, 2002).

The origins of this and other marked sex differences in control of GnRH secretion in the sheep are prenatal, and testosterone from the fetal testes programs the brain of the unborn male to defeminize the photoperiod control system timing puberty (Wood and Foster, 1998; Foster et al., 2002). Administration of testosterone to the developing female during a critical period before birth will irreversibly organize the brain such that androgenized genetic females no longer use photoperiod to time puberty, much like the normal male (Figure 11.6). Therefore, in the sheep the type of cue used to time puberty depends upon the sex of the brain. The male exclusively uses growth cues. The female uses photoperiod cues when nutrition is abundant and growth is rapid, but under circumstances of suboptimal food availability the female lambs use information about growth to time puberty. Growth cues become important only when the female is slow growing and all photoperiod requirements have been met. There is a practical lesson to be learned from this understanding and that is the developing male, not the female, is more useful as a model to

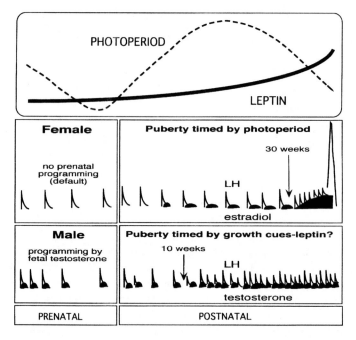

Figure 11.6. Sex differences in cues timing the pubertal increase in GnRH secretion in the sheep. In females, the pubertal increase in GnRH pulse frequency, as evidenced through pulsatile LH secretion, is timed by photoperiod (see Figure 11.5 for data on onset of cycles). In males, the default mechanism for photoperiod timing of puberty is abolished by testosterone secreted by the fetal testes, and the pubertal rise is timed by information about growth and metabolism. Leptin may be one of these growth cues/metabolic signals.

determine which cues, such as leptin, have the potential to relay information about energy balance to time the pubertal increase in GnRH secretion to the appropriate stage of growth. Studies of the developmental importance of leptin to control GnRH secretion are much more difficult in the prepubertal female because of the photoperiod block to high GnRH secretion during its prolonged period of sexual maturation.

11.5. CONSIDERATION OF SPECIES DIFFERENCES IN THE ROLE OF LEPTIN IN TIMING PUBERTY

The simple finding in the sheep model that even within a species there are differences between sexes in the type of cues used to initiate the pubertal GnRH rise, raises the obvious possibility that there may well be marked species differences in the types of cues that time puberty. This would arise, in part, because of differences in dependence on environmental and social information, but do these differences also arise because of species differences in rates of development and lifespans? Sometimes these differences and similarities are not considered in animals raised in laboratory conditions, and a one-hypothesis-fits-all approach is unrealistic. On the other hand, some mechanisms are highly conserved and only superficial species differences exist given the appropriate perspective. The reader is encouraged to read the other chapters in this volume on leptin and the timing of puberty in the rodent, subhuman primate, and human with these comparative considerations in mind.

The short-lived laboratory rodents are born in a relatively immature state neurologically and somatically (altricial). Early studies of the importance of leptin as a signal timing puberty led to conflicting conclusions. Perhaps this is because in the laboratory, the rate of development is so rapid that most experimental designs do not take into account that level of nutrition is an important variable timing puberty. Such considerations are particularly difficult when the animals are raised on an unlimited caloric diet where individuals are developing as rapidly as possible. However, an important role of leptin has been revealed when there are differences in energy intake because of diet, namely that leptin may serve as an important mediator of this information through energy balance. Thus, the current view is that leptin is a permissive cue. Subhuman primates are also altricial, and although the period of development seems long, it is relatively short in terms of total lifespan. Again, most studies have been conducted in animals on a standardized diet designed to promote optimal growth and their results do not provide evidence for an important role for leptin. Perhaps, as in the rodent, leptin is also important, but only during the final stages of sexual maturation when energy and other factors are being integrated by the brain in an attempt to determine when high-frequency GnRH secretion should first occur. Experimental evidence for this view is currently lacking. Lest one forget, puberty in some subhuman primates may be controlled by environmental variables, and this must be taken into account when assessing the role of leptin during its transition into adulthood. Finally, in the human, the precise role of leptin is difficult to assess because of the inability to design studies in normal children.

11.6. CONCLUSIONS

This chapter, when combined with others in this volume, should provide an overall impression about how complex the timing of puberty is. Many investigators have been searching for the "trigger" for puberty, but like the Seven Cites of Gold sought by Coronado, we are finding that this may well be a myth. There has been no unique mechanism noted to date to explain puberty that does not operate in the adult at either the integrative or cellular level. In the absence of such mechanisms, we are forced to view puberty as the first time the individual may use these mechanisms. We need to know much more about neurobiological mechanisms to provide a clear understanding of how internal and external information is integrated by the brain to regulate the control of GnRH secretion in the developing individual. Given that a positive energy balance is necessary for this activity, it is reasonable to deduce that puberty represents the first time an individual attains some type of genetically predetermined critical positive energy balance. During the pubertal process, leptin is likely to be one of the important cues used to relate energy balance to the brain to regulate GnRH secretion, and this same hormone seems to serve the same role in the adult. Thus, the view that leptin plays permissive role in the timing of puberty is reasonable. Other signals that are also permissive and that relate to the environment and society must be integrated with those of energy metabolism to ultimately control when reproductive activity occurs. In the context of these considerations, our current broad view of the mechanisms timing the pubertal increase in GnRH secretion is that there are likely to be several "permissive" mechanisms acting in parallel or in sequence that ultimately serve to time the first sustained increase in GnRH secretion. While these mechanisms are also used by the adult to time its periods of high GnRH secretion, there will certainly be pure developmental mechanisms that exist at earlier stages to pace the rate of sexual maturation. Only when these are discovered will we have a glimpse into what "nonpermissive" mechanisms are unique to development and underlie the transition into adulthood.

ACKNOWLEDGMENT

This work was supported by NIH-HD-18394.

REFERENCES

Blache, D., Celi, P., Blackberry, M. A., Dynes, R. A., and Martin, G. B. (2000). Decreases in voluntary feed intake and pulsatile luteinizing hormone secretion after intracerebroventricular infusion of recombinant bovine leptin in mature male sheep. *Reprod. Fertil. Dev., 12*, 373–381.

Bourguignon, J. P. and Plant, T. M. (2000). *Onset of Puberty in Perspective*. Elsevier Science Publishers, Amsterdam.

Claypool, L. E., Wood, R. I., Yellon, S. M., and Foster, D. L. (1989). The ontogeny of melatonin secretion in the lamb. *Endocrinology, 124*, 2135–2143.

Donovan, B. T. and Werff ten Bosch, J. J. van der (1965). *Physiology of Puberty*. Edward Arnold Publishers, London.

Ebling, F. J. P., Wood, R. I., Suttie, J. M., Adel, T. E., and Foster, D. L. (1989). Prenatal photoperiod influences neonatal prolactin secretion in the sheep. *Endocrinology, 125*, 384–391.

Elmquist, J. K., Bjørbæk, C., Ahima, R. S., Flier, J. S., and Saper, C. B. (1998). Distributions of leptin receptor mRNA isoforms in the rat brain. *J. Comp. Neurol., 395*, 535–547.

Finn, P. D., Cunningham, M. J., Pau, K.-Y. F., Spies, H. G., Clifton, D. K., and Steiner, R. A. (1998). The stimulatory effect of leptin on the neuroendocrine reproductive axis of the monkey. *Endocrinology, 139,* 4652–4662.

Foster, D. L. (1994). Puberty in the sheep. In E. Knobil and J. D. Neill (Eds.). *The Physiology of Reproduction* (Vol. 2), Raven Press, New York, pp. 411–451.

Foster, D. L. and Nagatani, S. (1999). Physiological perspectives of leptin as a regulator of reproduction: role in timing puberty. *Biol. Reprod., 60,* 205–215.

Foster, D. L. and Ryan, K. D. (1990). Puberty in the lamb: sexual maturation of a seasonal breeder in a changing environment. In P. C. Sizonenko, M. L. Aubert, and M. M. Grumbach (Eds.). *Control of the Onset of Puberty II,* William and Wilkins, Baltimore, pp. 108–142.

Foster, D. L., Ebling, F. J. P., Ryan, K. D., and Yellon, S. M. (1989). Mechanisms timing puberty: a comparative approach. In H. A. Delemarre-van de Waal, T. M. Plant, G. P. van Rees, and J. Schoemaker (Eds.). *Control of the Onset of Puberty III,* Elsevier Science Publishers B.V., Amsterdam, pp. 227–249.

Foster, D. L., Padmanabhan, V., Wood, R. I., and Robinson, J. E. (2002). Sexual differentiation of the neuroendocrine control of gonadotropin secretion: concepts derived from the sheep. In D. Skinner, N. Evans, and C. Doberska (Eds.). *Large Mammals as Neuroendocrine Models,* Reproduction Supplement 59, Cambridge University Press, Cambridge, pp. 83–99.

Håkansson, M.-L., Brown, H., Ghilardi, N., Skoda, R. C., and Meister B. (1998). Leptin receptor immunoreactivity in chemically defined target neurons of the hypothalamus. *J. Neurosci., 18,* 559–572.

Henry, B. A., Goding, J. W., Alexander, W. S., Tilbrook, A. J., Canny, B. J., Dunshea, F., Rao, A., Mansell, A., and Clarke, I. J. (1999). Central administration of leptin to ovariectomized ewes inhibits food intake without affecting the secretion of hormones from the pituitary gland: evidence for a dissociation of effects on appetite and neuroendocrine function. *Endocrinology, 140,* 1175–1182.

Henry, B. A., Goding, J. W., Tilbrook, A. J., Dunshea, F. R., and Clarke, I. J. (2001). Intracerebroventricular infusion of leptin elevates the secretion of luteinising hormone without affecting food intake in long-term food-restricted sheep, but increases growth hormone irrespective of bodyweight. *J. Endocrinol., 168,* 67–77.

Jackson, L. M., Ichimaru, T., Jaffe, C. A., and Foster, D. L. (2002). Effects of leptin on LH secretion during developmental and nutritional hypogonadotropism. *Program of the 84th Meeting of the Endocrine Society,* San Francisco. Abstract OR45-1.

Kennedy, G. C. and Mitra, J. (1963). Body weight and food intake as initiating factors for puberty in the rat. *J. Physiol., 166,* 408–418.

Lincoln, G. A. (2002). Neuroendocrine regulation of seasonal and gonadotropin rhythms: lessons learned from the Soay ram model. In D. Skinner, N. Evans, and C. Doberska (Eds.). *Large Mammals as Neuroendocrine Models,* Reproduction Supplement 59, Cambridge University Press, Cambridge, pp. 131–147.

Malpaux, B., Tricoire, H., Mailliet, Daveau, A., Migaud, M. Skinner, D. C., Pellietier, J., and Chimineau, P. (2002). Melatonin and seasonal reproduction: understanding the neuroendocrine mechanisms using the sheep as a model. In D. Skinner, N. Evans, and C. Doberska (Eds.). *Large Mammals as Neuroendocrine Models,* Reproduction Supplement 59, Cambridge University Press, Cambridge, pp. 131–147.

Mercer, J. G., Hoggard, N., Williams, L. M., Lawrence, C. B., Hannah, L. T., Morgan, P. J., and Trayhurn, P. (1996). Coexpression of leptin receptor and preproneuropeptide Y mRNA in arcuate nucleus of mouse hypothalamus. *J. Neuroendocrinol., 8,* 733–735.

Morrison, C. D., Daniel, J. A., Holmberg, B. J., Djiane, J., Raver, N., Gertler, A., and Keisler, D. H. (2001). Central infusion of leptin into well-fed and undernourished ewe lambs: effects on feed intake and serum concentrations of growth hormone and luteinizing hormone. *J. Endocrinol., 168,* 317–324.

Nagatani, S., Guthikonda, P., Thompson, R. C., Tsukamura, H. Maeda, K.-I., and Foster, D. L. (1998). Evidence for GnRH regulation by leptin: leptin administration prevents reduced pulsatile LH during fasting. *Neuroendocrinology, 67,* 370–376.

Nagatani, S., Guthikonda, P., and Foster, D. L. (2000a). Appearance of a nocturnal peak of leptin during puberty. *Horm. Behav., 37,* 345–352.

Nagatani, S., Zeng, Y., Keisler, D. H., Foster, D. L., and Jaffe, C. A. (2000b). Leptin regulates pulsatile luteinizing hormone and growth hormone secretion in the sheep. *Endocrinology, 141,* 3965–3975.

Ojeda, S. R. and Urbanski, H. F. (1994). Puberty in the rat. In E. Knobil and J. D. Neill (Eds.). *The Physiology of Reproduction* (Vol. 2), Raven Press, New York, pp. 363–409.

Olster, D. H. and Foster, D. L. (1986). Control of gonadotropin secretion in the male during puberty: a decrease in response to steroid inhibitory feedback in the absence of an increase in steroid-independent drive in the sheep. *Endocrinology, 118,* 2225–2234.

Plant, T. M. (1994). Puberty in primates. In E. Knobil, and J. D. Neill (Eds.). *The Physiology of Reproduction* (Vol. 2), Raven Press, New York, pp. 453–485.

Tanner, J. M. (1962). *Growth at Adolescence* (2nd ed.). Blackwell Scientific Publications, London.

Terasawa, E. (2001). Luteinizing hormone-releasing hormone (LHRH) neurons: mechanism of pulsatile LHRH release. *Vitam. Horm., 63*, 91–129.

Wood, R. E. and Foster, D. L. (1998). Sexual differentiation of reproductive neuroendocrine function. *Rev. Reprod., 3*, 130–140.

Wood, R. I., Newman, S. W., Lehman, M. N., and Foster, D. L. (1992). GnRH neurons in the fetal lamb hypothalamus are similar in males and females. *Neuroendocrinology, 55*, 427–433.

Wray, S. and Hoffman, G. E. (1986). Postnatal morphological changes in rat LH–RH neurons correlated with sexual maturation. *Neuroendocrinology, 43*, 93–97.

Yellon, S. M. and Newman, S. W. (1991). A developmental study of the GnRH neuronal system during sexual maturation in the male Djungarian hamster. *Biol. Reprod., 45*, 440–446.

IV

Pregnancy

12

Leptin and Fetal Growth and Development

HELEN CHRISTOU, SHANTI SERDY, AND CHRISTOS S. MANTZOROS

12.1. INTRODUCTION

Leptin, a 16 kDa adipocyte secreted protein, was initially discovered and characterized as a molecule which acts in the hypothalamus to signal adequacy of nutritional status, thereby regulating energy intake and expenditure, and thus weight gain and fat deposition. The role of leptin in conveying information to the hypothalamus regarding energy availability has recently been shown to have a broader significance, including leptin's role to regulate growth processes and to mediate the neuroendocrine response to fasting (Holness et al., 1999). In addition, experimental data in animals (Chehab et al., 1996; Mounzih et al., 1997) as well as observational studies in humans (Strobel et al., 1998; Mantzoros, 1999) indicate that leptin may be a factor that triggers the onset of puberty, ovulation, and reproduction.

Although several other actions of leptin have been proposed recently and are currently under intense investigation, the entire spectrum of physiological roles of leptin in addition to its classical role in energy homeostasis remains to be fully clarified. The discovery of (a) leptin production in non-adipose tissues including the placenta, and (b) leptin receptors in sites outside the central nervous system has triggered an enormous interest over the past several years in exploring the role of leptin in a number of physiologic and pathophysiologic processes, including its role in fetal growth. Specifically, the discovery of leptin production in the human placenta, and the presence of leptin receptors in fetal tissues has led to the hypothesis that leptin may be involved in developmental processes in

HELEN CHRISTOU • Division of Newborn Medicine, Children's Hospital, Harvard Medical School.
SHANTI SERDY, AND CHRISTOS S. MANTZOROS • Division of Endocrinology, Department of Internal Medicine, Beth Israel Deaconess Medical Center, Harvard Medical School.

Leptin and Reproduction. Edited by Henson and Castracane, Kluwer Academic/Plenum Publishers, 2003.

the fetus. Although significant insights in this area have been provided by several in vitro studies as well as animal studies in vivo, several questions regarding the functional role of leptin in the fetus remain unanswered. In this chapter, we discuss current knowledge regarding the role of leptin in fetal growth, its relationship to developmental processes in the fetus, and its potential role in pathologic conditions associated with these processes. Finally, we discuss areas that need to be further studied.

12.2. LEPTIN IS PRODUCED BY THE PLACENTA AND LEPTIN RECEPTOR IS EXPRESSED BY FETAL TISSUES

Masuzaki et al. (1997) reported that leptin is expressed in human placental trophoblasts and amnion cells as well as in a cultured human choriocarcinoma cell line. Work performed subsequently has confirmed the presence of leptin mRNA and protein in the human, rodent, ovine, and porcine placentas (Ashworth et al., 2000). Although significant species differences do exist, with human placenta producing much higher amounts of leptin compared to rodent placenta, the localization of leptin appears to be common in syncytiotrophoblast (in direct contact with maternal blood) and cytotrophoblast as well as in villous vascular endothelial cells (in direct contact with fetal blood).

Several splice variants of the leptin receptor have also been identified in the murine placenta, including OB-Rb (the signaling form of the leptin receptor), OB-Ra (the proposed transport form of the receptor), and OB-Re (the soluble form of the receptor). Leptin receptor mRNA is expressed in a number of mouse fetal tissues including cartilage and bone, lung, hair follicles, leptomeninges and choroid plexus (Hoggard et al., 1997), kidney and testes (Hoggard et al., 2001). In fetal rat pancreatic islets, leptin receptor isoforms have been detected by reverse transcriptase-polymerase chain reaction (RT-PCR) (Islam et al., 2000), and this has led investigators to propose a stimulatory role for leptin in pancreatic development and thus an indirect effect on fetal growth. Leptin mRNA and protein have been detected in adipose tissue from 20 to 38-week human fetuses (Lepercq et al., 2001), as well as in fetal heart and liver. The potential physiological significance of these observations remains to be determined and it remains unclear which fetal tissues produce sufficient amounts of leptin to have an endocrine effect and which tissues produce levels that are likely to have only local, autocrine or paracrine, effects (Reitman et al., 2001). Importantly, both fetal adipose tissue and the placenta are sources of cord blood leptin, the levels of which correlate with birth weight. The potential role of leptin in fetal growth is discussed below.

12.3. ROLE OF LEPTIN IN NORMAL REPRODUCTIVE FUNCTION

Several in vitro experiments as well as in vivo studies (including a leptin administration study to a leptin-deficient subject) indicate that leptin may be necessary for normal reproductive hormone levels and function (Reed et al., 1996; Moschos et al., 2002). More specifically, accumulating evidence from observational studies in humans, and more recently an interventional, leptin administration study in a leptin-deficient prepubertal child provides support to the notion that leptin plays an important role in human reproduction. The observational studies have clearly shown that leptin is secreted in a pulsatile fashion, and that circulating levels of this adipokine display a circadian rhythm in humans. Synchronicity of pulsatile luteinizing hormone (LH), estradiol, and leptin levels has been reported to occur during the mid-to-late follicular phase of the menstrual cycle in healthy women (Licinio et al., 1998). Further supportive evidence comes from several interventional studies in humans. For example, in clinical trials focusing on hypogonadotropic hypogonadism, it has been shown that pulsatile leptin secretion persists despite absence of gonadotropin-releasing hormone (GnRH)-LH pulsatility (Sir-Petermann et al., 1999, 2001) indicating that it is not the GnRH-LH pulsatility that influences leptin pulsatility. Most notably, leptin administration to a leptin-deficient girl with hypogonadotropic hypogonadism resulted in a pattern of gonadotropin secretion consistent with early puberty (Farooqi et al., 1999). Taken together, these data from observational and interventional studies in humans suggest that leptin plays a role not only in regulating physiologic levels but also the rhythmicity of LH and follicle-stimulating hormone (FSH) and possibly reproductive hormones. Ongoing interventional studies in humans aim to further explore the pathophysiologic significance of these observations and to investigate the role of leptin in the initiation of puberty as well as in reproductive function in healthy normal subjects. Similar studies are currently being performed in leptin-deficient states such as exercise-induced amenorrhea or anorexia nervosa.

12.4. ROLE OF LEPTIN IN NORMAL PREGNANCY AND IN PATHOLOGIC CONDITIONS DURING PREGNANCY

Circulating leptin levels are elevated during normal human pregnancy and leptin levels correlate significantly with serum levels of several other hormones (Sivan et al., 1998). Longitudinal studies in pregnant women have clearly shown that the increasing leptin levels peak during the second and third trimester and return to normal within 24 hr of delivery (Sivan et al., 1998). The changes of circulating leptin levels reported in these observational studies parallel the anabolic to catabolic shift in energy metabolism that occurs during pregnancy, and thus have been interpreted by many as suggesting that leptin may be one of the factors regulating maternal and fetal energy balance during normal gestation.

Both placental leptin and cord blood levels of leptin are increased in diabetic pregnancies (Lea et al., 2000) despite the fact that maternal levels remain essentially unaffected (Lepercq et al., 1998). Since insulin dependence during pregnancy is associated with relatively higher placental leptin levels, some authors have concluded that, similar to insulin's action in adipose tissue, insulin directly upregulates leptin expression in the placenta (Lepercq et al., 1998). Whether this hypothesis is correct as well as the exact mechanisms through which leptin may influence fetal growth in the diabetic state need to be studied further.

Increased placental production of leptin is also seen in preeclamptic pregnancies, and cord blood levels of leptin in preeclampsia correlate significantly with increased maternal serum levels (Mise et al., 1998; McCarthy et al., 1999). Several factors, including placental hypoxia, have been proposed as important regulators of placental leptin expression during preeclampsia (Mise et al., 1998) and recent studies have demonstrated a transcriptional effect of hypoxia on placental leptin production (Grosfeld et al., 2001). Hypoxia has also been proposed to be one of the factors possibly mediating the abnormally low leptin levels in small for gestational age infants. The complex interplay of factors influencing cord blood leptin levels and its effects on fetal growth remains to be fully elucidated.

12.5. ROLE OF LEPTIN IN EARLY EMBRYONIC DEVELOPMENT

Several converging lines of evidence emanating from independent groups suggest a role for leptin, mediated by JAK/STAT 3 signaling, in early embryonic development and implantation. Leptin has been detected in mouse and human oocytes and preimplantation embryos (Antczak et al., 1997). Leptin receptor is also present in mouse oocytes (Matsuoka et al., 1999). Finally, the polarized distribution of leptin/STAT 3 on the surface of the embryo at the morula stage, along with the observation that embryos from STAT 3-deficient mice degenerate after implantation (Takeda et al., 1997), has led several investigators to propose that leptin has an important role in early embryonic development. However, the exact role of leptin in implantation in the mouse remains to be fully elucidated (Mounzih et al., 1998).

A number of experimental data suggest a role for leptin during the preimplantation phase in humans. Leptin is produced by human blastocyst, endometrial epithelial cell cultures and cultured cytotrophoblasts (Gonzalez et al., 2000). Leptin's expression is regulated by interleukin (IL)-1β and 17β-estradiol (Chardonnens et al., 1999), and it appears that the IL-1 system plays an important role in the cross-talk between the preimplantation embryo and the receptive endometrium during the early phase of the human implantation process (Simon and Polan, 1994). Moreover, it has been proposed that IL-1β may induce leptin expression through an activation of PLE3, a placenta-specific enhancer element that was identified in the promoter region of the leptin gene (Wolf et al., 2000). Cytotrophoblast-derived leptin regulates protease synthesis (MMP-9) and integrin expression ($\alpha_6\beta_4$) in cytotrophoblast and may function as an autocrine/paracrine regulator of the invasive phenotype of cytotrophoblast during the invasion phase of human implantation (Gonzalez et al., 2000). In accordance with these in vitro observations, a clinical study focusing on serum and follicular leptin levels has shown that serum and most importantly follicular fluid leptin levels (and/or leptin resistance) play an important role in determining pregnancy rates and thus the likelihood of success of in vitro fertilization programs (Mantzoros et al., 2000).

12.6. ROLE OF LEPTIN IN FETAL GROWTH AND DEVELOPMENT

It is widely accepted that normal fetal growth and development reflect the interaction of maternal, placental, and fetal factors. Thus, in addition to genetic factors, nutritional, endocrine, and paracrine factors interact to regulate fetal growth. It is currently believed that insulin and the insulin-like growth factors (IGFs) are the main established endocrine regulators of fetal growth in humans and animals (Milner and Gluckman, 1996). Whereas IGF-I cord blood levels correlate well with birth weight (Verhaeghe et al., 1993), insulin's role is thought to be permissive in fetal growth and IGF-II levels correlate poorly with birth weight (Wiznitzer et al., 1998). Studies in null mutant mice have shown that IGF-I is the dominant growth-promoting factor during the rapid phase of somatic growth in late gestation (Baker et al., 1993; Liu et al., 1993) while IGF-II appears to be important in fetal growth in early gestation only (DeChiara et al., 1990).

Pregnancy is also a state of hyperleptinemia in both humans and rodents. In humans, the source of increased maternal leptin levels is thought to be the placenta (Lepercq et al., 2001), whereas in rodents it may also be oversecreted from adipose tissue (Kawai et al., 1997; Tomimatsu et al., 1997). Despite increased maternal leptin levels, pregnancy is also considered by many to be a leptin resistant state (Mounzih et al., 1998). It has also been proposed, although direct experimental evidence is still lacking, that leptin resistance during pregnancy may be indirectly influencing fetal growth by sustaining maternal food intake and thus providing sustained energy sources for fetal growth. Moreover, it has been shown that in mice, leptin-binding capacity in the serum rises approximately 40-fold by Day 18 of gestation (Gavrilova et al., 1997). This increase is most likely due to large amounts of the circulating OB-Re (short form) of the leptin receptor, which is produced in the murine placenta. Thus, although free leptin levels have not yet been measured directly, it is expected that the increase of free leptin levels will be relatively lower than that of total leptin levels in pregnancy. Given the above, it might have been expected that circulating maternal total leptin levels do not correlate with the offspring's birth weight. In contrast, numerous studies have shown a positive correlation between cord blood leptin levels and birth weight in human infants (Harigaya et al., 1997; Koistinen et al., 1997; Mantzoros et al., 1997; Ong et al., 1999; Varvarigou et al., 1999; Christou et al., 2001). These observations, together with the discovery of leptin production by the placenta and the presence of leptin receptors in a variety of fetal tissues has led to the speculation that fetal- and placental-derived leptin play a direct role in fetal growth and development.

Direct positive associations between cord blood leptin concentrations and fetal insulin or IGF-1 levels have been reported in normal infants (Varvarigou et al., 1999; Wolf et al., 2000) and cord blood leptin levels are also directly related to cord blood insulin levels in large for gestational age (LGA) infants (Coutant et al., 2001). In addition to observational studies in neonates, in vitro observations that insulin promotes leptin production in adipose tissue (Kolaczynski et al., 1996) lend further support to a possible interaction between fetal leptin and insulin in predicting fetal growth. Conversely, leptin has also been found to stimulate pancreatic islet cell development (Islam et al., 1997, 2000). Thus, to determine whether the observed associations between leptin and fetal growth are independent of other regulators of fetal growth, such as insulin levels, the relationship between leptin and the known hormonal regulators of fetal growth (insulin, IGFs, GH) has been examined in a number of studies. We have shown that the positive correlation between cord blood leptin concentrations and birth weight is independent of adiposity and insulin levels (Varvarigou et al., 1999) or IGF-I levels (Christou et al., 2001).

Experimental support of these observational data has recently been provided by an important study by Yamashita et al. (2001). These authors have demonstrated that alterations in placental leptin may contribute to the regulation of fetal growth independently of maternal glucose levels. Heterozygous mice deficient in the leptin receptor ($db/+$) develop gestational diabetes, which leads to fetal overgrowth associated with increased maternal weight gain during pregnancy, increased maternal glucose levels, and fetal hyperinsulinemia. The treatment of these mice with leptin reduced energy intake and maternal weight gain, improved maternal glucose tolerance, but had no effect on fetal overgrowth. Although fetal insulin levels were not directly measured in the study by Yamashita et al. (2001), it is unlikely that fetal overgrowth was secondary to hyperinsulinemia because glucose homeostasis was not disrupted. Instead, it was suggested that high placental leptin levels in $db/+$ pregnancies, which persisted after leptin treatment, may be etiologically related to fetal overgrowth. In agreement with this notion are data showing increased placental leptin levels in human diabetic pregnancies (Lepercq et al., 1998) and decreased placental and cord blood leptin levels in pregnancies complicated by intrauterine growth retardation (Lea et al., 2000). Thus, it has been proposed that leptin, either by affecting growth directly and/or by interacting with growth hormones, or by exerting specific effects on the feto-placenta-maternal unit to alter the partitioning of nutrients, plays an important role during normal fetal growth and its aberrations.

The precise mechanism underlying the involvement of leptin in fetal growth remains unclear and warrants further research, which is expected to define the molecular mechanisms by which leptin regulates fetal growth. It is considered highly likely, however, that leptin's role in fetal growth is only permissive since no major abnormalities in fetal growth were observed in leptin-deficient humans (Clement et al., 1998). Similarly, in *ob/ob* mice supplemented with leptin in doses sufficient to achieve fertility, withdrawal of leptin at any stage throughout pregnancy had no major effect on fetal growth (Mounzih et al., 1998). This finding has been interpreted as suggesting that in states of complete absence of leptin, alternative modulators of fetal growth are sufficient to allow normal fetal growth in the mouse. Finally, it should be kept in mind that the possibility exists that species differences in the placental production of leptin may be responsible for potential differences in the role of leptin in fetal growth in mice versus humans. This warrants careful interpretation of animal versus human studies, and the performance of more studies on the role of leptin in fetal growth in humans.

12.7. LEPTIN AND ANGIOGENESIS

Human umbilical venous endothelial cells express both the short and the long forms of leptin receptor (Bouloumie et al., 1998; Sierra-Honigmann et al., 1998). Treatment of these cells with leptin promotes angiogenic processes in a manner similar to vascular endothelial growth factor (VEGF) by activating a growth-related signaling pathway involving a tyrosine kinase intracellular pathway (most prominently the mitogen-activated protein kinases Erk1/2). The expression of leptin receptor in rat aortic smooth muscle cells further supports the possible role of leptin in vascular homeostasis. In these cells, leptin stimulated both proliferation and migration via activation of MAP kinase and PI3 kinase (Oda et al., 2001). As a result, it is reasonable to speculate that a possible angiogenic effect of leptin may be important in blood vessel growth in the maturing egg and early embryo. The clinical significance of these findings remains to be fully clarified.

12.8. ROLE OF LEPTIN IN HEMATOPOIESIS AND IMMUNE FUNCTION

Leptin receptor mRNA is expressed in hematopoietic tissues, including fetal liver, bone marrow, spleen, and human CD34+ progenitor cells as well as in two human hematopoietic cell lines in mice and humans (Gainsford et al., 1996). Although transcripts encoding both the long and short forms of the leptin receptor are co-expressed in human tissues, only the long form of leptin receptor can signal proliferation and differentiation in transfected mouse hematopoietic cells. In the mouse, leptin stimulates the proliferation of myelocytic progenitors and their differentiation into mature myelocytic cells (Umemoto et al., 1997). Leptin also enhances the production of cytokines, such as tumor necrosis factor (TNF)-α, IL-6, and IL-12 by macrophages (Loffreda et al., 1998) as well as mature macrophage function as indicated not only by increased cytokine production but also enhanced phagocytosis (Gainsford et al., 1996). Studies in leptin-deficient *ob/ob* and leptin-resistant *db/db* mice (Loffreda et al., 1998) have demonstrated that the absence of leptin or of a functional leptin receptor leads to impaired macrophage phagocytosis and abnormal cytokine gene expression, rendering these animals more susceptible to infection by various pathogens. Using the acute starvation model, Lord et al. (1998) have also reported that leptin administration to leptin-deficient mice reversed the immunosuppressive effects of acute nutrient deprivation. Together, these findings suggest a novel role for leptin in relating nutritional status to the regulation of cellular immune function, but direct evidence for a similar function in humans is still lacking. In addition, the full spectrum of leptin's immune modulating actions in humans remains unclear since human leptin failed to consistently influence a number of aspects of the biology of hematopoietic cells such as clonal proliferation, colony formation, and survival of progenitor cells (Gainsford et al., 1996).

In humans, only observational studies have focused on a role for leptin in the immune response. Ozata et al. (1999) reported that in a consanguineous Turkish family with leptin deficiency due to a leptin gene missense mutation, 7 of 11 affected family members died in childhood of infection. The authors implicated leptin deficiency as the factor responsible for altered immunity despite the fact that leptin-deficient survivors in the same extended family have no major immune deficiencies. Furthermore, in in vitro studies using human T lymphocytes, Lord et al. (1998) showed that similar to their previously mentioned data in mice, leptin differentially regulates the proliferation of naïve and memory T cells. However, the precise role of leptin in human immunity remains to be fully elucidated by interventional studies.

12.9. ROLE OF LEPTIN IN OSSIFICATION AND BONE DEVELOPMENT

Leptin is thought to play a role in fetal bone and cartilage development during intrauterine development of the skeleton. Leptin and leptin receptor expression have been detected in high levels in fetal bone and cartilage in the 14.5-day postcoitus mouse fetus (Hoggard et al., 1997) as well as in human articular chondrocytes (Figenschau et al., 2001).

An observational study in human fetuses (Ogueh et al., 2000) showed a gradual increase in fetal leptin levels with advancing gestation and a significant inverse correlation between fetal leptin levels and a marker of bone resorption. In agreement with these observations, bone mineral content and density are positively correlated with plasma leptin levels in premenarcheal children (Matkovic et al., 1997) as well as adolescents but not independently of weight, fat mass, and serum IGF-I levels (Manson et al., 1995; Roemmich et al., 2003). Treatment of bone marrow stromal cells with leptin in vitro led to increased expression of alkaline phosphatase, type I collagen, and osteocalcin and increased mineralized bone matrix formation (Thomas et al., 1999). These effects of leptin appear to be different from its negative effects on the central regulation of bone mass in the mature skeleton (Ducy et al., 2000), that is, that leptin inhibits osteoblastic activity in *ob/ob* and *db/db* mice via a centrally mediated inhibitory effect of leptin on osteoblast function (Ducy et al., 2000).

12.10. LEPTIN AND BRAIN DEVELOPMENT

Regulation of brain development by leptin in the mouse is supported by the observation of a number of brain abnormalities observed in the leptin-deficient (*ob/ob*) or leptin-resistant (*db/db*) mice such as decreased brain weight and DNA content, as well as structural neuronal abnormalities and impaired myelination (Bereiter and Jeanrenaud 1979, 1980; Rossier et al., 1979; Sena et al., 1985). However, the timing of leptin's effects on murine brain development appears to occur mostly in the postnatal period (Ahima et al., 1999). In contrast, in a very limited number of cases of human leptin deficiency, brain structures appear normal by CT scan and there is no association with psychomotor or learning abnormalities (Clement et al., 1998; Strobel et al., 1998). Thus, the role of leptin in human brain development remains uncertain at this time and needs to be further clarified.

12.11. CONCLUSIONS AND FUTURE DIRECTIONS

Leptin is a 16 kDa protein secreted by adipocytes in proportion to the size of energy stores in adipose tissue and its levels are also regulated by acute energy availability. Leptin conveys to the hypothalamus information on energy homeostasis and regulates reproductive function. In addition, other roles of leptin, including its role as a regulator of fetal growth and development are beginning to be explored. Accumulating evidence suggests that leptin produced by placental or fetal tissues acts through leptin receptors expressed in fetal tissues to regulate fetal growth and development. Although leptin may act in part by interacting with the insulin and GH–IGF-I system, observational studies in humans indicate that its effects on fetal growth are independent of these axes. The extent to which leptin per se mediates the fetal growth and developmental abnormalities associated with disease states in which leptin production is altered, such as diabetes, intrauterine growth retardation, and preeclampsia, remains to be fully clarified by future studies in humans. Elucidation of these mechanisms may provide novel therapeutic approaches for these conditions.

REFERENCES

Ahima, R. S., Bjorbaek, C., Osei, S., and Flier, J. S. (1999). Regulation of neuronal and glial proteins by leptin: implications for brain development. *Endocrinology, 140*(6), 2755–2762.

Antczak, M., Van Blerkom, J., and Clark, A. (1997). A novel mechanism of vascular endothelial growth factor, leptin and transforming growth factor-beta2 sequestration in a subpopulation of human ovarian follicle cells. *Hum. Reprod., 12*(10), 2226–2234.

Ashworth, C. J., Hoggard, N., Thomas, L., Mercer, J. G., Wallace, J. M., and Lea, R. G. (2000). Placental leptin. *Rev. Reprod., 5*(1), 18–24.

Baker, J., Liu, J. P., Robertson, E. J., and Efstratiadis, A. (1993). Role of insulin-like growth factors in embryonic and postnatal growth. *Cell, 75*(1), 73–82.

Bereiter, D. A. and Jeanrenaud, B. (1979). Altered neuroanatomical organization in the central nervous system of the genetically obese (ob/ob) mouse. *Brain Res., 165*(2), 249–260.

Bereiter, D. A. and Jeanrenaud, B. (1980). Altered dendritic orientation of hypothalamic neurons from genetically obese (ob/ob) mice. *Brain Res., 202*(1), 201–206.

Bouloumie, A., Drexler, H. C., Lafontan, M., and Busse, R. (1998). Leptin, the product of Ob gene, promotes angiogenesis. *Circ. Res., 83*(10), 1059–1066.

Chardonnens, D., Cameo, P., Aubert, M. L., Pralong, F. P., Islami, D., Campana, A., and Gaillard, R. C. (1999). Modulation of human cytotrophoblastic leptin secretion by interleukin-1alpha and 17beta-oestradiol and its effect on HCG secretion. *Mol. Hum. Reprod., 5*(11), 1077–1082.

Chehab, F. F., Lim, M. E., and Lu, R. (1996). Correction of the sterility defect in homozygous obese female mice by treatment with the human recombinant leptin. *Nat. Genet., 12*(3), 318–320.

Christou, H., Connors, J. M., Ziotopoulou, M., Hatzidakis, V., Papathanassoglou, E., Ringer, S. A., and Mantzoros, C. S. (2001). Cord blood leptin and insulin-like growth factor levels are independent predictors of fetal growth. *J. Clin. Endocrinol. Metab., 86*(2), 935–938.

Clement, K., Vaisse, C., Lahlou, N., Cabrol, S., Pelloux, V., Cassuto, D., Gourmelen, M., Dina, C., Chambaz, J., Lacorte, J. M., Basdevant, A., Bougneres, P., Lebouc, Y., Froguel, P., and Guy-Grand, B. (1998). A mutation in the human leptin receptor gene causes obesity and pituitary dysfunction. *Nature, 392*(6674), 398–401.

Coutant, R., Boux de Casson, F., Douay, O., Mathieu, E., Rouleau, S., Beringue, F., and Gillard, P. (2001). Relationships between placental GH concentration and maternal smoking, newborn gender, and maternal leptin: possible implications for birth weight. *J. Clin. Endocrinol. Metab., 86*(10), 4854–4859.

DeChiara, T. M., Efstratiadis, A., and Robertson, E. J. (1990). A growth-deficiency phenotype in heterozygous mice carrying an insulin-like growth factor II gene disrupted by targeting. *Nature, 345*(6270), 78–80.

Ducy, P., Amling, M., Takeda, S., Priemel, M., Schilling, A. F., Beil, F. T., Shen, J., Vinson, C., Rueger, J. M., and Karsenty, G. (2000). Leptin inhibits bone formation through a hypothalamic relay: a central control of bone mass. *Cell, 100*(2), 197–207.

Farooqi, I. S., Jebb, S. A., Langmack, G., Lawrence, E., Cheetham, C. H., Prentice, A. M., Hughes, I. A., McCamish, M. A., and O'Rahilly, S. (1999). Effects of recombinant leptin therapy in a child with congenital leptin deficiency. *N. Engl. J. Med., 341*(12), 879–884.

Figenschau, Y., Knutsen, G., Shahazeydi, S., Johansen, O., and Sveinbjornsson, B. (2001). Human articular chondrocytes express functional leptin receptors. *Biochem. Biophys. Res. Commun., 287*(1), 190–197.

Gainsford, T., Willson, T. A., Metcalf, D., Handman, E., McFarlane, C., Ng, A., Nicola, N. A., Alexander, W. S., and Hilton, D. J. (1996). Leptin can induce proliferation, differentiation, and functional activation of hemopoietic cells. *Proc. Natl. Acad. Sci. USA, 93*(25), 14564–14568.

Gavrilova, O., Barr, V., Marcus-Samuels, B., and Reitman, M. (1997). Hyperleptinemia of pregnancy associated with the appearance of a circulating form of the leptin receptor. *J. Biol. Chem., 272*(48), 30546–30551.

Gonzalez, R. R., Caballero-Campo, P., Jasper, M., Mercader, A., Devoto, L., Pellicer, A., and Simon, C. (2000). Leptin and leptin receptor are expressed in the human endometrium and endometrial leptin secretion is regulated by the human blastocyst. *J. Clin. Endocrinol. Metab., 85*(12), 4883–4888.

Grosfeld, A., Turban, S., Andre, J., Cauzac, M., Challier, J. C., Hauguel-de Mouzon, S., and Guerre-Millo, M. (2001). Transcriptional effect of hypoxia on placental leptin. *FEBS Lett., 502*(3), 122–126.

Harigaya, A., Nagashima, K., Nako, Y., and Morikawa, A. (1997). Relationship between concentration of serum leptin and fetal growth. *J. Clin. Endocrinol. Metab., 82*(10), 3281–3284.

Hoggard, N., Haggarty, P., Thomas, L., and Lea, R. G. (2001). Leptin expression in placental and fetal tissues: does leptin have a functional role? *Biochem. Soc. Trans., 29*(Pt 2), 57–63.

Hoggard, N., Hunter, L., Duncan, J. S., Williams, L. M., Trayhurn, P., and Mercer, J. G. (1997). Leptin and leptin receptor mRNA and protein expression in the murine fetus and placenta. *Proc. Natl. Acad. Sci. USA, 94*(20), 11073–11078.

Holness, M. J., Munns, M. J., and Sugden, M. C. (1999). Current concepts concerning the role of leptin in reproductive function. *Mol. Cell. Endocrinol., 157*(1–2), 11–20.

Islam, M. S., Morton, N. M., Hansson, A., and Emilsson, V. (1997). Rat insulinoma-derived pancreatic beta-cells express a functional leptin receptor that mediates a proliferative response. *Biochem. Biophys. Res. Commun., 238*(3), 851–85.

Islam, M. S., Sjoholm, A., and Emilsson, V. (2000). Fetal pancreatic islets express functional leptin receptors and leptin stimulates proliferation of fetal islet cells. *Int. J. Obes. Relat. Metab. Disord., 24*(10), 1246–1253.

Kawai, M., Yamaguchi, M., Murakami, T., Shima, K., Murata, Y., and Kishi, K. (1997). The placenta is not the main source of leptin production in pregnant rat: gestational profile of leptin in plasma and adipose tissues. *Biochem. Biophys. Res. Commun., 240*(3), 798–802.

Koistinen, H. A., Koivisto, V. A., Andersson, S., Karonen, S. L., Kontula, K., Oksanen, L., and Teramo, K. A. (1997). Leptin concentration in cord blood correlates with intrauterine growth. *J. Clin. Endocrinol. Metab., 82*(10), 3328–3330.

Kolaczynski, J. W., Nyce, M. R., Considine, R. V., Boden, G., Nolan, J. J., Henry, R., Mudaliar, S. R., Olefsky, J., and Caro, J. F. (1996). Acute and chronic effects of insulin on leptin production in humans: studies in vivo and in vitro. *Diabetes, 45*(5), 699–701.

Lea, R. G., Howe, D., Hannah, L. T., Bonneau, O., Hunter, L., and Hoggard, N. (2000). Placental leptin in normal, diabetic and fetal growth-retarded pregnancies. *Mol. Hum. Reprod., 6*(8), 763–769.

Lepercq, J., Cauzac, M., Lahlou, N., Timsit, J., Girard, J., Auwerx, J., and Haugel-de Mouzon, S. (1998). Overexpression of placental leptin in diabetic pregnancy: a critical role for insulin. *Diabetes, 47*(5), 847–850.

Lepercq, J., Challier, J. C., Guerre-Millo, M., Cauzac, M., Vidal, H., and Hauguel-de Mouzon, S. (2001). Prenatal leptin production: evidence that fetal adipose tissue produces leptin. *J. Clin. Endocrinol. Metab., 86*(6), 2409–2413.

Licinio, J., Negrao, A. B., Mantzoros, C., Kaklamani, V., Wong, M. L., Bongiorno, P. B., Mulla, A., Cearnal, L., Veldhuis, J. D., Flier, J. S., McCann, S. M., and Gold, P. W. (1998). Synchronicity of frequently sampled, 24-h concentrations of circulating leptin, luteinizing hormone, and estradiol in healthy women. *Proc. Natl. Acad. Sci. USA, 95*(5), 2541–2546.

Liu, J. P., Baker, J., Perkins, A. S., Robertson, E. J., and Efstratiadis, A. (1993). Mice carrying null mutations of the genes encoding insulin-like growth factor I (Igf-1) and type 1 IGF receptor (Igf1r). *Cell, 75*(1), 59–72.

Loffreda, S., Yang, S. Q., Lin, H. Z., Karp, C. L., Brengman, M. L., Wang, D. J., Klein, A. S., Bulkley, G. B., Bao, C., Noble, P. W., Lane, M. D., and Diehl, A. M. (1998). Leptin regulates proinflammatory immune responses. *FASEB J., 12*(1), 57–65.

Lord, G. M., Matarese, G., Howard, J. K., Baker, R. J., Bloom, S. R., and Lechler, R. I. (1998). Leptin modulates the T-cell immune response and reverses starvation-induced immunosuppression. *Nature, 394*(6696), 897–901.

Manson, J. E., Willett, W. C., Stampfer, M. J., Colditz, G. A., Hunter, D. J., Hankinson, S. E., Hennekens, C. H., and Speizer, F. E. (1995). Body weight and mortality among women. *N. Engl. J. Med., 333*(11), 677–685.

Mantzoros, C. S. (1999). The role of leptin in human obesity and disease: a review of current evidence. *Ann. Intern. Med., 130*(8), 671–680.

Mantzoros, C. S., Cramer, D. W., Liberman, R. F., and Barbieri, R. L. (2000). Predictive value of serum and follicular fluid leptin concentrations during assisted reproductive cycles in normal women and in women with the polycystic ovarian syndrome. *Hum. Reprod., 15*(3), 539–544.

Mantzoros, C. S., Varvarigou, A., Kaklamani, V. G., Beratis, N. G., and Flier, J. S. (1997). Effect of birth weight and maternal smoking on cord blood leptin concentrations of full-term and preterm newborns. *J. Clin. Endocrinol. Metab., 82*(9), 2856–2861.

Masuzaki, H., Ogawa, Y., Sagawa, N., Hosoda, K., Matsumoto, T., Mise, H., Nishimura, H., Yoshimasa, Y., Tanaka, I., Mori, T., and Nakao, K. (1997). Nonadipose tissue production of leptin: leptin as a novel placenta-derived hormone in humans. *Nat. Med., 3*(9), 1029–1033.

Matkovic, V., Ilich, J. Z., Skugor, M., Badenhop, N. E., Goel, P., Clairmont, A., Klisovich, D., Nahhas, R. W., and Landoll, J. D. (1997). Leptin is inversely related to age at menarche in human females. *J. Clin. Endocrinol. Metab., 82*(10), 3239–3245.

Matsuoka, T., Tahara, M., Yokoi, T., Masumoto, N., Takeda, T., Yamaguchi, M., Tasaka, K., Kurachi, H., and Murata, Y. (1999). Tyrosine phosphorylation of STAT3 by leptin through leptin receptor in mouse metaphase 2 stage oocyte. *Biochem. Biophys. Res. Commun., 256*(3), 480–484.

McCarthy, J. F., Misra, D. N., and Roberts, J. M. (1999). Maternal plasma leptin is increased in preeclampsia and positively correlates with fetal cord concentration. *Am. J. Obstet. Gynecol., 180*(3 Pt 1), 731–736.

Milner, R. D. G. and Gluckman, P. D. (1996). Regulation of intrauterine growth. In P. D. Gluckman and M. A. Heymann (Eds.). *Pediatrics and Perinatology, The Scientific Basis* (2nd Ed.), Arnold/Oxford University Press, pp. 284–289.

Mise, H., Sagawa, N., Matsumoto, T., Yura, S., Nanno, H., Itoh, H., Mori, T., Masuzaki, H., Hosoda, K., Ogawa, Y., and Nakao, K. (1998). Augmented placental production of leptin in preeclampsia: possible involvement of placental hypoxia. *J. Clin. Endocrinol. Metab., 83*(9), 3225–3229.

Moschos, S., Chan, J. L., and Mantzoros, C. S. (2002). Leptin and reproduction: a review. *Fertil. Steril., 77*(3), 433–444.

Mounzih, K., Lu, R., and Chehab, F. F. (1997). Leptin treatment rescues the sterility of genetically obese ob/ob males. *Endocrinology, 138*(3), 1190–1193.

Mounzih, K., Qiu, J., Ewart-Toland, A., and Chehab, F. F. (1998). Leptin is not necessary for gestation and parturition but regulates maternal nutrition via a leptin resistance state. *Endocrinology, 139*(12), 5259–5262.

Oda, A., Taniguchi, T., and Yokoyama, M. (2001). Leptin stimulates rat aortic smooth muscle cell proliferation and migration. *Kobe J. Med. Sci., 47*(3), 141–150.

Ogueh, O., Sooranna, S., Nicolaides, K. H., and Johnson, M. R. (2000). The relationship between leptin concentration and bone metabolism in the human fetus. *J. Clin. Endocrinol. Metab., 85*(5), 1997–1999.

Ong, K. K., Ahmed, M. L., Sherriff, A., Woods, K. A., Watts, A., Golding, J., and Dunger, D. B. (1999). Cord blood leptin is associated with size at birth and predicts infancy weight gain in humans. ALSPAC Study Team. Avon Longitudinal Study of Pregnancy and Childhood. *J. Clin. Endocrinol. Metab., 84*(3), 1145–1148.

Ozata, M., Ozdemir, I. C., and Licinio, J. (1999). Human leptin deficiency caused by a missense mutation: multiple endocrine defects, decreased sympathetic tone, and immune system dysfunction indicate new targets for leptin action, greater central than peripheral resistance to the effects of leptin, and spontaneous correction of leptin-mediated defects. *J. Clin. Endocrinol. Metab., 84*(10), 3686–3695.

Reed, D. R., Ding, Y., Xu, W., Cather, C., Green, E. D., and Price, R. A. (1996). Extreme obesity may be linked to markers flanking the human OB gene. *Diabetes, 45*(5), 691–694.

Reitman, M. L., Bi, S., Marcus-Samuels, B., and Gavrilova, O. (2001). Leptin and its role in pregnancy and fetal development—an overview. *Biochem. Soc. Trans., 29*(Pt 2), 68–72.

Roemmich, J. N., Clark, P. A., Mantzoros, C. S., Gurgol, C. M., Weltman, A., and Rogol, A. D. (2003). Relationship of Leptin to bone mineralization in children and adolescents. *J. Clin. Endocrinol. Metab.* (in press).

Rossier, J., Rogers, J., Shibasaki, T., Guillemin, R., and Bloom, F. E. (1979). Opioid peptides and alpha-melanocyte-stimulating hormone in genetically obese (ob/ob) mice during development. *Proc. Natl. Acad. Sci. USA, 76*(4), 2077–2080.

Sena, A., Sarlieve, L. L., and Rebel, G. (1985). Brain myelin of genetically obese mice. *J. Neurol. Sci., 68*(2–3), 233–243.

Sierra-Honigmann, M. R., Nath, A. K., Murakami, C., Garcia-Cardena, G., Papapetropoulos, A., Sessa, W. C., Madge, L. A., Schechner, J. S., Schwabb, M. B., Polverini, P. J., and Flores-Riveros, J. R. (1998). Biological action of leptin as an angiogenic factor. *Science, 281*(5383), 1683–1686.

Simon, C. and Polan, M. L. (1994). Cytokines and reproduction. *West. J. Med., 160*(5), 425–429.

Sir-Petermann, T., Maliqueo, M., Palomino, A., Vantman, D., Recabarren, S. E., and Wildt, L. (1999). Episodic leptin release is independent of luteinizing hormone secretion. *Hum. Reprod., 14*(11), 2695–2699.

Sir-Petermann, T., Recabarren, S. E., Lobos, A., Maliqueo, M., and Wildt, L. (2001). Secretory pattern of leptin and LH during lactational amenorrhoea in breastfeeding normal and polycystic ovarian syndrome women. *Hum. Reprod., 16*(2), 244–249.

Sivan, E., Whittaker, P. G., Sinha, D., Homko, C. J., Lin, M., Reece, E. A., and Boden, G. (1998). Leptin in human pregnancy: the relationship with gestational hormones. *Am. J. Obstet. Gynecol., 179*(5), 1128–1132.

Strobel, A., Issad, T., Camoin, L., Ozata, M., and Strosberg, A. D. (1998). A leptin missense mutation associated with hypogonadism and morbid obesity. *Nat. Genet., 18*(3), 213–215.

Takeda, K., Noguchi, K., Shi, W., Tanaka, T., Matsumoto, M., Yoshida, N., Kishimoto, T., and Akira, S. (1997). Targeted disruption of the mouse Stat3 gene leads to early embryonic lethality. *Proc. Natl. Acad. Sci. USA, 94*(8), 3801–3804.

Thomas, T., Gori, F., Khosla, S., Jensen, M. D., Burguera, B., and Riggs, B. L. (1999). Leptin acts on human marrow stromal cells to enhance differentiation to osteoblasts and to inhibit differentiation to adipocytes. *Endocrinology, 140*(4), 1630–1638.

Tomimatsu, T., Yamaguchi, M., Murakami, T., Ogura, K., Sakata, M., Mitsuda, N., Kanzaki, T., Kurachi, H., Irahara, M., Miyake, A., Shima, K., Aono, T., and Murata, Y. (1997). Increase of mouse leptin production by adipose tissue after midpregnancy: gestational profile of serum leptin concentration. *Biochem. Biophys. Res. Commun., 240*(1), 213–215.

Umemoto, Y., Tsuji, K., Yang, F. C., Ebihara, Y., Kaneko, A., Furukawa, S., and Nakahata, T. (1997). Leptin stimulates the proliferation of murine myelocytic and primitive hematopoietic progenitor cells. *Blood, 90*(9), 3438–3443.

Varvarigou, A., Mantzoros, C. S., and Beratis, N. G. (1999). Cord blood leptin concentrations in relation to intrauterine growth. *Clin. Endocrinol. (Oxf), 50*(2), 177–183.

Verhaeghe, J., Van Bree, R., Van Herck, E., Laureys, J., Bouillon, R., and Van Assche, F. A. (1993). C-peptide, insulin-like growth factors I and II and insulin-like growth factor binding protein-1 in umbilical cord serum: correlations with birth weight. *Am. J. Obstet. Gynecol., 169*(1), 89–97.

Wiznitzer, A., Reece, E. A., Homko, C., Furman, B., Mazor, M., and Levy, J. (1998). Insulin-like growth factors, their binding proteins, and fetal macrosomia in offspring of nondiabetic pregnant women. *Am. J. Perinatol., 15*(1), 23–28.

Wolf, H. J., Ebenbichler, C. F., Huter, O., Bodner, J., Lechleitner, M., Foger, B., Patsch, J. R., and Desoye, G. (2000). Fetal leptin and insulin levels only correlate in large-for-gestational age infants. *Eur. J. Endocrinol., 142*(6), 623–629.

Yamashita, H., Shao, J., Ishizuka, T., Klepcyk, P. J., Muhlenkamp, P., Qiao, L., Hoggard, N., and Friedman, J. E. (2001). Leptin administration prevents spontaneous gestational diabetes in heterozygous Lepr(db/+) mice: effects on placental leptin and fetal growth. *Endocrinology, 142*(7), 2888–2897.

13

Leptin in the Placenta

DORINA ISLAMI AND PAUL BISCHOF

13.1. ABSTRACT

Leptin is a polypeptide hormone that is clearly involved in the regulation of body mass by suppressing appetite and stimulating energy expenditure. The placenta is known to be an important source of circulating leptin, and leptin, also produced by primary cultures of human amnion cells is secreted into the amniotic fluid. Leptin concentration increases throughout human gestation, peaks in the second trimester and remains high until parturition. On the other hand, a decline in leptin mRNA abundance in syncytiotrophoblast is a common feature of advancing gestational age. In situ hybridization has indicated that leptin, OB-R_L, and OB-R_S mRNA transcript expression is limited to the endocrinologically active trophoblast, thus the syncytiotrophoblasts layer covering placental villi. Placental leptin is also released into the fetal circulation. Leptin is shown to act as a regulator of fetal weight and growth and as an important regulator of maternal and placental weight. Leptin may also be associated with mechanisms mediating lactation and neonatal growth. In the placenta, it could act in an autocrine way as a growth hormone (GH) and growth factor for angiogenesis. During the invasion phase of human implantation, leptin has been shown to play an important role in the regulation of hCG production by cytotrophoblastic cells (CTB), in the switching of the integrin repertoire and in stimulating metalloprotease activity. Leptin production is modulated by interleukin (IL)-1β and 17β-estradiol. A novel role for leptin as an autocrine/paracrine regulator of the invasion phase of human implantation may be proposed. Trophoblastic leptin could be an inducer of the secretion of metalloproteinases (MMP) and a modulator of the expression of integrins conferring an invasive phenotype to CTB. This review focuses on the role of leptin in the placenta and more specially in trophoblast invasion.

DORINA ISLAMI AND PAUL BISCHOF • Department of Obstetrics and Gynaecology, University Hospital of Geneva.

Leptin and Reproduction. Edited by Henson and Castracane, Kluwer Academic/Plenum Publishers, 2003.

13.2. INTRODUCTION

Leptin (from the Greek, *leptos*, means thin) is a small peptide product of the *ob* gene. It is a 16 KDa non-glycozylated polypeptide of 146 amino acids discovered in 1994 by Zhang et al. The precursor form of leptin contains 167 amino acids and is activated by cleavage of a 21 amino acid residue (Zhang et al., 1994; Ogawa et al., 1995). The amino acid sequence of leptin includes a putative amino-terminal leader sequence, but lacks an internal membrane-spanning domain (Zhang et al., 1994). Hassink et al. (1997) showed that the leptin molecule is a globular protein with a tertiary structure similar to that of helical cytokines. Nuclear magnetic resonance analysis revealed that it is a four-helix bundle cytokine, and helix length and disulfide pattern suggest that leptin is a member of the short-helix cytokine family (Kline et al., 1997).

The leptin receptor (OB-R) gene has been shown to have at least six splice variants OB-R (a–e) (Chen et al., 1996; Lee et al., 1996) and mu B219 (Cioffi et al., 1996). The OB-Rb variant encodes a receptor with a long intracellular signal transduction peptide (Tartaglia et al., 1995). The long form of the leptin receptor functions similarly to cytokine receptors and has been detected in human lung, kidney, liver, and skeletal muscle (Tartaglia et al., 1995), as well as heart, placenta, spleen, thymus, prostate, testis, ovary, small intestine, and colon (Cioffi et al., 1996). The presence of leptin receptors in numerous tissues supports the hypothesis that leptin is important for growth and development (Hassink et al., 1997). The identification of ob receptors in the hypothalamus raises the possibility that leptin is a regulator of hypothalamic releasing hormones in humans (Butte et al., 1997).

In humans and rodents, leptin is clearly involved in the regulation of body mass by suppressing appetite and stimulating energy expenditure (Campfield et al., 1995; Lönnqvist et al., 1995; Considine et al., 1996) (Figure 13.1), but additional functions of leptin at the feto–maternal interface have recently been reviewed (Bajoria et al., 2002).

13.3. PLACENTAL LEPTIN SECRETION

In 1997, Masuzaki et al., using Northern, Western blot, and immunohistochemistry, demonstrated the production of leptin in placental trophoblast and amnion cells, but not in the decidua or myometrium. These authors showed that in pregnant women, the placenta is an important source of circulating leptin, since after delivery leptin levels returned to nonpregnant levels within 24 hr. Leptin is also produced by primary cultures of human amnion cells and is secreted into the amniotic fluid (Figure 13.1).

Interestingly, the placenta probably serves both as a source of leptin and as a potential target for its action, since expression of the *ob* gene (Masuzaki et al., 1997) and its receptor (Luoh et al., 1997) have been detected in placental trophoblasts by Northern blot. Reverse transcriptase-polymerase chain reaction (RT-PCR) and immunohistochemistry showed the presence of leptin in the cytoplasm of syncytiotrophoblastic cells (Masuzaki et al., 1997). Leptin is not found in the mesenchyme of the villous core or in fetal blood vessels, suggesting that leptin is not taken up from the fetal circulation, but is rather synthesized and secreted locally. Placental leptin is synthesized as a single molecular variant identical to human recombinant leptin (Senaris et al., 1997).

There is still some controversy about the comparative capacity of placenta and adipose tissue to produce leptin. The placenta was found to express leptin mRNA at

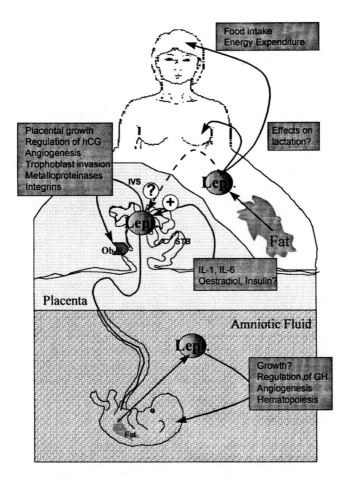

Figure 13.1. Maternal leptin is produced by adipocytes and stimulates food intake, energy expenditure, and is thought to affect lactation. During pregnancy, leptin is also produced by the placenta, and its secretion is stimulated by IL-1, IL-6, estradiol, and insulin. Placental leptin through its receptor (OB-R) modulates placental growth, regulates hCG, angiogenesis, trophoblast invasion, metalloproteinases, and integrins expression. Fetal leptin is produced by fetal adipocytes and is secreted in the amniotic fluid. Fetal leptin is thought to stimulate fetal growth and development, to regulate growth hormone (GH), fetal angiogenesis and hematopoiesis. STB—syncytiotrophoblast, IVS—intervillous space.

comparable or greater levels than adipose tissue (Hassink et al., 1997). In contrast, Green et al. (1995) and Bi et al. (1997), observed that leptin mRNA in human placenta was detected at much lower levels than in white adipose tissue. Irrespective of this difference, it is now admitted that trophoblastic cells are the main source of circulating leptin during pregnancy, particularly because after the delivery of the placenta, leptin levels return rapidly to normal (Masuzaki et al., 1997). This is further substantiated by the fact that human trophoblastic cell lines, JEG-3 and JAR (Hassink et al., 1997) and BeWo (Masuzaki et al., 1997), were found to express leptin. In their study Chardonnens et al. (1999), provided the first direct evidence of regulated leptin secretion by primary cultures of human CTB cells.

13.4. LEPTIN IN RODENT PREGNANCY

In rodents, leptin is probably not a critical molecule for implantation, gestation, and parturition. The pregnant rat placenta does not seem to be a major source of leptin production. The total amount of leptin mRNA was significantly increased only in rat maternal adipose tissue during pregnancy (Kawai et al., 1997; Tomimatsu et al., 1997). However, leptin receptor mRNA increases in the uterus of pregnant rats concomitantly with serum leptin (Chien et al., 1997) and the concentration of leptin protein increases in placenta, decidua, uterus, and adipose tissues with advancing gestation (Tomimatsu et al., 1997). In addition, female *ob/ob* mice previously treated with exogenous leptin and mated to similarly treated *ob/ob* males developed a normal pregnancy without further administration of leptin. Interestingly, in most of these pregnant mice a prolonged parturition has been observed (Mounzih et al., 1999). In contrast, other investigators have reported high levels of gene expression (mRNA) for leptin, its receptor, and the long spliced variant (OB-R_L) in the murine placenta and fetus (Hoggard et al., 1997). These authors observed that leptin, OB-R, and OB-Rb mRNA were expressed more abundantly in the placenta than in the fetus. Interestingly, they found also the presence of leptin gene expression and mature leptin protein in a number of tissues in the fetal mouse, thus suggesting a possible function of leptin as an autocrine or paracrine regulator in the fetus. This expression of leptin in fetal tissues suggests that the hormone be involved in the growth and development of the fetus with one possible function being a fetal growth factor or a signal to the fetus of maternal energy status. Alternatively, fetal leptin could provide a signal to the mother of fetal growth and development. Amico et al. (1998) found a significant increase of serum leptin concentrations from day 14 to 21 of rat gestation. This rise in leptin concentration was concomitant to a significant increase (4–5-fold) of placental leptin mRNA. Gavrilova et al. (1997) have reported a dramatic increase of circulating leptin levels during mouse gestation with, however, no detectable increase in placental leptin production. The serum levels of leptin-binding protein derived from the spliced leptin receptor increased in parallel to that of leptin. Since leptin is cleared by the kidney, at least partially by filtration (Cumin et al., 1996), binding of the 16 kDa leptin to the binding protein (which contains the extracellular region of the leptin receptor) should decrease its clearance. Thus, hyperleptinemia during pregnancy is probably due to reduced renal clearance of bound leptin. This study demonstrated that mice experience a profound increase in leptin and leptin binding capacity due to placental production of a soluble form of the leptin receptor.

13.5. LEPTIN SECRETION AND REGULATION DURING HUMAN PREGNANCY

13.5.1. Leptin in the Mother

The human situation is probably different. Leptin is a novel placental hormone in humans (Masuzaki et al., 1997), placental leptin release into the maternal circulation accounts for 14 percent of the total leptin production in normal adult women (Linnemann et al., 2000). Leptin concentration increases throughout gestation (Butte et al., 1997; Schubring et al., 1997), peaks in the second trimester, and remains high until parturition. In their study, Butte et al. (1997) have shown that serum leptin was significantly higher at 36 weeks of pregnancy than at 3 and 6 months postpartum. In more than 90 percent of these women, leptin declined from pregnancy to 3 months postpartum. Different factors may contribute to leptin changes during pregnancy and postpartum: (1) fat mass increases during pregnancy by about 15 percent (Lederman et al., 1999); this, however, cannot fully explain the up to 4-fold leptin rise; (2) in vitro experiments using trophoblast cell cultures have shown a significant leptin production by these cells (Hassink et al., 1997; Masuzaki et al., 1997; Chardonnens et al., 1999); (3) other pregnancy-specific tissues (decidua and amniochorion) may also produce leptin and with increasing gestational age, contribute progressively more to the peripheral circulation (Henson and Castracane, 2000); (4) an alteration in the placenta's contribution and/or that of maternal adipose stores, prompted by the elevated levels of hormones (most specifically, estrogen) typical of advancing primate pregnancy, might be at least partially responsible for increased serum leptin concentrations.

It appears, therefore, that neither fat accumulation nor placental leptin production alone, nor the sum of both can fully explain the marked rise in leptin levels in pregnancy or the prompt decrease immediately after delivery (Linnemann et al., 2000). Additional mechanisms may be involved:

1. Increased food intake is known to stimulate leptin levels (Kolaczynski et al., 1996). However, the elevated maternal leptin concentrations observed during pregnancy are not modified by a decreased food intake or a metabolic deficiency as might be expected. These discrepancies may be explicable in terms of a pregnancy-induced state of leptin resistance analogous to leptin resistance in obesity or, alternatively, a change in leptin bioavailability (Holness et al., 1999).

2. The increase in maternal leptin concentrations with advancing gestational age might be attributed to the action of a leptin-binding protein or soluble leptin receptor expressed in greater abundance, or perhaps exclusively, during pregnancy. Interaction with such a circulating receptor may be responsible, at least in part, for the hyperleptinemia typical of pregnancy and may serve to inhibit optimal interaction with hypothalamic receptors during pregnancy, thus promoting a state of leptin resistance, similar in effect to that conferred by the suppressor of cytokine signaling (SOCS-3) in the hypothalamus (Bjorbaek et al., 1999). The release of the soluble leptin receptor by placental membrane shedding and subsequent binding of free leptin to this receptor may impair the bioactivity of leptin and protect leptin from degradation or excretion (Gavrilova et al., 1997; Ashworth et al., 2000). Indeed, free leptin seems not to be elevated in pregnancy, in contrast to bound leptin (Lewandowski et al., 1999). Evidence has been found of a protein with definite leptin-binding capacity in human serum (Diamond et al., 1997; Gavrilova et al., 1997). As proposed by Heaney and Golde (1993), these binding proteins may take the form of soluble hormone receptors, available in peripheral serum. In a similar finding, a small increase in bound leptin is known to occur during human pregnancy (Stock and Bremme, 1998).

Leptin secretion from first trimester chorionic tissue has been shown to be 50-fold greater than that from term placental tissue (Yura et al., 1998a). This gestational age-related difference in leptin secretion is compatible with the marked difference in leptin gene expression between first trimester chorionic tissue and term placental tissue. Leptin production in explant cultures of first trimester chorionic tissue and term placental tissue is also stimulated by forskolin and phorblo-myristate-acetate (PMA) treatment. These findings suggest that leptin synthesis and secretion are augmented through the activation of protein kinase-A (PKA) and PKC in human trophoblasts in vitro, and probably also in vivo. CTB cells are markedly reduced in number in term placental tissue (Cunningham et al., 1999). In primary cultures of cytotrophoblasts from the term placenta, the rate of syncytial formation is not affected by forskolin (Strauss et al., 1992). Therefore, it has been suggested that the augmented leptin production through stimulation of PKA is caused by a functional activation of trophoblasts, independent of morphological changes (Yura et al., 1998b).

Henson et al. (1998) investigated the expression of leptin and leptin receptor mRNA transcripts with advancing human pregnancy, and using quantitative RT-PCR, reported that specific transcripts for leptin, as well as for $OB-R_L$ and $OB-R_S$ receptor isoforms, were expressed in placenta, both early (7–14 weeks) in gestation and at term (38 weeks). Although no changes in $OB-R_S$ and $OB-R_L$ transcript abundance were evident with respect to stage of gestation, the abundance of leptin mRNA transcripts in placental villous tissue decreased significantly. These authors suggest that both a significant increase in maternal leptin concentrations and a decline in leptin mRNA abundance in syncytiotrophoblast are common features of advancing gestational age in human pregnancy. Furthermore, in situ hybridization indicated that leptin, $OB-R_L$, and $OB-R_S$ mRNA transcript expression was limited to the endocrinologically active trophoblast, thus the syncytiotrophoblasts layer covering placental villi.

Although, in contrast with Hassink et al. (1997), who had suggested that placental tissue collected at term contains 1–2.5 times more leptin mRNA than maternal adipose tissue, Henson et al.'s (1998) observations give more credit to the notion that leptin-binding proteins are instrumental to the sustained increase in peripheral leptin concentrations during pregnancy.

Supportive of these findings, Ranganathan et al. (1998) reported that serum leptin levels in humans are not directly related to adipose tissue mRNA concentrations and that changes in leptin levels, while independent of fluctuations in mRNA abundance, are finely regulated by post-transcriptional mechanisms at the level of the adipocyte or by alteration of peripheral leptin degradation or clearance. Similarly, Kirchgessner et al. (1997) found that plasma leptin was decreased in tumor necrosis factor (TNF) knockout mice, with no corresponding decline in mRNA, and that 3T3-L1 cells evidenced a similar disparity between transcript abundance and leptin secretion.

Leptin was found to positively correlate with body weight and body mass index (BMI) at 36 weeks of gestation and at 3 and 6 months postpartum. Bi et al. (1997) have identified an enhancer of the human leptin gene and a placenta-specific element within the enhancer. This enhancer mediates placental expression of leptin that could explain the increased leptin levels during human pregnancy. Furthermore, leptin levels during pregnancy correlate with estradiol and human chorionic gonadotropin (hCG) (Hardie et al., 1997). To what extent these placental hormones regulate the placental enhancer of leptin expression is still unknown and remains to be explored. It has been reported (Lage et al.,

1999) that leptin levels in women suffering from spontaneous abortions in the first trimester have leptin concentrations that are 38 percent lower than women who successfully maintain their pregnancy. Leptin is thus needed for maintenance of pregnancy, maturation of the reproductive system, and for parturition and nursing (Chehab et al., 1996; Ahima et al., 1997). All these findings indicate that the regulation of leptin production in placental trophoblasts is different from that in adipocytes. Of particular importance in that respect is the identification of a novel placenta-specific transcription factor (Bi et al., 1997), not occurring in adipose tissue.

13.5.2. Leptin in the Fetus and Neonate

The nurture and survival of fetus in utero is dependent on the orchestration of endocrinological stimuli from surrounding gestational tissues. The localization of leptin in the placenta, amnion, amniotic fluid, and chorion leave suggests that leptin, in concert with the plethora of hormones already known to be produced from these tissues, may have a direct role in foeto-placental physiology. Placental leptin is also released into the fetal circulation (Linnemann et al., 2000). Fetal leptin concentrations, although lower than maternal levels, are detectable at term and are probably due to the production by fetal adipose tissue (Clapp and Kiess, 1998). However, it is not yet certain whether leptin of placental origin predominantly enters the maternal circulation, or is primarily directed toward the fetus (Holness et al., 1999). Immunocytochemical staining of developing subcutaneous tissues of human embryos, at 6–10 weeks of gestation, indicate that leptin is produced by developing fat cells from the beginning of lipidogenesis and differentiation (Atanassova and Popova, 2000). Although reasonable estimates of fetal fat mass are available (Fusch et al., 1997, 1999), leptin production rate by fetal adipose tissue are unknown. It remains, therefore, unclear to what extent fetal leptin originates from fetal adipose tissue or from the placenta. The proportion of total placental leptin released into the fetal circulation was found to be higher than the proportions of the classical placental hormones such as hCG and human placental lactogen (hPL) (Linnemann et al., 2000). This may be due to differences in diffusion rates as a reflection of differences in molecular weight (hCG, 39 kDa; hPL, 22 kDa; leptin, 16 kDa), or it may by a result of active transport with different rates of secretion. Another reason could be the production of leptin in villous endothelial cells, which is not described for hCG or hPL (Ashworth et al., 2000; Lea et al., 2000).

Schubring et al. (1997) have shown that ample concentrations of leptin are present in both maternal and cord blood at term, and it has been proposed that leptin in cord blood probably originates from both fetal and placental sources, as no correlation was found between maternal peripheral concentrations and cord levels. It is possible that high levels of leptin provide a signal of satiety around birth. Since leptin values in arterial cord blood are significantly higher than in venous cord blood, these authors and others (Jaquet et al., 1998) suggest that leptin synthesis by fetal tissues is higher that leptin synthesis by the placenta. In contrast to this study, Lin et al. (1999) have found higher concentrations of leptin in umbilical veins than in umbilical arteries, while Yura et al. (1998a) have shown a precipitous decline in neonatal levels immediately following birth, thus suggesting that the placenta may be an important source of leptin in the fetal circulation. Because leptin levels in cord blood correlate with birth weight, it is tempting to speculate that in the fetus, as in later life (Rink, 1994; Lindpaintner, 1995; Lönnqvist et al., 1995; Considine et al., 1996), leptin is signaling the expansion of fat stores toward the brain and other target tissues, thus leptin would act as a regulator of fetal weight and growth (Figure 13.1). This statement is further supported by the following observations: Leptin levels in cord blood increase according to the increase in the absolute weight and fat content of the newborn at a gestational age of 36–42 weeks (Matsuda et al., 1997). Absolute weight increases most sharply during the third trimester (Apte and Iyengar, 1972; Moore et al., 1988), and there is a change in body composition with regard to the percentages of both lipids (from <1 to 11 percent) and protein (from 9 to 12 percent). Matsuda et al. (1997) showed that serum leptin concentrations in newborns were higher than those in healthy non-obese adults. This difference did not appear to be due to body fat, since the percent body fat in newborns were about 10–15 percent, which are lower than those of non-obese adults (Apte and Iyengar, 1972). Furthermore, newborns show prominent brown adipose tissue (Aherne and Hull, 1966; Moragas and Toran, 1983), in which the expression of the *ob* gene has been shown to be lower than that in white adipose tissue (Masuzaki et al., 1995). In addition to an association between leptin levels and birth weight, Ong et al. (1997) noted that cord leptin levels were also correlated with infant length and head circumference. Thus, cord blood leptin, originating from the placenta (Hassink et al., 1997), and/or fetus (Butte et al., 1997; Sattar et al., 1998) may potentate growth by modulating GH secretion (Carro et al., 1997), as proposed in the rat (LaPaglia et al., 1998), via the regulation of GH-releasing hormone (Tannenbaum et al., 1998; Carro et al., 1999). Indeed, leptin receptor mRNA has been identified in human fetal anterior pituitary, and leptin specifically stimulated GH secretion from primary human fetal pituitary cell cultures without affecting adreno-corticophin hormone (ACTH), prolactin, or gonadotropin secretion (Shimon et al., 1998). Although placental GH, originating in the syncytiotrophoblast might also affect both placental development and overall conceptus growth by autocrine or paracrine mechanisms (Alsat et al., 1998), a direct effect of leptin on placental GH production has not yet been reported.

Besides a role as fetal growth regulator, leptin seems to also exert effects on the mother and the placenta. Maternal BMI and weight at delivery do not correlate with leptin levels in maternal serum. This might suggest that during gestation or at least at birth, the regulation of leptin levels differs from that in nonpregnant women, in whom leptin levels are highly correlated with BMI and fat mass (Considine et al., 1996; Kiess et al., 1996). Leptin levels in amniotic fluid are elevated at birth and correlate with levels in maternal serum but not in fetal blood. Placental weight correlates inversely with maternal leptin levels at birth, while leptin levels in arterial and venous cord blood are positively correlated with placental weight (Schubring et al., 1997). The relatively high concentrations in amniotic fluid and the negative correlation of maternal leptin levels and placental weight may point to a putative role for leptin as an important regulator of maternal and placental weight.

In addition to its proposed roles as a regulator of fetal and placental development, leptin may be associated with mechanisms mediating lactation and neonatal growth. Although, leptin is not significantly different between lactating and nonlactating women when measured at 3 and 6 months postpartum, it correlates negatively with prolactin when observed at these two moments after delivery (Butte et al., 1997). However, these authors did not observe any direct effect of leptin on milk production independent of prolactin. Leptin concentrations were shown by Mukherjea et al. (1999) to be higher in lactating women as compared to normal controls, leading the authors to suggest a role for leptin in mobilizing needed energy reserves. Additionally, leptin receptor mRNA transcripts have been determined to be specifically expressed in ovine mammary epithelial cells, strongly suggesting a role in mammary gland growth and development (Laud et al., 1999).

Hormonal regulation of leptin levels in the fetus and neonate might be different from the endocrine modulation of leptin levels seen throughout adult life; whereas insulin and glucocorticoids (Heiman et al., 1996; Rentsch and Chiesi, 1996; Sinha, 1997) are thought to modulate leptin levels throughout adult life, this might not be entirely relevant for the fetus and neonate. There is no sex difference of leptin levels in cord blood at term (Schubring et al., 1997). This is in contrast to the situation in the adult where consistently higher levels of leptin are found in serum from females than from males (Hebebrand et al., 1995; MacDougald et al., 1995; Considine et al., 1996). One possible explanation for the absence of a gender difference in leptin levels at birth might be that the percent body fat of a neonate is not gender specific. In contrast, Hassink et al. (1997) have speculated that leptin differentially affects growth and development in females, based on their study results, where they have found higher leptin concentrations in female as opposed to male newborns. During the last weeks of gestation, higher leptin concentrations have been observed in women carrying a female fetus as compared to those carrying males (Jaquet et al., 1998), suggesting that sexual differentiation may be involved in leptin regulation. Nevertheless, at the time of delivery, cord plasma leptin concentrations were not influenced by gender difference. During the early postnatal period, leptin levels decreased in both genders but higher concentrations were found in female babies (Matsuda et al., 1997; Hytinantti et al., 1999). The significance and the mechanisms of this gender difference in the serum leptin concentrations is not fully understood. It has been previously speculated that this phenomenon is due to a relatively greater body fat content in women, the difference in the body fat distribution between men and women, or reproductive hormone status (Murakami et al., 1995; Havel et al., 1996; Rosenbaum et al., 1996; Schwartz et al., 1996). In the newborn, the content and distribution of body fat do not differ between males and females (Apte and Iyengar, 1972). Furthermore, the serum concentrations of the reproductive hormones estrogen and testosterone do not differ and do not correlate with the leptin concentrations (Matsuda et al., 1997). These results suggest that this gender difference in the infant is unlikely to be due either to the body fat content or distribution or to the reproductive hormone status (estrogen and testosterone). However, different authors have hypothesized that the existence of a gender difference in the newborn as in adults, may depend on genetic differences. In addition, preterm neonates had significantly lower serum leptin concentration than full-term neonates. Thus, leptin derived either from placenta or fetal adipose tissue may regulate fetal growth and development. It has been suggested that testosterone could be a suppresser of leptin synthesis in preterm male infants (Ertl et al., 1999).

Leptin levels in neonates decrease rapidly following birth. It has been proposed that this rapid decline might be important as a stimulus of feeding behavior and the maintenance of energy homeostasis in early life (Schubring et al., 1998). Perhaps relatedly, growth-retarded babies exhibiting low leptin levels at birth, may indeed exhibit increased growth rates in early neonatal life, an effect still evident at 24 months of age (Ong et al., 1999). In another study (Harigaya et al., 1997), it has been shown that neonatal serum levels of leptin in large-for-gestational age infants at 6 hr after birth were significantly higher than average- and small-for-gestational age infants. Within 48 hr after birth, leptin levels of large-for-gestational age babies decline to the same level as small-for-gestational age babies, and remain constant thereafter. In their study of newborn infants,

Jaquet et al. (1998) found that neonatal serum leptin concentrations reflected adiposity. Thus, leptin may be a predictive indicator, rather than a determinant, of birth weight.

13.6. ROLE OF LEPTIN IN PREGNANCY

So far, no clear role for placental leptin secretion has been demonstrated. Placenta synthesizes leptin, thus, a role in intrauterine development has been proposed for this placental protein (Hassink et al., 1997). However, the maternal leptin concentration during pregnancy is not an accurate indicator of fetal growth (Tamura et al., 1998). One putative role for leptin could be related to follicular growth, egg, and embryo polarity (Antczak and van Blerkom, 1997; Antczak et al., 1997; Cioffi et al., 1997; Edwards and Beard, 1997). In pregnancy, the physiological function of high leptin levels is not clear. In the maternal body it may have effects on thermogenesis and mobilization of energy stores rather than on regulation of maternal food intake. In the placenta it could act in an autocrine way as a GH and growth factor for angiogenesis (Figure 13.1). The benefit of placental leptin to the fetus may thus be growth and angiogenesis, as leptin is considered an important growth factor in intrauterine and neonatal development (Hassink et al., 1997; Bouloumie et al., 1998; Ashworth et al., 2000). Leptin has been proposed as a regulator of both hematopoiesis (Bennett et al., 1996) and angiogenesis in various developmental models. Hirose et al. (1998) have reported a correlation between serum leptin levels and red blood cell counts in male adolescents. Similarly, transcripts of ObR/B219.1, a proposed hematological subtype of the leptin receptor, are detectable in blood cells within fetal vessels, but not in placental cells (Bodner et al., 1999). Leptin receptor has also been shown to be expressed in human vasculature and in primary cultures of human endothelial cells. Placental leptin may also function as an anti-inflammatory factor, because it counteracts proinflammatory cytokines such as TNF-α (Takahashi et al., 1999). This effect may be important, as successful pregnancy is associated with down-regulation of intrauterine proinflammatory cytokines (Wegmann et al., 1993; Sacks et al., 1999).

Leptin has also been related to *preeclampsia*, a hypertensive disorder of late pregnancy. Placental production of leptin is increased in severe preeclampsia, suggesting that leptin is a possible marker of placental hypoxia (Mise et al., 1998). In preeclamptic women, hypertension contributes to placental hypoxia, and plasma leptin levels are dramatically enhanced (McCarthy et al., 1999). This situation might be related to the degree of adiposity, as women with prepregnancy BMIs of >25 kg/m^2 exhibited lower peripheral leptin levels compared to normotensives of similar adiposity (Williams et al., 1999). In this capacity, leptin mRNA/beta-actin mRNA ratios were significantly higher in placental villous tissue from preeclamptics than in those of controls matched for gestational age. In pregnancies complicated by preeclampsia, certain angiogenic factors (vascular endothelial and placental growth factors) are depressed in maternal serum (Reuvekamp et al., 1999), potentially explaining the shallow placentation characteristic of this condition. Perhaps leptin, a proposed angiogenic regulator (Bouloumie et al., 1998), may act to support some degree of placentation in such patients.

Increased leptin mRNA and protein was found in placentas from *insulin-treated diabetic* women. Studies in normal and diabetic pregnancies indicate that fetal hyperleptinemia caused by hyperinsulinemia stimulates fetal growth (Koistinen et al., 1997; Lepercq et al., 1998). Because this increase in leptin mRNA resulted in a 3–5-fold increase in placental leptin protein, it was proposed that insulin might regulate fetal weight gain via an upregulation of leptin secretion. Divergently, in mild gestational diabetes, maternal leptin levels are significantly lower than in nonaffected controls, who exhibit similar adiposity and fasting insulin levels (Festa et al., 1999). Elevated leptin concentrations in cord blood are associated with macrosomia (Lepercq et al., 1999), although it was noted by Persson et al. (1999) that among insulin-treated diabetic mothers, cord blood leptin levels were not correlated with birth weight or altered significantly with respect to gender, as is typical in normal pregnancy.

Insulin is likely to play a critical role in leptin regulation (Lepercq et al., 1998) (Figure 13.1). During pregnancy, as well as at 3 and 6 months postpartum, leptin has been shown to correlate positively with serum insulin, but no independent effect of insulin on leptin was found after controlling for fat mass (Butte et al., 1997). Pregnancy is known to be characterized by an insulin resistance with advancing gestation. Obesity in humans is also characterized by hyperinsulinemia. The close association between hyperinsulinemia and hyperleptinemia suggests that *ob* gene expression may be mediated by insulin (Cusin et al., 1995; McGarry, 1995). Both insulin and leptin are suppressed during fasting and increase with refeeding. The relation between leptin and fat mass does not differ between pregnancy and the postpartum period, during the first 3 months after delivery, a mean 6 percent reduction in fat mass is associated with a mean 61 percent decrease in leptin (Butte et al., 1997). The decline in leptin might be partially explained by the decrease in insulin, but much of the variation (80 percent) remains to be explained. Reproductive hormones are also involved. Increased basal and glucose-stimulated levels of plasma insulin with advancing gestation parallel progressive increases in plasma progesterone, estrogen, and hPL, which may affect the expression of *ob* gene.

Although effects differ with regard to the tissue, species, and experimental paradigm employed, consensus opinion supports a role for estrogens in enhancing transcription/ production in leptin-producing tissues and a role for androgens in inhibiting the synthesis of the polypeptide (Henson and Castracane, 2000). Estrogen has been reported to enhance leptin secretion by cultured human placental cytotrophoblast cells in a dose-responsive manner (Chardonnens et al., 1999). Estradiol plays a role in the regulation of fetal growth (Abdul-Karim et al., 1971), onset of parturition (Chibbar et al., 1995), placental steroidogenesis (Petraglia et al., 1995; Grimes et al., 1996; Babischkin et al., 1997), release of neuropeptides (Petraglia et al., 1990), release of glycoproteins (Wilson et al., 1984; Petraglia et al., 1995), and leptin secretion (Sivan et al., 1998). Administration of estradiol antagonist reverses these effects of estradiol on the placenta, pointing toward a receptor-mediated effect. However, the exact mechanisms of estradiol action within the placenta are still a matter of debate. Our study demonstrates a concentration-dependent bimodal pattern of estradiol on the regulation of leptin secretion (Figure 13.1). We found an increase in leptin secretion of up to 100 nmol/L of estradiol, which is close to its physiological concentrations during late pregnancy (30–50 nmol/L) (Tulchinsky et al., 1972). Maternal serum leptin levels, however, are not directly correlated with peripheral estrogens (Schubring et al., 1998; Sivan et al., 1998), perhaps as a result of increased leptin clearance and/or degradation in late pregnancy. Certainly, the effects of estrogen on leptin transcription and synthesis may be tissue specific.

13.7. LEPTIN AND TROPHOBLAST INVASION

From the results of in vitro studies, it has been proposed that leptin plays an important role in the regulation of hCG production by CTB cells (Chardonnens et al., 1999) and in the switching of the integrin repertoire (Gonzalez et al., 2001) and metalloprotease activity (Castellucci et al., 2000) during the invasion phase of human implantation (Figure 13.1). We have demonstrated that human CTB cultured in vitro synthesized leptin and that its production is modulated by IL-1β and 17β-estradiol, providing evidence for an autocrine/paracrine regulation of leptin production in the human placenta (Chardonnens et al., 1999). Thus, IL-1β must be considered as a regulator of leptin secretion in first trimester CTB (Figure 13.1). Its effect seems to be specific, as neither TNF-α nor M-CSF changed leptin secretion. IL-1β is known to stimulate in vivo leptin secretion in humans (Janik et al., 1997) but it remains unclear whether this is a direct action of IL-1β on the adipocyte or via an endocrine cascade involving corticosteroids. In our study (Chardonnens et al., 1999), we provide evidence of a direct effect of IL-1β on CTB leptin secretion. The cellular mechanism of such an effect remains unclear. However, IL-1 receptors are present on the trophoblast (Simon et al., 1994). The placental leptin enhancer, or PLE, identified by Bi et al. (1997), has at least three protein binding sites. Protein binding to the PLE3 motif upregulates the transcription of the leptin gene in the human placenta. Since IL-1 exerts similar effects and since, in other cell systems, IL-1 effects are mediated through the NF-KB transcription factor (Friedman et al., 1996; Bird et al., 1997; Reddy et al., 1997), it is tempting to speculate that IL-1β may induce leptin expression through an NF-KB-dependent activation of PLE3. The specificity of IL-1β is interesting because the complete IL-1 system is present at the feto–maternal interface and is clearly involved in implantation and placentation (Simon et al., 1994). This raises the important question as to whether leptin is also involved in these crucial events.

In BeWo cells (a choriocarcinoma cell line), the trophoblast-specific transcription of the human leptin gene involved the promoter activity in the 208 bp region (Ebihara et al., 1997). In these cells, leptin secretion is increased by forskolin (an inducer of CTB differentiation to syncytium, Masuzaki et al., 1997). Plasma leptin levels were markedly elevated in patients with hydatiform mole or choriocarcinoma and were reduced after surgical treatment or chemotherapy. This study suggests a significant physiologic and pathophysiologic role of leptin in normal pregnancy and gestational trophoblastic neoplasms. Other investigators (Yura et al., 1998b) have also shown that leptin synthesis and secretion in BeWo cells are augmented by forskolin, an activator of PKA, and by PMA, an activator of PKC, in vitro. Augmentation of leptin production by forskolin is completely blocked by H89, an inhibitor of PKA (Chijiwa et al., 1990). PMA-induced augmentation of leptin production is also blocked by H7 or staurosporine, both of which are inhibitors of PKC (Hidaka et al., 1984; Tamaoki et al., 1986).

Secretion of proteases by maternal and fetal tissues and changes of the integrin repertoire in both maternal and fetal tissues characterize the invasion phase of embryo implantation. Gelatinase B (92 kDa) or MMP-9 is a major protease secreted by CTB cultured in vitro (Bischof et al., 1995a). Invasive CTB express the integrin $\alpha6\beta4$ (a laminin receptor) in a nonpolarized manner. A switch of integrin expression in CTB is believed to induce the invasion of trophoblast into decidual tissue. MMP-9 secretion is higher in $\alpha6\beta4$ positive CTB (Bischof et al., 1995b). Trophoectodermal cells, once they reach the endometrial basement membrane, express $\alpha6\beta4$ integrin and turn-on their secretion of gelatinases. These proteases digest the basement membrane, allowing the embryo to make contact with the stromal extracellular matrix (ECM). Integrin $\alpha5\beta1$ (a fibronectin receptor) anchors the embryo into ECM, and induces secretion of collagenases that digest the ECM allowing the embryo to burrow into the endometrium. Invasive trophoblast cells are thus characterized by protease secretion and $\alpha6\beta4$ integrin expression (Bischof and Campana, 1996). This process is under the paracrine control of endometrial cytokines and ECM glycoproteins.

CTB cultured in conditioned media by in vitro decidualized stromal cells shows a reduced gelatinolytic activity but an increased secretion of tissue inhibitor of MMP (TIMP-1) and fetal fibronectin. In contrast, IGFBP-1 (the main secretory product of decidualized stromal cells) increased the gelatinolytic activity of CTB. It has been suggested that the effects of IGFBP-1 are mediated through binding of this protein to the $\alpha5\beta1$ integrin through the Arg-Gly-Asp (RGD) integrin recognition sequence (Bischof et al., 1998).

The mechanism involved in the switching from non-invasive villous CTB to invasive extravillous CTB are still speculative (Bischof and Campana, 1996).

IL-1α and IL-6, transforming growth factor-β (TGF-β) and leptin have different effects on the expression of MMP in CTB cultures (Librach et al., 1994; Shimonovitz et al., 1996; Castellucci et al., 2000). TGF-β inhibits hCG production by CTB in a dose dependent manner (Song et al., 1996). Moreover, the TGF-β induced inhibition of CTB invasion is probably exerted through a stimulatory effect on trophoblastic TIMP synthesis (Graham and Lala, 1991) and through a down-regulation of urokinase plasminogen activator (uPA). Insulin-like growth factor-1 (IGF-1) has been recognized as an important regulator of proliferation and differentiation of trophoblast (Murata et al., 1994).

Leptin could be a regulator of protease synthesis and integrin expression in CTB. In vitro, leptin, TGF-β, and IL-6 did not affect $\alpha2\beta1$ integrin (laminin/collagen receptor) expression in cultured CTB from first trimester placenta. However, these factors upregulate the expression of $\alpha5\beta1$ integrin. Expression of $\alpha6\beta4$ integrin on CTB cultured in vitro was also upregulated by IL-1 and IL-6. A similar effect was observed for TGF-β and leptin, which upregulated $\alpha6\beta4$ integrin expression in 80 percent of CTB cultured in vitro. Moreover, leptin also increases MMP-9 activity in these cells. However, MMP-2 activity was not changed by leptin or TGF-β (Gonzalez et al., 2001). On the other hand, IL-6 induces a dose-dependent increase in the secretion of leptin, MMPs (2 and 9) activity, but not MMPs immunoreactivity, thus not their synthesis in trophoblastic cells (Meisser et al., 1999). Taking these data together, one could speculate that IL-6 activates trophoblastic MMP either through an increased secretion of leptin or a mechanism independent of leptin secretion. It is interesting to note that leptin and IL-6 share structural similarities and that the full-length leptin receptor has IL-6 signaling capabilities (Baumann et al., 1996; Tartaglia 1997). This IL-6 signaling capability might explain our results that recombinant human leptin activates MMPs in our primary trophoblastic cell cultures (Castellucci et al., 2000).

Thus, a novel role for leptin as autocrine/paracrine regulator of the invasion phase of human implantation may be proposed. Trophoblastic leptin could be an inducer of the secretion of invasion-inducing MMP and a modulator of the expression of integrins conferring an invasive phenotype to CTB. It is tempting to speculate that IL-1 stimulates leptin synthesis (Chardonnens et al., 1999) and that leptin in turn would contribute to switching on an invasive CTB phenotype. Further investigations should be performed to establish the autocrine/paracrine effects of leptin and cytokines on the invasive behavior of CTB during implantation and placentation.

REFERENCES

Abdul-Karim, R. W., Nesbitt, R. E. Jr., Drucker, M. H., and Rizk, P. T. (1971). The regulatory effect of estrogens on fetal growth. I. Placental and fetal body weight. *Am. J. Obstet. Gynecol., 109*, 656–661.

Aherne, W. and Hull, D. (1966). Brown adipose tissue and heat production in the newborn infant. *J. Pathol. Bacteriol., 91*, 223–234.

Ahima, R. S., Dushay, J., Flier, S. N., Prabakaran, D., and Flier, J. S. (1997). Leptin accelerates the onset of puberty in normal female mice. *J. Clin. Invest., 99*, 391–395.

Alsat, E., Guibourdenche, J., Coutourier, A., and Evain-Brion, D. (1998). Physiological role of human placental growth hormone. *Mol. Cell. Endocrinol., 140*, 121–127.

Amico, J. A., Thomas, A., Crowley, R. S., and Burmeister, L. A. (1998). Concentrations of leptin in the serum of pregnant, lactating and cycling rats and of leptin messenger ribonucleic acid in rat placental tissue. *Life Sci., 63*, 1387–1395.

Antczak, M. and van Blerkom, J. V. (1997). Oocyte influences on early development: the regulatory proteins leptin and STAT3 are polarized in mouse and human oocytes and differentially distributed within the cells of the preimplantatation stage embryo. *Mol. Hum. Reprod., 2*, 1067–1086.

Antczak, M., van Blerkom, J., and Clark, A. (1997). A novel mechanism of vascular endothelial growth factor, leptin and transforming growth factor-beta2 sequestration in a subpopulation of human ovarian follicle cells. *Hum. Reprod., 12*, 2226–2234.

Apte, S. V. and Iyengar, L. (1972). Composition of the human fetus. *Br. J. Nutr., 27*, 305–312.

Ashworth, C. J., Hoggard, N., Thomas, L., Mercer, J. G., Wallace, J. M., and Lea, R. G. (2000). Placental leptin. *Rev. Reprod., 5*, 18–24.

Atanassova, P. and Popova, L. (2000). Leptin expression during the differentiation of subcutaneous adipose cells of human embryos in situ. *Cells Tissues Organs, 166*, 15–19.

Babischkin, J. S., Grimes, R. W., Pepe, G. J., and Albrecht, E. D. (1997). Estrogen stimulation of P450 cholesterol side-chain cleavage activity in cultures of human placental syncytiotrophoblasts. *Biol. Reprod., 56*, 272–278.

Bajoria, R., Sooranna, S. R., Ward, B. S., and Chatterjee, R. (2002). Prospective function of placental leptin at maternal–fetal interface. *Placenta, 23*, 103–115.

Baumann, H., Morella, K. K., White, D. W., Dembski, M., Bailon, P. S., Kim, H., Lai, C. F., and Tartaglia, L. A. (1996). The full-length leptin receptor has signaling capabilities of interleukin 6-type cytokine receptors. *Proc. Natl. Acad. Sci. USA, 93*, 8374–8378.

Bennett, B. D., Solar, G. P., Yuan, J. Q., Mathias, J., Thomas, G. R., and Matthews, W. (1996). A role for leptin and its cognate in hematopoiesis. *Curr. Biol., 6*, 1170–1180.

Bi, S., Gavrilova, O., Gong, D. W., Mason, M. M., and Reitman, M. (1997). Identification of a placental enhancer for the human leptin gene. *J. Biol. Chem., 272*, 30583–30588.

Bird, T. A., Schooley, K., Dower, S. K., Hagen, H., and Virca, G. D. (1997). Activation of nuclear transcription factor NF-kappaB by interleukin-1 is accompanied by casein kinase II-mediated phosphorylation of the p65 subunit. *J. Biol. Chem., 272*, 32606–32612.

Bischof, P. and Campana, A. (1996). A model for implantation of the human blastocyste and early placentation. *Hum. Reprod. Update, 2*, 262–270.

Bischof, P., Haenggli, L., and Campana, A. (1995b). Gelatinase and oncofetal fibronectin secretion are dependent upon integrin expression on human cytotrophoblasts. *Hum. Reprod., 10*, 734–742.

Bischof, P., Martelli, M., Campana, A., Itoh, Y., Ogata, Y., and Nagase, H. (1995a). Importance of metalloproteinases (MMP) in human trophoblastic invasion. *Early Pregn. Biol. Med., 1*, 263–269.

Bischof, P., Meisser, A., Campana, A., and Tseng, L. (1998). Effects of decidua-conditioned medium and insulin-like growth factor binding protein-1 on trophoblastic matrix metalloproteinases and their inhibitors. *Placenta, 19*, 457–464.

Bjorbaek, C., El-Hascimi, K., Frantz, J. D., and Flier, J. S. (1999). The role of SOCS-3 in leptin signaling and leptin resistance. *J. Biol. Chem., 274*, 30059–30065.

Bodner, J., Ebenbichler, C. F., Wolf, H. J., Muller-Holzner, E., Stanzl, U., Gander, R., Huter, O., and Patsch, J. R. (1999). Leptin receptor in human term placenta: in situ hybridization and immunohistochemical localization. *Placenta, 20,* 677–682.

Bouloumie, A., Drexler, H. C., Lafontan, M., and Busse, R. (1998). Leptin, the product of Ob gene, promotes angiogenesis. *Circ. Res., 83,* 1059–1066.

Butte, N. F., Hopkinson, J. M., and Nicolson, M. A. (1997). Leptin in human reproduction: serum leptin levels in pregnant and lactating women. *J. Clin. Endocrinol. Metab., 82,* 585–589.

Campfield, L. A., Smith, F. J., Guisez, Y., Devos, R., and Burn, P. (1995). Recombinant mouse OB protein: evidence for a peripheral signal linking adiposity and central neural networks. *Science, 269,* 546–549.

Castellucci, M., De Matteis, R., Meisser, A., Cancello, R., Monsurro, V., Islami, D., Sarzani, R., Mirzioni, D., Cinti, S., and Bischof, P. (2000). Leptin modulates extracellular matrix molecules and metalloproteinases: possible implications for trophoblast invasion. *Mol. Hum. Reprod., 6,* 951–958.

Chardonnens, D., Cameo, P., Aubert, M. L., Pralong, F. P., Islami, D., Campana, A., Gaillard, R. C., and Bischof, P. (1999). Modulation of human cytotrophoblastic leptin secretion by interleukin-1alpha and 17beta-oestradiol and its effect on HCG secretion. *Mol. Hum. Reprod., 5,* 1077–1082.

Carro, E., Senaris, R., Considine, R. V., Casanueva, F. F., and Dieguez, C. (1997). Regulation of in vivo growth hormone secretion by leptin. *Endocrinology, 138,* 2203–2206.

Carro, E., Senaris, R., Seoane, L. M., Frohman, L. A., Arimura, A., Casanueva, F. F., and Dieguez, C. (1999). Role of growth hormone (GH)-releasing hormone and somatostatin on leptin-induced GH secretion. *Neuroendocrinology, 69,* 3–10.

Chehab, F., Lom, M., and Lu, R. (1996). Correction of the sterility defect in homozygous obese female mice by treatment with the human recombinant leptin. *Nat. Genet., 12,* 318–320.

Chen, H., Charlat, O., Tartaglia, L. A., Woolf, E. A., Weng, X., Ellis, S. J., Lakey, N. D., Culpepper, J., Moore, K. J., Breitbart, R. E., Duyk, G. M., Tepper, R. I., and Morgenstern, J. P. (1996). Evidence that the diabetes gene encodes the leptin receptor: identification of a mutation in the leptin receptor gene in db/db mice. *Cell, 84,* 491–495.

Chibbar, R., Wong, S., Miller, F. D., and Mitchell, B. F. (1995). Estrogen stimulates oxytocin gene expression in human chorio-decidua. *J. Clin. Endocrinol. Metab., 80,* 567–572.

Chien, E. K., Hara, M., Rouard, M., Yano, H., Philippe, M., Polonsky, K. S., and Bell, G. I. (1997). Increase in serum leptin and uterine leptin receptor messenger RNA levels during pregnancy rats. *Biochem. Biophys. Res. Commun., 237,* 476–480.

Chijiwa, T., Mishima, A., Hagiwara, M., Sano, M., Hayashi, K., Inoue, T., Naito, K., Toshioka, T., and Hidaka, H. (1990). Inhibition of forskolin-induced neurite outgrowth and protein phosphorylation by a newly synthesized selective inhibitor of cyclic AMP-dependent protein kinase, N-[2-(p-bromocinnamylamino)ethyl]-5-isoquinolinesulfonamide (H-89), of PC12D pheochromocytoma cells. *J. Biol. Chem., 265,* 5267–5272.

Cioffi, J. A., Shafer, A. W., Zupancic, T. J., Smith-Gbur, J., Mikhail, A., Platika, D., and Snodgrass, H. R. (1996). Novel B219/OB receptor isoforms: possible role of leptin in hematopoiesis and reproduction. *Nat. Med., 2,* 585–589.

Cioffi, J. A., Van Blerkom, J., Antczak, M., Shafer, A., Wittmer, S., and Snodgrass, H. R. (1997). The expression of leptin and its receptors in pre-ovulatory human follicles. *Mol. Hum. Reprod., 3,* 467–472.

Clapp, J. F. 3rd and Kiess, W. (1998). Cord blood leptin reflects fetal fat mass. *Soc. Gynecol. Invest., 5,* 300–303.

Considine, R. V., Sinha, M. K., Heiman, M. L., Kriauciunas, A., Stephens, T. W., Nyce, M. R., Ohannesian, J. P., Marco, C. C., McKee, L. J., and Bauer, T. L. (1996). Serum immunoreactive-leptin concentrations in normal-weight and obese humans. *N. Engl. J. Med., 334,* 292–295.

Cumin, F., Baum, H. P., and Levens, N. (1996). Leptin is cleared from the circulation primarily by the kidney. *Int. J. Obes. Relat. Metab. Disord., 20,* 1120–1126.

Cunningham, M. J., Clifton, D. K., and Steiner, R. A. (1999). Leptin's actions on the reproductive axis: perspectives and mechanisms. *Biol. Reprod., 60,* 216–222.

Cusin, I., Sainsbury, A., Doyle, P., Rohner-Jeanrenaud, F., and Jeanrenaud, B. (1995). The ob gene and insulin. A relationship leading to clues to the understanding of obesity. *Diabetes, 44,* 1467–1470.

Diamond, F. B. Jr., Eichler, D. C., Duckett, G., Jorgensen, E. V., Shulman, D., and Root, A. W. (1997). Demonstration of a leptin binding factor in human serum. *Biochem. Biophys. Res. Commun., 233,* 818–822.

Ebihara, K., Ogawa, Y., Isse, N., Mori, K., Tamura, N., Masuzaki, H., Kohno, K., Yura, S., Hosoda, K., Sagawa, N., and Nakao, K. (1997). Identification of the human leptin 5'-flanking sequences involved in the trophoblast-specific transcription. *Biochem. Biophys. Res. Commun., 241,* 658–663.

Edwards, R. G. and Beard, H. K. (1997). Oocyte polarity and cell determination in early mammalian embryos. *Mol. Hum. Reprod., 3,* 863–905.

Ertl, T., Funke, S., Sarkany, I., Szabo, I., Rascher, W., Blum, W. F., and Sulyok, E. (1999). Postnatal changes of leptin levels in full-term and preterm neonates: their relation to intrauterine growth, gender and testosterone. *Biol. Neonate, 75,* 167–176.

Festa, A., Shnawa, N., Krugluger, W., Hopmeier, P., Schernthaner, G., and Haffner, S. M. (1999). Relative hypoleptinemia in women with mild gestational diabetes mellitus. *Diabet. Med., 16,* 656–662.

Friedman, W. J., Thakur, S., Seidman, L., and Rabson, A. B. (1996). Regulation of nerve growth factor mRNA by interleukin-1 in rat hippocampal astrocytes is mediated by NFkappaB. *J. Biol. Chem., 271,* 31115–31120.

Fusch, C., Ozdoba, C., Kuhn, P., Durig, P., Remonda, L., Muller, C., Kaiser, G., Schroth, G., and Moessinger, A. C. (1997). Perinatal ultrasonography and magnetic resonance imaging findings in congenital hydrocephalus associated with fetal intraventricular hemorrhage. *Am. J. Obstet. Gynecol., 177,* 512–518.

Fusch, C., Slotboom, J., Fuehrer, U., Schumacher, R., Keisker, A., Zimmermann, W., Moessinger, A., Boesch, C., and Blum, J. (1999). Neonatal body composition: dual-energy X-ray absorptiometry, magnetic resonance imaging, and three-dimensional chemical shift imaging versus chemical analysis in piglets. *Pediatr. Res., 46,* 465–473.

Gavrilova, O., Barr, V., Marcus-Samuels, B., and Reitman, M. (1997). Hyperleptinemia of pregnancy associated with the appearance of a circulating form of the leptin receptor. *J. Biol. Chem., 272,* 30546–30551.

Gonzalez, R. R., Devoto, L., Campana, A., and Bischof, P. (2001). Effects of leptin, interleukin-1alpha, interleukin-6, and transforming growth factor-beta on markers of trophoblast invasive phenotype: integrins and metalloproteinases. *Endocrine, 15,* 157–164.

Graham, C. H. and Lala, P. K. (1991). Mechanism of control of trophoblast invasion in situ. *J. Cell. Physiol., 148,* 228–234.

Green, E. D., Maffei, M., Braden, V. V., Proenca, R., DeSilva, U., Zhang, Y., Chua, S. C. Jr, Leibel, R. L., Weissenbach, J., and Friedman, J. M. (1995). The human obese (OB) gene: RNA expression pattern and mapping on the physical, cytogenetic, and genetic maps of chromosome 7. *Genome Res., 5,* 5–12.

Grimes, R. W., Pepe, G. J., and Albrecht, E. D. (1996). Regulation of human placental trophoblast low-density lipoprotein uptake in vitro by estrogen. *J. Clin. Endocrinol. Metab., 81,* 2675–2679.

Hardie, L., Trayhurn, P., Abramovich, D., and Fowler, P. (1997). Circulating leptin in women: a longitudinal study in the menstrual cycle and during pregnancy. *J. Clin. Endocrinol., 47,* 101–106.

Harigaya, A., Nagashima, K., Nako, Y., and Morikawa, A. (1997). Relationship between concentration of serum leptin and fetal growth. *J. Clin. Endocrinol. Metab., 82,* 3281–3284.

Hassink, S. G., de Lancey, E., Sheslow, D. V., Smith-Kirwin, S. M., O'Connor, D. M., Considine, R. V., Opentanova, I., Dostal, K., Spear, M. L., Leef, K., Ash, M., Spitzer, A., and Funagane, V. L. (1997). Placental leptin: an important new growth factor in intrauterine and neonatal development? *Pediatrics, 100,* E1.

Havel, P. J., Kasim-Karakas, S., Dubuc, G. R., Mueller, W., and Phinney, S. D. (1996). Gender differences in plasma leptin concentrations. *Nat. Med., 2,* 949–950.

Heaney, M. L. and Golde, D. W. (1993). Soluble hormone receptors. *Blood, 82,* 1945–1948.

Hebebrand, J., van der Heyden, J., Devos, R., Kopp, W., Herpertz, S., Remschmidt, H., and Herzog, W. (1995). Plasma concentrations of obese protein in anorexia nervosa. *Lancet, 346,* 1624–1625.

Heiman, M. L., Ahima, R. S., Craft, L. S., Schoner, B., Stephens, T. W., and Flier, J. S. (1996). Leptin inhibition of the hypothalamic–pituitary–adrenal axis in response to stress. *Endocrinology, 138,* 3859–3863.

Henson, M. C. and Castracane, V. D. (2000). Leptin in pregnancy. *Biol. Reprod., 63,* 1219–1228.

Henson, M. C., Swan, K. F., and O'Neil, J. S. (1998). Expression of placental leptin and leptin receptor transcripts in early pregnancy and at term. *Obstet. Gynecol., 92,* 1020–1028.

Hidaka, H., Inagaki, M., Kawamoto, S., and Sasaki, Y. (1984). Isoquinolinesulfonamides, novel and potent inhibitors of cyclic nucleotide dependent protein kinase and protein kinase C. *Biochemistry, 23,* 5036–5041.

Hirose, H., Saito, I., Kawai, T., Nakamura, K., Maruyama, H., and Saruta, T. (1998). Serum leptin level: possible association with hematopoiesis in adolescents, independent of body mass index and serum insulin. *Clin. Sci. (Lond), 94,* 633–636.

Hoggard, N., Hunter, L., Duncan, J. S., Williams, L. M., Trayhurn, P., and Mercer, J. G. (1997). Leptin and leptin receptor mRNA and protein expression in the murine fetus and placenta. *Proc. Natl. Acad. Sci. USA, 94,* 11073–11078.

Holness, M. J., Munns, M. J., and Sugden, M. C. (1999). Current concepts concerning the role of leptin in reproductive function. *Mol. Cell. Endocrinol., 157*, 11–20.

Hytinantti, T., Koistinen, H. A., Koivisto, V. A., Karonen, S. L., and Andersson, S. (1999). Changes in the leptin concentration during the early postnatal period: adjustment to extrauterine life? *Pediatr. Res., 45*(2): 197–201.

Janik, J. E., Curti, B. D., Considine, R. V., Rager, H. C., Powers, G. C., Alvord, W. G., Smith, J. W., Gause, B. L., and Kopp, W. C. (1997). Interleukin-1 alpha increases serum leptin concentrations in humans. *J. Clin. Endocrinol. Metab., 82*, 3084–3086.

Jaquet, D., Leger, J., Levy-Marchal, C., Oury, J. F., and Czernichow, P. (1998). Ontogeny of leptin in human fetuses and newborns: effect of intrauterine growth retardation on serum leptin concentrations. *J. Clin. Endocrinol. Metab., 83*, 11243–11246.

Kawai, M., Yamaguchi, M., Murakami, T., Shima, K., Murata, Y., and Kishi, K. (1997). The placenta is not the main source of leptin production in pregnant rat: gestational profile of leptin in plasma and adipose tissues. *Biochem. Biophys. Res. Commun., 240*, 798–802.

Kiess, W., Englaro, P., Hanitsch, S., Rascher, W., Attanasio, A., and Blum, W. F. (1996). High leptin concentrations in serum of very obese children are further stimulated by dexamethasone. *Horm. Metab. Res., 28*, 708–710.

Kirchgessner, T. G., Uysal, K. T., Wiesbrock, S. M., Marino, M. W., and Hotamisligil, G. S. (1997). Tumor necrosis factor-alpha contributes to obesity-related hyperleptinemia by regulating leptin release from adipocytes. *J. Clin. Invest., 100*, 2777–2782.

Kline, A. D., Becker, G. W., Churgay, L. M., Landen, B. E., Martin, D. K., Muth, W. L., Rathnachalam, R., Richardson, J. M., Schoner, B., Ulmer, M., and Hale, J. E. (1997). Leptin is a four-helix bundle: secondary structure by NMR. *FEBS Lett., 407*, 239–242.

Koistinen, H. A., Koivisto, V. A., Andersson, S., Karonen, S. L., Kontula, K., Oksanen, L., and Teramo, K. A. (1997). Leptin concentration in cord blood correlates with intrauterine growth. *J. Clin. Endocrinol. Metab., 82*, 3328–3330.

Kolaczynski, J. W., Ohannesian, J. P., Considine, R. V., Marco, C. C., and Caro, J. F. (1996). Response of leptin to short-term and prolonged overfeeding in humans. *J. Clin. Endocrinol. Metab., 81*, 4162–4165.

Lage, M., Garcia-Mayor, R. V., Tome, M. A., Cordido, F., Valle-Inclan, F., Considine, R. V., Caro, J. F., Dieguez, C., and Casanueva, F. F. (1999). Serum leptin levels in women throughout pregnancy and the postpartum period and in women suffering spontaneous abortion. *Clin. Endocrinol. (Oxf), 50*, 211–216.

LaPaglia, N., Steiner, J., Kristeins, L., Emanuele, M., and Emanuele, N. (1998). Leptin alters the response of the growth hormone—insulin-like growth factor-I axis to fasting. *J. Endocrinol., 159*, 79–83.

Laud, K., Gourdou, I., Belair, L., Keisler, D. H., and Djiane, J. (1999). Detection and regulation of leptin receptor mRNA in ovine mammary epithelial cells during pregnancy and lactation. *FEBS Lett., 463*, 194–198.

Lea, R. G., Howe, D., Hannah, L. T., Bonneau, O., Hunter, L., and Hoggard, N. (2000). Placental leptin in normal, diabetic and fetal growth-retarded pregnancies. *Mol. Hum. Reprod., 6*, 763–769.

Lederman, S. A., Paxton, A., Heymsfield, S. B., Wang, J., Thornton, J., and Pierson, R. N. Jr. (1999). Maternal body fat and water during pregnancy: do they raise infant birth weight? *Am. J. Obstet. Gynecol., 180*(1 Pt 1), 235–240.

Lee, G. H., Proenca, R., Montez, J. M., Carroll, K. M., Darvishzadeh, J. G., Lee, J. I., and Friedman, J. M. (1996). Abnormal splicing of the leptin receptor in diabetic mice. *Nature, 379*, 632–635.

Lepercq, J., Cauzac, M., Lahlou, N., Timsit, J., Girard, J., Auwerx, J., and Hauguel-de-Mouzon, S. (1998). Overexpression of placental leptin in diabetic pregnancy: a critical role for insulin. *Diabetes, 47*, 847–850.

Lepercq, J., Lahlou, N., Timsit, J., Girard, J., and Mouzon, S. H. (1999). Macrosomia revisited: ponderal index and leptin delineate subtypes of fetal overgrowth. *Am. J. Obstet. Gynecol., 181*, 621–625.

Lewandowski, K., Horn, R., O'Callaghan, C. J., Dunlop, D., Medley, G. F., O'Hare, P., and Brabant, G. (1999). Free leptin, bound leptin, and soluble leptin receptor in normal and diabetic pregnancies. *J. Clin. Endocrinol. Metab., 84*, 300–306.

Librach, C. L., Feigenbaum, S. L., Bass, K. E., Cui, T. Y., Verastas, N., Sadovsky, Y., Quigley, J. P., French, D. L., and Fisher, S. J. (1994). Interleukin-1 beta regulates human cytotrophoblast metalloproteinase activity and invasion in vitro. *J. Biol. Chem., 269*, 17125–17131.

Lin, K. C., Hsu, S. C., Kuo, C. H., and Zhou, J. Y. (1999). Difference of plasma leptin levels in venous and arterial cord blood: relation to neonatal and placental weight. *Kaohsiung. J. Med. Sci., 15*, 679–685.

Lindpaintner, K. (1995). Finding an obesity gene—a tale of mice and man. *N. Engl. J. Med., 332*, 679–680.

Linnemann, K., Malek, A., Sager, R., Blum, W. F., Schneider, H., and Fusch, C. (2000). Leptin production and release in the dually in vitro perfused human placenta. *J. Clin. Endocrinol. Metab., 85,* 4298–4301.

Lönnqvist, F., Arner, P., Nordfors, L., and Schalling, M. (1995). Overexpression of the obese (ob) gene in adipose tissue of human obese subjects. *Nat. Med., 1,* 950–953.

Luoh, S. M., Di Marco, F., Levin, N., Armanini, M., Xie, M. H., Nelson, C., Bennett, G. L., Williams, M., Spencer, S. A., Gurney, A., and de Sauvage, F. J. (1997). Cloning and characterization of a human leptin receptor using a biologically active leptin immunoadhesion. *J. Mol. Endocrinol., 18,* 77–85.

MacDougald, O. A., Hwang, C. S., Fan, H., and Lane, M. D. (1995). Regulated expression of the obese gene product (leptin) in white adipose tissue and 3T3-L1 adipocytes. *Proc. Natl. Acad. Sci. USA, 91,* 9034–9037.

Masuzaki, H., Ogawa, Y., Isse, N., Satoh, N., Okazaki, T., Shigemoto, M., Mori, K., Tamura, N., Hosoda, K., and Yoshimasa, Y. (1995). Human obese gene expression. Adipocyte-specific expression and regional differences in the adipose tissue. *Diabetes, 44,* 855–858.

Masuzaki, H., Ogawa, Y., Sagawa, N., Hosoda, K., Matsumoto, T., Mise, H., Nishimura, H., Yoshimasa, Y., Tanaka, I., Mori, T., and Nakao, K. (1997). Nonadipose tissue production of leptin: leptin as a novel placenta-derived hormone in humans. *Nat. Med., 3,* 1029–1033.

Matsuda, J., Yokota, I., Iida, M., Murakami, T., Naito, E., Ito, M., Shima, K., and Kuroda, Y. (1997). Serum leptin concentration in cord blood: relationship to birth weight and gender. *J. Clin. Reprod. Metab., 82,* 1642–1644.

McCarthy, J. F., Misra, D. N., and Roberts, J. M. (1999). Maternal plasma leptin is increased in preeclampsia and positively correlates with fetal cord concentration. *Am. J. Obstet. Gynecol., 180*(3 Pt 1), 731–736.

McGarry, J. D. (1995). Appetite control: does leptin lighten the problem of obesity? *Curr. Biol., 5,* 1342–1344.

Meisser, A., Cameo, P., Islami, D., Campana, A., and Bischof, P. (1999). Effects of interleukin-6 (IL-6) on cytotrophoblastic cells. *Mol. Hum. Reprod., 5,* 1055–1058.

Mise, H., Sagawa, N., Matsumoto, T., Yura, S., Nanno, H., Itoh, H., Mori, T., Masuzaki, H., Hooda, K., Ogawa, Y., and Nakao, K. (1998). Augmented placental production of leptin in preeclampsia: possible involvement of placental hypoxia. *J. Clin. Endocrinol. Metab., 83,* 3225–3229.

Moore, R. M., Diamond, E. L., and Cavalieri, R. L. (1988). The relationship of birth weight and intrauterine diagnostic ultrasound exposure. *Obstet. Gynecol., 71,* 513–517.

Moragas, A. and Toran, N. (1983). Prenatal development of brown adipose tissue in man. A morphometric and biomathematical study. *Biol. Neonate, 43,* 80–85.

Mounzih, K., Qiu, J., Ewart-Toland, A., and Chehab, F. F. (1999). Leptin is not necessary for gestation and parturition but regulates maternal nutrition via a leptin resistance state. *Endocrinology, 139,* 5259–5262.

Mukherjea, R., Castonguay, T. W., Douglass, L. W., and Moser-Veillon, P. (1999). Elevated leptin concentrations in pregnancy and lactation: possible role as a modulator of substrate utilization. *Life Sci., 65,* 1183–1193.

Murakami, T., Iida, M., and Shima, K. (1995). Dexamethasone regulates obese expression in isolated rat adipocytes. *Biochem. Biophys. Res. Commun., 214,* 1260–1267.

Murata, K., Maruo, T., Matsuo, H., and Mochizuki, M. (1994). Insulin like growth factor 1 (IGF-1) as a local regulator of proliferation and differentiation of villous trophoblast in early pregnancy. *Nippon Sanka Fuynka Gakkai Zasshi, 46,* 87–94.

Ogawa, Y., Masuzaki, H., Isse, N., Okazaki, T., Mori, K., Shigemoto, M., Satoh, N., Tamura, N., Hosoda, K., and Yoshimasa, Y. (1995). Molecular cloning of rat obese cDNA and augmented gene expression in genetically obese Zucker fatty (fa/fa) rats. *J. Clin. Invest., 96,* 1647–1652.

Ong, K. K., Ahmed, M. L., Sheriff, A., Woods, K. A., Watts, A., Golding, J., and Dunger, D. B. (1999). Cord blood leptin is associated with size at birth and predicts infancy weight gain in humans. ALSPAC Study Team. Avon Longitudinal Study of Pregnancy and Childhood. *J. Clin. Endocrinol. Metab., 84,* 1145–1148.

Ong, L. C., Boo, N. Y., Chandran, V., Zamratol, S. M., Allison, L., Teoh, S. L., Nyein, M. K., and Lye, M. S. (1997). Relationship between head growth and neurodevelopmental outcome of Malaysian very low birth weight infants during the 1st year of life. *Ann. Trop. Pediatr., 17,* 209–216.

Persson, B., Westgren, M., Ceksi, G., Nord, E., and Ortqvist, E. (1999). Leptin concentrations in cord blood in normal newborn infants and offsprings of diabetic mothers. *Horm. Metab. Res., 31,* 467–471.

Petraglia, F., de Micheroux, A. A., Florio, P., Salvatori, M., Gallinelli, A., Cela, V., Palumbo, M. A., and Genazzani, A. R. (1995). Steroid-protein interaction in human placenta. *J. Steroid Biochem. Mol. Biol., 53,* 227–231.

Petraglia, F., Vaughan, J., and Vale, W. (1990). Steroid hormones modulate the release of immunoreactive gonadotropin-releasing hormone from cultured human placental cells. *J. Clin. Endocrinol. Metab., 70*, 1173–1178.

Ranganathan, S., Maffei, M., and Kern, P. A. (1998). Adipose tissue ob mRNA expression in humans: discordance with plasma leptin and relationship with adipose TNFalpha expression. *J. Lipid. Res., 39*, 724–730.

Reddy, S. A., Huang, J. H., and Liao, W. S. (1997). Phosphatidylinositol 3-kinase in interleukin 1 signaling. Physical interaction with the interleukin 1 receptor and requirement in NFkappaB and AP-1 activation. *J. Biol. Chem., 272*, 29167–29173.

Rentsch, J. and Chiesi, M. (1996). Regulation of ob gene mRNA levels in cultured adipocytes. *FEBS Lett., 379*(1), 55–59.

Reuvekamp, A., Velsing-Aarts, F. V., Poulina, I. E., Capello, J. J., and Duits, A. J. (1999). Selective deficit of angiogenic growth factors characterizes pregnancies complicated by pre-eclampsia. *Br. J. Obstet. Gynaecol., 106*, 1019–1022.

Rink, T. J. (1994). Genetics. In search of a satiety factor. *Nature, 372*, 406–407.

Rosenbaum, M., Nicolson, M., Hirsch, J., Heymsfield, S. B., Gallagher, D., Chu, F., and Leibel, R. L. (1996). Effects of gender, body composition, and menopause on plasma concentrations of leptin. *J. Clin. Endocrinol. Metab., 81*, 3424–3427.

Sacks, G., Sargent, I., and Redman, C. (1999). An innate view of human pregnancy. *Immunol. Today, 20*, 114–118.

Sattar, N., Greer, I. A., Pirwani, I., Gibson, J., and Wallace, A. M. (1998). Leptin levels in pregnancy: marker for fat accumulation and mobilization? *Acta Obstet. Gynecol. Scand., 77*, 278–283.

Schubring, C., Kiess, W., Englaro, P., Rascher, W., Dotsch, J., Hanitsch, S., Attanasio, A., and Blum, W. F. (1997). Levels of leptin in maternal serum, amniotic fluid, and arterial and venous cord blood: relation to neonatal and placental weight. *J. Clin. Endocrinol. Metab., 82*, 1480–1483.

Schubring, C., Siebler, T., Englaro, P., Blum, W. F., Kratusch, J., Triep, K., and Kiess, W. (1998). Rapid decline of serum leptin levels in healthy neonates after birth. *Eur. J. Pediatr., 157*, 263–264.

Schwartz, M. W., Peskind, E., Raskind, M., Boyko, E. J., and Porte, D. Jr. (1996). Cerebrospinal fluid leptin levels: relationship to plasma levels and to adiposity in humans. *Nat. Med., 2*, 589–593.

Senaris, R., Garcia-Caballero, T., Casabiell, X., Gallego, R., Castro, R., Considine, R. V., Dieguez, C., and Casanueva, F. F. (1997). Synthesis of leptin in human placenta. *Endocrinology, 138*, 4501–4504.

Shimon, I., Yan, X., and Melmed, S. (1998). Human fetal pituitary expresses functional growth hormone-releasing peptide receptors. *J. Clin. Endocrinol. Metab., 83*, 174–178.

Shimonovitz, S., Hurwitz, A., Barak, V., Dushnik, M., Adashi, E. Y., Anteby, E., and Yagel, S. (1996). Cytokine-mediated regulation of type IV collagenase expression and production in human trophoblast cells. *J. Clin. Endocrinol. Metab., 81*, 3091–3096.

Simon, C., Francés, A., Piquette, G., el Danasouri, I., Zurawski, G., Dang, W., and Polan, M. L. (1994). Embryonic implantation in mice is blocked by interleukin-1 receptor antagonist. *Endocrinology, 134*, 521–528.

Sinha, M. K. (1997). Human leptin: the hormone of adipose tissue. *Eur. J. Endocrinol., 136*, 461–464.

Sivan, E., Whittaker, P. G., Sinha, D., Homko, C. J., Lin, M., Reece, E. A., and Boden, G. (1998). Leptin in human pregnancy: the relationship with gestational hormones. *Am. J. Obstet. Gynecol., 179*, 1128–1132.

Song, Y., Keelan, J., and Frame, J. T. (1996). Activin A stimulates, while transforming growth factor beta 1 inhibits chorionic gonadotropin production and aromatase activity in cultured human placental trophoblast. *Placenta, 17*, 603–610.

Stock, S. M. and Bremme, K. A. (1998). Elevation of plasma leptin levels during pregnancy in normal and diabetic women. *Metabolism, 47*, 840–843.

Strauss, J. F. 3rd, Kido, S., Sayegh, R., Sakuragi, N., and Gafvels, M. E. (1992). The cAMP signaling system and human trophoblast function. *Placenta, 13*, 389–403.

Takahashi, N., Waelput, W., and Guisez, Y. (1999). Leptin is an endogenous protective protein against the toxicity exerted by tumor necrosis factor. *J. Exp. Med., 189*(1): 207–212.

Tamaoki, T., Nomoto, H., Takahashi, I., Kato, Y., Morimoto, M., and Tomita, F. (1986). Staurosporine, a potent inhibitor of phospholipid/Ca++ dependent protein kinase. *Biochem. Biophys. Res. Commun., 135*, 397–402.

Tamura, T., Goldenberg, R. L., Johnston, K. E., and Cliver, S. P. (1998). Serum leptin concentrations during pregnancy and their relationship to fetal growth. *Obstet. Gynecol., 91*, 389–395.

Tannenbaum, G. S., Gurd, W., and Lapointe, M. (1998). Leptin is a potent stimulator of spontaneous pulsatile growth hormone (GH) secretion and the GH response to GH-releasing hormone. *Endocrinology, 139*, 3871–3875.

Tartaglia, L. A. (1997). The leptin receptor. *J. Biol. Chem., 272*, 6093–6096.

Tartaglia, L. A., Dembski, M., Weng, X., Deng, N., Culpepper, J., Devos, R., Richards, G. J., Campfield, L. A., Clark, F. T., and Deeds, J. (1995). Identification and expression cloning of a leptin receptor, OB-R. *Cell, 83*, 1263–1271.

Tomimatsu, T., Yamaguchi, M., Murakami, T., Ogura, K., Sakata, M., Mitsuda, N., Kanzaki, T., Kurachi, H., Irahara, M., Miyake, A., Shima, K., Aono, T., and Murata, Y. (1997). Increase of mouse leptin production by adipose tissue after midpregnancy: gestational profile of serum leptin concentration. *Biochem. Biophys. Res. Commun., 240*, 213–215.

Tulchinsky, D., Hobel, C. J., Yeager, E., and Marshall, J. R. (1972). Plasma estrone, estradiol, estriol, progesterone, and 17-hydroxyprogesterone in human pregnancy. I. Normal pregnancy. *Am. J. Obstet. Gynecol., 112*, 1095–1100.

Wegmann, T. G., Lin, H., Guilbert, L., and Mosmann, T. R. (1993). Bi-directional cytokine interactions in the maternal–fetal relationship: is successful pregnancy a TH2 phenomenon? *Immunol. Today, 14*, 353–356.

Williams, M. A., Havel, P. J., Schwartz, M. W., Leisenring, W. M., King, I. B., Zingheim, R. W., Zebelman, A. M., and Luthy, D. A. (1999). Pre-eclampsia disrupts the normal relationship between serum leptin concentrations and adiposity in pregnant women. *Pediatr. Perinat. Epidemiol., 13*, 190–204.

Wilson, E. A., Jawad, M. J., and Powell, D. E. (1984). Effect of estradiol and progesterone on human chorionic gonadotropin secretion in vitro. *Am. J. Obstet. Gynecol., 149*, 143–148.

Yura, S., Sagawa, N., Mise, H., Mori, T., Masuzaki, H., Ogawa, Y., and Nakao, K. (1998a). A positive umbilical venous-arterial difference of leptin level and its rapid decline after birth. *Am. J. Obstet. Gynecol., 178*, 926–930.

Yura, S., Sagawa, N., Ogawa, Y., Masuzaki, H., Mise, H., Matsumoto, T., Ebihara, K., Fujii, S., and Nakao, K. (1998b). Augmentation of leptin synthesis and secretion through activation of protein kinases A and C in cultured human trophoblastic cells. *J. Clin. Endocrinol. Metab., 83*, 3609–3614.

Zhang, Y., Proenca, R., Maffei, M., Barone, M., Leopold, L., and Friedman, J. M. (1994). Positional cloning of the mouse obese gene and its human homologue. *Nature, 372*, 425–432.

Leptin in Rodent Pregnancy

BRENDAN J. WADDELL AND JEREMY T. SMITH

14.1. INTRODUCTION

Leptin, the peptide hormone product of the *ob* gene, is produced predominantly by adipocytes and acts to regulate food intake and energy expenditure at the hypothalamus (van Dijk, 2001). In addition, it is now recognized that leptin impacts on several aspects of reproduction including puberty onset, gonadal function, and pregnancy (Clarke and Henry, 1999). Accordingly, total deficiency of leptin in mice due to a mutation in the *ob* gene prevents sexual maturation, with fertility restored only after administration of exogenous leptin (Zhang et al., 1994; Chehab et al., 1996). Of particular interest to the present review, recent work by Malik et al. (2001) demonstrates an obligatory role for leptin in rodent pregnancy. Potentially, leptin could influence a range of gestational adaptations, including maternal metabolism, function of the placenta and related tissues, and the promotion of fetal growth. In this review we examine the biology of leptin during rodent pregnancy with a particular emphasis on the central role played by the placenta. Initially, the key features of gestation in the rat and mouse are presented as necessary background for the subsequent discussion of leptin in pregnancy.

14.2. PREGNANCY IN THE RAT AND MOUSE

Estrous cycles in the rat and mouse are normally of 4–5 days duration, with ovulation, formation of corpora lutea, and mating occurring early on estrus (Rugh, 1968; Freeman, 1988). The mating stimulus initiates twice-daily surges in pituitary prolactin secretion that, in turn, promotes growth of the corpora lutea and associated progesterone

---

BRENDAN J. WADDELL AND JEREMY T. SMITH • School of Anatomy and Human Biology, The University of Western Australia.

Leptin and Reproduction. Edited by Henson and Castracane, Kluwer Academic/Plenum Publishers, 2003.

secretion (Freeman, 1988). Thus, endocrine adaptations of rodent pregnancy are initiated directly by the mating stimulus and thereby interrupt estrous cyclicity. In both species implantation and the associated decidualization of the endometrium begin at around days 5–6 and the chorio-vitelline placenta is established (Enders and Schlafke, 1969; Rugh, 1968; Steven and Morriss, 1975). The definitive, chorio-allantoic placenta subsequently forms from germinal trophoblast cells and becomes the primary site of nutrient and gaseous exchange from about midgestation. This definitive placenta is of hemochorial form, is discoid in shape, and composed of two morphologically and functionally distinct regions termed the basal (or junctional) and labyrinth zones. The basal zone, consisting of spongiotrophoblasts, glycogen cells, giant trophoblast cells and maternal (but not fetal) blood vessels, is located adjacent to the decidua basalis and is the major site of hormone production (Enders et al., 1965; Davies and Glasser, 1968). The labyrinth zone is the major site of maternal–fetal exchange after midgestation, and comprises fetal blood vessels and mesenchyme surrounded by trophoblast epithelium (Enders et al., 1965; Davies and Glasser, 1968; Rugh, 1968). Rapid growth of this placental labyrinth zone supports a dramatic increase in fetal mass over the final week of pregnancy. Parturition normally occurs from days 18–21 in the mouse depending on strain (Rugh, 1968), whereas gestation length in the rat is generally 22–23 days (Lincoln and Porter, 1976; Moore, 1990).

14.3. PLASMA LEPTIN LEVELS THROUGHOUT GESTATION

Maternal plasma leptin concentrations remain relatively stable during the first half of rodent pregnancy but then increase dramatically (approximately 40-fold) from midgestation in the mouse (Gavrilova et al., 1997; Tomimatsu et al., 1997; Yamaguchi et al., 1998) (see Figure 14.1a). Most reports indicate that a more modest yet still substantial increase occurs in the rat from midgestation, and this is followed by a clear pre-partum decline (Kawai et al., 1997; Amico et al., 1998; Garcia et al., 2000; Seeber et al., 2002) (see Figure 14.1b). One study reported that pregnancy-induced increases in plasma leptin are not evident if compared with age-matched, nonpregnant control rats, but in this study the initial plasma leptin values appeared unusually high (Terada et al., 1998). The rise in plasma leptin from midpregnancy appears due to the combined effects of increased plasma leptin binding activity (Gavrilova et al., 1997; Seeber et al., 2002) (see Figure 14.2) and enhanced maternal adipocyte expression of leptin mRNA (Kawai et al., 1997; Tomimatsu et al., 1997; Garcia et al., 2000). Importantly, the rodent placenta appears to contribute little, if any, leptin to the maternal circulation (Kawai et al., 1997; Amico et al., 1998; Terada et al., 1998; Hoggard et al., 2000), unlike the significant contribution made by the placenta in human and nonhuman primates (Masuzaki et al., 1997; Henson et al., 1998). The increase in plasma leptin binding activity, which is particularly striking during mouse pregnancy (Gavrilova et al., 1997), raises some uncertainty as to the biological significance of increased plasma leptin in rodent pregnancy. In particular it is unclear whether plasma leptin binding restricts access of leptin to target tissues. This issue is addressed below in relation to the effects of the soluble isoform of the leptin receptor on leptin transport and metabolism in pregnancy. Interestingly, the pre-partum decline in plasma leptin levels in the rat occurs despite a concomitant rise in plasma leptin binding activity (Seeber et al., 2002). This apparent inconsistency may reflect reduced fat mass late in pregnancy (Herrera et al., 2000),

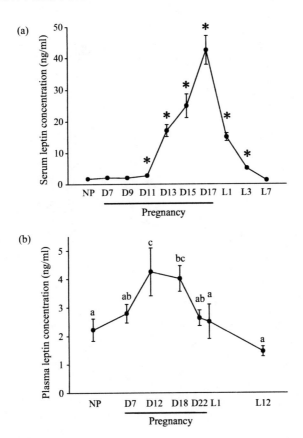

Figure 14.1. (a) Mouse serum leptin concentration (ng/ml) before pregnancy (NP), on day 7 (D7), 9 (D9), 11 (D11), 13 (D13), 15 (D15), and 17 (D17) of pregnancy, and on day 1 (L1), 3 (L3), and 7 (L7) of lactation. $*p < 0.05$ versus nonpregnant. Figure redrawn from Tomimatsu et al. (1997) with permission. (b) Rat plasma leptin concentration before pregnancy (NP), on day 7 (D7), 12 (D12), 18 (D18), and 22 (D22) of pregnancy, and on day 1 (L1) and 12 (L12) of lactation. Values are the mean $\pm$ SEM. Values without common notations differ significantly ($p < 0.05$, one-way ANOVA and LSD-test). Figure reproduced from Seeber et al. (2002).

a fall in adipocyte leptin mRNA expression (Terada et al., 1998), or a rise in placental clearance of leptin (see below).

14.4. LEPTIN RECEPTOR EXPRESSION IN PREGNANCY

14.4.1. Leptin Receptor Isoforms and their Functional Properties

Leptin actions are exerted via a receptor (Ob-R) that is a member of the class 1 cytokine receptor family (for review see Tartaglia, 1997). Six splice variants of Ob-R have been identified (Wang et al., 1996; Tartaglia, 1997) and these fall into three categories: a long

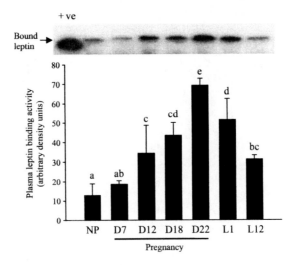

Figure 14.2. Autoradiograph showing binding of ^{125}I-leptin in plasma from rats before pregnancy (NP) and at days 7 (D7), 12 (D12), 18 (D18), and 22 (D22) of pregnancy and days 1 (L1) and 12 (L12) of lactation. A serum sample from a day 17 pregnant mouse is included as a positive control (+ve). The graph shows plasma leptin-binding activity, and values are the mean $\pm$ SEM. There was significant variation among groups ($p < 0.05$, two-way ANOVA) and those without common notations differ significantly ($p < 0.05$, LSD-test). Figure reproduced from Seeber et al. (2002).

form (Ob-Rb) capable of full signal transduction, a number of C-terminally truncated forms (Ob-Ra, Ob-Rc, Ob-Rd, Ob-Rf) of which Ob-Ra is the most physiologically significant and may act as a transporter of leptin through physiological barriers, and Ob-Re which contains only the extracellular domain and acts as a leptin binding protein. Signal transduction via Ob-Rb involves activation of the Jak-STAT (signal transducer and activator of transcription) and mitogen-activated protein kinase pathways (Ghilardi et al., 1996; Bjorbaek et al., 1997). Hypothalamic expression of Ob-Rb enables leptin to influence energy balance centrally through suppression of orexigenic signals and stimulation of metabolic rate (Halaas et al., 1995; Pelleymounter et al., 1995; Mantzoros, 1999). Hypothalamic expression of Ob-Rb in pregnancy is of particular interest since the normal mechanisms controlling energy balance appear to be altered as part of the maternal adaptation. For example, food intake increases progressively during rat pregnancy despite a corresponding rise in plasma leptin, suggesting that pregnancy may be a state of leptin resistance (Seeber et al., 2002). In addition to the hypothalamus, understanding the role of leptin in pregnancy requires knowledge of Ob-R isoform expression in various reproductive tissues including the uterus and placenta, and each of these is discussed separately below.

14.4.2. Hypothalamus

Potentially, the development of apparent leptin resistance in pregnancy could result from a fall in hypothalamic expression of Ob-Rb. Indeed, in the nonpregnant rat, hypothalamic Ob-Rb expression falls in response to estrogen (Bennett et al., 1998), suggesting that the progressive rise in estrogen characteristic of rodent pregnancy may contribute to leptin resistance. Therefore, we profiled hypothalamic Ob-Rb mRNA (by real-time quantitative Reverse Transcriptase-Polymerase Chain Reaction [RT-PCR]) and protein expression across pregnancy in the rat (days 0, 7, 12, 18, and 22), and although a transient peak was observed at day 7, expression remained generally stable. Most importantly, hypothalamic Ob-Rb levels were unchanged at the time when apparent leptin resistance was most prevalent after midgestation (Seeber et al., 2002). This contrasts with a recent report by Garcia et al. (2000) in which a slight fall in Ob-Rb mRNA occurred at day 18 compared with nonpregnant controls, although this was measured only by semiquantitative RT-PCR. Hypothalamic expression of all other Ob-R isoforms was also unaffected by pregnancy (Garcia et al., 2000).

14.4.3. Uterus and Ovary

The rat uterus expresses multiple Ob-R isoforms but there is some uncertainty as to their relative expression and distribution. Chien et al. (1997) detected both Ob-Ra and Ob-Rc mRNAs by RT-PCR analysis of pregnant myometrium, but expression of the major signal transduction isoform, Ob-Rb, was not observed. Subsequently, however, we reported mRNA expression for all three major Ob-R isoforms (Ob-Ra, Ob-Rb, and Ob-Re) in rat uterus (by RT-PCR analysis) (Plastow and Waddell, 2002), consistent with observations in the human nonpregnant uterus (Alfer et al., 2000). We also observed intense Ob-R immunostaining in both endometrium and myometrium during the estrous cycle and pseudopregnancy. The antibody employed in our studies (Santa Cruz, K-20) potentially recognizes all three isoforms, and so in situ hybridization studies are required to ascertain the specific expression patterns for each isoform. Nevertheless, immunolocalization of Ob-R exhibited a dynamic pattern over the 4-day estrous cycle, being localized predominantly to glandular and luminal epithelial cells when estrogen was high (i.e., at proestrus), but to the myometrium when estrogen was minimal (i.e., at postestrus). Moreover, while endometrial stromal cells were clearly Ob-R immunonegative during the cycle, artificial induction of decidualization in pseudopregnant rats resulted in a rapid induction of Ob-R immunoreactivity (Plastow and Waddell, 2002). Consistent with this upregulation in endometrial Ob-R expression and the increase in myometrial Ob-R mRNA reported by Chien et al. (1997), we recently observed that retention of ^{125}I-leptin by the uterus is markedly increased in pregnancy (see Figure 14.3). Further studies are required to establish the functional significance of this leptin binding in both the endometrium and myometrium.

The ovary of the nonpregnant rat expresses multiple Ob-R isoforms (Zachow et al., 1999) and ovulation rate is reduced in response to leptin (Duggal et al., 2000). It is not certain whether the ovary of the pregnant rat expresses any of the Ob-R isoforms, although we observed an increase in ovarian retention of ^{125}I-leptin in vivo during late pregnancy (see Figure 14.3). This raises the possibility that the corpus luteum (CL), the major constituent of the rat ovary at this time, expresses Ob-R and is responsive to leptin. Indeed, Ruiz-Cortes et al. (2000) demonstrated that porcine granulosa cells express Ob-R mRNA and that this expression increases with luteinization in vitro. Moreover, luteal expression of Ob-R in vivo was maximal in functional CL but then declined during regression (Ruiz-Cortes et al., 2000). Further studies are required to establish whether leptin influences luteal function during rodent pregnancy.

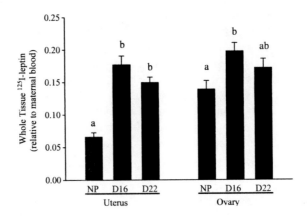

Figure 14.3. ^{125}I-leptin in the uterus and ovary from rats before pregnancy (NP) and at days 16 (D16) and 22 (D22) of pregnancy. Values are the tissue concentrations of trichloroacetic acid (TCA)-precipitable radioactivity (cpm) expressed as a proportion of the value in maternal plasma. Tissues and maternal plasma were obtained 90 min after the administration of ^{125}I-leptin. For both the uterus and ovary, ^{125}I-leptin activity varied with the stage of pregnancy (uterus, $p < 0.001$; ovary, $p < 0.05$, one-way ANOVAs) and for each tissue values without common notations (a, b) differ significantly ($p < 0.05$, LSD-test).

14.4.4. Placenta

The rodent placenta expresses all three major Ob-R isoforms, Ob-Ra, Ob-Rb, and Ob-Re (Gavrilova et al., 1997; Kawai et al., 1999; Hoggard et al., 2000). Thus, expression of these receptors could influence leptin actions in the placenta and fetus by mediating leptin transport (via Ob-Ra), direct leptin actions within the placenta (via Ob-Rb), and leptin bioavailability (via Ob-Re). We recently reported marked changes in the expression of all three Ob-R isoforms over the final week of rat pregnancy, as well as important spatial differences in their expression between the two placental zones (Smith and Waddell, 2002). Thus, Ob-Re mRNA was by far the most highly expressed among the three isoforms, with maximal levels observed in the labyrinth zone just prior to term (see Figure 14.4). This coincides with the time of maximal plasma leptin binding activity measured in maternal plasma (Seeber et al., 2002), consistent with the suggestion that plasma leptin binding activity is due to placental secretion of Ob-Re (Yamaguchi et al., 1998; Gavrilova et al., 1997). Ob-Ra mRNA was also expressed in both placental zones at day 16 of rat pregnancy, and by day 22 this had increased in the labyrinth zone by around 5-fold (Smith and Waddell, 2002). Because this placental zone is the major site of maternal–fetal exchange and Ob-Ra is thought to mediate leptin transport, we propose that placental expression of Ob-Ra may be the key determinant of transplacental leptin passage. This possibility is discussed in relation to leptin transport and metabolism in pregnancy (see Section 14.5.1).

Placental expression of Ob-Rb mRNA and protein were readily detected in both placental zones over the final week of rat pregnancy, suggesting that the placenta is a leptin target tissue. Immunolocalization of Ob-R confirmed its presence in both zones, although staining appeared more intense in the labyrinth zone (Smith and Waddell, 2002) consistent with a previous report of Ob-R mRNA expression measured by in situ hybridization

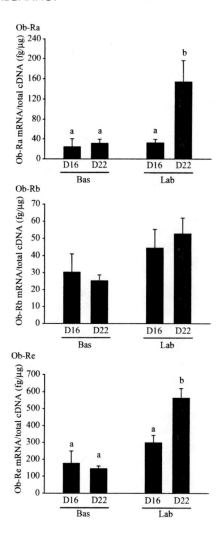

Figure 14.4. Quantification of placental Ob-Ra, Ob-Rb, and Ob-Re mRNA in the basal (Bas) and labyrinth (Lab) zones of the rat placenta at day 16 (D16) and day 22 (D22) of pregnancy by real-time RT-PCR. Values are expressed as fg RNA/μg total cDNA and are the mean ± SEM. Ob-Ra and Ob-Re varied with gestational age ($p < 0.05$) and placental zone ($p < 0.02$), both with significant interaction ($p < 0.05$, two-way ANOVA). Values without common notations differ significantly ($p < 0.05$, LSD-test). Figure reproduced from Smith and Waddell (2002).

(Kawai et al., 1999). Within the labyrinth zone, Ob-R immunoreactivity was specifically localized to trophoblast tissue as previously reported for the human placenta (Henson et al., 1998; Bodner et al., 1999). In addition, however, Ob-R staining was particularly intense within the adventitial layer of fetal blood vessels, possibly related to the proposed role of leptin in promoting angiogenesis (see Section 14.6.1).

14.5. LEPTIN TRANSPORT AND METABOLISM IN PREGNANCY

14.5.1. Effects of Placental Ob-Re on Maternal Leptin

Placental secretion of Ob-Re appears to have a major impact on the transport and metabolism of maternal leptin in rodent pregnancy. Thus, circulating Ob-Re is thought to account for the marked rise in plasma leptin binding activity in both the mouse (Gavrilova et al., 1997) and rat (Seeber et al., 2002), with approximately 95 percent of circulating leptin bound to plasma proteins by day 18 of pregnancy in the mouse (Gavrilova et al., 1997). As observed for several other peptides such as corticotropin releasing hormone (Seasholtz et al., 2001) and the insulin-like growth factors (Baxter, 1994), such binding would be expected to restrict the exit of plasma leptin into extravascular tissues, thus reducing its metabolic clearance rate (Cumin et al., 1997) and access to target cells. To test this possibility we compared the fate of ^{125}I-leptin injected intravenously into nonpregnant and pregnant rats and showed that leptin disappears from the circulation more slowly during pregnancy (see Figure 14.5). These data suggest that access of leptin to target tissues, including the hypothalamus, is reduced in pregnancy.

14.5.2. Placental Ob-Ra and Transplacental Passage of Maternal Leptin to the Fetus

Because the rodent placenta and fetus synthesize relatively little leptin (Dessolin et al., 1997; Kawai et al., 1997; Amico et al., 1998; Terada et al., 1998; Hoggard et al., 2000), transplacental passage of maternal leptin is likely to be the major source of fetal

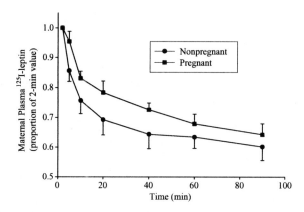

Figure 14.5. Disappearance rates of ^{125}I-leptin administered to nonpregnant and pregnant (day 22) rats. Values are the TCA-precipitable activity (cpm) of plasma expressed as a proportion of the 2-min value. There was significant variation with both time ($p < 0.001$) and treatment ($p < 0.01$, two-way ANOVA).

leptin in the rat and mouse. Expression of Ob-Ra by the placental labyrinth zone may provide the mechanism for this transfer of maternal leptin, since Ob-Ra is thought to mediate leptin transport across physiological barriers such as the choroid plexus (Kastin et al., 1999) and exhibits transcellular leptin transport when transfected into epithelial cells in vitro (Hileman et al., 2000). The labyrinth zone is the major site of maternal–fetal exchange and its expression of Ob-Ra increases by around 5-fold in the final days of gestation (Smith and Waddell, 2002). We surmised that this increase in placental Ob-Ra may improve the capacity for transplacental passage of maternal leptin to the fetus, and to test this possibility we examined the fate of ^{125}I-leptin administered to the maternal circulation at days 16 and 22 of pregnancy, in particular, its passage to the fetus. As predicted, there was a marked increase in the amount of ^{125}I-leptin reaching the fetus at day 22 compared with day 16, consistent with a role for Ob-Ra in leptin transport. Moreover, inhibition of endogenous glucocorticoid synthesis by maternal treatment with metyrapone, an 11β-hydroxylase inhibitor, enhanced both placental Ob-Ra expression (Smith and Waddell, 2002) and transplacental passage of ^{125}I-leptin in vivo (see Figure 14.6). Fetal leptin levels remained similar to those of control fetuses after maternal metyrapone treatment, despite suppression of maternal leptin in these animals (see Figure 14.7). Collectively, these data indicate that placental Ob-Ra expression correlates positively with the capacity of the placenta to transport maternal leptin to the fetus, and is tonically suppressed by endogenous glucocorticoids. Interestingly, maternal treatment with either dexamethasone or carbenoxolone (an 11β-hydroxysteroid dehydrogenase inhibitor that enhances local levels of active glucocorticoid within the placenta) did not affect Ob-Ra expression, but did inhibit placental Ob-Rb expression and appeared to limit passage of endogenous leptin from the mother to the fetus. Specifically, while dexamethasone treatment stimulated maternal leptin by 2.7-fold, it resulted in a 5.5-fold reduction in fetal leptin, similar to recent observations of Sugden et al. (2001). Fetal leptin levels were similarly reduced by carbenoxolone treatment, but this occurred without any effect on maternal leptin, further supporting the contention that glucocorticoids suppress placental

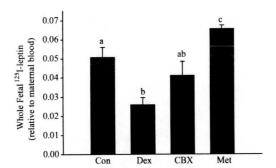

Figure 14.6. Fetal ^{125}I-leptin concentration at day 22 of pregnancy in untreated rats (Con) and in rats after maternal treatment with dexamethasone (Dex), carbenoxolone (CBX), or metyrapone (Met) from day 13 to day 22. Values are the tissue concentrations of TCA-precipitable radioactivity (cpm) expressed as a proportion of the value in maternal plasma. Tissues and maternal plasma were obtained 90 min after the administration of ^{125}I-leptin. ^{125}I-leptin activity varied with treatment ($p < 0.001$, one-way ANOVA) and values without common notations differ significantly ($p < 0.05$, LSD-test).

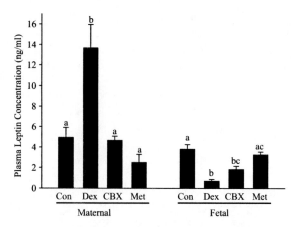

Figure 14.7. Plasma leptin concentrations in the maternal and fetal circulation at day 22 of pregnancy in untreated rats (Con) and in rats after maternal treatment with dexamethasone (Dex), carbenoxolone (CBX), or metyrapone (Met) from day 13 to day 22. Values are the mean $\pm$ SEM. Within each compartment (maternal or fetal), values without common notations differ significantly ($p < 0.01$, one-way ANOVAs, LSD-tests). Figure taken from Smith and Waddell (2002).

leptin transport (Smith and Waddell, 2002). The latter was confirmed by the demonstration that transplacental passage of ^{125}I-leptin to the fetus was reduced after dexamethasone treatment (see Figure 14.6). Because this effect of dexamethasone was accompanied by reduced placental expression of Ob-Rb, but not Ob-Ra, it raises the possibility that Ob-Rb also contributes to placental leptin transport. Such a role would be consistent with observations of Ob-Rb-mediated internalization of leptin in transfected CHO cells (Uotani et al., 1999). An improved placental capacity for transfer of maternal leptin to the fetus late in pregnancy may be important with respect to fetal development, since leptin has been proposed as a key component of the complex hormonal regulation of fetal growth (see Section 14.6.2). Moreover, increased transplacental passage of maternal leptin to the fetus may contribute to the pre-partum decline in maternal plasma leptin levels, despite the continued increase in plasma leptin binding activity at this time (Seeber et al., 2002). Indeed, a role for increased placental clearance near term is supported by the recent observation that the pre-partum decline in maternal leptin does not occur in pregnant rats that have a small litter size (Butterstein and Castracane, 2002).

14.6. BIOLOGICAL ACTIONS OF LEPTIN IN PREGNANCY AND LACTATION

14.6.1. Implantation and Placentation

Although leptin is essential for normal fertility in rodents (Zhang et al., 1994; Chehab et al., 1996), the requirement for leptin during pregnancy per se has been more difficult to establish. Studies by Mounzih et al. (1998) had suggested that after conception, *ob/ob* mice could successfully establish and maintain a pregnancy in the absence of leptin. More recently, however, Malik et al. (2001) demonstrated that exogenous leptin is indeed required for successful implantation, but not for the subsequent maintenance of pregnancy beyond 6.5 days. It was suggested that the discrepancy between these reports relates to the high dose of exogenous leptin used in the initial study and the consequent residual effect of this leptin well into pregnancy (Malik et al., 2001). Thus, leptin appears to play a significant role in the preparation for implantation or in the implantation process per se, and this possibly relates to actions of leptin in the decidua. Indeed, as discussed above, we observed a striking induction of Ob-R expression in endometrial stromal cells following artificial induction of decidualization (Plastow and Waddell, 2002). Potentially, leptin could exert several important effects in the decidua, including provision of an angiogenic stimulus either directly (Bouloumie et al., 1998; Sierra-Honigman et al., 1998) or indirectly via interactions with vascular endothelial growth factor (Ashworth et al., 2000; Cao et al., 2001). Leptin may also promote vascular permeability (Cao et al., 2001), a key event in the process of decidualization and implantation (Rabbani and Rogers, 2001), and could directly influence the function of the invading embryo. For example, leptin may stimulate mitogenesis in the rapidly growing trophoblast, comparable to its effect on human epidermal cells (Frank et al., 2000). Finally, leptin could promote the invasive capacity of placental trophoblast via stimulation of matrix metalloproteinase production or activity (Castellucci et al., 2000).

14.6.2. Fetal and Placental Growth

Leptin does not appear to play an indispensable role in rodent pregnancy after implantation (Malik et al., 2001), but it may still have a major impact on several physiological processes. Of particular interest is the clinical evidence showing that fetal leptin levels are positively correlated with placental and fetal weights (Hassink et al., 1997; Yamashita et al., 2001), an association that appears to be independent of insulin-like growth factors (Christou et al., 2001). Fetal weight is also correlated with placental leptin levels following experimental manipulations in wild type and leptin-receptor deficient mice (Yamashita et al., 2001). Leptin may promote fetal growth directly since Ob-R expression has been identified in several fetal tissues including bone, cartilage, heart, liver, and testis (Hoggard et al., 1997, 2000; El-Hefnawy et al., 2000). Leptin may provide a particularly important developmental stimulus to the central nervous system, since decreased brain weight and impaired expression of synaptic and glial proteins have been observed in both *ob/ob* and *db/db* mice (Ahima et al., 1999). Consistent with such a role, Ob-R expression has been identified in the developing cerebral cortex of the fetal rat (Matsuda et al., 1999). In addition to actions within the fetus, leptin could promote fetal growth indirectly via mitogenic and angiogenic effects in the placenta as discussed above. We have explored the latter possibility in our models of glucocorticoid-induced fetal growth retardation (dexamethasone and carbenoxolone treatments), and shown that reduced fetal and placental growth is associated with a marked reduction in placental Ob-Rb expression (Smith and Waddell, 2002). Conversely, suppression of endogenous maternal glucocorticoid synthesis by metyrapone enhances both fetal growth and placental expression of Ob-Rb, as well as Ob-Ra and Ob-Re (Smith and Waddell, 2002). Collectively these data suggest that glucocorticoid-induced fetal growth retardation may be due, in part, to inhibition of leptin actions in the placenta, and/or a reduction in leptin transport to the fetus as discussed above (see Section 14.5.2).

14.6.3. Maternal Metabolism

Changes in maternal plasma leptin during rodent pregnancy are part of the complex endocrine adaptations that maintain maternal metabolic homeostasis, yet ensure provision of energy to meet the demands of fetal growth (for review see Holness et al., 1999). Maternal metabolism in rat pregnancy is characterized by two distinct phases: an initial anabolic phase followed by a catabolic phase with the switch between the two occurring over days 12–16 (Knopp et al., 1973). During the first two thirds of pregnancy the mother is hyperphagic, yet the energy demands of the conceptus are minimal. While blood glucose remains constant there is an increase in plasma insulin, a decrease in fatty acids, and an increase in adipocyte insulin receptors (Flint et al., 1983). Thus, the additional energy consumed is laid down in peripheral fat depots and the mother is in a net anabolic condition (Herrera et al., 1991; Lopez-Luna et al., 1991). Maternal plasma leptin levels increase prior to the end of this anabolic phase (Seeber et al., 2002) presumably due to the additional fat depots and placental Ob-Re secretion. During the final third of pregnancy there is an exponential increase in the energy demands of the growing fetuses, and this is met by a switch to a catabolic environment in the mother. Thus, while hyperphagia continues, there is an increase in fat mobilization, and so concentrations of fatty acids and triglycerides rise (Montes et al., 1978). Although insulin levels continue to increase during this period the mother becomes relatively insulin resistant and the catabolic environment is maintained (Knopp et al., 1973; Herrera et al., 1994; Holness et al., 1999). Leptin has been shown to inhibit insulin stimulation of glucose utilization and lipogenesis (Muller et al., 1997), suggesting that it may contribute to the increased fuel mobilization after midpregnancy by increasing insulin resistance (Holness et al., 1999). Increased levels of maternal glucocorticoids after midpregnancy (Atkinson and Waddell, 1995) are also likely to promote fuel mobilization by direct effects on insulin sensitivity (Jeanrenaud and Rohner-Jeanrenaud, 2000) and indirectly via stimulation of adipocyte leptin synthesis (De Vos et al., 1998; Elimam et al., 1998). This capacity for glucocorticoid stimulation of adipocyte leptin synthesis is clearly maintained in late pregnancy (Sugden et al., 2001; Smith and Waddell, 2002).

The regulation of food intake by leptin appears to be altered during pregnancy, since hyperphagia is maintained despite progressive increases in plasma leptin. While this likely reflects the increased plasma binding activity resulting from placental Ob-Re (Gavrilova et al., 1997; Seeber et al., 2002), the pre-partum decline in maternal plasma leptin does not result in increased food consumption; indeed, food consumption after the pre-partum decline in leptin is the lowest observed at any stage of pregnancy (Seeber et al., 2002). This suggests that the dynamic hormonal milieu of late pregnancy, which includes rapid increases in pituitary secretion of prolactin and luteinizing hormone (Morishige et al., 1973), and a marked reduction in ovarian secretion of progesterone (Waddell et al., 1989), further modulates leptin regulation of food intake. Because progesterone is a known appetite stimulant (Hervey and Hervey, 1967) its pre-partum decline may directly impact on food consumption.

14.6.4. Lactation

Plasma leptin levels continue to fall after birth in the rat (Herrera et al., 2000; Woodside et al., 2000) consistent with the sustained hyperphagia required to meet the increased metabolic demands of lactation (Seeber et al., 2002). Interestingly, plasma leptin and hypothalamic Ob-Rb expression both remain comparable to levels observed in non-pregnant rats (Brogan et al., 2000; Seeber et al., 2002), suggesting that the hyperphagia of lactation is driven primarily by signals other than leptin. Among a large number of potential stimuli, direct activation of orexigenic signals, such as neuropeptide Y, by the suckling stimulus (Li et al., 1998; Brogan et al., 1999) and the sustained increase in prolactin secretion (Noel and Woodside, 1993) are likely to be of particular importance.

Although pregnancy maintenance beyond day 6.5 in *ob/ob* mice does not require administration of exogenous leptin, newborn pups fail to survive unless leptin treatment is continued throughout pregnancy and after birth (Malik et al., 2001). Even when leptin administration is recommenced after birth, pups still fail to survive (Malik et al., 2001). These observations indicate that leptin plays an indispensable role in mammary gland development and/or function. Accordingly, mammary gland epithelial cells have been shown to express Ob-R in the sheep (Laud et al., 1999), and although we are unaware of Ob-R expression in the rodent mammary gland, leptin itself appears to be produced locally in the mouse mammary epithelial cells (Aoki et al., 1999). Further studies are required to assess the specific role of leptin in mammary gland development.

14.7. SUMMARY AND CONCLUSIONS

Leptin is clearly required for the establishment of rodent pregnancy, with likely effects on decidualization, implantation, and placental formation. Increased maternal plasma leptin appears to be an important part of the maternal adaptation to pregnancy, and leptin is essential for mammary gland development and lactation. Pregnancy-induced changes in plasma leptin and its effects on the mother and fetus appear to be regulated by the placenta. Specifically, the rodent placenta mediates leptin effects in the maternal compartment via secretion of Ob-Re, and appears to regulate access of maternal leptin to the fetus via expression of Ob-Ra. Because the placenta also expresses Ob-Rb, leptin is likely to exert a direct, positive influence on placental function, possibly in relation to growth and angiogenesis. Therefore, we suggest that these actions of leptin in the placenta contribute to its proposed role as a fetal growth factor.

REFERENCES

Ahima, R. S., Bjorbaek, C., Osei, S., and Flier, J. S. (1999). Regulation of neuronal and glial proteins by leptin: implications for brain development. *Endocrinology, 140*, 2755–2762.

Alfer, J., Muller-Schottle, F., Classen-Linke, I., von Rango, U., Happel, L., Beier-Hellwig, K., Rath, W., and Beier, H. M. (2000). The endometrium as a novel target for leptin: differences in fertility and subfertility. *Mol. Hum. Reprod., 6*, 595–601.

Amico, J. A., Thomas, A., Crowley, R. S., and Burmeister, L. A. (1998). Concentrations of leptin in the serum of pregnant, lactating, and cycling rats and of leptin messenger ribonucleic acid in rat placental tissue. *Life Sci., 63*, 1387–1395.

Aoki, N., Kawamura, M., and Matsuda, T. (1999). Lactation-dependent down regulation of leptin production in mouse mammary gland. *Biochim. Biophys. Acta.*, *1427*, 298–306.

Ashworth, C. J., Hoggard, N., Thomas, L., Mercer, J. G., Wallace, J. M., and Lea, R. G. (2000). Placental leptin. *Rev. Reprod.*, *5*, 18–24.

Atkinson, H. C. and Waddell, B. J. (1995). The hypothalalmic–pituitary–adrenal axis in rat pregnancy and lactation: circadian variation and interrelationship of plasma adrenocorticotropin and corticosterone. *Endocrinology*, *136*, 512–520.

Baxter, R. C. (1994). Insulin-like growth factor binding proteins in the human circulation: a review. *Horm. Res.*, *42*, 140–144.

Bennett, P. A., Lindell, K., Karlsson, C., Robinson, I. C., Carlsson, L. M., and Carlsson, B. (1998). Differential expression and regulation of leptin receptor isoforms in the rat brain: effects of fasting and oestrogen. *Neuroendocrinology*, *67*, 29–36.

Bjorbaek, C., Uotani, S., da Silva, B., and Flier, J. S. (1997). Divergent signaling capacities of the long and short isoforms of the leptin receptor. *J. Biol. Chem.*, *272*, 32686–32695.

Bodner, J., Ebenbichler, C. F., Wolf, H. J., Muller-Holzner, E., Stanzl, U., Gander, R., Huter, O., and Patsch, J. R. (1999). Leptin receptor in human term placenta: in situ hybridization and immunohistochemical localization. *Placenta*, *20*, 677–682.

Bouloumie, A., Drexler, H. C. A., Lafontan, M., and Busse, R. (1998). Leptin, the product of the *ob* gene, promotes angiogenesis. *Circ. Res.*, *83*, 1059–1066.

Brogan, R. S., Grove, K. L., and Smith, M. S. (2000). Differential regulation of leptin receptor but not orexin in the hypothalamus of the lactating rat. *J. Neuroendocrinol.*, *12*, 1077–1086.

Brogan, R. S., Mitchell, S. E., Trayhurn, P., and Smith, M. S. (1999). Suppression of leptin during lactation: contribution of the suckling stimulus versus milk production. *Endocrinology*, *140*, 2621–2627.

Butterstein, G. M., and Castracane, V. D. (2002). The influence of the number of fetal/placental units on maternal serum leptin in the pregnant rat. *Proc. Soc. Study Reprod.*, *35*.

Cao, R., Brakenhielm, E., Wahlestedt, C., Thyberg, J., and Cao, Y. (2001). Leptin induces vascular permeability and synergistically stimulates angiogenesis with FGF-2 and VEGF. *Proc. Natl. Acad. Sci. USA*, *98*, 6390–6395.

Castellucci, M., De Matteis, R., Meisser, A., Cancella, R., Monsurro, V., Islami, D., Sarzani, R., Marzioni, D., Cinti, S., and Bischof, P. (2000). Leptin modulates extracellular matrix molecules and metalloproteinases: possible implications for trophoblast invasion. *Mol. Hum. Reprod.*, *6*, 951–958.

Chehab, F. F., Lim, M. E., and Lu, R. (1996). Correction of the sterility defect in homozygous obese female mice by treatment with the human recombinant leptin. *Nat. Genet.*, *12*, 318–320.

Chien, E. K., Hara, M., Rouard, M., Yano, H., Phillippe, M., Polonsky, K. S., and Bell, G. I. (1997). Increase in serum leptin and uterine leptin receptor messenger RNA levels during pregnancy in rats. *Biochem. Biophys. Res. Commun.*, *237*, 476–480.

Christou, H., Connors, J. M., Ziotopoulou, M., Hatzidakis, V., Papathanassoglou, E., Ringer, S. A., and Mantzoros, C. S. (2001). Cord blood leptin and insulin-like growth factor levels are independent predictors of fetal growth. *J. Clin. Endocrinol. Metab.*, *86*, 935–938.

Clarke, I. J. and Henry, B. A. (1999). Leptin and reproduction. *Rev. Reprod.*, *4*, 48–55.

Cumin, F., Baum, H. P., and Levens, N. (1997). Mechanism of leptin removal from the circulation by the kidney. *J. Endocrinol.*, *155*, 577–585.

Davies, J. and Glasser, S. R. (1968). Histological and fine structural observations on the placenta of the rat. *Acta. Anat.*, *69*, 542–608.

De Vos, P., Lefebvre, A. M., Shrivo, I., Fruchart, J. C., and Auwerx, J. (1998). Glucocorticoids induce the expression of the leptin gene through a non-classical mechanism of transcriptional activation. *Eur. J. Biochem.*, *253*, 619–626.

Dessolin, S., Schalling, M., Champigny, O., Lonnqvist, F., Ailhaud, G., Dani, C., and Ricquier, D. (1997). Leptin gene is expressed in rat brown adipose tissue at birth. *Faseb. J.*, *11*, 382–387.

Duggal, P. S., Van Der Hoek, K. H., Milner, C. R., Ryan, N. K., Armstrong, D. T., Magoffin, D. A., and Norman, R. J. (2000). The in vivo and in vitro effects of exogenous leptin on ovulation in the rat. *Endocrinology*, *141*, 1971–1976.

El-Hefnawy, T., Ioffe, S., and Dym, M. (2000). Expression of the leptin receptor during germ cell development in the mouse testis. *Endocrinology*, *141*, 2624–2630.

Elimam, A., Knutsson, U., Bronnegard, M., Stierna, P., Albertsson-Wikland, K., and Marcus, C. (1998). Variations in glucocorticoid levels within the physiological range affect plasma leptin levels. *Eur. J. Endocrinol.*, *139*, 615–620.

Enders, A. C. and Schlafke, S. (1969). Cytological aspects of trophoblast–uterine interaction in early implantation. *Am. J. Anat.*, *125*, 1–29.

Enders, J. F., Gunalp, A., Gresser, I., Diamandopoulos, G. T., and Shein, H. M. (1965). Extensive polynucleate giant cell formation after x-irradiation or addition of colchicine in SV40 transformed Syrian hamster cells. *Arch. Gesamte. Virusforsch.*, *17*, 347–373.

Flint, D. J., Clegg, R. A., and Vernon, R. G. (1983). Adipose tissue metabolism during early pregnancy in the rat: temporal relationships of changes in the metabolic activity, number of insulin receptors, and serum hormone concentrations. *Arch. Biochem. Biophys.*, *224*, 677–681.

Frank, S., Stallmeyer, B., Kampfer, H., Kolb, N., and Pfeilschifter, J. (2000). Leptin enhances wound re-epithelialization and constitutes a direct function of leptin in skin repair. *J. Clin. Invest.*, *106*, 501–509.

Freeman, M. E. (1988). The ovarian cycle of the rat. In E. Knobil and J. Neill (Eds.). *The Physiology of Reproduction* (Vol. 29), Raven Press Ltd., New York, pp.1893–1928.

Garcia, M. D., Casanueva, F. F., Dieguez, C., and Senaris, R. M. (2000). Gestational profile of leptin messenger ribonucleic acid (mRNA) content in the placenta and adipose tissue in the rat, and regulation of the mRNA levels of the leptin receptor subtypes in the hypothalamus during pregnancy and lactation. *Biol. Reprod.*, *62*, 698–703.

Gavrilova, O., Barr, V., Marcus-Samuels, B., and Reitman, M. (1997). Hyperleptinemia of pregnancy associated with the appearance of a circulating form of the leptin receptor. *J. Biol. Chem.*, *272*, 30546–30551.

Ghilardi, N., Ziegler, S., Wiestner, A., Stoffel, R., Heim, M. H., and Skoda, R. C. (1996). Defective STAT signaling by the leptin receptor in diabetic mice. *Proc. Natl. Acad. Sci., USA*, *93*, 6231–6235.

Halaas, J. L., Gajiwala, K. S., Maffei, M., Cohen, S. L., Chait, B. T., Rabinowitz, D., Lallone, R. L., Burley, S. K., and Friedman, J. M. (1995). Weight-reducing effects of the plasma protein encoded by the obese gene. *Science*, *269*, 543–546.

Hassink, S. G., de Lancey, E., Sheslow, D. V., Smith-Kirwin, S. M., O'Connor, D. M., Considine, R. V., Opentanova, I., Dostal, K., Spear, M. L., Leef, K., Ash, M., Spitzer, A. R., and Funanage, V. L. (1997). Placental leptin: an important new growth factor in intrauterine and neonatal development? *Pediatrics*, *100*, E1.

Henson, M. C., Swan, K. F., and O'Neil, J. S. (1998). Expression of placental leptin and leptin receptor transcripts in early pregnancy and at term. *Obstet. Gynecol.*, *92*, 1020–1028.

Herrera, E., Lasuncion, M. A., Huerta, L., and Martin-Hidalgo, A. (2000). Plasma leptin levels in rat mother and offspring during pregnancy and lactation. *Biol. Neonate*, *78*, 315–320.

Herrera, E., Lasuncion, M. A., Palacin, M., Zorzano, A., and Bonet, B. (1991). Intermediary metabolism in pregnancy. First theme of the Freinkel era. *Diabetes*, *40*(Suppl. 2), 83–88.

Herrera, E., Munoz, C., Lopez-Luna, P., and Ramos, P. (1994). Carbohydrate–lipid interactions during gestation and their control by insulin. *Braz. J. Med. Biol. Res.*, *27*, 2499–2519.

Hervey, E. and Hervey, G. R. (1967). The effects of progesterone on body weight and composition in the rat. *J. Endocrinol.*, *37*, 361–381.

Hileman, S. M., Tornoe, J., Flier, J. S., and Bjorbaek, C. (2000). Transcellular transport of leptin by the short leptin receptor isoform ObRa in Madin-Darby Canine Kidney cells. *Endocrinology*, *141*, 1955–1961.

Hoggard, N., Hunter, L., Duncan, J. S., Williams, L. M., Trayhurn, P., and Mercer, J. G. (1997). Leptin and leptin receptor mRNA and protein expression in the murine fetus and placenta. *Proc. Nat. Acad. Sci., USA*, *94*, 11073–11078.

Hoggard, N., Hunter, L., Lea, R. G., Trayhurn, P., and Mercer, J. G. (2000). Ontogeny of the expression of leptin and its receptor in the murine fetus and placenta. *Brit. J. Nut.*, *83*, 317–326.

Holness, M. J., Munns, M. J., and Sugden, M. C. (1999). Current concepts concerning the role of leptin in reproductive function. *Mol. Cell. Endocrinol*, *157*, 11–20.

Jeanrenaud, B. and Rohner-Jeanrenaud, F. (2000). CNS–periphery relationships and body weight homeostasis: influence of the glucocorticoid status. *Int. J. Obes. Relat. Metab. Disord.*, *24*(Suppl. 2), S74–76.

Kastin, A. J., Pan, W., Maness, L. M., Koletsky, R. J., and Ernsberger, P. (1999). Decreased transport of leptin across the blood–brain barrier in rats lacking the short form of the leptin receptor. *Peptides*, *20*, 1449–1453.

Kawai, M., Murakami, T., Otani, S., Shima, K., Yamaguchi, M., and Kishi, K. (1999). Colocalization of leptin receptor (OB-R) mRNA and placental lactogen-II in rat trophoblast cells: gestational profile of OB-R mRNA expression in placentae. *Biochem. Biophys. Res. Commun.*, *257*, 425–430.

Kawai, M., Yamaguchi, M., Murakami, T., Shima, K., Murata, Y., and Kishi, K. (1997). The placenta is not the main source of leptin production in pregnant rat: gestational profile of leptin in plasma and adipose tissues. *Biochem. Biophys. Res. Commun.*, *240*, 798–802.

Knopp, R. H., Saudek, C. D., Arky, R. A., and O'Sullivan, J. B. (1973). Two phases of adipose tissue metabolism in pregnancy: maternal adaptations for fetal growth. *Endocrinology*, *92*, 984–988.

Laud, K., Gourdou, I., Belair, L., Keisler, D. H., and Djiane, J. (1999). Detection and regulation of leptin receptor mRNA in ovine mammary epithelial cells during pregnancy and lactation. *FEBS. Lett.*, *463*, 194–198.

Li, C., Chen, P., and Smith, M. S. (1998). The acute suckling stimulus induces expression of neuropeptide Y (NPY) in cells in the dorsomedial hypothalamus and increases NPY expression in the arcuate nucleus. *Endocrinology*, *139*, 1645–1652.

Lincoln, D. W. and Porter, D. G. (1976). Timing of the photoperiod and hour of birth in rats. *Nature*, *260*, 780–781.

Lopez-Luna, P., Maier, I., and Herrera, E. (1991). Carcass and tissue fat content in the pregnant rat. *Biol. Neonate*, *60*, 29–38.

Malik, N. M., Carter, N. D., Murray, J. F., Scaramuzzi, R. J., Wilson, C. A., and Stock, M. J. (2001). Leptin requirement for conception, implantation, and gestation in the mouse. *Endocrinology*, *142*, 5198–5202.

Mantzoros, C. S. (1999). Leptin and the hypothalamus: neuroendocrine regulation of food intake. *Mol. Psychiatry*, *4*, 8–12, 6–7.

Masuzaki, H., Ogawa, Y., Sagawa, N., Hosoda, K., Matsumoto, T., Mise, H., Nishimura, H., Yoshimasa, Y., Tanaka, I., Mori, T., and Nakao, K. (1997). Nonadipose tissue production of leptin: leptin as a novel placenta-derived hormone in humans. *Nat. Med.*, *3*, 1029–1033.

Matsuda, J., Yokota, I., Tsuruo, Y., Murakami, T., Ishimura, K., Shima, K., and Kuroda, Y. (1999). Developmental changes in long-form leptin receptor expression and localization in rat brain. *Endocrinology*, *140*, 5233–5238.

Montes, A., Humphrey, J., Knopp, R. H., and Childs, M. T. (1978). Lipid metabolism in pregnancy. VI. Lipoprotein composition and hepatic lipids in the fed pregnant rat. *Endocrinology*, *103*, 1031–1038.

Moore, D. M. (1990). Rats. In B. E. Rollin and L. Kesel (Eds.). *The Experimental Animal in Biomedical Research* (Vol. 2), CRC Press, Boca Raton, pp. 251–277.

Morishige, W. K., Pepe, G. J., and Rothchild, I. (1973). Serum luteinizing hormone, prolactin and progesterone levels during pregnancy in the rat. *Endocrinology*, *92*, 1527–1530.

Mounzih, K., Qiu, J., Ewart-Toland, A., and Chehab, F. F. (1998). Leptin is not necessary for gestation and parturition but regulates maternal nutrition via a leptin resistance state. *Endocrinology*, *139*, 5259–5262.

Muller, G., Ertl, J., Gerl, M., and Preibisch, G. (1997). Leptin impairs metabolic actions of insulin in isolated rat adipocytes. *J. Biol. Chem.*, *272*, 10585–10593.

Noel, M. B., and Woodside, B. (1993). Effects of systemic and central prolactin injections on food intake, weight gain, and estrous cyclicity in female rats. *Physiol. Behav.*, *54*, 151–154.

Pelleymounter, M. A., Cullen, M. J., Baker, M. B., Hecht, R., Winters, D., Boone, T., and Collins, F. (1995). Effects of the obese gene product on body weight regulation in ob/ob mice. *Science*, *269*, 540–543.

Plastow, K. P., and Waddell, B. J. (2002). Leptin receptor expression in the rat uterus: variation across the estrous cycle and with decidualization. *Proc. Endocr. Soc.*, *84*.

Rabbani, M. L. and Rogers, P. A. (2001). Role of vascular endothelial growth factor in endometrial vascular events before implantation in rats. *Reproduction*, *122*, 85–90.

Rugh, R. (1968). *The Mouse: Its Reproduction and Development*. Burgess Publishing Company, Minneapolis.

Ruiz-Cortes, Z. T., Men, T., Palin, M. F., Downey, B. R., Lacroix, D. A., and Murphy, B. D. (2000). Porcine leptin receptor: molecular structure and expression in the ovary. *Mol. Reprod. Dev.*, *56*, 465–474.

Seasholtz, A. F., Burrows, H. L., Karolyi, I. J., and Camper, S. A. (2001). Mouse models of altered CRH-binding protein expression. *Peptides*, *22*, 743–751.

Seeber, R. M., Smith, J. T., and Waddell, B. J. (2002). Plasma leptin binding activity and hypothalamic leptin receptor expression during pregnancy and lactation. *Biol. Reprod.*, *66*, 1762–1767.

Sierra-Honigmann, M. R., Nath, A. K., Murakami, C., Garcia-Cardana, G., Papapetropoulos, A., Sessa, W. C., Madge, L. A., Schechner, J. S., Schwabb, M. B., Polverini, P. J., and Flores-Riveros, J. R. (1998). Biological action of leptin as an angiogenic factor. *Science*, *281*, 1683–1686.

Smith, J. T., and Waddell, B. J. (2002). Leptin receptor expression in the rat placenta: changes in Ob-Ra, Ob-Rb and Ob-Re with gestational age and suppression by glucocorticoids. *Biol. Reprod. 67*, 1204–1210.

Steven, D. and Morriss, G. (1975). Development of foetal membranes. In D. H. Steven (Ed.). *Comparative Placentation, Essays in Structure and Function*, Academic Press, London, pp. 75–84.

Sugden, M. C., Langdown, M. L., Munns, M. J., and Holness, M. J. (2001). Maternal glucocorticoid treatment modulates placental leptin and leptin receptor expression and materno-fetal leptin physiology during late pregnancy, and elicits hypertension associated with hyperleptinaemia in the early-growth-retarded adult offspring. *Eur. J. Endocrinol., 145*, 529–539.

Tartaglia, L. A. (1997). The leptin receptor. *J. Biol. Chem., 272*, 6093–6096.

Terada, Y., Yamakawa, K., Sugaya, A., and Toyoda, N. (1998). Serum leptin levels do not rise during pregnancy in age-matched rats. *Biochem. Biophys. Res. Commun., 253*, 841–844.

Tomimatsu, T., Yamaguchi, M., Murakami, T., Ogura, K., Sakata, M., Mitsuda, N., Kanzaki, T., Kurachi, H., Irahara, M., Miyake, A., Shima, K., Aono, T., and Murata, Y. (1997). Increase of mouse leptin production by adipose tissue after midpregnancy: gestational profile of serum leptin concentration. *Biochem. Biophys. Res. Commun., 240*, 213–215.

Uotani, S., Bjorbaek, C., Tornoe, J., and Flier, J. S. (1999). Functional properties of leptin receptor isoforms: internalization and degradation of leptin and ligand-induced receptor downregulation. *Diabetes, 48*, 279–286.

van Dijk, G. (2001). The role of leptin in the regulation of energy balance and adiposity. *J. Neuroendocrinol., 13*, 913–921.

Waddell, B. J., Bruce, N. W., and Dharmarajan, A. M. (1989). Changes in ovarian blood flow and secretion of progesterone and 20 alpha-hydroxypregn-4-en-3-one on day 16 and the morning and afternoon of day 22 of pregnancy in the rat. *Biol. Reprod., 41*, 990–996.

Wang, M. Y., Zhou, Y. T., Newgard, C. B., and Unger, R. H. (1996). A novel leptin receptor isoform in rat. *FEBS. Lett., 392*, 87–90.

Woodside, B., Abizaid, A., and Walker, C. (2000). Changes in leptin levels during lactation: implications for lactational hyperphagia and anovulation. *Horm. Behav., 37*, 353–365.

Yamaguchi, M., Murakami, T., Yasui, Y., Otani, S., Kawai, M., Kishi, K., Kurachi, H., Shima, K., Aono, T., and Murata, Y. (1998). Mouse placental cells secrete soluble leptin receptor (sOB-R): cAMP inhibits sOB-R production. *Biochem. Biophys. Res. Commun., 252*, 363–367.

Yamashita, H., Shao, J., Ishizuka, T., Klepcyk, P. J., Muhlenkamp, P., Qiao, L., Hoggard, N., and Friedman, J. E. (2001). Leptin administration prevents spontaneous gestational diabetes in heterozygous Lepr(db/+) mice: effects on placental leptin and fetal growth. *Endocrinology, 142*, 2888–2897.

Zachow, R. J., Weitsman, S. R., and Magoffin, D. A. (1999). Leptin impairs the synergistic stimulation by transforming growth factor-beta of follicle-stimulating hormone-dependent aromatase activity and messenger ribonucleic acid expression in rat ovarian granulosa cells. *Biol. Reprod., 61*, 1104–1109.

Zhang, Y., Proenca, R., Maffei, M., Barone, M., Leopold, L., and Friedman, J. M. (1994). Position cloning of the mouse obese gene and its human homologue. *Nature, 372*, 425–432.

15

Leptin in Primate Pregnancy

MICHAEL C. HENSON AND V. DANIEL CASTRACANE

15.1. INTRODUCTION

Leptin is a polypeptide hormone once considered to be an exclusive product of adipose tissue, it's sole function being to aid in the maintenance of energy homeostasis through the induction of satiety (Considine and Caro, 1997; Weigle, 1997). Mutations in the leptin *(ob)* gene are responsible for a lack of circulating leptin and for obesity in homozygous *(ob/ob)* mice (Zhang et al., 1994), with some cases of human obesity potentially corresponding to a similar mechanism (Montague et al., 1997; Rosenbaum and Leibel, 1998). Most cases of human obesity exhibit a state of "leptin resistance," which may result from any of a number of potential causes (Ahima and Flier, 2000). Regulatory mechanisms remain incompletely defined, although interactions with various hormones are known to affect leptin's ability to regulate adipose energy reserves (Jequier and Tappy, 1999; Moschos et al., 2002).

Identification of both leptin and its receptor in the fetus also implies a role for leptin as a regulator of conceptus development, while localization in the placental trophoblast may relate to autocrine and/or paracrine regulatory functions vital to pregnancy maintenance (Jequier and Tappy, 1999; Ashworth et al., 2000; Gonzalez et al., 2000a,b; Mantzoros, 2000; Bajoria et al., 2001; Hoggard et al., 2001a). Further evidence suggests that cross-talk occurs between the placenta, fetus, and maternal adipose tissue and that placental estrogen, present in increasingly higher concentrations throughout gestation, influences leptin synthesis in a tissue- and cell type- specific fashion (Henson and Castracane, 2000, 2002). It is clear that a better understanding of regulatory mechanisms will have

MICHAEL C. HENSON • Departments of Obstetrics and Gynecology, Physiology, and Structural and Cellular Biology, and the Tulane National Primate Research Center; Tulane University Health Sciences Center.

V. DANIEL CASTRACANE • Department of Obstetrics and Gynecology, and the Women's Health Research Institute of Amarillo; Texas Tech University Health Sciences Center—Amarillo; Diagnostic Systems Laboratories, Inc., Webster, Texas.

Leptin and Reproduction. Edited by Henson and Castracane, Kluwer Academic/Plenum Publishers, 2003.

direct clinical significance, as leptin has been proposed to impact on those causes of human perinatal morbidity and mortality that are exacerbated by anomalies of implantation, fetal maturity and development, general conceptus growth, trophoblast endocrinology, and placental sufficiency.

15.2. LEPTIN DYNAMICS IN MAMMALIAN PREGNANCY

Maternal leptin concentrations rise with those of estrogen and are correlated in early human pregnancy with those of human chorionic gonadotropin (hCG; Hardie et al., 1997). Although not directly correlated with peripheral estrogens (Schubring et al., 1998; Sivan et al., 1998), perhaps due to increased leptin clearance and/or degradation in late pregnancy, leptin release by adipocytes increased dramatically when cultured with estradiol (E_2; Sivan et al., 1998). Throughout pregnancy, maternal peripheral leptin concentrations are enhanced over those in nonpregnant women (Butte et al., 1997; Hardie et al., 1997; Schubring et al., 1997). This enhancement, which occurs prior to any significant increase in body weight due to the pregnancy, suggests that factors other than adiposity affect maternal leptin levels (Highman et al., 1998). Fetal leptin concentrations, although lower than maternal levels, may be owed largely to production by fetal adipose tissue (Clapp and Kiess, 1998; Lepercq et al., 2001). Fetal levels (Butte et al., 1997; Sivan et al., 1997) and those in umbilical cord blood at term are correlated with birth weight (Matsuda et al., 1997; Sivan et al., 1997; Tamura et al., 1998; Varvarigou et al., 1999), although most agree that maternal and cord levels are unrelated to one another (Schubring et al., 1997; Tamas et al., 1998). Significant concentrations in the umbilical vasculature, combined with an immediate decline in neonatal levels at birth, further suggest that the placenta is an important source of fetal leptin (Yura et al., 1998; Reitman et al., 2001). Localization of leptin mRNA transcripts (Hassink et al., 1997; Masuzaki et al., 1997; Senaris et al., 1997; Yura et al., 1998) and leptin protein (Masuzaki et al., 1997; Senaris et al., 1997) to the syncytiotrophoblast further suggests, that in addition to the leptin produced by adipocytes, some of the increase observed in the maternal periphery with advancing gestation is placental. In pigs, leptin levels were not correlated with leptin mRNA expression in fetal adipose tissue, leading the authors to conclude that the placenta is the major source of that leptin associated with fetal growth in that species (Chen et al., 2000). Similarly, in dually perfused human placental villous tissue, it was determined that a higher proportion of leptin is released into the fetal (rather than the maternal) circulation than other placental polypeptides, such as hCG and human placental lactogen (Linneman et al., 2000). This observation, along with those of Laml et al. (2001a), who reported a lack of correlation between maternal and cord leptin concentrations, support the concept of divergently regulated fetal, placental, and maternal compartments.

In the pregnant rodent (Chien et al., 1997; Hoggard et al., 1997; Kawai et al., 1997; Tomimatsu et al., 1997) significant differences in leptin ontogeny exist in comparison to both the human (Henson et al., 1998) and nonhuman primate (Henson et al., 1999; Reitman et al., 2001). In this capacity, although it is commonly accepted that maternal leptin levels are enhanced during pregnancy in the rat, declining just prior to parturition (Chien et al., 1997), others have determined that maternal leptin concentrations were unchanged with advancing rat pregnancy (Terada et al., 1998). Further, although we have determined that leptin mRNA in human placental villous tissue was greater in the first trimester than at term (Henson et al., 1998), Amico et al. (1998) reported that placental leptin mRNA increased 4–5-fold over the final one third of rat pregnancy, while Garcia et al. (2000) reported that leptin mRNA concentrations in both placenta and maternal adipose tissue increased throughout gestation in the rat. Conversely, Kawai et al. (1997) reported, although the possibility of assay insensitivity might be advanced, that the rat placenta does not express leptin mRNA (Kawai et al., 1997), but does express leptin receptor in late gestation (Kawai et al., 1999). In the mouse, although leptin transcripts were abundant in maternal adipose tissue and placenta (Hoggard et al., 1997), as well as in numerous fetal tissues (Hoggard et al., 2000), leptin does not appear to be directly associated with conceptus growth or development (Mounzih et al., 1998). From most accounts, therefore, it appears that the regulation and roles of leptin during pregnancy may be species-specific. Thus, although an assortment of species manifest pregnancy-associated hyperleptinemia, they accomplish it by diverse mechanisms. The highly conserved nature of this trait, however, strongly implies fundamental roles for leptin in mammalian pregnancy.

Because of species differences in regard to leptin regulation and function, the use of an accepted nonhuman primate model of human pregnancy (Castracane and Goldzieher, 1986a,b; Pepe and Albrecht, 1995; Henson, 1998) was required to adequately investigate leptin as a gestational hormone (Henson et al., 1999). Therefore, female baboons (*Papio* sp.) were individually housed, maintained, and mated as we have previously described (Henson et al., 1988, 1991, 1992, 1997). Venous blood samples were obtained and serum leptin concentrations determined by radioimmunoassay (RIA). Serum leptin concentrations in pregnant baboons, between days 60 and 160 of gestation, were approximately 10-fold greater than in nonpregnant animals in the luteal phase of the menstrual cycle. Maternal leptin concentrations increased approximately 2.5-fold between day 60 and day 160 of gestation. Normal term in the baboon is approximately 184 days. Following cesarean delivery in both early pregnancy (approximately day 60) and near term (approximately day 160), maternal leptin levels declined within 15 days to levels equivalent to those of nonpregnant, cycling females. Leptin transcripts in placental villous tissue declined between early and late gestation, while maternal serum leptin levels increased over the same period. The progressive decline in placental leptin mRNA transcripts during baboon pregnancy was reminiscent of our previous investigation (Henson et al., 1998) of human placental tissue collected both early in gestation and at term.

The leptin receptor is manifested in alternatively spliced isoforms, which are identified by various nomenclatures and differ in the length of their intracellular domains. OB-R_L is a long form, which has JAK/STAT signaling capabilities and is in greatest abundance in the hypothalamus (Tartaglia, 1997). OB-R_S, which is a short form, has signaling capabilities involving mitogen activated protein kinase (MAPK; Bjorbaek et al., 1997) and is expressed in the brain (Couce et al., 1997; Fei et al., 1997) and other organs (Tartaglia et al., 1995; Luoh et al., 1997; Weigle, 1997). A soluble isoform appears to function as a circulating leptin receptor (Moschos et al., 2002). As is leptin, the leptin receptor is also present in the trophoblast and we have reported that transcripts for leptin, as well as OB-R_L and OB-R_S receptor isoforms, were expressed in human placenta both early (7–14 weeks) in gestation and at term (38 weeks of gestation) (Henson et al., 1998). No quantitative changes in OB-R_L and OB-R_S mRNAs were apparent in respect to stage of gestation, but the abundance of leptin mRNA transcripts in placental villous tissue was significantly greater in early gestation than at term. Although leptin receptor transcripts were previously identified in term placenta (Amato et al., 1997; Luoh et al., 1997), this study was the first to identify both OB-R_L and OB-R_S isoforms in early pregnancy. Although the mRNA encoding OB-R_L is expressed as the predominant isoform in hypothalamus, it is generally less abundant in peripheral tissues than the transcripts encoding the shorter isoform. Our findings indicate that the placenta is similar in regard to OB-R_S abundance (Henson et al., 1998), with in situ hybridization suggesting that transcripts for leptin, OB-R_L, and OB-R_S were exclusive to the trophoblast.

In the interest of further characterizing our nonhuman primate model, sections of baboon placental villous tissue were examined to localize mRNA transcripts for leptin,

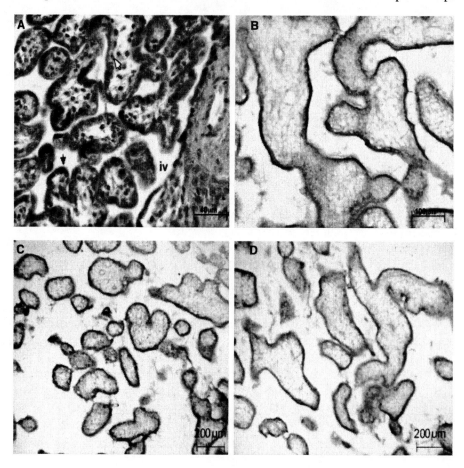

Figure 15.1. Representative phase contrast photomicrographs (400×) of baboon placental villous tissue. A photomicrograph depicting hematoxylin–eosin staining (A) serves as a histological reference for latter photomicrographs depicting in situ hybridization. Multinucleated syncytiotrophoblast cells (black arrowhead), mononucleated cytotrophoblast cells (white arrowhead), and intervillous space (iv) are identified. Photomicrographs depict in situ hybridization positive for leptin (B), OB-R_L (C), and OB-R_S (D) in baboon placental villous tissue. Taken from Henson et al. (1999) and Green et al. (2000). Reproduced with the permission of The Endocrine Society and The Society for Experimental Biology and Medicine.

$OB-R_L$, and $OB-R_S$ in specific placental cell types. As shown in Figure 15.1A (Henson et al., 1999; Green et al., 2000), multinucleated syncytiotrophoblasts formed a continuous surface covering of placental villi (black arrowhead) in early pregnancy, as determined by hematoxylin and eosin staining. Few mononucleated cytotrophoblasts (white arrowhead) were present. Specific in situ hybridization to leptin (Figure 15.1B) determined transcript expression to be predominantly localized in trophoblast cells surrounding villi. Subjectively, expression intensity appeared be greatest in early pregnancy, which agreed with the enhanced abundance of leptin transcripts at that time (Henson et al., 1999). In situ hybridization to $OB-R_L$ and $OB-R_S$ also determined transcript expression for long (Figure 15.1C) and short (Figure 15.1D) isoforms of the leptin receptor. Therefore, as in human placental villous tissue (Henson et al., 1998), mRNA transcripts for both leptin and the leptin receptor in the baboon were located mainly in the trophoblast, stressing the potential for autocrine and/or paracrine signaling mechanisms at the maternal–fetal interface. In addition, omental and subcutaneous adipose tissues were collected at cesarean section from baboons on days 60, 100, and 160 of pregnancy (Green et al., 2000), with a resurgent corpus luteum (CL), decidual tissue, and a sample of amniochorion also collected on day 160. In order to establish the potential of leptin as a regulator of fetal development, fetal brain (hypothalamic region) was also collected. Utilizing human leptin receptor primers, transcripts for both $OB-R_L$ and $OB-R_S$ mRNAs were detected by Reverse Transcriptase- Polymerase Chain Reaction (RT-PCR) in omental and subcutaneous adipose tissue throughout pregnancy and in CL, decidua, amniochorion, and fetal brain near term. Semiquantitative comparison of transcript abundance, in relation to constitutively expressed glyceraldehyde-3-phosphate dehydrogenase (GAPDH), suggested no differences in transcript abundance, in placenta or adipose tissue, with advancing gestation. Transcripts for the short form were expressed in greater abundance than the long form in all tissues, which agreed with prior studies that concluded that $OB-R_S$ was present in greater abundance in peripheral tissues (lung, kidney, etc.) than $OB-R_L$ (Tartaglia, 1997).

15.3. LEPTIN REGULATION BY STEROID HORMONES

As in rodents, leptin appears to help regulate the onset of human reproductive function (Friedman and Halaas, 1998; Foster and Nagatani, 1999), as increases in peripheral levels are associated with maturity (Ahmed et al., 1999; Clayton and Trueman, 2000) and the onset of menarche (Matkovic et al., 1997). Even when relative adiposities are controlled for (Fruehwald-Schultes et al., 1998), significant differences in peripheral leptin concentrations occur simply as a consequence of gender, with women of reproductive age evidencing higher levels than men (Hickey et al., 1996; Castracane et al., 1998; Moschos et al., 2002). Gender-linked differences in leptin production might result from the enhanced potential of tissues taken from females (Williams et al., 2000), to synthesize and release leptin (Menendez et al., 2000). In addition, serum leptin levels in young women are greater than in postmenopausal women (Shimizu et al., 1997), and levels in female neonates (Matsuda et al., 1997; Helland et al., 1998; Gomez et al., 1999) and fetuses in utero (Jaquet et al., 1998; Yang and Kim, 2000) are higher than in male counterparts, perhaps suggesting a link with estrogen (Butte et al., 1997; Hardie et al., 1997). It might follow, therefore, that the maternal hyperleptinemia noted in human pregnancy could result from the stimulation of maternal adipose tissue by gestational hormones (Sivan et al., 1998; Kronfeld-Schor et al., 2000; Linnemann et al., 2001). Thus, placental estrogen biosynthesis increases with advancing gestation and is responsible for the regulation of hormones important for pregnancy maintenance (Pepe and Albrecht, 1995; Henson, 1998), while E_2 administration has been shown to enhance the expression of leptin mRNA transcripts and leptin secretion by adipocytes, both in vitro (Murakami et al., 1995; Slieker et al., 1996; Kronfeld-Schor et al., 2000) and in vivo (Brann et al., 2002). Similarly, E_2-stimulated expression of leptin in isolated rat adipocytes was inhibited by an estrogen receptor antagonist, while coincubation with the transcriptional inhibitor, actionomycin D, prevented E_2-induced increases in leptin mRNA (Machinal et al., 1999). Leptin and E_2 levels rise and fall at similar times during the human menstrual cycle (Mannucci et al., 1998; Cella et al., 2000) and E_2, perhaps in cooperation with progesterone (Messinis et al., 2001), has been reported to enhance serum leptin concentrations in women. In rats, ovariectomy diminished leptin gene expression in subcutaneous and retroperitoneal white adipose tissue and caused a decline in serum leptin levels (Shimizu et al., 1997; Yoneda et al., 1998; Chu et al., 1999; Machinal et al., 1999) while leptin expression in mesenteric white adipose tissue was enhanced (Shimizu et al., 1997). E_2 administration reversed these effects. As in the rodent, ovariectomy reduced serum leptin levels in humans (Messinis et al., 1999).

The disparate effects of estrogen on leptin gene transcription and leptin synthesis, therefore, appear to be both tissue- and species-specific, as estrogen does not enhance *ob* gene expression in ovariectomized swine (Qian et al., 1999) at all and may even precipitate a general transcriptional decline among avians (Ashwell et al., 1999). The apparently divergent effects of estrogen have been further addressed by reports (Lindheim et al., 2000; Yamada et al., 2000; Unkila-Kallio et al., 2001) that leptin levels in women were not affected by the relatively small changes in estrogen concentrations associated with normal menstrual cycles. However, large estrogen increases resulting from ovulation induction served to up-regulate leptin synthesis and it has been proposed that estrogen replacement therapy (ERT) may act to enhance serum leptin levels in postmenopausal women. In this capacity, some investigators (Elbers et al., 1999; Lavoie et al., 1999; Konukoglu et al., 2000) have reported that serum leptin levels increase in postmenopausal women receiving ERT, while others (Kohrt et al., 1996; Haffern et al., 1997; Castracane et al., 1998; Gower et al., 2000) have failed to detect an estrogen effect. Different estrogen dosages, administration regimens, or differences in the ages and relative adiposities of patients may account for differences in reported results.

During pregnancy, estrogen has been suggested to regulate leptin expression by acting on a portion of the estrogen response element (ERE; Kumar and Chambon, 1998) in the leptin promoter (Machinal et al., 1999; O'Neil et al., 2001a). It may be by this mechanism that leptin secretion by human cytotrophoblasts was dose-dependently potentiated by E_2 in vitro (Chardonnens et al., 1999). The presence of estrogen receptor in primate trophoblast (Albrecht et al., 2000) suggests that, as in adipose tissue (Mystkowski and Schwartz, 2000), this probably constitutes an estrogen receptor-mediated effect.

Estrogen may affect leptin transcription as a consequence of tissue type (Machinal et al., 1999; Villafuerte et al., 2000; Li et al., 2001) or in respect to the endocrine environment from which it was collected (Casabiel et al., 2001; Martin et al., 2002). Therefore, we hypothesized that enhanced maternal estrogen levels (Henson et al., 1996) in baboon pregnancy would act to enhance leptin transcription in adipose tissue and result in elevated leptin concentrations. In order to investigate this, we collected venous blood samples and subcutaneous (SAT) and omental (OAT) adipose tissues from pregnant baboons in early (day 60, $n = 5$), mid (day 100, $n = 5$), and late (day 160, $n = 5$) gestation (O'Neil et al., 2001b). Venous blood and SAT was also collected from nonpregnant baboons in the mid luteal phase of the menstrual cycle ($n = 5$). E_2 was quantitated in serum by RIA and leptin mRNA transcripts were quantitated in adipose tissue by competitive RT-PCR. Leptin concentrations in SAT and OAT homogenates were quantitated by RIA. As expected, E_2 concentrations were lowest in cycling animals (0.06 ± 0.02 ng/ml) and increased ($p < 0.01$) with advancing gestational age (4.17 ± 0.87 ng/ml on day 160). However, the abundance of leptin transcripts in both SAT and OAT (data not shown) was relatively unchanged ($p > 0.05$) in regard to the change in pregnancy status or to advancing gestation. Although leptin mRNA concentrations were not different, tissue leptin concentrations (Figure 15.2) in SAT were higher ($p < 0.02$) throughout pregnancy than in nonpregnant baboons. Leptin increased in both SAT and OAT with advancing gestation, further suggesting the potential for an estrogen effect.

The baboon possesses a true maternal-fetoplacental unit, which as in the human, relies on androgen precursors from the fetal adrenal gland for optimal placental estrogen synthesis (Pepe and Albrecht, 1995; Henson, 1998). The surgical removal of the fetus (fetectomy), inhibits placental estrogen production by the syncytiotrophoblast and reduces maternal serum E_2 concentrations dramatically. Although in the rhesus monkey fetectomy may not significantly affect maternal estrogen levels, possibly due to an enhancement in maternal adrenal function in late gestation (Nathanielsz et al., 1992), removal of the fetus in mid baboon pregnancy clearly results in a dramatic decline in maternal estrogen concentrations (Albrecht et al., 1980, 1991; Albrecht and Pepe, 1985; Henson, 1998). Therefore, placental villous tissue, OAT, and SAT were collected from baboons in late (day 160, $n = 5$) pregnancy (O'Neil et al., 2001b). Estrogen production was inhibited in

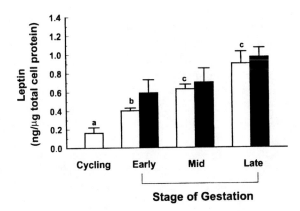

Figure 15.2. Leptin levels (mean $\pm$ SEM) in subcutaneous adipose tissue from nonpregnant cycling baboons ($n = 5$); and in subcutaneous (open bars) and omental (closed bars) tissue at early ($n = 5$), mid ($n = 5$), and late ($n = 5$) baboon gestation. Different lowercase letters indicate significant differences (ab, $p < 0.02$; ac, $p < 0.01$). Reproduced from O'Neil et al. (2001b), with the permission of The Endocrine Society.

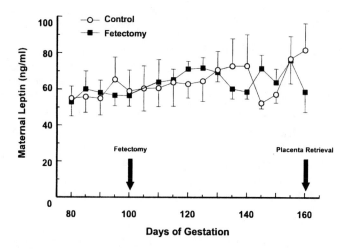

Figure 15.3. Maternal serum leptin concentrations (mean $\pm$ SEM) with advancing gestation in both intact ($n = 5$) and fetectomized ($n = 5$) baboons. Reproduced from O'Neil et al. (2001b), with the permission of The Endocrine Society.

some pregnant baboons at day 100 ($n = 5$) by fetectomy. Placentae were left in situ until day 160 of gestation when, following laparotomy and hysterotomy, they were retrieved and SAT and OAT were again collected. Maternal peripheral blood samples were drawn at regular intervals from days 80 to 160 of gestation. As we have previously reported (Albrecht et al., 1991; Putney et al., 1991), fetectomy was again associated with a marked decrease in maternal E_2 concentrations to a mean of 0.49 ± 0.04 ng/ml following fetectomy, or 13 percent of that in intact controls (3.72 ± 0.25 ng/ml). Maternal serum leptin levels, however, did not differ ($p > 0.05$) between intact and fetectomized baboons (Figure 15.3). Leptin mRNA transcripts in placental villous and maternal adipose tissues were quantitated by competitive RT-PCR. Leptin transcript abundance declined only modestly ($p > 0.05$) in OAT (Figure 15.4B), while in SAT (Figure 15.4C), leptin mRNA declined from 49.8 ± 9.8 to 5.8 ± 2.6 attomoles/μg total RNA as a consequence of fetectomy. Divergently, the abundance of leptin mRNA transcripts increased almost 3-fold in placental villous tissue (Figure 15.4A) from 0.4 ± 0.1 to 1.1 ± 0.3 as a result of fetectomy. Additionally, leptin protein was quantitated by RIA in adipose and placental tissue homogenates, from tissues collected on day 160. As also shown in Figure 15.4, leptin levels were similar in OAT, 0.60 ± 0.15 ng leptin/μg total cell protein in fetectomized baboons, as compared to 0.97 ± 0.10 ng leptin/μg total cell protein in pregnancy-intact controls. In SAT, leptin levels in fetectomized baboons were approximately one half that ($p < 0.05$) of controls (0.90 ± 0.13 ng leptin/μg total cell protein), while in placental villous tissue, levels were 3-fold higher in fetectomized animals (0.28 ± 0.08 ng leptin/μg total cell protein) compared to baboons with intact pregnancies (0.09 ± 0.01 ng leptin/μg total cell protein).

Collectively, results in the baboon strongly suggested the existence of (1) tissue-specific regulation of the leptin gene by estrogen and/or androgens, as evidenced by the decline in leptin mRNA in adipose tissue and the increase in placenta, following fetectomy; and (2) the existence of post-transcriptional (translational) mechanisms, possibly

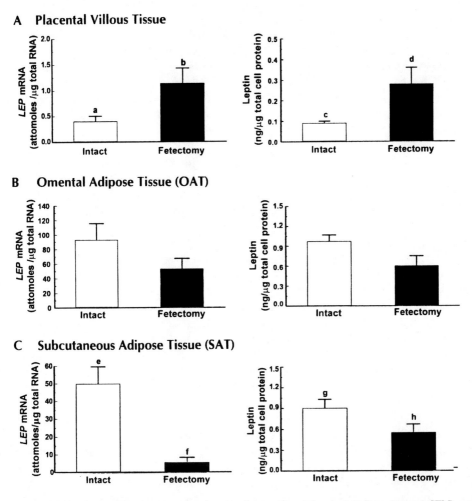

Figure 15.4. Leptin (LEP) mRNA transcript abundance and protein concentrations (mean ± SEM) upon placental retrieval (day 160) in placental villous tissue (A), omental adipose tissue (B), and subcutaneous adipose tissue (C) in intact control ($n = 5$) and fetectomized ($n = 5$) baboons. Significant differences (ab, cd, gh, $p < 0.05$; ef, $p < 0.003$). Reproduced from O'Neil et al. (2001b), with the permission of The Endocrine Society.

mediated by estrogen, resulting in enhanced leptin production. Although, following fetectomy, we noted quantitative, tissue-specific changes in leptin production following fetectomy, maternal serum leptin levels remained the same as pregnancy-intact controls. Therefore, although leptin expression declined in SAT, increased placental expression suggested a compensatory mechanism. Results reveal a tissue-specific regulatory mechanism, with estrogen stimulating leptin production in adipose tissue and inhibiting it in the trophoblast. Of course, the potential for divergent transcriptional regulation in placenta and adipose tissue is known to exist, as evidenced by the presence of a functional enhancer for the leptin gene in choriocarcinoma cells, that is not active in adipocytes (Bi et al., 1997; Ebihara et al., 1997).

Characteristics inherent in the fetectomized baboon model, however, make it impossible to rule out the possible contributions of a viable fetus. Future studies, featuring the administration of estrogen to fetectomized baboons and the inhibition of estrogen synthesis in pregnancy-intact baboons, will be needed to fully elucidate the mechanism(s) involved. Our studies also identify leptin receptor in a number of pregnancy-specific tissues and suggest that they could play a role in regulating gestational leptin dynamics. Perhaps relatedly, a recent report by Lindell et al. (2001) revealed that a putative estrogen response element, close to the most frequently used transcriptional start sites of the leptin receptor gene in the rat hypothalamus, provided a possible mechanism by which estrogen may regulate the leptin receptor.

In vitro studies have also proved useful in elucidating regulatory mechanisms. Therefore, it is possible to directly determine gene activation in a cell line that is similar in many respects to the first trimester human cytotrophoblast; a cell type in which leptin release may be potentiated by estrogen (Chardonnens et al., 1999). In this capacity, we transfected JEG-3 choriocarcinoma cells with the pERE-2-luciferase reporter or p1774 leptin-luciferase reporter and co-transfected with estrogen receptor (ER)α to elicit estrogen responsiveness (O'Neil et al., 2001a). Because we have proposed that estrogen exerts divergent, tissue-specific effects, we also examined the effects of estrogen on *LEP* promoter activation in MCF-7 breast cancer cells, a cell type that naturally expresses the estrogen receptor. Because a portion of the ERE consensus sequence occurs in the *LEP* promoter region (Machinal et al., 1999; O'Neil et al., 2001a), we hypothesized that *LEP* promoter activation by estrogen could be inhibited by an estrogen receptor antagonist. In the first experiment, we transfected JEG-3 choriocarcinoma cells with both leptin luciferase (p1774) and ERα. In order to ensure that cells were not estrogen responsive without the addition of ER, cells were also transfected with ERE-luciferase and a vector plasmid, without ERα. In this negative control, no luciferase activity was detected. Figure 15.5A illustrates the effects of phorbol myristate acetate (PMA), E_2, and the ER antagonists, ICI_{182780} (ICI) and 4-hydroxytamoxifen (OHT) on luciferase activity in cells co-transfected with pERE-2-luciferase and ERα. As expected, E_2 stimulated luciferase activity ($p < 0.05$) and was inhibited by both ICI and OHT. Figure 15.5B shows the effects of PMA, E_2, and the estrogen receptor antagonists on JEG-3 cells co-transfected with p1774 and ERα. E_2 stimulated luciferase activity ($p < 0.05$), which was inhibited by the antiestrogens. This demonstrated the potential for estrogen to enhance leptin transcription, as well as for this effect to be blocked by an antiestrogen. The results of this experiment represent an estrogen-induced transcriptional enhancement of leptin in a cell type that is similar in many respects to normal cytotrophoblast, but different in many structural and endocrine respects from normal syncytiotrophoblast (Ringler and Strauss, 1990). Figure 15.5C shows the effects of PMA, E_2, and ICI on pERE-2-luciferase activity MCF-7 cells. As expected, E_2 stimulated pERE-2-luciferase activity ($p < 0.05$), which was inhibited by co-incubation with ICI. Figure 15.5D shows the effects of PMA, E_2, and ICI on p1774 activity. Although the LEP promoter was stimulated ($p < 0.05$) by PMA, E_2 and ICI had no effect. Taken together, results indicate that estrogen can activate the LEP promoter in choriocarcinoma cells through ERα and suggest that the regulation of leptin biosynthesis may depend upon a functional ER. Results also imply that leptin gene promoter activation may depend upon co-activators present in only some types of cells, further attesting to the tissue-specificity of mechanisms regulating leptin production.

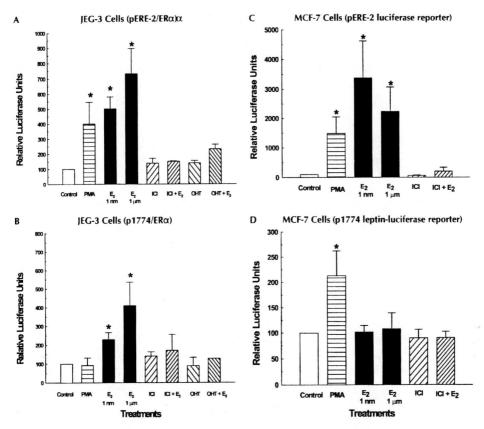

Figure 15.5. Panel A—Relative luciferase activity in JEG-3 cells co-transfected with the ERE-luciferase reporter and ERα, treated with 20 ng/ml PMA, 1 nM or 1 μM E_2, 100 nM ICI, 100 nM ICI + 1 nM E_2, 100 nM OHT or 100 nM OHT + 1 nM E_2. Panel B—Relative luciferase activity in JEG-3 cells co-transfected with the ERE-luciferase reporter and ERα (same treatments as in Panel A). Panel C—Relative luciferase activity in MCF-7 cells transfected with the ERE-luciferase reporter, treated with 20 ng/ml PMA, 1 nM or 1 μM E_2, 100 nM ICI, or 100 nM ICI + 1 nM E_2. Panel D—Relative luciferase activity in MCF-7 cells transfected with the leptin-luciferase reporter (same treatments as in Panel C). *Values (means ± SEM of 3 cultures) are different from that of control ($P < 0.05$). Reproduced from O'Neil et al. (2001a), with permission from Elsevier Science.

As estrogen is associated in many experimental models with the enhancement of leptin transcription, synthesis, and release, androgens are associated with leptin inhibition (Mystkowski and Schwartz, 2000; Reidy and Weber, 2000). Thus, in healthy men testosterone concentrations were negatively correlated with leptin in serum (Thomas et al., 2000; Soderberg et al., 2001), while in prostate cancer patients treated with a nonsteroidal antiandrogen, leptin levels rose dramatically (Nowicki et al., 2001). Leptin levels in untreated patients were no different from that of healthy controls. These findings are in keeping with previous in vivo human studies (Luukkaa et al., 1998; Hislop et al., 1999; Watanobe and Suda, 1999) and in vitro experiments in the rat (Kristensen et al., 1999) that indicated a negative effect of testosterone on leptin production. In addition to studies documenting the negative association between androgen levels and leptin, testosterone administration directly inhibited leptin concentrations in both prepubertal boys (Adan et al., 1999) and hypogonadal men (Jockenhovel et al., 1997), an effect that may be mimicked to some degree by the "weaker" androgens. Thus, administration of 50 mg dehydroepiandrosterone (DHEA)/day to women exhibiting adrenal insufficiency led to a significant decline in serum leptin levels compared to placebo treated controls (Callies et al., 2001). The relative effects (stimulatory or inhibitory) of the high levels of aromatizable fetal androgens in primate pregnancy (Henson, 1998; Albrecht et al., 2000) are unknown.

15.4. ROLES OF LEPTIN IN PREGNANCY

As we (Henson and Castracane, 2000, 2002) and others (Kratzsch et al., 2000; Caprio et al., 2001; Hoggard et al., 2001a; Linnemann et al., 2001; Reitman et al., 2001; Sagawa et al., 2002) have reviewed, a number of physiological roles are indicated for leptin in human pregnancy. Among these is the regulation of fetal and placental growth (Kiess et al., 1998; Mantzoros and Moschos, 1998). Harigaya et al. (1997), reported that leptin levels in infants whose birth weights were classed as large for gestational age were significantly higher than those whose body weights were appropriate for gestational age, but still greater than for infants that were small for gestational age. Others (Koistinen et al., 1997; Gomez et al., 1999; Hoggard et al., 2001a) reported similar results and added that leptin concentrations in umbilical cord blood were correlated with both birth weights and placental weights, although apparently unrelated to levels of other polypeptide regulators of fetal growth (Christou et al., 2001). Similarly, decreases in placental leptin mRNA were associated with decreased leptin concentrations in umbilical vein blood from fetal growth-restricted pregnancies (Lea et al., 2000). Most specifically, as proposed by Hoggard and coworkers, placental leptin directly influences fetal adiposity in response to a fetal demand that is relative to placental supply (Hoggard et al., 2001a). These observations, in combination with the knowledge that venous cord blood leptin concentrations are higher than arterial concentrations (Yura et al., 1998; Reitman et al., 2001), underscore a role for placental leptin in regulating normal fetal development directly. Although early reports of leptin as a fetal growth regulator were of a mostly associative nature, subsequent investigations have provided better and more direct evidence. Therefore, studies of a third trimester twin pregnancy reported that a growth-restricted twin evidenced lower placental and cord blood leptin than its normal size sibling (Lea et al., 2000). This study also indicated that low levels of cord blood leptin in low birth weight infants were a direct reflection of low concentrations in placenta. Subsequently, observations from second trimester monochorionic twin pregnancies revealed that fetal and cord leptin levels in normal size fetuses were twice that of their growth restricted twins (Soorana et al., 2001a,b). It is interesting to note that children born with intrauterine growth restriction (IUGR) retain low leptin levels into adulthood, suggesting a permanent alteration in adipose tissue (Jaquet et al., 2001). Leptin, originating in the placenta (Masuzaki et al., 1997) and fetus (Schubring et al., 1997; Clapp and Kiess, 1998; Atanassova and Popova, 2000; Lepercq et al., 2001), may also potentiate growth indirectly by modulating growth hormone (GH) secretion (Carro et al., 1997), as proposed in the rat (LaPaglia et al., 1998), via hypothalamic regulation of GH-releasing hormone (Carro et al., 1999). Therefore, because leptin receptor mRNA has been identified in the human fetal anterior pituitary and administration of recombinant human leptin stimulated GH secretion from fetal pituitary cultures (Shimon et al., 1998), it might be postulated that placental GH, stimulated by leptin, could facilitate fetoplacental growth (Alsat et al., 1998) via this mechanism. Reports associating the polypeptide with fetal adiposity are joined by those correlating leptin levels with infant length and head circumference (Ong et al., 1999), those stating that postnatal administration restores the depressed brain weights of leptin-deficient *ob/ob* neonates (Steppan and Swick, 1999; Ahima and Flier, 2000; Hoggard et al., 2001b), and those reporting that leptin directly stimulates fetal (Ogueh et al., 2000) and adult (Burguera et al., 2001) bone growth. As proposed by Gordeladze et al. (2001), this latter leptin-mediated effect could proceed via numerous mechanisms affecting bone cell growth and differentiation. Indeed, leptin administration is known to dramatically stimulate bone growth in leptin-deficient *ob/ob* mice (Steppan et al., 2000) and a correlation between leptin and bone density has been observed in nonobese women (Pasco et al., 2001).

Physiological roles for the polypeptide are not limited to those regulating conceptus growth, as leptin has also been proposed to regulate angiogenesis in genetically obese rats (Sierra-Honigmann et al., 1998) and cultured human and porcine vascular endothelial cells (Bouloumie et al., 1998). Furthermore, Buchbinder et al. (2000) reported that leptin receptor mRNA was enhanced in the uterine artery of ovariectomized sheep treated with estrogen. These observations support the proposal that leptin stimulates blood vessel formation in developing tissues, either alone or in concert with other angiogenic factors, such as fibroblast growth factor II and vascular endothelial growth factor (Cao et al., 2001). Additionally, leptin may also directly affect the proliferation of fetal pancreatic islet cells, which express functional receptors, and may play a role in determining islet cell mass at birth (Islam et al., 2000). Finally, leptin could have direct effects on general organ maturation in the fetus, as leptin infusion in chronically instrumented ovine fetuses is associated with increased swallowing and urine output in utero (Roberts et al., 2001).

In addition to leptin's effects on conceptus growth and development, identification of leptin and leptin receptor in the syncytiotrophoblast may also relate to the placenta's function as an endocrine organ (Ashworth et al., 2000; Gonzalez et al., 2000a,b; Henson and Castracane, 2000; Kratzsch et al., 2000; Caprio et al., 2001), suggesting an assortment of regulatory mechanisms in an organ that produces an array of vital hormones throughout gestation (Reitman et al., 2001). A well-defined interrelationship of trophoblastic polypeptide and steroid hormones is required to maintain primate pregnancy (Pepe and Albrecht, 1995). Thus, hCG and enhanced concentrations of estrogens and progesterone in the maternal peripheral circulation are of placental origin (Solomon, 1997), with leptin, of both placental and maternal adipose origin, also increasing with advancing gestation. Because cultured first trimester trophoblast cells produce leptin in considerable amounts (Meisser et al., 1999) and the addition of recombinant leptin enhanced hCG release, (Chardonnens et al., 1999), the fact that leptin levels in women suffering spontaneous first trimester abortions were abnormally low (Lage et al., 1999; Laird et al., 2001) may have special significance. Finally, recombinant leptin infused into the fetal circulation inhibited activation of the hypothalamic–pituitary–adrenal (HPA) axis in late ovine pregnancy, suggesting that mechanisms controlling the initiation of labor might be fine-tuned by such a metabolic trigger that originates in a maturing placenta or fetus (Howe et al., 2000). It has been speculated that the depressed leptin levels seen in premature human infants might similarly reflect an immaturity of the HPA axis (Spear et al., 2001), an effect potentially reflected in ovine pregnancy, in which an ontogenic rise in fetal plasma leptin concentrations was attributed, in part, to the normal pre-partum rise in plasma cortisol (Forhead et al., 2002). It is interesting to note, therefore, that glucocorticoid administration enhances leptin mRNA transcript abundance, as well as leptin secretion, in cultured human trophoblasts (Coya et al., 2001).

Leptin and leptin receptor expression in human placenta and uterine endometrium, and the suggestion that endometrial leptin secretion is enhanced in the presence of a viable blastocyst, also implicates leptin in implantation (Gonzalez et al., 2000a; Dominguez et al., 2002; Kawamura et al., 2002). This process, which could involve the regulation of matrix metalloproteinases in controlling the invasive properties of extravillous cytotrophoblasts (Castellucci et al., 2000), represents yet another potential role for leptin (Gonzalez et al., 2000a). In pre-eclampsia, a condition that is characterized by shallow endometrial invasion and the sudden onset of maternal hypertension, maternal leptin concentrations are often elevated. Even when pre-eclampsia is not manifested, shallow invasion is intimately associated with IUGR (Hoang et al., 2001). Although it is generally agreed that maternal and fetal leptin concentrations are enhanced in pre-eclampsia (Mise et al., 1998; McCarthy et al., 1999; Williams et al., 1999; Anim-Nyame et al., 2000; Teppa et al., 2000; Linnemann et al., 2001; Odegard et al., 2002), some investigators have reported no significant changes in mean leptin levels (Martinez-Abundis et al., 2000; Laml et al., 2001b). This disparity may reflect differences in the stage of pregnancy when women develop the condition, or differences in the criteria used for diagnosis. Nevertheless, onset of pre-eclampsia is a sequelae of cellular hypoxia (Mise et al., 1998; Hytinantti et al., 2000a; Grosfeld et al., 2001), as is pregnancy-associated diabetes (Hytinantti et al., 2000 b), a condition which is itself characterized by enhanced leptin synthesis (Stock and Bremme, 1998; Lepercq et al., 1999; Lewandowski et al., 1999).

In the mouse, a soluble isoform of the leptin receptor ($OB-R_e$) is released into the maternal circulation where it may potentiate leptin resistance by binding leptin and rendering it biologically unavailable (Gavrilova et al., 1997; Hoggard et al., 1997). A substantial enhancement in the concentration of this protein has been proposed to explain the increase in leptin, an appetite suppressant, which occurs during human pregnancy, a period normally perceived to be one of enhanced nutritional need (Hoggard et al., 1998). Although studies have detected a protein in the human (Diamond et al., 1997) of identical size as one we have observed in the baboon (Edwards et al., 2002), a comparably large increase has not been reported to occur in pregnancy, prompting speculation that human pregnancy may more closely resemble pregnancy in the rat, a species that exhibits only a modest gestational increase (Gavrilova et al., 1997). At least two isoforms of the soluble leptin receptor may functionally bind leptin in the human peripheral circulation (Lambert et al., 2001) and we have identified similar candidates in the pregnant baboon (Edwards et al., 2002). Increasing concentrations of soluble leptin receptor that are generated by tissues specific to pregnancy (placenta, amniochorion, decidua) may be at least partially responsible for maternal hyperleptinemia. In addition to the circulating receptor potentially acting to inhibit leptin clearance (Huang et al., 2001) from the maternal circulation, Schulz et al. (2000) have proposed the receptor to be an active participant in the regulation of fetal growth. Thus, in respect to their characterization of leptin and soluble receptor in maternal, fetal, and umbilical sera and their identification of soluble receptor in placenta by immunocytochemistry and immunoblotting, they proposed that while membrane-bound leptin receptors are actively involved in the autocrine regulation of placental leptin production, the soluble form of the receptor serves as an active transport vehicle for leptin to developing fetal tissues.

15.5. SUMMARY OF HYPOTHESES CONCERNING THE REGULATION OF LEPTIN AND ITS ROLES IN PRIMATE PREGNANCY

At this stage of investigation, it is still difficult to accurately predict all of leptin's roles, or characterize completely the mechanisms regulating its production throughout gestation. However, in respect to animal studies, observations made during human pregnancy and the insight provided by in vitro models, we can propose certain possible relationships (summarized in Figure 15.6). Androgen precursor (of both maternal and fetal adrenal origin) facilitates estrogen production in the placental trophoblast. As pregnancy progresses and the influence of the fetal adrenal grows, fetal androgens may down-regulate leptin production in fetal adipose tissue. Placental estrogen, however, appears to stimulate leptin production in early pregnancy trophoblast. Although this may account for enhanced placental transcript abundance at this time, this increase is not directly reflected by commensurately elevated leptin concentrations in the maternal peripheral circulation; as it is possible that much placental leptin is shunted to the fetus via the umbilical vasculature. This influx of placental leptin may be responsible for promoting fetal growth and facilitating angiogenesis in early gestation. Net increases in fetal leptin levels with advancing gestational age may occur simply as a result of increasing adipose mass, or as a result of the enhanced contribution of a growing trophoblast. Placental leptin production, specifically stimulated in cytotrophoblast cells early in pregnancy helps to stimulate hCG synthesis/release, which wanes in the second trimester. Placental estrogen, produced in increasingly greater amounts with advancing pregnancy, stimulates maternal adipose

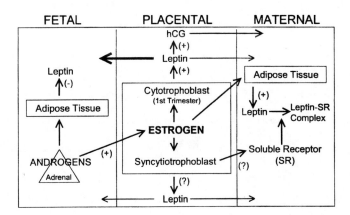

Figure 15.6. Schematic representation of potential leptin-associated pathways in the primate maternal–fetoplacental unit.

tissue, perhaps bearing primary responsibility for the maternal hyperleptinemia that is characteristic of pregnancy. Changing levels of estrogen, that are commensurate with the progressive maturation of the placental trophoblast, may affect leptin production by the syncytiotrophoblast, but to date this has not been conclusively demonstrated. Circulating leptin receptor, a product of the syncytiotrophoblast, binds leptin in the maternal peripheral circulation and may promote a state of pregnancy-specific leptin resistance and/or influence leptin transport to the developing fetus. Whether the production of this protein is regulated by estrogen is also subject to debate. Finally, maternal estrogen levels decrease rapidly with delivery of the placenta, and leptin production by maternal adipose tissue declines commensurately, perhaps stimulating maternal appetite in preparation for lactation. Although much has been learned in a comparatively short time about leptin as a gestational hormone, it is anticipated that ongoing research will continue to delineate regulatory mechanisms and better define the physiological roles of leptin in pregnancy.

ACKNOWLEDGMENTS

The support of NIH RR00164 to the Tulane National Primate Research Center is gratefully acknowledged. We also express our sincere appreciation to Dr. April G. O'Quinn, Chair, Elease Bradford, Nathlynn Dellande, Dr. Deborah E. Edwards, Dr. Amy E. Green, Dr. Jennifer S. O'Neil, Dr. Wenliang Shi, and Kenneth F. Swan of the Department of Obstetrics and Gynecology, Tulane University Health Sciences Center; Dr. Rudolf S. Bohm, Dr. Jeanette Purcell, and Dr. Marion Ratterree of the Tulane National Primate Research Center; and Terry Gimpel of the Department of Obstetrics and Gynecology, Texas Tech University Health Sciences Center, for their significant contributions to studies performed in our laboratories.

REFERENCES

Adan, L., Bussieres, L., Trivin, C., Souberbielle, J. C., and Brauner, R. (1999). Effect of short-term testosterone treatment on leptin concentrations in boys with pubertal delay. *Horm. Res., 52,* 109–112.

Ahima, R. S., and Flier, J. S. (2000). Leptin. *Ann. Rev. Physiol., 62,* 413–437.

Ahmed, M. L., Ong, K. K. L., Morrell, D. J., Cox, L., Drayer, N., Perry, L., Precce, M. A., and Dunger, D. B. (1999). Longitudinal study of leptin concentrations during puberty: sex differences and relationships to changes in body composition. *J. Clin. Endocrinol. Metab., 84,* 899–905.

Albrecht, E. D. and Pepe, G. J. (1985). The placenta remains functional following fetectomy in baboons. *Endocrinology, 116,* 843–845.

Albrecht, E. D., Haskins, A. L., and Pepe, G. J. (1980). The influence of fetectomy upon the serum concentrations of progesterone, estrone, and estradiol in baboons. *Endocrinology, 107,* 766–770.

Albrecht, E. D., Henson, M. C., and Pepe, G. J. (1991). Regulation of placental low density lipoprotein uptake in baboons by estrogen. *Endocrinology, 128,* 450–458.

Albrecht, E. D., Aberdeen, G. W., and Pepe, G. J. (2000). The role of estrogen in the maintenance of primate pregnancy. *Am. J. Obstet. Gynecol., 182,* 432–438.

Alsat, E., Guibourdenche, J., Couturier, A., and Evain-Brion, D. (1998). Physiological role of human placental growth hormone. *Molec. Cell. Endocrinol., 140,* 120–127.

Amato, P., Zwain, I. H., and Yen, S. S. C. (1997). Identification and expression of leptin receptor in human placenta: a possible role in reproduction. *Proceedings of the 79th Annual Meeting of the Endocrine Society,* Minneapolis, MN, p. 396.

Amico, J. A., Thomas, A., Crowley, R. S., and Burmeister, L. A. (1998). Concentrations of leptin in the serum of pregnant, lactating, and cycling rats and of leptin messenger ribonucleic acid in rat placental tissue. *Life Sci.*, *63*, 1387–1395.

Anim-Nyame, N., Soorana, S. R., Steer, P. J., and Johnson, M. R. (2000). Longitudinal analysis of maternal plasma leptin concentrations during normal pregnancy and pre-eclampsia. *Hum. Reprod.*, *15*, 2033–2036.

Ashwell, C. M., Czerwinski, S. M., Brocht, D. M., and McMurtry, J. P. (1999). Hormonal regulation of leptin expression in broiler chickens. *Am. J. Physiol.*, *276*, R226–R232.

Ashworth, C. J., Hoggard, N., Thomas, L., Mercer, J. G., Wallace, J. M., and Lea, R. G. (2000). Placental leptin. *Rev. Reprod.*, 5, 18–24.

Atanassova, P. and Popova, L. (2000). Leptin expression during the differentiation of subcutaneous adipose cells of human embryos in situ. *Cell. Tiss. Org.*, *166*, 15–19.

Bajoria, R., Soorana, S. R., Ward, B. S., and Chatterjee, R. (2001). Prospective function of placental leptin at maternal–fetal interface. *Placenta*, *23*, 103–115.

Bi, S., Gavrilova, O., Gong, D. W., Mason, M. M., and Reitman, M. (1997). Identification of a placental enhancer for the human leptin gene. *J. Biol. Chem.*, *272*, 30583–30588.

Bjorbaek, C., Uotani, S., da Silva, B., and Flier, J. S. (1997). Divergent signaling capacities of the long and short isoforms of the leptin receptor. *J. Biol. Chem.*, *272*, 32686–32695.

Bouloumie, A., Drexler, H. C. A., Lafontan, M., and Busse, R. (1998). Leptin, the product of *ob* gene, promotes angiogenesis. *Circ. Res.*, *83*, 1059–1066.

Brann, D. W., Wade, M. F., Dhandapani, M., Mahesh, V. B., and Buchanan, C. D. (2002). Leptin and reproduction. *Steroids*, *67*, 95–104

Buchbinder, A., Baker, R. S., Hirth, J. A., Mershon, J. L., and Clark, K. I. (2000). Expression of leptin receptor in the ovine uterine artery and its regulation by estrogen. *J. Soc. Gynecol. Invest.*, *7*(Suppl.), Abstract #749.

Burguera, B., Hofbauer, L., Thomas, T., Gori, F., Evans, G. L., Khosla, S., Riggs, B. L., and Turner, R. T. (2001). Leptin reduces ovariectomy-induced bone loss in rats. *Endocrinology*, *142*, 3546–3553.

Butte, N. F., Hopkinson, J. M., and Nicolson, M. A. (1997). Leptin in human reproduction: serum leptin levels in pregnant and lactating women. *J. Clin. Endocrinol. Metab.*, *82*, 585–589.

Callies, F., Fassnacht, M., Christoph van Vlijmen, J., Koehler, I., Huebler, D., Seibel, M. J., Ar, H. W., and Allolio, B. (2001). Dehydroepiandrosterone replacement in women with adrenal insufficiency: effects on body composition, serum leptin, bone turnover, and exercise capacity. *J. Clin. Endocrinol. Metab.*, *86*, 1968–1972.

Cao, R., Brakenheim, E., Wahlestedt, C., Thyberg, J., and Cao, Y. (2001). Leptin induces vascular permeability and synergistically stimulates angiogenesis with FGF-2 and VEGF. *Proc. Natl. Acad. Sci. USA*, *98*, 6390–6395.

Caprio, M., Fabbrini, E., Isidori, A. M., Avevsa, A., and Fabbri, A. (2001). Leptin in reproduction. *Trends Endocrinol. Metab.*, *12*, 65–72.

Carro, E., Senaris, R., Considine, R. V., Casanneva, F. F., and Dieguez, C. (1997). Regulation of in vivo growth hormone secretion by leptin. *Endocrinology*, *138*, 2202–2206.

Carro, E., Senaris, R. M., Seoane, L. M., Frohman, L. A., Arimura, A., Casanueva, F. F., and Dieguez, C. (1999). Regulation of growth hormone (GH)-releasing hormone and somatostatin on leptin-induced GH secretion. *Neuroendocrinology*, *69*, 3–10.

Casabiel, X., Pineiro, V., Vega, F., De La Cruz, L. F., Dieguez, C., and Casanueva, F. F. (2001). Leptin, reproduction and sex steroids. *Pituitary*, *4*, 93–99.

Castellucci, M., De Matteis, R., Meisser, A., Cancello, R., Monsurro, V., Islami, D., Sarzani, R., Marzioni, D., Cinti, S., and Bischof, P. (2000). Leptin modulates extracellular matrix molecules and metalloproteinases: possible implications for trophoblast invasion. *Molec. Hum. Reprod.*, *6*, 951–958.

Castracane, V. D. and Goldzieher, J. W. (1986a). Timing of the luteal-placental shift in the baboon (*Papio cynocephalus*). *Endocrinology*, *118*, 506–512.

Castracane, V. D. and Goldzieher, J. W. (1986b). The relationship of estrogen to placental steroidogenesis in the baboon. *J. Clin. Endocrinol. Metab.*, *62*, 1163–1166.

Castracane, V. D., Kraemer, R. R., Franken, M. A., Kraemer, G. R., and Gimpel, T. (1998). Serum leptin concentration in women: effect of age, obesity and estrogen administration. *Fertil. Steril.*, *70*, 472–477.

Cella, F., Giordano, G., and Cordera, R. (2000). Serum leptin concentrations during the menstrual cycle in normal-weight women: effects of an oral triphasic estrogen–progestin medication. *Eur. J. Endocrinol.*, *142*, 174–178.

Chardonnens, D., Cameo, P., Aubert, M. L., Pralong, F. P., Islami, D., Campana, A., Gaillard, R. C., and Bischof, P. (1999). Modulation of human cytotrophoblastic leptin section by interleukin-1 alpha and 17 beta-oestradiol and its effect on hCG secretion. *Mol. Hum. Reprod.*, 5, 1077–1082.

Chen, X., Lin, J., Hausman, D. B., Martin, R. J., Dean, R. G., and Hausman, G. J. (2000). Alterations in fetal adipose tissue leptin expression correlate with the development of adipose tissue. *Biol. Neonate*, 78, 41–47.

Chien, E. K., Hara, M., Rouard, M., Yano, H., Phillippe, M., Polonsky, K. S., and Bell, G. I. (1997). Increase in serum leptin and uterine leptin receptor messenger RNA levels during pregnancy in rats. *Biochem. Biophys. Res. Comm.*, 237, 476–480.

Christou, H., Connors, J. M., Ziotopoulou, M., Hatzidakis, V., Papathanassoglou, E., Ringer, S. A., and Matzoros, C. S. (2001). Cord blood leptin and insulin-like growth factor levels are independent predictors of fetal growth. *J. Clin. Endocrinol. Metab.*, 86, 935–938.

Chu, S.-C., Chou, Y.-C., Lin, J.-Y., Chen, C.-H., Shyu, J.-C., and Chou, F.-P. (1999). Fluctuation of serum leptin levels in rats after ovariectomy and the influence of estrogen supplement. *Life Sci.*, 64, 2299–2306.

Clapp, III, J. F. and Kiess, W. (1998). Cord blood leptin reflects fetal fat mass. *J. Soc. Gynecol. Invest.*, 5, 300–303.

Clayton, P. E. and Trueman, J. A. (2000). Leptin and puberty. *Arch. Dis. Child.*, 83, 1–4.

Considine, R. V. and Caro, J. F. (1997). Leptin and the regulation of body weight. *Int. J. Biochem. Cell. Biol.*, 29, 1255–1272.

Couce, M. E., Burquera, B., Paresi, J. E., Jensen, M. D., and Lloyd, R. V. (1997). Localization of leptin receptor in the human brain. *Neuroendocrinology*, 66, 145–150.

Coya, R., Gualillo, O., Pineda, J., del Carmen Garcia, M., de los Angeles Busturia, M., Ariel-Quiroga, A., Martul, P., and Senaris, R. M. (2001). Effects of cyclic 3′,5′-adenosine monophosphate, glucocorticoids, and insulin on leptin messenger RNA levels and leptin secretion in cultured human trophoblast. *Biol. Reprod.*, 65, 814–819.

Diamond, F. B., Jr., Eclair, D. C., Duckett, G., Jergensen, E. V., Shulman, D., and Root, A. W. (1997). Demonstration of a leptin binding factor in human serum. *Biochem. Biophys. Res. Commun.*, 233, 818–822.

Dominguez, F., Pellicer, A., and Simon, C. (2002). Paracrine dialogue in implantation. *Mol. Cell. Endocrinol.*, 186, 175–181.

Ebihara, K., Ogawa, Y., Isse, N., Mori, K., Tamura, N., Masuzaki, H., Kohno, K., Yura, S., Hosoda, K., Sagawa, N., and Nakao, K. (1997). Identification of the human leptin 5′ flanking sequences involved in the trophoblast-specific transcription. *Biochem. Biophys. Res. Commun.*, 241, 658–663.

Edwards, D. E., Swan, K. F., Castracane, V. D., and Henson, M. C. (2002). Effects of fetectomy and estrogen supplementation on soluble leptin receptor in baboon pregnancy. *Annual Meeting, Society for the Study of Reproduction*. Baltimore, MD. *Biol. Reprod.*, 66(Suppl. 1), Abstract #322.

Elbers, J. M., de Valk-de Roo, G. W., Popp-Snijders, C., Nicolaas-Merkus, A., Westerveen, E., Joenje, B. W., and Netelenbos, J. C. (1999). Effects of administration of 17 β-oestradiol on serum leptin levels in healthy postmenopausal women. *Clin. Endocrinol. (Oxf)*, 51, 449–454.

Fei, H., Okano, H. J., Li, C., Lee, G. H., Zhao, C., Darnell, R., and Friedman, J. M. (1997). Anatomic localization of alternatively spliced leptin receptors (Ob-R) in mouse brain and other tissues. *Proc. Natl. Acad. Sci., USA*, 94, 7001–7005.

Forhead, A. J., Thomas, L., Crabtree, J., Hoggard, N., Gardner, D. S., Giussani, D. A., and Fowden, A. L. (2002). Plasma leptin concentration in fetal sheep during late gestation: ontogeny and effect of glucocorticoids. *Endocrinology*, 143, 1166–1173.

Foster, D. L. and Nagatani, S. (1999). Physiological perspectives on leptin as a regulator of reproduction: role in timing of puberty. *Biol. Reprod.*, 60, 205–215.

Friedman, J. M. and Halaas, J. L. (1998). Leptin and the regulation of body weight in mammals. *Nature*, 395, 763–770.

Fruehwald-Schultes, B., Peters, A., Kern, W., Beyer, J., and Pfützner, A. (1998). Influence of sex differences in subcutaneous fat mass on serum leptin concentrations. *Diabetes Care*, 21, 1204–1205.

Garcia, M. D., Casanueva, F. F., Diguez, C., and Senaris, R. M. (2000). Gestational profile of leptin messenger ribonucleic acid (mRNA) content in the placenta and adipose tissues in the rat, and regulation of the mRNA levels of the leptin receptor subtypes in the hypothalamus during pregnancy and lactation. *Biol. Reprod.*, 62, 698–703.

Gavrilova, O., Barr, V., Marcus-Samuels, B., and Reitman, M. (1997). Hyperleptinemia of pregnancy associated with the appearance of a circulating form of the leptin receptor. *J. Biol. Chem.*, 272, 30546–30551.

Gomez, L., Carrascosa, A., Yeste, D., Potau, N., Rique, S., Ruiz-Cuevas, P., and Almar, J. (1999). Leptin values in placental cord blood of human newborns with normal intrauterine growth after 30–42 weeks of gestation. *Horm. Res.*, *51*, 10–14.

Gonzalez, R. R., Caballero-Campo, P., Jasper, M., Mercader, A., Devote, L., Pellicer, A., and Simon, C. (2000a). Leptin and leptin receptor are expressed in the human endometrium and endometrial leptin secretion is regulated by the human blastocyst. *J. Clin. Endocrinol. Metab.*, *85*, 4883–4888.

Gonzalez, R. R., Simon, C., Caballero-Campo, P., Norman, R., Chardonnens, D., Devote, L., and Bischof, P. (2000b). Leptin and reproduction. *Human Reprod. Update*, *6*, 290–300.

Gordeladze, J. O., Reseland, J. E., and Drevon, C. A. (2001). Pharmacological interference with transcriptional control of osteoblasts: a possible role for leptin and fatty acids in maintaining bone strength and body lean mass. *Curr. Pharm. Design*, 7, 275–290.

Gower, B. A., Nagy, T. R., Goran, M. I., Smith, A., and Kent, E. (2000). Leptin in postmenopausal women: influence of hormone therapy, insulin, and fat distribution. *J. Clin. Endocrinol. Metab.*, *85*, 1770–1775.

Green, A. E., O'Neil, J. S., Swan, K. F., Bohm, R. P., Ratterree, M. S., and Henson, M. C. (2000). Leptin receptor transcripts are constitutively expressed in placenta and adipose tissue with advancing baboon pregnancy. *Proc. Soc. Exp. Biol. Med.*, *223*, 362–366.

Grosfeld, A., Turban, S., Andre, J., Cauzac, M., Challier, J. C., Hauguel-de Mouzon, S., and Guerre-Millo, M. (2001). Transcriptional effect of hypoxia on placental leptin. *FEBS. Lett.*, *502*, 122–126.

Haffern, S. M., Mykkanen, L., and Stern, M. P. (1997). Leptin concentrations in women in the San Antonio heart study: effect of menopausal status and postmenopausal hormone replacement therapy. *Am. J. Epidemiol.*, *146*, 581–585.

Hardie, L., Trayhurn, P., Abramovich, D., and Fowler, P. (1997). Circulating leptin in women: a longitudinal study in the menstrual cycle and during pregnancy. *Clin. Endocrinol.*, *47*, 101–106.

Harigaya, A., Nagashima, K., Nako, Y., and Morikawa, A. (1997). Relationship between concentrations of serum leptin and fetal growth. *J. Clin. Endocrinol. Metab.*, *82*, 3281–3284.

Hassink, S. G., de Lancey, E., Sheslow, D. V., Smith-Kirwin, S. M., O'Connor, D. M., Considine, R. V., Opentanova, I., Dostal, K., Spear, M. L., Leef, K., Ash, M., Spitzer, A. R., and Funanage, V. L. (1997). Placental leptin: an important new growth factor in intrauterine and neonatal development? *Pediatrics*, *100/1/e1*, 1–6.

Helland, I. B., Reseland, J. E., Saugstad, O. D., and Drevon, C. A. (1998). Leptin levels in pregnant women and newborn infants: gender differences and reduction during the neonatal period. *Pediatrics*, *101/3/e12*, 1–5.

Henson, M. C. (1998). Pregnancy maintenance and the regulation of placental progesterone biosynthesis in baboon pregnancy. *Hum. Reprod. Update*, *4*, 389–405.

Henson, M. C. and Castracane, V. D. (2000). Leptin in pregnancy. *Biol. Reprod.*, *63*, 1219–1228.

Henson, M. C., and Castracane, V. D. (2002). Leptin: Roles and regulation in primate pregnancy. *Sem. Reprod. Med.*, *20*, 113–121.

Henson, M. C., Babischkin, J. S., Pepe, G. J., and Albrecht, E. D. (1988). Effect of the antiestrogen ethamoxytriphetol (MER-25) on placental low density lipoprotein uptake and degradation in baboons. *Endocrinology*, *122*, 2019–2026.

Henson, M. C., Pepe, G. J., and Albrecht, E. D. (1991). Regulation of placental low density lipoprotein uptake in baboons by estrogen: dose dependent effects of the anti-estrogen ethamoxytriphetol (MER-25). *Biol. Reprod.*, *45*, 43–48.

Henson, M. C., Pepe, G. J., and Albrecht, E. D. (1992). Developmental increase in placental low density lipoprotein uptake during baboon pregnancy. *Endocrinology*, *130*, 1698–1706.

Henson, M. C., Shi, W., Greene, S. J., and Reggio, B. C. (1996). Effects of pregnant human, nonpregnant human, and fetal bovine sera on human chorionic gonadotropin, estradiol, and progesterone release by cultured human trophoblast cells. *Endocrinology*, *137*, 2067–2074.

Henson, M. C., Greene, S. J., Reggio, B. C., Shi, W., and Swan, K. F. (1997). Effects of reduced maternal lipoprotein-cholesterol availability on placental progesterone biosynthesis in the baboon. *Endocrinology*, *138*, 1385–1397.

Henson, M. C., Swan, K. F., and O'Neil, J. S. (1998). Expression of placental leptin and leptin receptor transcripts in early pregnancy and at term. *Obstet. Gynecol.*, *92*, 1020–1028.

Henson, M. C., Castracane, V. D., O'Neil, J. S., Gimpel, T., Swan, K. F., Green, A. E., and Shi, W. (1999). Serum leptin concentrations and expression of leptin transcripts in placental trophoblast with advancing baboon pregnancy. *J. Clin. Endocrinol. Metab.*, *84*, 2543–2549.

Hickey, M. S., Israel, R. G., Gardiner, S. N., Considine, R. V., McCammon, M. R., Tyndall, G. L., Houmard, J. A., Marks, R. H., and Caro, J. F. (1996). Gender differences in serum leptin levels in humans. *Biochem. Mole. Med.*, *59*, 1–6.

Highman, T. J., Friedman, J. E., Huston, L. P., Wong, W. W., Catalano, P. M. (1998). Longitudinal changes in maternal serum leptin concentrations, body composition, and resting metabolic rate in pregnancy. *Am. J. Obstet. Gynecol.*, *178*, 1010–1015.

Hislop, M. S., Ratanjee, B. D., Soule, S. G., and Marais, A. D. (1999). Effects of anabolic-androgenic steroid use or gonadal testosterone suppression on serum leptin concentration in men. *Eur. J. Endocrinol.*, *141*, 40–46.

Hoang, V. M., Foulk, R., Claussar, K., Burlingame, A., Gibson, B. W., and Fisher, S. J. (2001). Functional proteonomics: examining the effects of hypoxia on the cytotrophoblast protein repertoire. *Biochemistry*, *40*, 4077–4086.

Hoggard, N., Hunter, L., Duncan, J. S., Williams, L. M., Trayhurn, P., and Mercer, J. G. (1997). Leptin and leptin receptor mRNA and protein expression in the murine fetus and placenta. *Proc. Natl. Acad. Sci. USA*, *94*, 11073–11078.

Hoggard, N., Hunter, L., Trayhurn, P., Williams, L. M., and Mercer, J. G. (1998). Leptin and reproduction. *Proc. Nutr. Soc.*, *57*, 421–427.

Hoggard, N., Hunter, L., Lea, R. G., Trayhurn, P., and Mercer, J. G. (2000). Ontogeny of the expression of leptin and its receptor in the murine fetus and placenta. *Br. J. Nutr.*, *83*, 317–326.

Hoggard, N., Haggarty, P., Thomas, L., and Rea, G. (2001a). Leptin expression in placental and fetal tissues: does leptin have a functional role? *Biochem. Soc. Trans.*, *29*, 57–63.

Hoggard, N., Crabtree, J., Allstaff, S., Abramovich, D. R., and Haggarty, P. (2001b). Leptin secretion to both the maternal and fetal circulations in the *ex vivo* perfused human term placenta. *Placenta*, *22*, 347–352.

Howe, D. C., Gertler, A., and Challis, J. R. G. (2000). Leptin suppresses the increase in ACTH secretion in the late gestation fetal sheep: metabolic cues and the timing of parturition. *J. Soc. Gynecol. Invest.*, *7*(Suppl.), Abstract #35.

Huang, L., Wang, Z., and Li, C. (2001). Modulation of circulating leptin levels by its soluble leptin receptor. *J. Biol. Chem.* 276, 6343–6349.

Hytinantti, T., Koistinen, H. A., Koivisto, V. A., Karonen, S. L., Rutanen, E. M., and Andersson, S. (2000a). Increased leptin concentration in preterm infants of pre-eclamptic mothers. *Arch. Dis. Child. Fetal. Neonatal. Ed.*, *83*, F13–F16.

Hytinantti, T., Koistinen, H. A., Teramo, K., Karonen, S. L., Koivisto, V. A., and Andersson, S. (2000b). Increased fetal leptin in type I diabetes mellitus pregnancies complicated by chronic hypoxia. *Diabetologia*, *43*, 709–713.

Islam, M. S., Sjoholm, A., and Emilsson, V. (2000). Fetal pancreatic islets express functional leptin receptors and leptin stimulates proliferation of fetal islet cells. *Int. J. Obes.*, *24*, 1246–1253.

Jaquet, D., Leger, J., Levy-Marchal, C., Oury, J. F., and Czernichow, P. (1998). Ontogeny of leptin in human fetuses and newborns: effect of intrauterine growth retardation on serum leptin concentrations. *J. Clin. Endocrinol. Metab.*, *83*, 1243–1246.

Jaquet, D., Gaboriau, A., Czernichow, P., and Levy-Marchal, C. (2001). Relatively low serum leptin levels in adults born with intra-uterine growth retardation. *Int. J. Obes. Relat. Metab. Disord.*, *25*, 491–495.

Jequier, E. and Tappy, L. (1999). Regulation of body weight in humans. *Physiol. Rev.*, *79*, 451–480.

Jockenhovel, F., Blum, W. F., Vogel, E., Englaro, P., Muller-Wieland, D., Reinwein, D., Rascher, W., and Krone, W. (1997). Testosterone substitution normalizes elevated serum leptin levels in hypogonadal men. *J. Clin. Endocrinol. Metab.*, *82*, 2510–2513.

Kawai, M., Yamaguchi, M., Murakami, T., Shima, K., Murata, Y., and Kishi, K. (1997). The placenta is not the main source of leptin production in pregnant rat: gestational profile of leptin in plasma and adipose tissues. *Biochem. Biophys. Res. Comm.*, *240*, 798–802.

Kawai, M., Murakawi, T., Otani, S., Shima, K., Yamaguchi, A., and Kishi, K. (1999). Colocalization of leptin receptor (OB-R) mRNA and placental lactogen-II in rat trophoblast cells: gestational profile of OB-R mRNA expression in placentae. *Biochem. Biophys. Res. Comm.*, *257*, 425–430.

Kawamura, K., Sato, N., Fukuda, J., Kodama, H., Kumagai, J., Tanikawa, H., Nakamura, A., and Tanaka, T. (2002). Leptin promotes the development of mouse preimplantation embryos in vitro. *Endocrinology*, *143*, 1922–1931.

Kiess, W., Siebler, T., Inglaro, P., Kratzsch, J., Deutscher, J., Meyer, K., Gallaher, B., Blum, W. F. (1998). Leptin as a metabolic regulator during fetal and neonatal life and in childhood and adolescence. *J. Ped. Endocrinol. Metab.*, *11*, 483–496.

Kohrt, W. M., Landt, M., and Birge, Jr., S. J. (1996). Serum leptin levels are reduced in response to exercise training, but not hormone replacement therapy, in older women. *J. Clin. Endocrinol. Metab.*, *81*, 3980–3985.

Koistinen, H. A., Koivisto, V. A., Andersson, S., Karonen, S. L., Kontula, K., Oksanen, L., and Teramo, K. A. (1997). Leptin concentration in cord blood correlates with intrauterine growth. *J. Clin. Endocrinol. Metab.*, *82*, 3328–3330.

Konukoglu, D., Serin, O., and Erean, M. (2000). Plasma leptin levels in obese and non-obese postmenopausal women before and after hormone replacement therapy. *Maturitas*, *36*, 203–207.

Kratzsch, J., Hockel, M., and Kiess, W. (2000). Leptin and pregnancy outcome. *Curr. Opin. Obstet. Gynecol.*, *12*, 501–505.

Kristensen, K., Pederson, S. B., and Richelsen, B. (1999). Regulation of leptin by steroid hormones in rat adipose tissue. *Biochem. Biophys. Res. Comm.*, *259*, 624–630.

Kronfeld-Schor, N., Zhao, J., Silvia, B. A., Bicer, E., Matthews, P. T., Urban, R., Zimmerman, S., Kunz, T. H., and Widmaier, E. D. (2000). Steroid-dependent up-regulation of adipose leptin secretion in vitro during pregnancy in mice. *Biol. Reprod.*, *63*, 274–280.

Kumar, V. and Chambon, P. (1998). The estrogen receptor binds tightly to its responsive element as a ligand-induced homodimer. *Cell*, *55*, 145–146.

Lage, M., Garcia-Mayo, R. V., Tome, M. A., Cordido, F., Valle-Inclan, F., Considine, R. V., Caro, J. F., Dieguez, C., and Casanueva, F. F. (1999). Serum leptin levels in women throughout pregnancy and the postpartum period and in women suffering spontaneous abortion. *Clin. Endocrinol.*, *50*, 211–216.

Laird, S. M., Quinton, N. D., Anstie, B., Li, T. C., and Blakemore, A. I. (2001). Leptin and leptin-binding activity in women with recurrent miscarriage: correlation with pregnancy outcome. *Hum. Reprod.*, *16*, 2008–2013.

Lambert, A., Kiess, W., Bottler, A., Glasow, A., and Kratzsch, J. (2001). Soluble leptin receptor represents the main leptin binding activity in human blood. *Biochem. Biophys. Res. Commun.*, *283*, 982–988.

Laml, T., Hartmann, B. W., Ruecklinger, E., Preyer, O., Soregi, G., and Wagenbichler, P. (2001a). Maternal serum leptin concentrations do not correlate with cord blood leptin concentrations in normal pregnancy. *J. Soc. Gynecol. Investig.*, *8*, 43–47.

Laml, T., Preyer, O., Hartmann, B. W., Ruecklinger, E., Soregi, G., and Wagenbichler, P. (2001b). Decreased maternal leptin in pregnancies complicated by preeclampsia. *J. Soc. Gynecol. Investig.*, *8*, 89–93.

LaPaglia, N., Steiner, J., Kirsteins, L., Emanuele, M., and Emanuele, N. (1998). Leptin alters the response of the growth hormone releasing factor–growth hormone–insulin-like growth factor-I axis to fasting. *J. Endocrinol.*, *159*, 79–83.

Lavoie, H. B., Taylor, A. E., Sharpless, J. L., Anderson, E. J., Strauss, C. C., and Hall, J. E. (1999). Effects of short-term hormone replacement on serum leptin levels in postmenopausal women. *Clin. Endocrinol. (Oxf)*, *51*, 415–422.

Lea, R. G., Howe, D., Hannah, L. T., Bonneau, O., Hunter, L., and Hoggard, N. (2000). Placental leptin in normal, diabetic and fetal growth-retarded pregnancies. *Mole. Hum. Reprod.*, *6*, 763–769.

Lepercq, J., Cauzac, M., Lahlon, N., Timsit, J., Girard, J., Auwerx, J., and Hanguel-de Mouzon, S. (1999). Overexpression of placental leptin in diabetic pregnancy: a critical role for insulin. *Diabetes*, *47*, 847–850.

Lepercq, J., Challier, J.-C., Guerre-Millo, M., Cauzac, M., Vidal, H., and Hauguel-de Mouzon, S. (2001). Prenatal leptin production: evidence that fetal adipose tissue produces leptin. *J. Clin. Endocrinol. Metab.*, *86*, 2409–2413.

Lewandowski, K., Horn, R., O'Callaghan, C. J., Dunlop, D., Medley, G. F., O'Hare, P., and Brabant, G. (1999). Free leptin, bound leptin, and soluble leptin receptor in normal and diabetic pregnancies. *J. Clin. Endocrinol. Metab.*, *84*, 300–306.

Li, A. W., Morash, B., Hollenberg, A. N., Ur, E., Wilkinson, M., and Murphy, P. R. (2001). Transcriptional regulation of the leptin gene promoter in rat GH3 pituitary and C6 glioma cells. *Mole. Cell. Endocrinol.*, *176*, 57–65.

Lindell, K., Bennett, P. A., Itoh, Y., Robinson, I. C. A. F., Carlsson, L. M. S., and Carlsson, B. (2001). Leptin receptor 5′ untranslated regions in the rat: relative abundance, genomic organization and relation to putative response elements. *Mole. Cell. Endocrinol.*, *172*, 37–45.

Lindheim, S. R., Sauer, M. V., Carmina, F., Chang, P. L., Zimmerman, R., and Lobo, R. A. (2000). Circulating leptin levels during ovulation induction: relation to adiposity and ovarian morphology. *Fertil. Steril.*, *73*, 493–498.

Linnemann, K., Malek, A., Sager, R., Blum, W. F., Schneider, H., and Fusch, C. (2000). Leptin production and release in the dually in vitro perfused human placenta. *J. Clin. Endocrinol. Metab.*, *85*, 4298–4301.

Linnemann, K., Malek, A., Schneider, H., and Fusch, C. (2001). Physiological and pathological regulation of feto/placento/maternal leptin expression. *Biochem. Soc. Trans.*, *29*, 86–90.

Luoh, S. M., Di Marco, F., Levin, N., Armanini, M., Xie, M. H., Nelson, C., Bennett, G. L., Williams, M., Spencer, S. A., Gurney, A., and de Sauvage, F. J. (1997). Cloning and characterization of a human leptin receptor using a biologically active leptin immunoadhesin. *J. Molec. Endocrinol.*, *18*, 77–85.

Luukkaa, V., Pesonen, U., Huhtaniemi, I., Lehtonen, A., Tilvis, R., Tuomilehto, J., Koulu, M., and Huupponen, R. (1998). Inverse correlation between serum testosterone and leptin in men. *J. Clin. Endocrinol. Metab.*, *83*, 3243–3246.

Machinal, F., Dieudonne, M. N., Leneveu, M. C., Pecquery, R., and Giudicelli, Y. (1999). In vivo and in vitro *ob* gene expression and leptin secretion in rat adipocytes: evidence for a regional specific regulation by sex steroid hormones. *Endocrinology*, *140*, 1567–1574.

Mannucci, E., Ognibene, A., Becorpi, A., Cremasco, F., Pellegrini, S., Otanelli, S., Rizello, S. M., Massi, G., Messeri, G., and Rotella, C. M. (1998). Relationship between leptin and oestrogens in healthy women. *Eur. J. Endocrinol.*, *139*, 198–201.

Mantzoros, C. S. (2000). Role of leptin in reproduction. *Ann. NY. Acad. Sci.*, *900*, 174–183.

Mantzoros, C. S. and Moschos, S. J. (1998). Leptin: in search of role(s) in human physiology and pathophysiology. *Clin. Endocrinol.*, *49*, 551–567.

Martin, L. J., Mahaney, M. C., Almasy, L., MacCluer, J. W., Blangero, J., Jaquish, C. E., and Comuzzie, A. G. (2002). Leptin's sexual dimorphism results from genotype by sex interactions mediated by testosterone. *Obes. Res.*, *10*, 14–21.

Martinez-Abundis, E., Gonzalez-Ortiz, M., and Pascoe-Gonzales, S. (2000). Serum leptin levels and the severity of preeclampsia. *Arch. Gynecol. Obstet.*, *264*, 71–73.

Masuzaki, H., Ogawa, Y., Sagawa, N., Hosoda, K., Matsumoto, T., Mise, H., Nishimura, H., Yoshimasa, Y., Tanaka, I., Mori, T., and Nakao, K. (1997). Non-adipose tissue production of leptin: leptin as a novel placenta-derived hormone in humans. *Nat. Med.*, 3, 1,029–1,033.

Matkovic, V., Ilich, J. Z., Skugor, M., Badenhop, N. E., Goel, P., Clairmont, A., Klisovic, D., Nahhas, R. W., and Landoll, J. D. (1997). Leptin is inversely related to age at menarche in human females. *J. Clin. Endocrinol. Metab.*, *82*, 3239–3245.

Matsuda, J., Yokota, I., Iida, M., Murakami, T., Naito, E., Ito, M., Shima, K., and Kuroda, Y. (1997). Serum leptin concentration in cord blood: relationship to birth weight and gender. *J. Clin. Endocrinol. Metab.*, *82*, 1642–1644.

McCarthy, J. F., Misra, D. N., and Roberts, J. M. (1999). Maternal plasma leptin is increased in preeclampsia and positively correlates with fetal cord concentration. *Am. J. Obstet. Gynecol.*, *180*, 731–736.

Meisser, A., Cameo, P., Islami, D., Campana, A., and Bischof, P. (1999). Effects of interleukin-6 (Il-6) on trophoblast cells. *Mol. Hum. Reprod.*, *5*, 1055–1058.

Menendez, C., Baldelli, R., Lage, M., Casabiell, X., Pinero, V., Solar, J., Diguez, C., and Casanueva, F. F. (2000). The in vitro secretion of human leptin is gender-dependent but independent of the body mass of the donors. *Eur. J. Endocrinol.*, *143*, 711–714.

Messinis, I. E., Milingos, S. D., Alexandris, E., Kariotis, G., and Seferiadis, K. (1999). Leptin concentrations in normal women following bilateral ovariectomy. *Hum. Reprod.*, *14*, 913–918.

Messinis, I. E., Papageorgiou, I., Milingos, S., Asprodini, E., Kollios, G., and Seferiadis, K. (2001). Oestradiol plus progesterone treatment increases serum leptin concentrations in normal women. *Hum. Reprod.*, *16*, 1827–1832.

Mise, H., Sagawa, N., Matsumoto, T., Yura, S., Nanno, H., Itoh, H., Mori, T., Mazuzaki, H., Hoshoda, K., Ogawa, Y., and Nakao, K. (1998). Augmented placental production of leptin in preeclampsia: possible involvement of placental hypoxia. *J. Clin. Endocrinol. Metab.*, *83*, 3225–3229.

Montague, C. T., Farooqi, I. S., Whitehead, J. P., Soos, M. A., Rau, H., Wareham, N. J., Sewter, C. P., Digby, J. E., Mohammed, S. W., Hurst, J. A., Cheetham, C. H., Earley, A. R., Barnett, A. H., Prins, J. B., and O'Rahilly, S. (1997). Congenital leptin deficiency is associated with severe early-onset obesity in humans. *Nature*, *387*, 903–907.

Moschos, S., Chan, J. L., and Mantzoros, C. S. (2002). Leptin and reproduction: a review. *Fertil. Steril.*, *77*, 433–444.

Mounzih, K., Qiu, J., Ewart-Toland, A., and Chehab, F. F. (1998). Leptin is not necessary for gestation and parturition but regulates maternal nutrition via a leptin resistance state. *Endocrinology*, *139*, 5259–5262.

Murakami, T., Iida, M., and Shima, K. (1995). Dexamethasone regulates obese expression in isolated rat adipocytes. *Biochem. Biophys. Res. Commun.*, *241*, 1260–1267.

Mystkowski, P. and Schwartz, M. W. (2000). Gonadal steroids and energy homeostasis in the leptin era. *Nutrition.*, *16*, 937–946.

Nathanielsz, P. W., Figueroa, J. P., and Honnebier, M. B. (1992). In the rhesus monkey placental retention after fetectomy at 121 to 130 days gestation outlasts the normal duration of pregnancy. *Am. J. Obstet. Gynecol.*, *166*, 1529–1535.

Nowicki, M., Brye, W., and Kokot, F. (2001). Hormonal regulation of appetite and body mass in patients with advanced prostate cancer treated with combined androgen blockade. *J. Endocrinol. Invest.*, *24*, 31–36.

Odegard, R. A., Vatten, L. J., Nilsen, S. T., Salveson, K. A., and Austgulen, R. (2002). Umbilical cord leptin is increased in preeclampsia. *Am. J. Obstet. Gynecol.*, *186*, 427–432.

Ogueh, O., Sooranna, S., Nicolaides, K. H., and Johnson, M. R. (2000). The relationship between leptin concentration and bone metabolism in the human fetus. *J. Clin. Endocrinol. Metab.*, *85*, 1997–1999.

O'Neil, J. S., Burow, M. E., Green, A. E., McLachlan, J. A., and Henson, M. C. (2001a). Effects of estrogen on leptin gene promoter activation in MCF-7 breast cancer and JEG-3 choriocarcinoma cells: selective regulation via estrogen receptors α and β *Molec. Cell. Endocrinol.*, *176*, 67–75.

O'Neil, J. S., Green, A. E., Edwards, D. E., Swan, K. F., Gimpel, T., Castracane, V. D., and Henson, M. C. (2001b). Regulation of leptin and leptin receptor in baboon pregnancy: effects of advancing gestation and fetectomy. *J. Clin. Endocrinol. Metab.*, *86*, 2518–2524.

Ong, K. K. L., Ahmed, M. L., Sheriff, A., Woods, K. A., Watts, A., Golding, J., Dunger, D. B. (1999). Cord blood leptin is associated with size at birth and predicts infancy weight gain in humans. *J. Clin. Endocrinol. Metab.*, *84*, 1145–1148.

Pasco, J. A., Henry, M. J., Kotowiez, M. A., Collier, G. R., Ball, M. J., Ugoni, A. M., Nicholson, G. C. (2001). Serum leptin levels are associated with bone mass in nonobese women. *J. Clin. Endocrinol. Metab.*, *86*, 1884–1887.

Pepe, G. J. and Albrecht, E. D. (1995). Actions of placental and fetal adrenal steroid hormones in primate pregnancy. *Endocr. Rev.*, *16*, 608–648.

Putney, D. J., Henson, M. C., and Pepe, G. J. (1991). Influence of the fetus and estrogen on maternal serum growth hormone, insulin-like growth factor-II, and epidermal growth factor concentrations during baboon pregnancy. *Endocrinology*, *129*, 3109–3117.

Qian, H., Barb, C. R., Compton, M. M., Hausman, G. J., Azain, M. J., Kraeling, R. R., and Baile, C. A. (1999). Leptin mRNA expression and serum leptin concentrations as influenced by age, weight and estradiol in pigs. *Dom. Anim. Endocrinol.*, *16*, 135–143.

Reidy, S. P. and Weber, J. M. (2000). Leptin: an essential regulator of lipid metabolism. *Comp. Biochem. Physiol. (Part A)*, *125*, 285–297.

Reitman, M. L., Bi, S., Marcus-Samuels, B., and Gavrilova, O. (2001). Leptin and its role in pregnancy and fetal development—an overview. *Biochem. Soc. Trans.*, *29*, 68–72.

Ringler, G. E. and Strauss, III, J. F. (1990). In vitro systems for the study of placental endocrine function. *Endocr. Rev.*, *11*, 105–123.

Roberts, T. J., Nijland, M. J., Caston-Balderrama, A., and Ross, M. G. (2001). Central leptin stimulates ingestive behavior and urine flow in the near term ovine fetus. *Horm. Metab. Res.*, *33*, 144–150.

Rosenbaum, M. and Leibel, R. L. (1998). Leptin: a molecule integrating somatic energy stores, energy expenditure and fertility. *Trends Endocrinol. Metab.*, *9*, 117–124.

Sagawa, N., Yura, S., Itoh, H., Mise, H., Kakui, K., Korita, D., Takemura, M., Nuamah, M. A., Ogawa, Y., Masuzaki, H., Nakao, K., and Fujii, S. (2002). Role of leptin in pregnancy—a review. *Placenta*, *23*(Suppl. A), S80–S86.

Schubring, C., Kiess, W., Englaro, P., Rascher, W., Dotsch, J., Hanitsch, S., Ahanasio, A., and Blum, W. F. (1997). Levels of leptin in maternal serum, amniotic fluid, and arterial and venous cord blood: relation to neonatal and placental weight. *J. Clin. Endocrinol. Metab.*, *82*, 1480–1483.

Schubring, C., Englaro, P., Siebler, T., Blum, W. F., Demiraka, T., Kratzsh, J., and Kiess, W. (1998). Longitudinal analysis of maternal serum leptin levels during pregnancy, at birth and up to six weeks after birth: relation to body mass index, skinfolds, sex steroids and umbilical cord blood leptin levels. *Horm. Res.*, *50*, 276–283.

Schulz, S., Hockel, C., and Weise, W. (2000). Hormonal regulation of neonatal weight: placental leptin and leptin receptors. *Br. J. Obstet. Gynecol.*, *107*, 1486–1491.

Senaris, R., Garcia-Caballaro, T., Casabiell, X., Gallego, R., Castro, R., Considine, R. V., Dieguez, C., and Casanueva, F. F. (1997). Synthesis of leptin in human placenta. *Endocrinology*, *138*, 4501–4504.

Shimizu, H., Shimomura, Y., Nakanishi, Y., Futawatari, T., Ohtani, K., Sato, N., and Mori, M. (1997). Estrogen increases in vivo leptin production in rats and human subjects. *J. Endocrinol.*, *154*, 285–292.

Shimon, I., Yan, X., Magoffin, D. A., Friedman, T. C., and Melmed, S. (1998). Intact leptin receptor is selectively expressed in human fetal pituitary and pituitary adenomas and signals human fetal pituitary growth hormone secretion. *J. Clin. Endocrinol. Metab.*, *83*, 4059–4064.

Sierra-Honigmann, M. R., Nath, A. K., Murakami, C., Garcia-Cardena, G., Papapetropoulos, A., Sessa, W. C., Madge, L. A., Schechner, J. S., Schwabb, M. B., Polverini, P. J., and Flores-Riveros, J. R. (1998). Biological action of leptin as an angiogenic factor. *Science*, *281*, 1683–1686.

Sivan, E., Lin, W. M., Homko, C. J., Reece, E. A., and Boden, G. (1997). Leptin is present in human cord blood. *Diabetes*, *46*, 917–919.

Sivan, E., Whittaker, P. G., Sinha, D., Homko, C. J., Lin, M., Reece, E. A., and Boden, G. (1998). Leptin in human pregnancy: the relationship with gestational hormones. *Am. J. Obstet. Gynecol.*, *179*, 1128–1132.

Slieker, L. J., Sloop, K. W., and Surface, P. L. (1996). Regulation of expression of *ob* mRNA and proteins by glucocorticoids and cAMP. *J. Biol. Chem.*, *271*, 5301–5304.

Soderberg, S., Olsson, T., Eliasson, M., Johnson, O., Brismar, K., Carlstrom, K., and Ahren, B. (2001). A strong association between biologically active testosterone and leptin in non-obese men and women is lost with increasing (central) adiposity. *Int. J. Obes. Relat. Metab. Disord.*, *25*, 98–105.

Solomon, S. (1997). The primate placenta as an endocrine organ: steroids. In E. Knobil, and J. D. Neill (Eds.). *The Physiology of Reproduction* (2nd ed.), Raven Press, New York, pp. 863–873.

Soorana, S. R., Ward, S., and Bajoria, R. (2001a). Discordant fetal leptin levels in monochorionic twin with chronic midtrimester twin–twin transfusion syndrome. *Placenta*, *22*, 392–398.

Soorana, S. R., Ward, S., and Bajoria, R. (2001b). Fetal leptin influences birth weight in twins with discordant growth. *Pediatr. Res.*, *49*, 667–672.

Spear, M. L., Hassink, S. G., Leef, K., O'Conner, D. M., Kirwin, S. M., Locke, R., Gorman, R., and Funanage, V. L. (2001). Immaturity or starvation? Longitudinal study of leptin levels in premature infants. *Biol. Neonate*, *80*, 35–40.

Steppan, C. M. and Swick, A. G. (1999). A role for leptin in brain development. *Biochem. Biophys. Res. Comm.*, *256*, 600–602.

Steppan, C. M., Crawford, D. T., Chidsey-Frink, K. L., Ke, H., and Swick, A. G. (2000). Leptin is a potent regulator of fetal growth in *ob/ob* mice. *Regul. Pept.*, *92*, 73–78.

Stock, S. M. and Bremme, K. A. (1998). Elevation of plasma leptin levels during pregnancy in normal and diabetic women. *Metabolism*, *47*, 840–843.

Tamas, P., Sulyok, E., Szabo, I., Vizer, M., Ertl, T., Rascher, W., and Blum, W. F. (1998). Changes of maternal serum leptin levels during pregnancy. *Gynecol. Obstet. Invest.*, *46*, 169–171.

Tamura, T., Goldenberg, R. L., Johnston, K. E., and Cliver, S. P. (1998). Serum leptin concentrations during pregnancy and their relationship to fetal growth. *Obstet. Gynecol.*, *91*, 389–395.

Tartaglia, L. A. (1997). The leptin receptor. *J. Biol. Chem.*, *272*, 6093–6096.

Tartaglia, L. A., Dembski, M., Weng, X., Deng, N., Culpepper, J., Devos, R., Richards, G. J., Campfield, L. A., Clark, F. T., Deeds, J., Muir, C., Sanker, S., Moriarity, A., Moore, K. J., Smutko, J. S., Mays, G. G., Woolf, E. A., Monroe, C. A., and Tepper, R. I. (1995). Identification and expression cloning of a leptin receptor, OB-R. *Cell*, *83*, 1263–1271.

Teppa, R. J., Ness, R. B., Crombleholme, W. R., and Roberts, J. M. (2000). Free leptin is increased in normal pregnancy and further increased in preeclampsia. *Metabolism*, *49*, 1043–1048.

Terada, Y., Yamakawa, K., Sugaya, A., and Toyoda, N. (1998). Serum leptin levels do not rise during pregnancy in age-matched rats. *Biochem. Biophys. Res. Commun.*, *253*, 841–844.

Thomas, T., Burguera, B., Melton, III L. J., Atkinson, E. J., O'Fallon, W. M., Riggs, B. L., and Khosla, S. (2000). Relationship of serum leptin levels with body composition and sex steroid and insulin levels in men. *Metabolism.*, *49*, 1278–1284.

Tomimatsu, T., Yamaguchi, M., Murakami, T., Ogura, K., Sakata, M., Mitsuda, N., Kanzaki, T., Kurachi, H., Irahara, M., Miyake, A., Shima, K., Aono, T., and Murata, Y. (1997). Increase of mouse leptin production by adipose tissue after midpregnancy: gestation profile of serum leptin concentration. *Biochem. Biophys. Res. Commun.*, *240*, 213–245.

Unkila-Kallio, L., Andersson, S., Koistinen, H. A., Karonen, S. L., Ylikorkala, O., and Tiitinen, A. (2001). Leptin during assisted reproductive cycles: the effect of ovarian stimulation and of very early pregnancy. *Hum. Reprod.*, *16*, 657–662.

Varvarigou, A., Mantzoros, C. S., and Beratis, N. G. (1999). Cord blood leptin concentrations in relation to intrauterine growth. *Clin. Endocrinol.*, *50*, 177–183.

Villafuerte, B. C., Fine, J. B., Bai, Y., Zhao, W., Fleming, S., and DiGirolamo, M. (2000). Expressions of leptin and insulin-like growth factor-I are highly correlated and region-specific in adipose tissue of growing rats. *Obesity Res.*, *8*, 646–655.

Watanobe, H. and Suda, T. (1999). A detailed study on the role of sex steroid milieu in determining plasma leptin concentrations in adult male and female rats. *Biochem. Biophys. Res. Commun.*, *259*, 56–59.

Weigle, D. S. (1997). Leptin and other secretory products of adipocytes modulate multiple physiological functions. *Annales d'Endocrinologie (Paris)*, *58*, 132–136.

Williams, M. A., Havel, P. J., Schuartz, M. W., Schwartz, M. W., Leisewring, W. B., King, I. B., Zingheim, R. M., Zebelman, A. M., and Luthy, D. A. (1999). Pre-eclampsia disrupts the normal relationship between serum leptin concentrations and adiposity in pregnant women. *Paediatr. Perinat. Epidemiol.*, *3*, 190–204.

Williams, C. B., Fawcett, R. L., Waechter, A. S., Zhang, P., Kogon, B. E., Jones, R., Inman, M., Huse, J., and Considine, R. V. (2000). Leptin production in adipocytes from morbidly obese subjects: stimulation by dexamethasone, inhibition with troglitazone, and influence of gender. *J. Clin. Endocrinol. Metab.*, *85*, 2678–2684.

Yamada, M., Irahara, M., Tezuka, M., Murakami, T., Shima, K., and Aono, T. (2000). Serum leptin profiles in the normal menstrual cycles and gonadotropin treatment cycles. *Gynecol. Obstet. Invest.*, *49*, 119–123.

Yang, S. W. and Kim, S. Y. (2000). The relationship of the levels of leptin, insulin-like growth factor-I and insulin in cord blood with birth size, ponderal index, and gender difference. *J. Pediat. Endocrinol. Metab.*, *13*, 289–296.

Yoneda, N., Saito, S., Kimura, M., Yamada, M., Iida, M., Murakami, T., Irahara, M., Shima, K., and Aono, T. (1998). The influence of ovariectomy on *ob* gene expression in rats. *Horm. Metab. Res.*, *30*, 263–265.

Yura, S., Sagawa, N., Mise, H., Mori, T., Masuzaki, H., Ogawa, Y., and Nakao, K. (1998). A positive umbilical venous–arterial difference of leptin level and its rapid decline after birth. *Am. J. Obstet. Gynecol.*, *178*, 926–930.

Zhang, Y., Proenca, R., Maffei, M., Barone, M., Leopold, L., and Friedmand, J. M. (1994). Positional cloning of the mouse obese gene and its human homologue. *Nature*, *372*, 425–432.

V

Genetics

16

Genetics and Physiology Link Leptin to the Reproductive System

FARID F. CHEHAB, AMANDA EWART-TOLAND, KHALID MOUNZIH, JUN QIU, AND SCOTT OGUS

16.1. INTRODUCTION

Studies linking leptin to the reproductive system have drastically expanded since the initial study that reported the rescuing effects of leptin on the reproductive system of *ob/ob* mice (Chehab et al., 1996). A search of the current literature revealed that this topic has evolved in various directions; however, the broader issue remains that of nutrition and reproduction. This subject continues to draw upon the seminal observations of Kennedy and Mitra (1969) in rodents and of Frisch and McArthur in humans (1974) whose experiments laid the foundations for this field of investigation. Thus, it is without surprise that this area has witnessed multiple interesting studies over the years with undeniably the major breakthrough being the discovery of the hormone leptin (Zhang et al., 1994). This chapter will review in the pre-leptin era, the most pertinent reports that have paved the way for understanding the relevance and importance of studies carried out in the post-leptin era. Together, the integration of these investigations will offer the reader a broad insight into this subject.

FARID F. CHEHAB, AMANDA EWART-TOLAND, KHALID MOUNZIH, JUN QIU, AND SCOTT OGUS • Department of Laboratory Medicine, University of California—San Francisco.

Leptin and Reproduction. Edited by Henson and Castracane, Kluwer Academic/Plenum Publishers, 2003.

16.2. ADIPOSE TISSUE AND REPRODUCTION

The onset of reproductive function is initially a central nervous system event. While, in females, the pituitary, ovary, and uterus are poised well before the triggering of this event, the pulsatile release of the gonadotropin-releasing hormone (GnRH) is the critical event that mediates reproductive maturation. Although there is some variation among different mouse strains, it takes approximately 4 weeks for vaginal opening, an easily detectable pubertal index to manifest in female mice. Vaginal opening is followed by the first estrous cycle, which although somewhat inefficient for pregnancy, highlights the firing of the reproductive system. Pregnancy, a potential end result of these key events imposes on the organism a nutritional demand, which is largely stored in adipose tissue. It is logical to assume that the adequacy and attainment of these energy stores could be relayed to appropriate neuroendocrine pathways via a peripheral signal. In seminal studies, Gordon Kennedy (Kennedy and Mitra, 1969) found that puberty was closely related to body weight, more specifically to food intake per unit body weight. Since growth is a prerequisite for breeding, Kennedy asked about the nature of the synchronization signal between growth and the initiation of reproduction. He postulated that the brain receives a metabolic factor related to food intake that signals the adequacy of energy stores necessary for reproduction. In prehistoric times, when food reserves were scarce or fluctuated seasonally, stored fat became an essential component of reproduction. Since fat is the most labile component of body weight, a system that informs neuroendocrine pathways about these fluctuations in energy stored in the adipose mass is therefore essential for the maintenance of reproduction. The sexual dimorphism of fat storage is exemplified by the fact that the overall fat masses of a well-nourished 57 kg 18-year-old girl and 57 kg 15-year-old boy measuring both 165 cm are approximately 16 kg and 7 kg, representing respectively, 28 and 12 percent of body weight (Frisch, 1981). In terms of calories, the 16 kg of fat in the female is equivalent to 144,000 calories. Since pregnancy requires approximately 50,000 calories above normal metabolic requirements and lactation needs about 1,000 calories per day, the stored calories in fat could provide energy for a pregnancy and about 3 months lactation (Frisch, 1990). It is evident that if such energy reserves could not be made available for pregnancy, then there would be no need for the reproductive system to maintain its function and waste resources that can be utilized for other critical body functions. Thus, a factor that links the state of energy reserves in the adipose mass to the central control of reproduction would only allow firing of the reproductive system when energy reserves are adequate, thus shunting and saving energy resources for other functions that are necessary for survival.

16.3. ROLE OF FAT IN REPRODUCTION AND IN SIGNALING PUBERTY

Extensive literature exists about the association between nutritional state and reproductive maturity. Pioneering studies by Kennedy and Mitra (1969) proposed that puberty is somehow linked to body weight and more specifically to fat storage, which as they conclude, is one of the signals responsible for the initiation of hypothalamic control of ovarian function (Kennedy, 1957). Later studies by Frisch and McArthur (1974) related the loss or restoration of menstrual cycles in young girls to a minimum weight for height. In addition, these investigators find that normal girls become relatively fatter from menarche to reproductive maturity. These findings established a relationship between reproduction and adipose tissue prompting additional studies that attributed dysfunctional reproduction in female marathon runners, young ballet dancers, and college rowers to a substantial loss of adipose mass (Frisch et al., 1980, 1981) and a certain degree of hypothalamic dysfunction (Vigersky, 1977). Therefore, the reassertion of the link between adipose tissue and reproduction has led to the "critical weight" hypothesis (Johnston et al., 1971), which essentially stipulates that a functional reproductive system requires a minimum amount of stored adipose mass. Another characteristic of adipose tissue is that it can convert androgens to estrogens via aromatization (Siiteri, 1981) thereby contributing additional estrogens outside the ovarian axis. Overall these findings established a strong association between reproduction and fat reinforcing the hypothesis that a metabolic signal may be responsible for the initiation of reproduction.

16.4. STERILITY OF *ob/ob* AND *db/db* MICE

Male and female homozygous *ob/ob* and *db/db* mice show obesity with marked hypogonadism and infertility. Seminal parabiosis experiments performed by Lane (1959) demonstrated that the ovary of an *ob/ob* female was capable of releasing gonadal hormones in response to the excess gonadotropin hormones secreted by an ovariectomized lean female when both mice were joined in a parabiotic union. The findings and implications of these studies was that the pituitary of the obese female was incapable of releasing gonadotropins (follicle-stimulating hormone [FSH], luteinizing hormone [LH]) but that its reproductive tract could respond to gonadotropins suggesting that the reproductive defects of *ob/ob* mice was central in origin rather than peripheral. In a detailed study of gonadotropins and gonadal function in *ob/ob* males, Swerdloff et al. (1976) found that the levels of FSH, LH, and testosterone were reduced in comparison with lean animals. Since low levels of gonadal steroids tend to increase gonadotropin secretion in the presence of functioning feedback mechanisms at the hypothalamus–pituitary level, the *ob/ob* mouse is therefore deficient in this respect. Along the same lines, intact male *ob/ob* mice were found to be less responsive than normal mice to a bolus of GnRH, and when castrated, exhibited an attenuated pituitary response to testosterone (Swerdloff et al., 1976, 1978). Thus, it appears that the immaturity of the hypothalamic–pituitary axis of *ob/ob* mice resembles in some aspects that of prepubertal mice suggesting that *ob/ob* mice cannot complete the pubertal process even though they exhibit normal early sexual development. Hypofunction of the female *ob/ob* pituitary was demonstrated by showing that pituitary extracts administered to *ob/ob* females induced ovulation and conception, but not implantation (Runner, 1954), which was achieved following treatment with gonadotropic hormones (Smithberg and Runner, 1956). Furthermore, the administration of high doses of progesterone maintained pregnancy for 19 days p.c., but did not enable the mothers to deliver the fetuses except after administration of relaxin, which stimulated parturition and lactation

(Smithberg and Runner, 1957). These findings demonstrated that the sterility of the *ob/ob* female was caused by an insufficiency of hormones at the hypothalamic–pituitary level rather than by physical hindrance of copulatory activity due to excessive adipose tissue. However, although the pituitary and plasma concentrations of LH were reduced or near normal in *ob/ob* mice, it has been suggested that the *ob/ob* females have a basal tonic release of LH but that they lack the pre-ovulatory surge of gonadotropin secretion, which characterizes the cycle of lean females.

In males, the testes of *ob/ob* and *db/db* animals are smaller than their corresponding lean siblings and often remain undescended. Testes histology revealed that *ob/ob* mice had abnormal spermatogenesis and multinucleated spermatids along with atrophied Leydig cells (Hellman et al., 1963; Hellman, 1965). The failure of homozygous *ob/ob* and *db/db* females to reproduce results from a deficiency in estrous cyclicity resulting from atrophied ovaries. However, these ovaries are capable of producing viable eggs when transplanted into lean female recipients (Hummel, 1957; Johnson and Sidman, 1979) suggesting that their infertility is not due to any ovarian defect. These findings allowed the implementation of a strategy based on *ob/ob* and *db/db* ovarian transplants into normal females and subsequent fertilization with normal males to generate obligate heterozygous mutant mice, which can be used in crosses to generate homozygous *ob/ob* and *db/db* mice.

As previously noted, it is often perceived that the sterility of *ob/ob* and *db/db* mice could be caused by a physical barrier of adipose mass. To address this issue, efforts were aimed at inducing fertility by thinning down the body weight of *ob/ob* mice via food restriction. Although these approaches largely failed (Mayer, 1953), there were occasional reports of success only in *ob/ob* males (Lane and Dickie, 1954; Eastcott, 1972), which upon careful inspection of the literature were derived from an outbred stock and thus with different genetic backgrounds (an issue addressed below). Therefore, thinning down the body weight of *ob/ob* mice by food restriction did not largely improve their impaired reproduction. Another observation that supports the idea that excessive fat does not necessarily block reproduction in rodents is the fact that other rodent models of obesity are indeed fertile such as mice that have mutations at the *Fat*, A^y and *tub* loci. Additionally, the obesity exhibited by the gold-thioglucose (GTG) and the monosodium glutamate (MSG) nongenetic rodent models of obesity are not associated with any histological abnormality in reproductive tissues as these animals are fertile and reproduce normally despite their obese state. Overall, it is concluded that the infertility of *ob/ob* mice is closely related to a genetic deficiency lying at the *ob* locus on mouse chromosome 6.

16.5. LEPTIN AND ITS RECEPTOR

The cloning and identification of the obese (*ob*) gene and its encoded protein product, leptin, raised tremendous interest in the scientific and medical communities. The uncovering of the mouse *ob* gene necessitated positional cloning strategies coupled with genetic mapping and exon-trapping techniques to reveal the underlying molecular lesion responsible for the sterile-obese phenotype of the *ob/ob* mouse (Zhang et al., 1994). The mutation, R105X, consisted of a single base substitution at codon 105 of the cloned *ob* gene converting arginine to a stop codon thus leading to a truncated protein, with no biological activity. Treatment of *ob/ob* mice with the recombinant *ob* protein reduced their massive body weight (Campfield et al., 1995; Halaas et al., 1995; Pelleymounter et al., 1995; Weigle et al., 1995), which was mostly attributed mostly to excess adipose tissue mass. The discovery of leptin provided an entry point to begin to unravel the biology of obesity and to develop modalities for its treatment. Subsequent cloning of the *ob* receptor (Ob-R), from the *db/db* mouse (Tartaglia et al. 1995) heightened additional interest into this nascent leptin pathway. Expression studies revealed that the Ob-R gene expresses various mRNA isoforms via alternate mRNA splicing. However, a point mutation in an Ob-R intron generates a donor splice site that prevents alternate mRNA splicing of a specific protein isoform. This isoform encodes an Ob-R isoform that spans a cytoplasmic domain characteristic of signaling proteins (Chen et al., 1996; Lee et al., 1996). The absence of this long form of Ob-R, termed Ob-Rb, is the underlying cause of the *db/db* phenotype, which exhibits similar to *ob/ob* mice, an obese-sterile phenotype.

16.6. CORRECTION OF THE STERILITY OF *ob/ob* MICE

Classical parabiosis experiments between *ob/ob, db/db*, and their lean counterparts suggested that the *ob* and *db* proteins are, respectively, a circulating ligand and the receptor for this ligand (Coleman and Hummel, 1969; Coleman, 1973). What this seminal experiment also suggested was that the *ob/ob* mouse is deficient in the production of the *ob* protein whereas the *db/db* mouse overexpresses it but is resistant to its effect due to a mutation in the *ob* receptor. Cloning of the *ob* and *db* genes vindicated these original findings as shown by leptin treatment and subsequent weight loss of *ob/ob* and normal mice but not of *db/db* mice, which are resistant to leptin due to the absence of Ob-Rb. However, because these experiments did not address the reproductive phenotype of *ob/ob* mice, we turned our interests to the effects of leptin on the reproductive axis of *ob/ob* mice.

16.6.1. *ob/ob* Females

The studies outlined below have largely been previously described (Chehab et al., 1996). We expressed the secreted form of leptin in *Escherichia coli*, purified and refolded it before injecting it daily into *ob/ob* females. Treating an initial group of *ob/ob* females first tested the efficacy of the leptin preparation (Figure 16.1A). Leptin treatment of these obese mice consisted of three phases. In the first phase, which lasted 8 days, the *ob/ob* mice received two doses per day of either phosphate-buffered saline (PBS) or recombinant leptin. At the end of the first phase, body weights and food intake of leptin-treated *ob/ob* mice were markedly decreased versus control mice. Withdrawal of leptin injections in the second phase resulted in a rapid increase of food consumption and body weight, demonstrating the need for repeated injections to maintain a biological effect. The third phase, which consisted of single daily injections, produced a similar but slower biological response than the first phase. Leptin treatment was continued for 42 days until the declining body weights of the *ob/ob* mice stabilized approximately to 47 percent. In a second set of *ob/ob* females treated for 30 days, saline injected mice increased their body weight by 13 percent whereas leptin-treated and food-restricted mice lost, respectively, 40 percent and 38 percent of their body weight (Figure 16.1B). In parallel, cumulative food

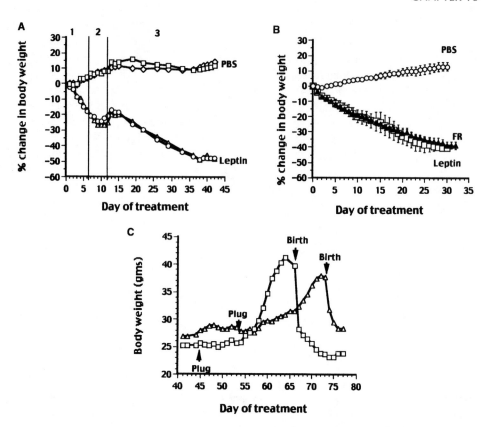

Figure 16.1. Effect of leptin treatment and withdrawal on *ob/ob* females. (A). Treatment of *ob/ob* females with vehicle (PBS) or recombinant leptin during three phases. In the 1st phase, leptin treatment was initiated, withdrawn in the second phase, and resumed in the third phase. (B). Treatment of *ob/ob* females with vehicle (PBS), recombinant leptin or food restriction (FR). (C). Body weight of leptin-treated females after exposure to stud males. Presence of the copulatory plug is indicative of copulation and ensuing pregnancy. (Reprinted with permission from Chehab, F. F., Lim, M. E., and Lu, R. [1997]. Correction of the sterility defect in homozygous obese female mice by treatment with the human recombinant leptin. *Nat. Genet., 12*, 318–320. Copyright Nature Publishing group, Macmillan Publishers Ltd.)

consumption in the leptin-treated mice was reduced to 35 percent of the PBS controls. After 42 and 30 days of daily leptin treatment in the first and second sets of *ob/ob* females, vehicle, food-restricted, and leptin-treated *ob/ob* mice were mated with normal C57BL/6J males. Copulatory plugs, which indicate mating, were detected in all the leptin-treated *ob/ob* females, 1–6 days after introduction of the males, demonstrating indirectly estrous cyclicity and ovulation. Neither PBS-control nor food-restricted mice showed any copulatory plugs even after 2 weeks following introduction of the males. Plugged leptin-treated females showed an increase in body weight at around day 12 p.c. and delivered newborn pups timely between 19 and 20 days p.c. (Figure 16.1C). None of the pups survived beyond 2 days of age apparently due to a lack of lactation as evidenced by a lack or reduced milk intake. While under continuous leptin treatment, the *ob/ob* mothers went through a second round of pregnancies. The resulting pups were fostered to a wild-type mother who lactated them showing that they did not suffer from suckling but suggested rather that the *ob/ob* mothers were largely incapable of lactation, although one *ob/ob* mother did lactate her pups. We interpreted these experiments to suggest that the pharmacological leptin treatment of the *ob/ob* mothers may have significantly disrupted the breast fat pads, thus inhibiting development of the breast alveolar system in anticipation of lactation. In a third round of *ob/ob* pregnancies, withdrawal of leptin treatment at day 13 p.c. did not prevent the mothers from carrying out their pregnancies to term and delivering the pups suggesting that leptin may not have been required for gestation and parturition, an issue that is addressed later in this chapter. However, the cessation of leptin treatment following delivery led to a restoration of the sterile-obese phenotype characteristic of *ob/ob* mice. Although we have not measured the levels of reproductive hormones in the pregnant females, it is evident that the entire reproductive cascade of events could not have occurred in the absence of these hormones. Nevertheless, increased secretion of FSH, LH, and estradiol were subsequently confirmed following leptin treatment of *ob/ob* mice (Barash et al., 1996). Overall, it is apparent that leptin-induced weight loss and not food-restricted weight loss corrected the sterility of the *ob/ob* females suggesting a stimulatory role of leptin on the reproductive system.

16.6.2. *ob/ob* Males

Male *ob/ob* mice are also sterile and their infertility has been attributed to a deficiency in GnRH secretion (Swerdloff et al., 1976). Similar to *ob/ob* females, *ob/ob* males were either food restricted or treated with recombinant leptin (Figure 16.2A) (Mounzih et al., 1997). After 12 days of treatment, at a time where body weight had decreased to 24 percent and 32 percent in food-restricted and leptin-treated *ob/ob* mice, respectively, exposure of lean females to *ob/ob* males from both groups resulted in pregnancies from all the leptin-treated *ob/ob* males, but none of the food-restricted *ob/ob* males. These results show that leptin treatment of homozygous male *ob/ob* mice induces fertility as demonstrated by their ability to fertilize lean female mice. It is interesting to note that reversal of the sterility occurred despite an initial body weight of approximately $65.6 + 4.9$ g indicating that accumulation of excessive fat does not permanently block the immaturity of the hypothalamic–pituitary–gonadal axis, which only becomes functional following leptin treatment. To gain further insights into the effects of leptin treatment, histology sections of testes from normal lean, untreated *ob/ob*, food-restricted *ob/ob* and leptin-treated *ob/ob* mice were examined. Previous studies had shown that *ob/ob* testis contain multinucleated spermatids, few spermatozoa, and a small amount of interstitial Leydig tissue (Jones and Ainsworth-Harrisson, 1957), which was reduced by about 50 percent (Hellman, 1965). In untreated and food-restricted *ob/ob* males, the lumen of their seminiferous tubules contained strikingly less sperm than lean mice and their interstitial Leydig cells appeared atrophied. However, the seminiferous tubules of the leptin-treated *ob/ob* males were abundant with mature sperms and their Leydig cells regained their usual morphology and clustering characteristic (Figure 16.2B). Therefore, leptin treatment of the *ob/ob* males stimulated spermatogenesis and allowed regeneration of the testosterone producing Leydig cells resulting in correction of their sterility as effectively as in the *ob/ob* females, implicating leptin in the reproductive physiology of the male as well.

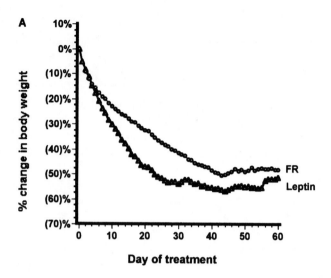

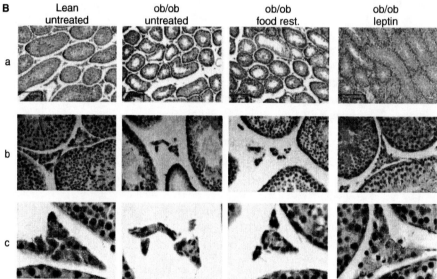

Figure 16.2. Effect of leptin treatment and food restriction (FR) on the body weight of *ob/ob* males. (A). Treatment of *ob/ob* males by FR or recombinant leptin. (B). Hematoxylin and eosin stained section across the seminiferous tubules of untreated lean and *ob/ob* males and FR or leptin-treated *ob/ob* males. The sections show in (a) (magnification 100×) the seminiferous tubules and their sperm content and in (b) (magnification 400×) and (c) (magnification 1000×), the morphology of the Leydig cells, which regenerate upon leptin treatment. (Reprinted with permission from Mounzih, K., Lu, R., and Chehab, F. F. [1997]. Leptin rescues the sterility of genetically obese *ob/ob* males. *Endocrinology, 138*(3), 1190–1193. Copyright owner, The Endocrine Society.)

16.7. CORRECTION OF THE STERILITY IN *db/db* MICE

Induction of fertility in *db/db* mice by manipulation of the leptin/leptin receptor system is more difficult than in *ob/ob* mice owing to the transmembrane nature of Ob-Rb and its localization in the hypothalamus. Nevertheless, transgenic expression of Ob-Rb under the control of the neuron specific enolase promoter, rescued the sterility of *db/db* males but intriguingly not that of *db/db* females (Kowalski et al., 2001) suggesting that perhaps over-expression of Ob-Rb in the hypothalamus or elsewhere in the brain might be detrimental to the female reproductive system.

16.8. IMPACT OF MODIFIER GENES ON THE *ob/ob* PHENOTYPE

Variation of a particular phenotype is often accounted for by the genetic diversity, which in rodents can easily be achieved by crossing different inbred strains together, thus introducing modifier genes into a particular trait. In the context of the *ob/ob* phenotype, we asked whether introduction of the Balb/cJ genetic background would modify the sterile-obese phenotype of C57BL/6J *ob/ob* mice. Toward this end, we generated an F2 intercross between the C57BL/6J and Bab/cJ genomes and studied their interaction on *ob/ob* mice (Ewart-Toland et al., 1999). In addition, we recently generated a backcross to determine the effects of the Balb/cJ genome alone on the *ob/ob* phenotype (Qiu et al., 2001). These two studies will be briefly reviewed in this section.

In the F2 intercross, C57BL/6J-Balb/cJ *ob/ob* (herein referred to as F2 *ob/ob*) mice remained morbidly obese; however, their obesity was variable. For example, at 23 weeks of age, the body weights of F2 *ob/ob* males, but not F2 *ob/ob* females, were significantly less than those of C57BL/6J *ob/ob* males (Figure 16.3A and B). A 15 week mating period revealed that approximately 40 percent of male F2 *ob/ob* mice were capable of inducing 1–6 pregnancies in normal female mice (Figure 16.3C). The lack of a body weight difference between F2 fertile *ob/ob* males and F2 infertile *ob/ob* males showed that body weight and adiposity were neither a contributing nor an inhibitory factor to the onset of fertility in F2 *ob/ob* males (Figure 16.3D). Thus, the modifier genes from either or both genetic backgrounds acted to decrease the obesity and enhance the fertility of male F2 *ob/ob* mice. It is not known at the present time whether these modifier genes are activated or enhanced by the absence of leptin but it appears to suggest that they partially compensate for a defective leptin pathway. Surprisingly, because none of the F2 *ob/ob* females showed any sign of fertility, it is reasonable to speculate that some of these modifier genes may be X-linked and thus subject to Iyonization. The powerful effect of the modifier genes on the reproductive system of *ob/ob* males warranted a search for candidate genes. Although cloning modifier genes is a daunting task with limited success and little precedent, we undertook a genome scan using 123 microsatellites spaced approximately 10 cm apart. Chromosomal regions with LOD scores of 5 on chromosome 5 and 5.6 on chromosome 1 were associated, respectively, with traits such as body weight and the number of pregnancies that a male F2 *ob/ob* mouse could induce in a normal female. DNA sequencing of the coding

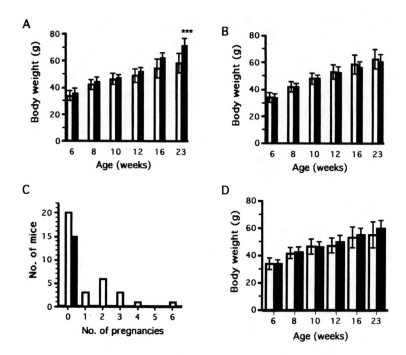

Figure 16.3. Effect of the mixed C57BL/6J-Balb/cJ genetic background on the *ob/ob* phenotype. Body weights of F2 *ob/ob* (white bars) and C57BL/6J (black bars) males (A) and females (B) from 6 to 23 weeks of age. (C) Number of pregnancies that F2 *ob/ob* (white bars) and C57BL/6J (black bars) males could induce in lean females. (D) Body weights of fertile (white bars) and infertile (black bars) F2 *ob/ob* males from 6 to 23 weeks of age. (Reprinted with permission from Ewart-Toland, A., Mounzih, K., Qiu, J., and Chehab, F. F. [1999]. Effect of the genetic background on the reproduction of leptin-deficient obese mice. *Endocrinology, 140*(2), 732–738. Copyright owner, The Endocrine Society.)

sequences of candidate genes in these regions (such as TGFβ2, Ptx2, and the GnRH receptor) did not reveal any DNA polymorphism between the C57BL/6J and Balb/cJ strains (unpublished data) concluding that the regulatory regions of these genes or of other candidate genes in the vicinity may account for the significant LOD scores.

In the second cross, which actually consisted of a backcross, the aim was to investigate the effect of the Balb/cJ genome only on the *ob/ob* phenotype (Qiu et al., 2001). Derivation of the Balb/cJ *ob/ob* line necessitated 10 successive rounds of backcrossing onto the Balb/cJ genetic background before the resulting phenotype could be investigated. It was found that the body weights and adiposities of Balb/cJ *ob/ob* mice were drastically reduced compared to those of C57BL/6J *ob/ob* mice. For example, at 25 weeks of age, Balb/cJ *ob/ob* males and females reached a body weight plateau of approximately 50 g whereas C57BL/6J *ob/ob* mice weighed over 75 g (Figure 16.4A and B), the difference being attributed to the adipose mass (Figure 16.4C). Food intake was similar and not significantly different between *ob/ob* mice from both of these strains demonstrating that metabolism rather than reduced food intake accounted for the reduced adiposity in the *ob/ob* Balb/cJ strain. Furthermore, insulin and triglycerides levels were significantly elevated in fed Balb/cJ *ob/ob* mice suggesting a more severe diabetic course than their

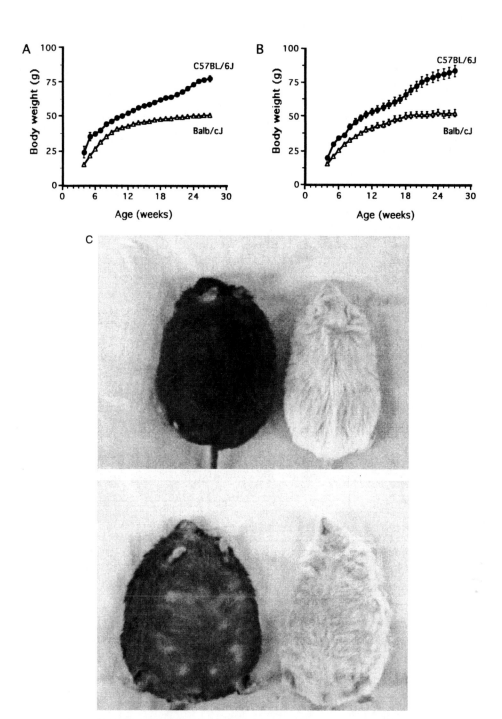

Figure 16.4. Effect of the Balb/cJ genome on the *ob/ob* phenotype. Body weights of *ob/ob* males (A) and females (B) bred either on the C57BL/6J (circles) or Balb/cJ (triangles) congenic genetic backgrounds. (C) Dorsal and ventral views of photographs taken at 23 weeks of age of a C57BL/6J *ob/ob* (black fur) and Balb/cJ *ob/ob* (white fur) mice, showing the reduced adiposity of the Balb/cJ *ob/ob* mouse. (Reprinted with permission from Qiu, J., Ogus, S., Mounzih, K., Ewart-Toland, A., and Chehab, F. F. [2001]. Leptin-deficient mice backcrossed to the Balb/cJ genetic background have reduced adiposity, enhanced fertility, normal body temperature, and severe diabetes. *Endocrinology, 142*(8), 3421–3425. Copyright owner, The Endocrine Society.)

C57BL/6J *ob/ob* counterparts. This observation raised the possibility that their reduced obese state might be secondary to an uncontrolled severe diabetic state similar to that of C57BL/KsJ *ob/ob* mice, which were severely diabetic and lost weight progressively before dying of β-cell exhaustion and pancreatic failure at approximately 6 months of age (Coleman and Hummel, 1973). However, unlike C57BL/KsJ *ob/ob* mice, Balb/cJ *ob/ob* mice lived well over a year of age and did not show any evidence of pancreatic failure despite massive islet cell hyperplasia. Therefore, the modifier genes from the C57BL/KsJ and Balb/cJ genetic backgrounds have profound but distinctive effects on the progression of the diabetes of leptin-deficient mice. On the reproductive side, 6 out of 10 Balb/cJ *ob/ob* males and 4 out of 4 Balb/cJ *ob/ob* females were capable of inducing pregnancies during a 16 week mating period despite their obese state (Table 16.1). It is to be noted that some of the Balb/cJ *ob/ob* males that could induce multiple pregnancies in females, had body weights in the range of 30.7–52.1 g while those of the Balb/cJ *ob/ob* females were 35.7–63.6 g showing again that the body weights and consequently adiposities of these obese mice were not an obstacle to their fertility. Overall, it can be concluded that this congenic line of Balb/cJ *ob/ob* mice is capable of reproducing without leptin treatment, albeit at a reduced rate. Again, the action of Balb/cJ modifier genes is powerful and appears to complement the effect seen in the F2 intercross. A combination of genetic and genomic strategies, are currently being pursued to uncover the nature of these modifier genes. Although it is expected that other genetic backgrounds will also likely alleviate the phenotype of C57BL/6J *ob/ob* mice, it will be interesting to uncover the genetic backgrounds that will exacerbate the penetrance of the *ob* mutation even more than the

Table 16.1. Fertility of ob/ob males and ob/ob females on the Balb/cJ genetic background. The body weight and age of each mouse at the beginning of the mating period and near each mating time is shown. (Reprinted with permission from Qiu, J., Ogus, S., Mounzih, K., Ewart-Toland, A., and Chehab, F. F. [2001]. Leptin-deficient mice backcrossed to the Balb/cJ genetic background have reduced adiposity, enhanced fertility, normal body temperature, and severe diabetes. *Endocrinology, 142*(8), 3421–3425. Copyright owner, The Endocrine Society)

	Body weight (g)	Age (weeks)	Pregnancies	Age at mating(s) (weeks)	Body weight at mating (g)
Males					
1	39.7	12	0	—	—
2	43.0	11	3	21, 22, 26	47.5, 48.7, 50.5
3	45.7	11	2	11, 24	45.7, 52.1
4	43.3	10	0	—	—
5	42.9	9	0	—	—
6	33.1	8	1	13	41.7
7	33.5	7	0	—	—
8	33.7	7	1	18	49.8
9	25.9	6	2	8, 19	34.3, 45.7
10	25.3	6	2	7, 13	30.7, 45.2
Females					
1	39.0	10	1	11	41.7
2	33.6	8	2	13, 39	47.3, 63.6
3	29.3	8	2	10, 21	38.8, 52.6
4	23.4	6	1	9	35.7

C57BL/6J background. Our results also caution that modification of the *ob/ob* phenotype via the genetic background may confound investigations of *ob/ob* mice that fail to take into account their genetic background.

16.9. LEPTIN AND PUBERTY

Puberty is a broad term that refers to the stage of becoming physiologically capable of sexual reproduction. Some of the visible changes that highlight this period are genital maturation, development of secondary sexual characteristics; in females, the first occurrence of ovulation and in males, sperm ejaculation. It is also evident that hormonal signals at the neuroendocrine level must precede peripheral or external changes. However, the timing and nature of these signals is debatable. Because the reproductive axis can be activated prior to puberty by infusion of GnRH (Hotchkiss et al., 1971), showing that the pituitary and the ovaries are poised for function, the questions then are: What are the critical signals that trigger GnRH release? Do these signals remove an inhibition that is placed on GnRH neurons or do they exert positive stimulatory effects? Despite signal candidates such as growth hormone, neuropeptide Y, and insulin-like growth factor I, these questions have remained largely unanswered. In this section, we will address the effects of leptin, the latest of these signal candidates on the initiation of reproductive function.

As previously noted, the studies of Kennedy and Mitra (1963) and those of Frisch and Revelle (1970) proposed that body fat storage and food rather than age could be initiating factors for puberty. Thus, adiposity was suggested to be a critical determinant for the onset of puberty and the maintenance of menstrual cycles (Frisch and McArthur, 1974). This hypothesis was mostly based on the findings that loss of adipose tissue mass due to excessive exercise led to an interruption of menses (Frisch et al., 1980, 1981). The importance of the work outlined by Kennedy and Frisch lay in the fact that they brought forward the importance of adipose tissue as an organ that may significantly contribute to the onset of reproductive function by secreting a metabolic factor. So it was apparent that when the hormone leptin was found to correct the sterility of *ob/ob* mice, this association could be investigated.

The reasoning behind our puberty experiment (Chehab et al., 1997) was that if leptin was an informant to neuroendocrine pathways about the attainment of adequate fat stores, then it would be possible via leptin treatment to trick these pathways into believing that the animal had enough adipose tissue stores, when in fact it had very little. With this hypothesis in mind, we treated normal mice with pharmacological doses of recombinant leptin, about a week prior to the time they would have started to exhibit vaginal opening, which is one of the first external signs of puberty in the mouse. The leptin treatment, as expected, caused a decrease in growth and body weight of the prepubertal females (Figure 16.5A), but when they were housed with normal males, they mated earlier than the PBS treated females (Figure 16.5B) demonstrating an earlier onset of fertility. In another experiment, the leptin-treated females exhibited vaginal opening 1–3 days earlier than the vehicle-treated prepubertal females implying that they started cycling and their reproductive organs had matured earlier than PBS treated mice (Figure 16.5D). Overall, this experiment was significant, confirmed the original hypothesis, and was confirmed by independent studies (Ahima et al., 1996), thus strengthening these findings. In another

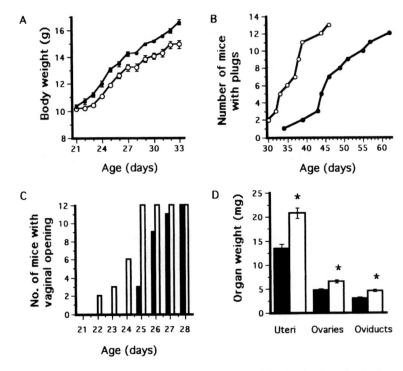

Figure 16.5. Effect of leptin on the onset of reproduction in normal female mice. Prepuberal mice were treated with PBS (black circles or black bars) or recombinant leptin (white circles or white bars) showing effects on (A) body weight, (B) copulatory plugs, (C) vaginal opening, and (D) weights of reproductive organs. (Reprinted with permission from Chehab, F. F., Mounzih, K., Lu, R., and Lim, M. E. [1997, January 3]. Early onset of reproductive function in normal female mice treated with leptin. *Science, 275*, 88–90. Copyright 1997 American Association for the Advancement of Science.)

experiment, transgenic mice overexpressing 12-fold leptin under the control of a liver-specific promoter, were found to exhibit early signs of puberty (Yura et al., 2000). These results are also consistent with the original hypothesis, namely that leptin plays an important role in the onset of reproductive function. Furthermore, we found that transgenic mice overexpressing 5-fold more leptin than endogenous levels undergo early puberty only when supplemented with exogenous leptin (Ogus and Chehab, unpublished observations) suggesting a quantitative effect of leptin on the reproductive axis at puberty. Further strength was added to this hypothesis when it was found that a transgene expressing very low leptin level rescued the reproductive phenotype of *ob/ob* mice without significantly affecting their morbid obesity (Ioffe et al., 1998).

A search for a leptin spike, presumably analogous to the LH surge, was found to occur in mice 10 days postnatally and was followed by rising levels of estradiol (Ahima et al., 1998) while in rats, a leptin spike was found a day after vaginal opening (Nagatini et al., 2000). In monkeys, a modest leptin spike was evident only during nighttime (Suter et al., 2000) but not daytime (Plant and Durrant, 1997) blood samplings. From these studies stems the notion that leptin may in fact act similar to LH in terms of its amplitude prior to the critical events that occur around puberty.

Overall, it is evident that leptin is a serious participant in the pubertal process and a key player to the reproductive system as vindicated by the sterility rescue of the leptin-deficient *ob/ob* mouse and the pubertal leptin studies. What is still missing though is a mechanism of action by which leptin stimulates the reproductive system. It already appears that such a mechanism would have to diverge from the leptin melanocortin neuronal circuit since melanocortin-4 receptor knockout mice are obese but fertile (Huszar et al., 1997). The ability of leptin to stimulate directly GnRH has been reported (Magni et al., 1999) although not confirmed in our hands using leptin-stimulated GT-1 cells either directly or following transfection with the long form OB-Rb receptor (Qiu and Chehab, unpublished observations). Nevertheless an upstream activator of GnRH is a prerequisite and would fit best in a mechanism of leptin action on the initiation of the reproductive system.

16.10. LEPTIN IN PREGNANCY

Early reports showed that pregnancy is associated with rising leptin levels (Butte et al., 1997; Schubring et al., 1997) suggesting that leptin may play a significant role in the biology of pregnancy. Although leptin is normally secreted from adipose tissue, studies have shown that in humans, it was also secreted from the human placenta (Masuzaki et al., 1997). In mice, the elevated leptin levels during pregnancy were proposed to result from the expression of the soluble leptin receptor, Ob-Re, which could bind circulating leptin and reduce its renal clearance (Gavrilova et al., 1997). We found that in vitro, recombinant Ob-Re falsely elevates leptin levels in a radioimmunoassay presumably by binding to ^{125}I-leptin and preventing the competition reaction between leptin and the primary antibody (Qiu and Chehab, unpublished observations). Therefore, leptin levels in the presence of Ob-Re may be erroneously elevated when quantitated by radioimmunoassay. In addition, we found that pregnant *ob/ob* mice have greatly elevated "leptin-like" levels in their circulation, due to Ob-Re levels as well (Qiu and Chehab, unpublished observations). Nevertheless, because leptin and Ob-Re may together play an important role in the biology of pregnancy, we asked whether a mouse pregnancy could progress with little or no leptin (Mounzih et al., 1998). Again, we turned to the sterile and leptin-deficient *ob/ob* mouse model, which gains its fertility only after being treated with leptin. Male and female *ob/ob* mice were treated with leptin to induce fertility and were then mated together to generate only *ob/ob* fetuses. Such pregnancies were deficient in endogenous leptin production, but could be exposed to leptin only via exogenous treatment. Thus, after copulation, exogenous leptin treatment was withdrawn at different timepoints to assess its impact on the continuation of pregnancy. Briefly, it was found that the absence of leptin at Day 1 or any day thereafter did not affect the successful outcome of the *ob/ob* pregnancies (Figure 16.6). More specifically, the lack of leptin did not prevent implantation and development of the fetuses despite occasional complications of delayed gestation, extended parturition, and lack of lactation. Thus, in this animal model, leptin did not play an important role in the critical events of pregnancy. Interestingly, monitoring of food intake in leptin-treated pregnant *ob/ob* females throughout gestation revealed that around mid-gestation, the food reducing effects of leptin appeared to fade as a state of leptin resistance was established (Figure 16.7). While a leptin resistant state in pregnancy makes a lot of sense for the build-up of nutrient reserves necessary for late fetal development and in anticipation of lactation,

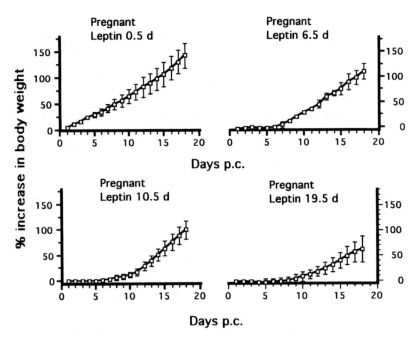

Figure 16.6. Effect of leptin treatment on the progression of pregnancy in *ob/ob* females mated with *ob/ob* males. The graphs depict the increase in body weight of pregnant *ob/ob* females that were treated with recombinant leptin until 0.5, 6.5, 10.5, or 19.5 days following detection of the copulatory plug at day 0.5 of the pregnancy. (Reprinted with permission from Mounzih, K., Qiu, J., Ewart-Toland, A., Chehab, F. F. [1998]. Leptin is not necessary for gestation and parturition but regulates maternal nutrition via a leptin resistance state. *Endocrinology, 139*(12), 5259–5262. Copyright owner, The Endocrine Society.)

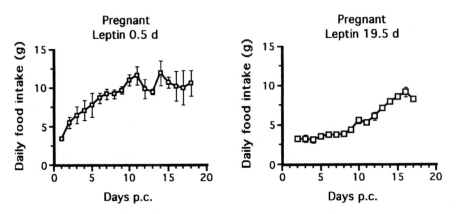

Figure 16.7. Food intake of pregnant *ob/ob* females treated with recombinant leptin until 0.5 and 19.5 days after detection of the copulatory plug. Treatment of leptin throughout gestation demonstrates that food intake is initially suppressed but then increased around mid-gestation demonstrating a leptin resistance effect during pregnancy. (Reprinted with permission from Mounzih, K., Qiu, J., Ewart-Toland, A., Chehab, F. F. [1998]. Leptin is not necessary for gestation and parturition but regulates maternal nutrition via a leptin resistance state. *Endocrinology, 139*(12), 5259–5262. Copyright owner, The Endocrine Society.)

the underlying mechanisms of this leptin resistant state may involve central and peripheral pathways, which may be similar to those involved in obesity.

16.11. CONCLUDING REMARKS

The cataclysm that the fat-derived hormone leptin brought to the biology of obesity has also been extended to reproductive biology. Although leptin has occasionally been hailed as a reproductive hormone, its role in reproduction is a peculiar one and certainly does not simulate any of the established reproductive hormones. Rather, leptin appears to act as a peripheral messenger conveying nutritional information to central pathways controlling energy expenditure and reproduction, two processes that are essential and necessary for the survival of an organism and of the species. The notion and concept that fat plays an important role in reproduction is attractive, logical, substantiated in the literature and has even recently been extended to fish (Peyon et al., 2001). However, skepticism and uncertainty about the role of fat and leptin in reproduction (Bronson, 2001) is disturbing and exemplifies some of the controversies that revolve around this subject. Although there is undoubtedly much to be learned from both sides, the current literature clearly favors a role for leptin in reproduction. Whether leptin acts as a critical, permissive, or facilitator factor may be semantic and will necessitate further investigation. What is undeniable however is that leptin-deficient mice (and humans) are obese and fail to complete the activation of their reproductive system unless exogenous leptin is provided. What is also important is that leptin can prevent the fasting-induced shutdown of the reproductive system (Ahima et al., 1996). Altogether, these seminal findings support the importance of the adipose tissue and more specifically of leptin in the triggering and maintenance of an active reproductive state.

REFERENCES

Ahima, R. S., Prabakaran, D., Mantzoros, C., Qu, D., Lowell, B., Maratos-Flier, E., and Flier, J. S. (1996). Role of leptin in the neuroendocrine response to fasting. *Nature, 382,* 250–252.

Ahima, R. S., Dushay, J., Flier, S. N., Prabakaran, D., and Flier, J. S. (1997). Leptin accelerates the onset of puberty in normal female mice. *J. Clin. Invest., 99,* 391–395.

Ahima, R. S., Prabakaran, D., and Flier, J. S. (1998). Postnatal leptin surge and regulation of circadian rhythm of leptin by feeding *J. Clin. Invest., 101,* 1020–1027.

Barash, I. A., Cheung, C. C., Weigle, D. S., Ren, H., Kabingting, E. B., Kuijper, J. L., Clifton, D. K., and Steiner, R. A. (1996). Leptin is a metabolic signal to the reproductive system. *Endocrinology, 137,* 3144–3147.

Bronson, F. H. (2001). Puberty in female mice is not associated with increases in either body fat or leptin. *Endocrinology, 142,* 4758–4761.

Butte, N. F., Hopkinson, J. M., and Nicolson, M. A. (1997). Leptin in human reproduction: serum leptin levels in pregnant and lactating women. *J. Clin. Endocrinol. Metab., 82,* 585–589.

Campfield, L. A., Smith, F. J., Guisez, Y., Devos, R., and Burn, P. (1995). Recombinant mouse OB protein: evidence for a peripheral signal linking adiposity and central neural networks. *Science, 269,* 546–549.

Chehab, F. F., Lim, M. E., and Lu, R. (1996). Correction of the sterility defect in homozygous obese female mice by treatment with the human recombinant leptin. *Nat. Gen., 12,* 318–320.

Chehab, F. F., Mounzih, K., Lu, R., and Lim, M. E. (1997). Early onset of reproductive function in normal female mice treated with leptin. *Science, 275,* 88–90.

Chen, H., Charlat, O., Tartaglia, L. A., Woolf, E. A., Weng, X., Ellis, S. J., Lakey, N. D., Culpepper, J., Moore, K. J., Breitbart, R. E., Duyk, G. M., Tepper, R. I., and Morgenstern, J. P. (1996). Evidence that the diabetes gene encodes the leptin receptor: identification of a mutation in the leptin receptor gene in *db/db* mice. *Cell, 84,* 491–495.

Coleman, D. J. (1973). Effects of parabiosis of obese with diabetes and normal mice. *Diabetologia, 9,* 294–298.

Coleman, D. J. and Hummel, K. H. (1969). Effect of parabiosis of normal with genetically diabetic mice. *Am. J. Phyiol., 217,* 1298–1304.

Coleman, D. L. and Hummel, K. P. (1973). The influence of the genetic background on the expression of the obese gene (ob) in the mouse. *Diabetologia, 9,* 287–293.

Eastcott, A. (1972). Influence of restricting the diet on breeding in genetically obese mice. *Lab. Anim., 6,* 9–18.

Ewart-Toland, A., Mounzih, K., Qiu, J., and Chehab, F. F. (1999). Effect of the genetic background on the reproduction of leptin-deficient obese mice. *Endocrinology, 140,* 732–738.

Frisch, R. E. (1981). What's below the surface? *N. Engl. J. Med., 305,* 1019–1020.

Frisch, R. E. (1990). Body fat, menarche, fitness and fertility. In R. E. Frisch (Ed.). *Progress in Reproductive Biology and Medicine* (vol. 14), Karger Press, Basel, pp. 1–26.

Frisch, R. E., Gotz-Welbergen, A. V., McArthur, J. W., Albright, T., Witschi, J., Bullen, B., Birnholz, J., Reed, R. B., and Hermann, H. (1981). Delayed menarche and amenorrhea of college athletes in relation to age of onset of training. *JAMA, 246,* 1559–1563.

Frisch, R. E. and McArthur, J. W. (1974). Menstrual cycles: fatness as a determinant of minimum weight for height necessary for their maintenance or onset. *Science, 185,* 949–951.

Frisch, R. E., Wyshak, G., and Vincent, L. (1980). Delayed menarche and amenorrhea in ballet dancers. *N. Engl. J. Med., 303,* 17–19.

Gavrilova, O., Barr, V., Marcus-Samuels, B., and Reitman, M. (1997). Hyperleptinemia of pregnancy associated with the appearance of a circulating form of the leptin receptor. *J. Biol. Chem., 272,* 30546–30551.

Halaas, J. L., Gajiwala, K. S., Maffei, M., Cohen, S. L., Chait, B. T., Rabinowitz, D., Lallone, R. L., Burley, S. K., and Friedman, J. M. (1995). Weight-reducing effects of the plasma protein encoded by the obese gene. *Science, 269,* 543–546.

Hellman, B. (1965). Studies in obese-hyperglycemic mice. *Ann. NY Acad. Sci., 131,* 541–558.

Hellman, B., Jacobsen, L., and Taljedal, I. B. (1963). Endocrine activity of the testis in obese-hyperglycemic mice. *Acta Endocrinol., 44,* 20–26.

Hotchkiss, J., Atkinson, L. E., and Knobil, E. (1971). Time course of serum estrogen and luteinizing hormone (LH) concentrations during the menstrual cycle of the rhesus monkey. *Endocrinology, 89,* 177–183.

Hummel, K. P. (1957). Transplantation of ovaries of the obese mouse. *Anat. Rec., 128,* 569.

Huszar, D., Lynch, C. A., Fairchild-Huntress, V., Dunmore, J. H., Fang, Q., Berkemeier, L. R., Gu, W., Kesterson, R. A., Boston, B. A., Cone, R. D., Smith, F. J., Campfield, L. A., Burn, P., and Lee, F. (1997). Targeted disruption of the melanocortin-4 receptor results in obesity in mice. *Cell, 88,* 131–141.

Ioffe, E., Moon, B., Connolly, E., and Friedman, J. M. (1998). Abnormal regulation of the leptin gene in the pathogenesis of obesity. *Proc. Natl. Acad. Sci. USA, 195,* 11852–11857.

Johnson, M. J. and Sidman, R. (1979). A reproductive endocrine profile in the diabetes (db) mutant mouse. *Biol. Reprod., 20,* 552–559.

Johnston, F. E., Malina, R. M., Galbraith, M. A., Frisch, R. E., Revelle, R., and Cook, S. (1971). Height, weight and age at menarche and the "critical weight" hypothesis. *Science, 174,* 1148–1149.

Jones, N. and Ainsworth-Harrisson, G. (1957). Genetically determined obesity and sterility in the mouse. *Stud. Fert., 9,* 51–64.

Kennedy, G. C. and Mitra, J. (1969). Body weight and food intake as initiating factors for puberty in the rat. *J. Physiol., 166,* 408–418.

Kennedy, G. C. (1957). The development with age of hypothalamic restraint upon the appetite of the rat. *J. Endocrinol., 16,* 9–17.

Kowalski, T. J., Liu, S. M., Leibel, R. L., and Chua, S. C. Jr. (2001). Transgenic complementation of leptin-receptor deficiency. I. Rescue of the obesity/diabetes phenotype of LEPR-null mice expressing a LEPR-B transgene. *Diabetes, 50,* 425–435.

Lane, P. (1959). The pituitary-gonad response of genetically obese mice in parabiosis with thin and obese siblings. *Endocrinology, 65,* 863–868.

Lane, P. W. and Dickie, M. M. (1954). The effect of restricted food intake on the lifespan of genetically obese mice *J. Nutr.*, *64*, 549.

Lee, G. H., Proenca, R., Montez, J. M., Carroll, K. M., Darvishzadeh, J. G., Lee, J. I., and Friedman, J. M. (1996). Abnormal splicing of the leptin receptor in diabetic mice. *Nature*, *379*, 632–635.

Magni, P., Vettor, R., Pagano, C., Calcagno, A., Beretta, E., Messi, E., Zanisi, M., Martini, L., and Motta, M. (1999). Expression of a leptin receptor in immortalized gonadotropin-releasing hormone-secreting neurons. *Endocrinology*, *140*, 1581–1585.

Mayer, J. (1953). Genetic, traumatic and environmental factors in the etiology of obesity. *Physiol. Rev.*, *33*, 472–508.

Mounzih, K., Lu, R., and Chehab, F. F. (1997) Leptin treatment rescues the sterility of genetically obese *ob/ob* males. *Endocrinology*, *138*, 1190–1193.

Mounzih, K., Qiu, J., Ewart-Toland, A., and Chehab, F. F. (1998). Leptin is not necessary for gestation and parturition but regulates maternal nutrition via a leptin resistance state *Endocrinology*, *139*, 5259–5262.

Nagatani, S., Guthikonda, P., and Foster, D. L. (2000). Appearance of a nocturnal peak of leptin secretion in the pubertal rat. *Horm. Behav.*, *37*, 345–352.

Pelleymounter, M. A., Cullen, M. J., Baker, M. B., Hecht, R., Winters, D., Boone, T., and Collins, F. (1995). Effects of the obese gene product on body weight regulation in ob/ob mice. *Science*, *269*, 540–543.

Peyon, P., Zanuy, S., and Carrillo, M. (2001). Action of leptin on in vitro luteinizing hormone release in the European sea bass (Dicentrarchus labrax). *Biol. Reprod.*, *65*, 1573–1578.

Qiu, J., Ogus, S., Mounzih, K., and Chehab, F. F. (2001). Leptin-deficient mice backcrossed to the BALB/cJ genetic background have reduced adiposity, enhanced fertility, normal body temperature, and severe diabetes. *Endocrinology*, *142*, 3421–3425.

Masuzaki, H., Ogawa, Y., Sagawa, N., Hosoda, K., Matsumoto, T., Mise, H., Nishimura, H., Yoshimasa, Y., Tanaka, I., Mori, T., and Nakao, K. (1997). Non-adipose tissue production of leptin: leptin as a novel placenta-derived hormone in humans. *Nat. Med.*, *3*, 1029–1033.

Plant, T. M. and Durrant, A. R. (1997). Circulating leptin does not appear to provide a signal for triggering the initiation of puberty in the male rhesus monkey (*Macaca mulatta*). *Endocrinology*, *138*, 4505–4508.

Runner, M. N. (1954). Inherited hypofunction of the female pituitary in the sterile-obese syndrome in the mouse. *Genetics*, *39*, 990–991.

Schubring, C., Kiess, W., Englaro, P., Rascher, W., Dotsch, J., Hanitsch, S., Attanasio, A., and Blum, W. F. (1997). Levels of leptin in maternal serum, amniotic fluid, and arterial and venous cord blood: relation to neonatal and placental weight. *J. Clin. Endocrinol. Metab.*, *82*, 1480–1483.

Siiteri, P. K. (1981). Extraglandular estrogen formation and serum binding of estradiol. Relationship to cancer. *J. Endocrinol.*, *89*, 119–129.

Smithberg, M. and Runner, M. N. (1957). Pregnancy induced in genetically sterile mice. *J. Hered.*, *48*, 97–100.

Smithberg, M. and Runner, M. N. (1956). The induction and maintenance of pregnancy in prepubertal mice. *J. Exp. Zool.*, *133*, 441–458.

Suter, K. J., Pohl, C. R., and Wilson, M. E. (2000). Circulating concentrations of nocturnal leptin, growth hormone, and insulin-like growth factor-I increase before the onset of puberty in agonadal male monkeys: potential signals for the initiation of puberty. *J. Clin. Endocrinol. Metab.*, *85*, 808–814.

Swerdloff, R. S., Batt, R. A., and Bray, G. A. (1976). Reproductive hormonal function in the genetically obese (*ob/ob*) mouse. *Endocrinology*, *98*, 1359–1364.

Swerdloff, R. S., Peterson, M., Vera, A., Batt, R. A. L., Heber, D., and Bray, G. A. (1978). The hypothalamic–pituitary axis in genetically obese (ob/ob) mice: response to luteinizing horomone releasing hormone. *Endocrinology*, *103*, 542–547.

Tartaglia, L. A., Dembski, M., Weng, X., Deng, N., Culpepper, J., Devos, R., Richards, G. J., Campfield, L. A., Clark, F. T., Deeds, J., Muir, C., Sanker, S., Moriarty, A., Moore, K. J., Smutlo, J. S., Mays, G. G., Woolf, E. A., Monroe, C. A., and Tepper, R. I. (1995). Identification and expression cloning of a leptin receptor, OB-R. *Cell*, *83*, 1263–1271.

Vigersky, R. A., Andersen, A. E., Thompson, R. H., and Loriaux, L. (1977). Hypothalamic dysfunction in secondary amenorrhea associated with simple weight loss. *N. Engl. J. Med.*, *297*, 1141–1145.

Weigle, D. S., Bukowski, T. R., Foster, D. C., Holderman, S., Kramer, J. M., Lasser, G., Lofton-Day, C. E., Prunkard, D. E., Raymond, C., and Kuijper, J. L. (1995). Recombinant ob protein reduces feeding and body weight in the *ob/ob* mouse. *J. Clin. Invest.*, *96*, 2065–2070.

Yura, S., Ogawa, Y., Sagawa, N., Masuzaki, H., Itoh, H., Ebihara, K., Aizawa-Abe, M., Fujii, S., and Nakao, K. (2000). Accelerated puberty and late-onset hypothalamic hypogonadism in female transgenic skinny mice overexpressing leptin. *J. Clin. Inves.*, *105*, 749–755.

Zhang, Y., Proenca, R., Maffei, M., Barone, M., Leopold, L., and Friedman, J. M. (1994). Positional cloning of the mouse obese gene and its human homologue. *Nature*, *372*, 425–432.

Leptin and the Onset of Puberty

Insights from Rodent and Human Genetics

I. SADAF FAROOQI

17.1. ABSTRACT

Deficiency of the adipocyte-derived hormone leptin in *ob/ob* mice results in severe, early-onset obesity and infertility. Administration of leptin results in complete reversal of the phenotype suggesting that leptin is needed for the development of puberty in rodents. In humans, mutations in the genes encoding leptin and the leptin receptor result in obesity syndromes and hypogonadotropic hypogonadism. We have shown that administration of recombinant human leptin results in the onset of puberty at an appropriate developmental age in human congenital leptin deficiency. This work suggests that leptin is a metabolic gate for the onset of puberty in humans. Leptin's actions may be mediated by central pathways and by direct action on peripheral organs.

17.2. INTRODUCTION

There is considerable evidence that nutrition influences pubertal development and it is well-recognized that alterations in diet and exercise, fasting and starvation, exert dramatic influences on pubertal maturation in many species, including the rat, sheep, and

I. SADAF FAROOQI • University Departments of Medicine and Clinical Biochemistry, Addenbrooke's Hospital—Cambridge.

Leptin and Reproduction. Edited by Henson and Castracane, Kluwer Academic/Plenum Publishers, 2003.

humans (Warren, 1980; Bronson and Rissman, 1986; Ebling et al., 1990). Classical studies performed by Kennedy and Mitra (1963) illustrated how the onset of puberty in the female rat is influenced by body size and food intake and led the authors to suggest that "such information might be conveyed by fairly direct chemical feedbacks". From an evolutionary perspective, it is readily plausible to assume that sexual maturation is only initiated when energy reserves are adequate to meet the demands of mating, pregnancy, and lactation. Although such studies have suggested that a signal related to energy stores or adipose tissue mass may be involved in the timing of puberty, the existence of such a signal has only recently emerged frrom studies in genetic models of rodent and human obesity.

17.3. *Ob/Ob* MICE AND THE DISCOVERY OF LEPTIN

The severely obese *ob/ob* mouse, first described by Ingalls et al. (1950), inherits its early-onset obesity in an autosomal recessive pattern and weighs three times more than normal mice by maturity. A decrease in thermogenesis, and hyperphagia after weaning, are followed by the development of severe obesity due to the preferential deposition of fat. In 1994, Jeffrey Friedman and colleagues cloned and characterized the *ob* gene which is expressed predominantly in white adipose tissue and encodes the 167 amino acid secreted protein, leptin (Zhang et al., 1994). The ob transcript is mutant in two strains of *ob/ob* mice which are leptin deficient as a result. Whereas *ob/ob* males can occasionally reproduce if maintained on a restricted diet, *ob/ob* females are always sterile and weight loss induced by dietary restriction fails to correct their sterility (Lane and Dickie, 1954). Early sexual development is normal in *ob/ob* females and their ovaries are capable of producing viable eggs when transplanted into lean female recipients (Hummel, 1957). However, ovulation never follows and the mice remain prepuberal indefinitely with no occurrence of estrus cycles. Gonadotropin concentrations are reduced in *ob/ob* animals but increase in response to stimulation with exogenous gonadotropin releasing hormone (GnRH) in vitro and in vivo implying that the sterility of *ob/ob* mice is caused by hypogonadotropic hypogonadism rather than physical hindrance of copulatory activity, pregnancy, and parturition caused by excess adipose tissue (Swerdloff et al., 1976, 1978). Female *ob/ob* mice can reproduce if ovulation and pregnancy are induced and maintained artificially with gonadotropins and progesterone, suggesting a lack of gonadotropin stimulation as the basis of their infertility (Hummel, 1957). The phenotypic characteristics of the *ob/ob* mouse are shared by another strain of severely obese mice, *db/db*, which are identical when expressed on the same genetic background (Coleman, 1978). The *db/db* mouse was subsequently shown to harbour a mutation in the signalling isoform of the leptin receptor (ObRb) (Chen et al., 1996), which is mainly expressed in the hypothalamus and thus lack of functioning leptin receptors also results in infertility.

17.3.1. Responses to Leptin Administration

Administration of recombinant leptin to leptin-deficient *ob/ob* mice not only restores normal body weight, but also corrects the sterility of adult female (Barash et al., 1996) and male ob/ob mice (Chehab et al., 1996) compared to saline-treated pair fed ob/ob controls. No such effect is seen in db/db mice which lack functioning leptin receptors.

In fasting rodents, reduced luteinizing hormone (LH) pulse frequency is seen, which is restored with the administration of leptin (Nagatani et al., 1998). Thus, in sexually mature mice, leptin is a key link between nutrition and the reproductive axis. These findings were extended by the observation that leptin treatment of normal, prepubertal female mice accelerates the onset of puberty as determined by vaginal opening, estrus, and cyclicity (Ahima et al., 1997). Importantly, the ability of exogenous leptin to accelerate puberty without affecting body weight in normal mice suggests that leptin may be the factor that normally links body weight and energy stores to the timing of puberty. As these observations in normal mice are seen at doses that do not change body weight, this work suggests that the actions of leptin to regulate neuroendocrine and reproductive function in normal mice are not secondary to effects on energy balance.

17.4. TRANSGENIC MODELS

The effects of leptin on the reproductive axis may, however, extend beyond the paradigm of starvation and feeding. Detailed physiological studies by Leibel and colleagues showed that fat mass, adjusted for age and sex was 27 percent higher in the Lep $ob/+$ than $+/+$ animals ($p = 0.03$) (Chung et al., 1998). Plasma leptin concentration adjusted for fat mass was 32 percent lower in the Lep $ob/+$ than wild-type mice, suggesting that induction of the single normal leptin allele in Lep $ob/+$ mice was not sufficient to compensate completely for the defective allele. However, despite an effect on fat deposition, Lep $ob/+$ mice are fertile suggesting that the lower concentration of leptin found in these animals is sufficient for pubertal development.

Ioffe et al. (1998) further tested the possibility that a relative decrease in leptin production could lead to obesity, by breeding mice expressing a weak human leptin transgene to ob/ob mice. Constitutive expression of leptin at a low level, equivalent to half that of wild-type mice, resulted in a moderate obesity phenotype, with a 3-fold increase in fat mass compared to wild-type mice. Several of the neuroendocrine features associated with leptin deficiency were abolished in low level leptin transgenic mice, which were fertile. However, low level leptin transgenic mice still exhibited abnormal thermoregulation in response to cold exposure and had mildly elevated plasma insulin concentrations suggesting that different thresholds exist for the different biological responses elicited by changes in serum leptin concentration and that these could be reversed by leptin administration (Ioffe et al., 1998).

Nakao and colleagues have generated transgenic skinny mice overexpressing leptin under the control of the liver-specific human serum amyloid P component (SAP) promoter (Ogawa et al., 1999). With no apparent adipose tissue and 10-fold higher serum leptin concentrations than wild-type littermates, these mice provide the opportunity to study the effects of chronic hyperleptinemia. The female skinny transgenic mice exhibit accelerated vaginal opening and ovarian and uterine maturation compared to wild-type littermates (Yura et al., 2000). The fertility rate of skinny mice was comparable to wild-type mice at 8 weeks of age. Reproductive function at older ages was characterized by prolonged estrus, atrophic ovaries, and reduced GnRH and LH secretion. However, no impairment of function was seen in male transgenic skinny mice (Yura et al., 2000). Hyperleptinemia in this rodent model seems to facilitate the onset of puberty but if chronically persistent, appears to subsequently downregulate the central leptin signals that stimulate reproductive function or interfere with gonadotropin stimulation of peripheral target organs.

In conclusion, the evidence from rodent models suggests that there may be multiple sites of leptin action and different thresholds for the various effects of leptin on the reproductive axis.

17.5. LEPTIN ACTION IN HUMANS

The identification of leptin was followed by an explosion of interest in clarifying its role as a regulator of body weight in humans. In general, the amount of *ob* mRNA in human adipocytes is positively correlated with weight and adipose tissue mass and is upregulated in obese humans (Hamilton et al., 1995). Despite mutational screening of a significant number of obese individuals (Considine et al., 1995; Maffei et al., 1996), no subjects with mutations in the *ob* gene had been identified and the importance of leptin and leptin-mediated pathways in human energy homeostasis was uncertain.

In 1997, we identified 2 cousins within a highly consanguineous family of Pakistani origin with severe obesity of early-onset; an 8-year-old girl weighing 86 kg (Child A) and a 2-year-old boy weighing 29 kg (Child B) (Montague et al., 1997). These children were noted to be severely hyperphagic, constantly demanding food, with an intense drive to eat which was never satisfied. They were found to have undetectable levels of serum leptin and were homozygous for a deletion of a single guanine nucleotide at codon 133 which results in a truncated protein. We have subsequently identified 2 further children with the same condition (Children C and D, aged 3 years). Prepubertal concentrations of follicle-stimulating hormone (FSH), LH, and gonadal steroids were observed in Children B, C, and D. Despite advanced skeletal maturation, with a bone age of 12.5 years, Child A was clinically prepubertal (Tanner stage 1) and levels of estradiol, FSH, and LH and androgens were at prepubertal concentrations before and after GnRH stimulation, which would be consistent with hypogonadotropic hypogonadism. An ultrasound scan of the pelvis showed an infantile uterus and ovaries.

Strobel et al. (1998) subsequently described a Turkish family in whom two severely obese adults were found to be leptin-deficient and were homozygous for a missense mutation in the *ob* gene. As well as severe, early-onset obesity, it was notable that both adults had failed to undergo pubertal development with low serum FSH and LH concentrations consistent with hypogonadotropic hypogonadism (Strobel et al., 1998). Additionally, Clement et al. (1998) identified three morbidly obese sisters who were found to be homozygous for a mutation in the leptin receptor. Clinically, these subjects failed to develop secondary sexual characteristics and remained amenorrhoeic at ages 13.5 and 19 years, with biochemical features of central hypogonadism. Whilst these monogenic obesity disorders are rare, they do provide unique insights into the role of leptin in human biology and it is clear that leptin is required for the onset of puberty in humans.

17.5.1. Heterozygotes for Leptin Mutation

We have studied heterozygotes for the leptin mutation and shown that inheriting one defective allele of the leptin gene appears to result in a state of relative hypoleptinaemia which is associated with the development of a moderate increase in body fat mass (Farooqi et al., 2001). These observations suggest that the modulation of energy balance by leptin levels may be more subtle than a simple on/off switch operating at a threshold level close to starvation. However, this partial leptin deficiency did not appear to be associated with any defect in gonadotropin secretion, with all heterozygotes having undergone appropriate development of secondary sexual characteristics, and adults having regular menstrual cycles, and fertility suggesting that a low level of leptin is sufficient to preserve these functions.

17.5.2. Response to Leptin Administration

We have undertaken a clinical trial of recombinant human leptin therapy in these four children with congenital leptin deficiency. We demonstrated that treatment of human congenital leptin deficiency with recombinant leptin can lead to a significant and sustained reduction in weight (Farooqi et al., 1999). As in leptin-treated *ob/ob* mice, this was predominantly due to loss of fat. During the first twelve months of leptin therapy, there was a gradual increase in basal and GnRH stimulated gonadotropin secretion, although there was no development of secondary sexual characteristics, and pelvic ultrasonography revealed an infantile uterus and ovaries. To examine the pulsatile secretion of LH and FSH, blood samples were taken every 10 min for 6 hr during the day and 12 hr overnight. After 12 months of leptin therapy, Child A had pulsatile secretion of FSH and LH at night consistent with the early stages of puberty (Figure 17.1). During the second year of treatment, it is notable that the increase in basal and stimulated gonadotropins was arrested at 18–20 months at a time when the child appeared to have relapsed and gained weight. Furthermore, when 24 hr LH and FSH pulsatility studies were repeated at 22 months, there was a complete regression of the pulsatile secretion that had previously been seen at night (Figure 17.2). With an increase in the dose of recombinant human leptin and a regain of the suppressive effect on appetite, the basal and stimulated gonadotropins increased and there was a restoration of pulsatility with increased amplitude of nocturnal secretion. Subsequently, Child A has gone on to progress through the stages of pubertal development, with clinical development of secondary sexual characteristics. This has been associated with a growth spurt, behavioral

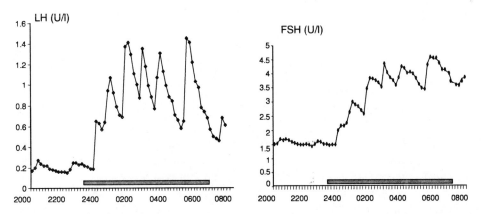

Figure 17.1. Pulsatile secretion of LH and FSH after 12 months of leptin therapy in congenital leptin deficiency.

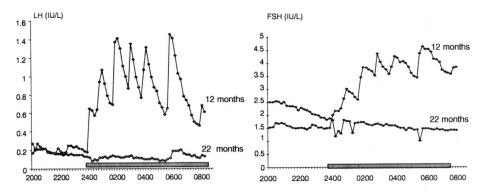

Figure 17.2. Regression of pulsatile gonadotropin secretion at 22 months of leptin treatment. This pattern was synchronous with a period of clinical relapse and weight gain.

changes associated with pubertal development, enlargement of the ovaries on ultrasound with observation of follicles, and an increase in uterine size. Recently Child A has had her first menstrual period and now has regular menstrual cycles.

For Child B, basal LH and FSH concentrations remained in the prepubertal range after 12 months of treatment. At 12 months, 24 hr FSH and LH pulsatility was measured, which showed that there was absolutely no change in LH or FSH concentrations with 10-min sampling, confirming the prepubertal state. Similarly there has been no evidence of premature pubertal development in Children C and D (age 4 and 6 years).

These studies in human congenital leptin deficiency suggest that leptin is permissive for the onset of puberty in humans, but only at an appropriate developmental stage. The observation that pituitary reserves and gonadotropin pulsatility parallel the effects of leptin on appetite, suggests that these two processes are closely coupled in humans.

Whether adequate serum leptin concentrations are permissive for normal pubertal development or, alternatively, that leptin plays a more active role in the initiation of puberty is not known. Based on our data we cannot currently establish whether leptin activates the reproductive axis or serves as a permissive signal that maintains reproductive function when circulating levels are above a certain threshold. It is likely that in these subjects, as in *ob/ob* mice, absent circulating levels of leptin indicate that the resources necessary for successful reproduction are unavailable.

17.6. DOWNSTREAM MEDIATORS OF LEPTIN ACTION

Although the precise mechanisms by which leptin mediates its effects on the timing of puberty are unknown, it is plausible that leptin exerts direct actions on the GnRH neurone, such as modifying its inherent pulsatility or capacity to express GnRH. However, there is little evidence for coexpression of the leptin receptor and GnRH at either the mRNA or protein level in the rat or monkey. The long form leptin receptor (Ob-Rb) is coexpressed with neuropeptide Y (NPY), agouti-related peptide (AgRP), proopiomelanocortin (POMC), and cocaine- and amphetamine-regulated transcript (CART) in the arcuate hypothalamic nucleus (Elmquist et al., 1998).

17.6.1. Melanocortin Pathways

NPY and AgRP are expressed in the same neurones in the medial arcuate nucleus, whereas POMC and CART are coexpressed in the lateral arcuate nucleus. Intracerebral injection of NPY stimulates feeding (i.e., orexigenic). In contrast, αMSH (a product of POMC) and CART inhibit feeding (i.e., anorexigenic). αMSH is thought to regulate feeding through melanocortin 4 (MC4) receptors in the hypothalamus. AgRP antagonizes the actions of αMSH at MC4 receptors and is therefore orexigenic. We have shown that mutations in the melanocortin 4 receptor occur in 5 percent of our cohort of severely obese children, making this the commonest monogenic form of human obesity thus far described (Farooqi et al., 2000). Most of the mutations appear to be dominantly inherited with variable penetrance in some subjects. These mutations are associated with either a complete or partial loss of function in an invitro assay, and co-transfection studies of mutant and wild-type receptors indicate that the likely mechanism for a pathological effect is haploinsufficiency, rather than dominant negativity. Studies of patients with MC4 receptor mutations reveal that this obesity syndrome is characterized by hyperphagia, increased linear growth throughout childhood, and severe hyperinsulinemia. These features are similar to those seen in MC4R knockout mice and in agouti mice, suggesting the preservation of melanocortin pathways in rodents and humans. Importantly in both rodent and humans with genetic lesions in the melanocortin pathway, reproductive function is preserved with appropriate pubertal development and normal fertility.

Thus, it is more likely that leptin's effects are mediated by other neurotransmitters such as glutamate, gamma amino butyric acid (GABA) and NPY, which have been proposed as potential regulators of GnRH activity during puberty (Aubert et al., 1998).

17.6.2. Neuropeptide Y

Leptin suppresses hypothalamic NPY gene expression and release, and could potentially lead to activation of the GnRH neurone as widespread coexpression of leptin receptor and preproNPY mRNA has been observed in the arcuate nucleus of the mouse hypothalamus (Mercer et al., 1996). Furthermore, NPY has been shown to stimulate GnRH release acutely, and acts in an inhibitory fashion when given chronically or when given to animals with low estrogen levels (Gruaz et al., 1998). However, NPY cannot be the sole mediator of leptin's effects as NPY knockout mice are not infertile (Erickson et al., 1996) and have normal serum gonadal steroid concentrations when fed ad libitum (Erickson et al., 1997).

Studies in rodents and humans with complete deficiency of the adipocyte derived hormone leptin suggest that leptin may serve to be the molecular explanation for the observed physiological relationship between body weight and the onset of puberty in humans. However, the mechanisms through which these effects are mediated are unclear. Continuation of recombinant human leptin therapy in leptin deficient subjects will provide a unique opportunity to examine whether leptin does indeed act as a metabolic gate for the onset of puberty in male and female humans and whether leptin is also the signal for the maintenance of reproductive function in adult life.

REFERENCES

Ahima, R. S., Dushay, J., Flier, S. N. et al. (1997). Leptin accelerates the onset of puberty in normal female mice. *J. Clin. Invest.*, *99*(3), 391–395.

Aubert, M. L., Pierroz, D. D., Gruaz, N. M. et al. (1998). Metabolic control of sexual function and growth: role of neuropeptide Y and leptin. *Mol. Cell. Endocrinol.*, *140*(1–2), 107–113.

Barash, I. A., Cheung, C. C., Weigle, D. S. et al. (1996). Leptin is a metabolic signal to the reproductive system. *Endocrinology*, *137*(7), 3144–3147.

Bronson, F. H. and Rissman, E. F. (1986). The biology of puberty. *Biol. Rev. Camb. Philos. Soc.*, *61*(2), 157–195.

Chehab, F. F., Lim, M. E., and Lu, R. (1996). Correction of the sterility defect in homozygous obese female mice by treatment with the human recombinant leptin. *Nat. Genet.*, *12*(3), 318–320.

Chen, H., Charlat, O., Tartaglia, L. A. et al. (1996). Evidence that the diabetes gene encodes the leptin receptor: identification of a mutation in the leptin receptor gene in db/db mice. *Cell*, *84*(3), 491–495.

Chung, W. K., Belfi, K., Chua, M. et al. (1998). Heterozygosity for Lep(ob) or Lep(rdb) affects body composition and leptin homeostasis in adult mice. *Am. J. Physiol.*, *274*(4 Pt 2), R985–990.

Clement, K., Vaisse, C., Lahlou, N. et al. (1998). A mutation in the human leptin receptor gene causes obesity and pituitary dysfunction. *Nature*, *392*(6674), 398–401.

Coleman, D. L. (1978). Obese and diabetes: two mutant genes causing diabetes-obesity syndromes in mice. *Diabetologia*, *14*(3), 141–148.

Considine, R. V., Considine, E. L., Williams, C. J. et al. (1995). Evidence against either a premature stop codon or the absence of obese gene mRNA in human obesity [see comments]. *J. Clin. Invest.*, *95*(6), 2986–2988.

Ebling, F. J., Wood, R. I., Karsch, F. J. et al. (1990). Metabolic interfaces between growth and reproduction. III. Central mechanisms controlling pulsatile luteinizing hormone secretion in the nutritionally growth-limited female lamb. *Endocrinology*, *126*(5), 2719–2727.

Elmquist, J. K., Bjorbaek, C., Ahima, R. S. et al. (1998). Distributions of leptin receptor mRNA isoforms in the rat brain. *J. Comp. Neurol.*, *395*(4), 535–547.

Erickson, J. C., Ahima, R. S., Hollopeter, G. et al. (1997). Endocrine function of neuropeptide Y knockout mice. *Regul. Pept.*, *70*(2–3), 199–202.

Erickson, J. C., Clegg, K. E., and Palmiter, R. D. (1996). Sensitivity to leptin and susceptibility to seizures of mice lacking neuropeptide Y. *Nature*, *381*(6581), 415–421.

Farooqi, I. S., Jebb, S. A., Langmack, G. et al. (1999). Effects of recombinant leptin therapy in a child with congenital leptin deficiency. *N. Engl. J. Med.*, *341*(12), 879–884.

Farooqi, I. S., Keogh, J. M., Kamath, S. et al. (2001). Partial leptin deficiency and human adiposity. *Nature*, *414*(6859), 34–35.

Farooqi, I. S., Yeo G. S., Keogh, J. M. et al. (2000). Dominant and recessive inheritance of morbid obesity associated with melanocortin 4 receptor deficiency. *J. Clin. Invest.*, *106*(2), 271–279.

Frisch, R. E. and Renelle, R. (1971). Height and weight at menarche and a hypothesis of menarche. *Arch. Dis. Child.*, *46*, 695–701.

Gruaz, N. M., Lalaoui, M., Pierroz, D. D. et al. (1998). Chronic administration of leptin into the lateral ventricle induces sexual maturation in severely food-restricted female rats. *J. Neuroendocrinol.*, *10*(8), 627–633.

Hamilton, B. S., Paglia, D., Kwan, A. Y., and Deitel, M. (1995). Increased obese mRNA expression in omental fat cells from massively obese humans. *Nat. Med.*, *1*(9), 953–956.

Hummel, K. P. (1957). Transplantation of ovaries of the obese mouse. *Anat. Rec.*, *128*, 569.

Ingalls, A. M., Dickie, M. M., and Snell, G. D. (1950). Obese, a new mutation in the house mouse. *Obes. Res.*, *4*(1), 101.

Ioffe, E., Moon, B., Connolly, E., and Friedman, J. M. (1998). Abnormal regulation of the leptin gene in the pathogenesis of obesity. *Proc. Natl. Acad. Sci. USA*, *95*(20), 11852–11857.

Kennedy, G. C. and Mitra, J. (1963). Body weight and food intake as initiating factors for puberty in the rat. *J. Physiol.*, *166*, 408–418.

Lane, P. W. and Dickie, M. M. (1954). Fertile obese male mice. Relative sterility in obese males corrected by dietary restriction. *J. Hered.*, *45*, 56–58.

Maffei, M., Stoffel, M., Barone, M. et al. (1996). Absence of mutations in the human OB gene in obese/diabetic subjects. *Diabetes*, *45*(5), 679–682.

Mercer, J. G., Hoggard, N., Williams, L. M. et al. (1996). Coexpression of leptin receptor and prepronneuropeptide Y mRNA in arcuate nucleus of mouse hypothalamus. *J. Neuroendocrinol.*, *8*(10), 733–735.

Montague, C. T., Farooqi, I. S., Whitehead, J. P. et al. (1997). Congenital leptin deficiency is associated with severe early-onset obesity in humans. *Nature*, *387*(6636), 903–908.

Nagatani, S., Guthikonda, P., Thompson, R. C. et al. (1998). Evidence for GnRH regulation by leptin: leptin administration prevents reduced pulsatile LH secretion during fasting. *Neuroendocrinology*, *67*(6), 370–376.

Ogawa, Y., Masuzaki, H., Hosoda, K. et al. (1999). Increased glucose metabolism and insulin sensitivity in transgenic skinny mice overexpressing leptin. *Diabetes*, *48*(9), 1822–1829.

Strobel, A., Issad, T., Camoin, L. et al. (1998). A leptin missense mutation associated with hypogonadism and morbid obesity. *Nat Genet.*, *18*(3), 213–215.

Swerdloff, R. S., Batt, R. A., and Bray, G. A. (1976). Reproductive hormonal function in the genetically obese (ob/ob) mouse. *Endocrinology*, *98*(6), 1359–1364.

Swerdloff, R. S., Peterson, M., Vera, A. et al. (1978). The hypothalamic–pituitary axis in genetically obese (ob/ob) mice: response to luteinizing hormone-releasing hormone. *Endocrinology*, *103*(2), 542–547.

Warren, M. P. (1980). The effects of exercise on pubertal progression and reproductive function in girls. *J. Clin. Endocrinol. Metab.*, *51*(5), 1150–1157.

Yura, S., Ogawa, Y., Sagawa, N. et al. (2000). Accelerated puberty and late-onset hypothalamic hypogonadism in female transgenic skinny mice overexpressing leptin. *J. Clin. Invest.*, *105*(6), 749–755.

Zhang, Y., Proenca, R., Maffei, M. et al. (1994). Positional cloning of the mouse obese gene and its human homologue. *Nature*, *372*(6505), 425–432.

VI

Clinical Challenges

18

Does Leptin Play a Role in Preeclampsia?

LUCILLA POSTON

18.1. THE SYNDROME OF PREECLAMPSIA

Preeclampsia affects between 2.6 and 7.3 percent of all pregnancies (Levine et al., 1997). It is defined as a gestational increase in blood pressure with superimposed proteinuria (Davey and McGillivray, 1988) but these are outward symptoms of a what is now known to be a complex multisystem disorder (Roberts and Redman, 1993; Higgins and de Sweit, 2001). Women with preeclampsia may have fits (eclampsia), or may develop renal and liver complications. Oedema is common and pulmonary edema or stroke may lead to maternal death. Preeclampsia, one of the commonest disorders of pregnancy, therefore remains a major cause of maternal and fetal morbidity and mortality (de Swiet, 2000; CESDI, 2001; Dekker and Sibai, 1999). Neonatal morbidity is also high, largely because the only cure for preeclampsia is delivery which may be indicated from as early as 26 weeks of gestation. The disorder is, therefore, a major cause of premature birth and its attendant problems of immaturity. The origins of preeclampsia remain poorly defined although the placenta undoubtedly plays a pivotal role. Poor placentation resulting from impairment of trophoblast invasion of the spiral arteries (Dekker and Sibai, 1999) and maternal "susceptibility" factors are proposed to interact to promote vasoconstriction leading to hypertension and potentially to seizures and to hepatic and renal dysfunction (Roberts and Hubel, 1999). The disorder is associated with intense vasoconstriction and maternal vascular endothelial cell activation (Roberts and Redman, 1993; Roberts and Cooper, 2001; Roberts and Hubel, 1999). However, activation of the endothelium has now been ascertained to be part of a genereralized inflammatory state, involving substantial lecucocyte activation (Redman et al., 1999). Poor perfusion of the placenta and the inflammatory response are considered to

LUCILLA POSTON • Maternal and Fetal Research Unit, Guy's Kings and St. Thomas' School of Medicine, King's College—London.

Leptin and Reproduction. Edited by Henson and Castracane, Kluwer Academic/Plenum Publishers, 2003.

299

contribute to a state of oxidative stress and increased generation of lipid peroxides, facilitated by maternal dyslipidaemia (Hubel and Roberts, 1999; Poston, 2001). Lipid peroxides are very damaging to endothelial cells and may contribute to leucocyte adhesion and platelet activation and to reduced synthesis of endothelium derived vasodilators, so increasing the risk of hypercoagulation and hypertension. These in turn may give rise to all the signs and symptoms of the syndrome.

18.2. LEPTIN IN NORMAL PREGNANCY

The leptin profile observed during normal gestation has been considered in other chapters and will not be addressed in any detail here. Maternal blood concentrations rise 2 to 4-fold (Schubring et al., 1998; Geary et al., 1999; McCarthy et al., 1999), peaking in the second trimester and falling rapidly after delivery (Masuzaki et al., 1997). The role of leptin in normal pregnancy is not known but it may play a role in successful pregnancy outcome since women suffering from recurrent miscarriage have low blood leptin concentrations (Laird et al., 2001). The human placenta is a source of leptin (Green et al., 1995; Masuzaki et al., 1997; Lea et al., 2000) and leptin mRNA, leptin and leptin receptors are present in the placental syncytiotrophoblast, the site of nutrient and gas exchange (Lea et al., 2000). The human placental leptin gene also has a promoter region with a different sequence to that of adipocytes which suggests that placental leptin may be differentially regulated compared to that synthesized in adipocytes (Bi et al., 1997). The placenta, therefore, synthesizes leptin and can be also be a target organ.

18.3. WHY INVESTIGATE LEPTIN IN PREECLAMPSIA?

Leptin has been implicated in lipid metabolism, obesity and insulin resistance. All three are associated with preeclampsia. Since the placenta also plays a pivotal and undisputed contribution to the origin of preeclampsia, there is a good rationale for investigation of leptin in preeclampsia. Many investigations of preeclampsia have been compromised by inaccurate classification of the disorder; in this review only those studies which have classified preeclampsia by internationally accepted criteria have been included.

18.4. LEPTIN IN MATERNAL BLOOD IN PREECLAMPSIA

Most investigations of leptin in preeclampsia have simply involved measurement of leptin in the maternal blood (Mise et al., 1998; Sattar et al., 1998; McCarthy et al., 1999; Williams et al., 1999; Anim-Nyame et al., 2000; Laivuori et al., 2000; Teppa et al., 2000; Laml et al., 2001; Vitoratos et al., 2001; Chappell et al., 2002a,b). Protocols show little conformity. Blood has been sampled before the onset of clinical disease (Williams et al.,

1999; Anim-Nyame et al., 2000; Chappell et al., 2002a,b), or just before delivery but before labour onset (Mise et al., 1998; McCarthy et al., 1999) and one study (Laml et al., 2001) measured leptin in samples taken at delivery. Disease severity also varies as indicated by differences in the mean gestational age at delivery in the patient groups studied. Most investigators have measured total (i.e., bound and free) concentrations of leptin and the majority has used the same source of radioimmunoassay (RIA, Linco, MO, USA). The majority has studied carefully matched controls and PE patients for gestational age.

Elevation of the maternal plasma or serum leptin concentration has almost uniformly been observed in the maternal blood of women with preeclampsia or in those who later developed the disorder (Mise et al., 1998; McCarthy et al., 1999; Williams et al., 1999; Anim-Nyame et al., 2000; Laivouri et al., 2000; Teppa et al., 2000; Bartha et al., 2001; Vitoratos et al., 2001; Chappell et al., 2002a). The lower leptin concentration reported by Laml et al. (2001) is an unexplained different finding, although preeclampsia may have been more mild in the women in this study compared with others, as gestational age at delivery was 40 weeks. Also, as the women were investigated at delivery leptin concentrations may have been affected by the process of labour. An early study by Sattar et al. (1998) reported no difference in preeclamptic and normal pregnancies, but this was a small study with a trend towards higher values in the preeclamptic group.

Several authors have reported an increase in leptin rises before the onset of the clinical disease (Williams et al., 1999; Anim-Nyame et al., 2000; Chappell et al., 2002a) although a recent preliminary report (Laivouri and Roberts, 2002) observed that leptin increased in the second trimester only in those women who later developed preeclampsia near term. No rise occurred in those with early onset disease. However, this was a retrospective investigation which included samples from women taking a low dose of aspirin. There appeared to be no effect of aspirin on the leptin concentrations but it would be of interest to verify this observation in a prospectively studied group not taking aspirin prophylaxis. The early elevation of leptin in women destined to develop preeclampsia might suggest a pathophysiological role rather than failure of excretion through reduced renal leptin clearance, which might occur in women with established disease in whom renal function is often compromised. This observation has led to interest in the potential of leptin as an early predictor of the disorder. Indeed, in a study of 81 women at risk of preeclampsia, we have shown that serum leptin was effective in prediction of the disease in 21 women who later developed it (Chappell et al., 2002a) (Figure 18.1). Using receiver operation characteristic (ROC) curves to assess the predictive value, the area under the curve for leptin at 20 weeks gestation was 0.71, where chance $= 0.5$ and perfect value $= 1.00$, and at 24 weeks' gestation was 0.77 (Figure 18.2). It was of interest that leptin concentrations did not rise in women who delivered babies small for gestational age but without associated preeclampsia, which would agree with reports of reduced leptin concentrations in pregnancies associated with growth restricted fetuses (Lea et al., 2000; Linnemann et al., 2001). We also found that the concentration of placenta growth factor (PlGF) failed to show the normal pregnancy related rise in women who developed preeclampsia and when the serum leptin concentration was combined with the concentration of placenta growth factor PlGF (leptin/PlGF ratio), predictive power was improved. Similarly an algorithm combing leptin and the plasminogen activator inhibitor-1 : plasminogen activator inhibitor 2 ratio (PAI-1:PAI-2 ratio) provided good predictive power at 20 and 24 weeks' gestation. In our study, risk was assessed on the basis of an abnormal

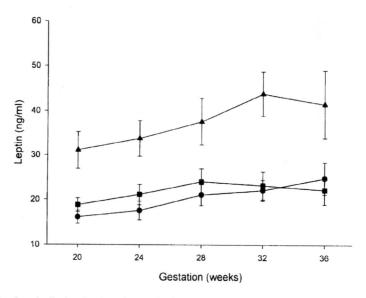

Gestation (weeks)

Figure 18.1. Longitudinal evaluation of serum leptin concentrations in low risk women (■) and from women in the high risk group who developed preeclampsia (▲) or who did not develop preeclampsia but delivered small for gestational age infants (●). Data are given as geometric mean ± SEM. From Chappell et al., 2002a (with permission, *Am. J. Obstet. Gynecol.*).

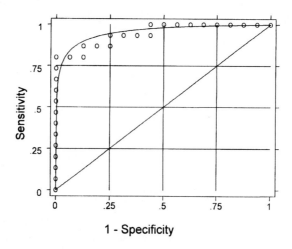

1 - Specificity

Figure 18.2. ROC curve for the prediction of preeclampsia at 24 weeks' gestation using the algorithm $Log_e PlGF-3.0\{PAI-1/PAI-2 ratio\}$ in comparison with low risk women with normal outcome. This algorithm achieves a combination of high sensitivity and specificity for the later prediction of preeclampsia. From Chappell et al., 2002a (with permission, *Am. J. Obstet. Gynecol.*).

uterine artery waveform or because of previous preeclampsia. Further prospective studies are now required to determine whether these makers alone, or in combination, may provide adequate sensitivity and specificity for prediction of preeclampsia in women in whom risk is assessed by standard clinical criteria (e.g., diabetes, underlying essential hypertension), or indeed in low risk women.

18.5. THE POSSIBLE ORIGINS OF RAISED MATERNAL BLOOD LEPTIN CONCENTRATIONS IN PREECLAMPSIA

The association between insulin resistance and plasma leptin in non-pregnant subjects is well recognized and an association has often been observed between insulin resistance and preeclampsia (Solomon and Seely, 2001). However, no association between leptin and insulin resistance, as assessed by an oral glucose tolerance test was found in a study of leptin and insulin resistance in women with preeclampsia (Laivuori et al., 2000) but this isolated investigation is worthy of repetition. In non-pregnant subjects, blood concentrations of leptin seem to be a reflection of adiposity as there is a strong correlation the body mass index (BMI). In pregnancy, a relationship between maternal BMI and leptin has also generally been found (Butte et al., 1997; Hartmann et al., 1997), but with exceptions (Schubring et al., 1997; McCarthy et al., 1999). In preeclampsia, plasma leptin increases independently of the BMI, an observation supported by several different studies. In one, leptin was raised compared to controls despite there being no difference in pre-pregnancy BMI between PE and control groups (McCarthy et al., 1999). Similarly case control studies in which BMI has been carefully matched (Mise et al., 1998; Williams et al., 1999) have shown differences in leptin concentrations between PE and control groups. Williams et al. (1999) found a weak and non-significant association with pre-pregnancy BMI and leptin in preeclampsia whereas there was a strong assocation in normal pregnancy. It was of interest that only those women with a low or normal BMI who later developed preeclampsia had a raised second trimester serum leptin concentration. Overall, therefore, an increase in BMI and associated increase in adipocyte mass would not provide an obvious explanation for the increase of maternal blood leptin concentrations in preeclampsia. It could be hypothesized, nonetheless, that increased stimulation of leptin synthesis from existing adipocytes might be responsible for the rise in the maternal leptin concentration, a possibility addressed recently in a preliminary report (Laivouri et al., 2002). These authors found that leptin mRNA expression was similar in adipocytes from women with preeclampsia as in those from normal women. Moreover the expression of two target enzymes for leptin; fatty acid synthase and fatty acid translocase, were no different in adipocytes from the two subject groups, leading the authors to conclude that the placenta rather than adipocytes was the most likely source of leptin in women with preeclampsia.

18.5.1. The Placenta as a Source of Leptin

Leptin messenger RNA levels are increased in placentas from preeclamptic women compared with those from normal women, an observation first reported by Mise et al. (1998). To further implicate the placenta, this study also reported that, post delivery, concentrations of leptin in the maternal circulation were found to fall to those which would be anticipated from the woman's BMI. Dötsch et al. (1999) later reported an increase in leptin message in the placentas of women with preeclampsia compared to gestation matched controls, when expressed as the ratio of leptin/βactin (a housekeeping gene). As placental mass is often reduced in preeclampsia, a simple relationship between placental mass and leptin synthesis is an unlikely explanation for the rise in maternal circulating concentrations, and other explanations must be sought.

18.5.2. The Role of Placental Hypoxia

Incomplete placentation in preeclampsia may lead to underperfusion of the placenta and to relative hypoxia. A relationship between hypoxia and leptin was first suggested in non-pregnant subjects by the demonstration of raised leptin concentrations at high altitude, thus providing a potential explanation for appetite loss and weight loss (Tschop et al., 1998). Mise et al. (1998) subsequently showed that hypoxia stimulates leptin synthesis in BeWo cells, a cell line derived from placental trophoblast and Grosfeld et al. (2001) have since identified a hypoxia response element on the promoter region of the ob gene in BeWo cells. Hypoxia may, therefore upregulate placental leptin expression through a transcriptional mechanism likely to involved in distinct sequences on the promoter region of the gene. The specificity of the placental gene promoter was highlighted by the recent observation that hypoxia *reduces* the synthesis of leptin from cultured rat adipocytes (Yasumasu et al., 2002). Incidentally, this finding also argues against hyperleptinaemia at altitude being the result of a direct effect of hypoxia on adipocytes.

18.5.3. Interactions with Inflammatory Mediators

Normal pregnancy is a state of mild inflammation as evidenced by leucocyte activation (Sacks et al., 1998), and from a recent study in our laboratory (Chappell et al., 2002a), by modest endothelial cell activation. In women with preeclampsia aggressive leucocyte and endothelial cell activation occur and there is an increase in the circulating concentrations of a range of cytokines, for example, TNFα and IL-6 (Redman et al., 1999). Leptin is a member of the cytokine family (Zhange et al., 1997) and although a cytokine itself, leptin synthesis is stimulated by cytokines (Grunfeld et al., 1996). For example, leptin production in rodents is increased after adminstration of TNFα (Grunfeld et al., 1996) and leptin has been implicated in the anorexia of acute inflammatory conditions (Sarraf et al., 1997). The elevation of the maternal blood leptin concentration in preeclampsia could, therefore, reflect an increase in adipocyte/trophoblast leptin synthesis via cytokine stimulation. In normal pregnancy, Clapp and Kiess (2000) have previously reported a relationship between leptin and TNFα in early and mid pregnancy but not in late pregnancy. More recently, Bartha et al. (2001) in a study of 27 women with preeclampsia and 25 controls, investigated relationships between leptin, BMI, and the cytokines TNFα and IL-6. In agreement with earlier studies they found a relationship between BMI and leptin in normal pregnancy and similarly to Clapp and Kiess (2000), found a relationship with leptin and TNFα. For the first time these authors also reported a correlation with IL-6 and leptin. In preeclamptic women, leptin was raised, the relationship with BMI and IL-6 was lost but the association with TNFα was maintained. TNFα has also been linked to insulin resistance (Hotamisligil, 1999), raising the intriguing possibility that it may be a common factor linking increased leptin concentrations to the insulin resistance of preeclampsia.

18.6. POTENTIAL BENEFITS OF RAISED LEPTIN IN PREECLAMPTIC PREGNANCIES

Leptin normally decreases the appetite, but this cannot be the case in normal pregnancy in which concentrations are raised and the appetite increases. Pregnancy would appear therefore, at least in this respect, to be a leptin resistant state. Indeed, animal studies have suggested that pregnancy is associated with resistance to the satiety induced by leptin through the hypothalamic Ob-re receptor (Garcia et al., 2000). In preeclampsia, it has been suggested that the further rise in leptin could overcome leptin resistance at the level of the placenta in order to improve placental nutrient availability in the face of adversity. The rise in leptin in preeclampsia could also be an attempt to increase placental growth, since leptin may be an angiogenic factor (Sierra-Honigmann et al., 1998; Park et al., 2001). However, there appear to be no leptin receptors on vascular endothelial cells in the placenta (Hoggard et al., 2001) and if leptin has an angiogenic influence it must occur indirectly through other mediators. Placental apoptosis is recognized to occur in the placenta in preeclampsia (Leung et al., 2001), and leptin may blunt responses to apoptotic stimuli, for example, hypoxia. Stimulation of nitric oxide synthesis by leptin (Kimura et al., 2000) could also play a beneficial role by counteracting endothelial dysfunction, but it is debatable whether physiological concentrations of leptin are adequate to stimulate NO synthesis (Mitchell et al., 2001).

The stimulation of leptin synthesis by cytokines could be part of a protective response to preeclampsia. Leptin deficiency is known to increase susceptibility to LPS and TNF induced lethality (Takahashi et al., 1999) and liver damage (Yang et al., 1997), and leptin deficient (ob/ob mice) have a greater number of monocytes and dysfunctional macrophages together with higher levels of pro-inflammatory cytokines and lower concentrations of anti-inflammatory cytokines (for review see Faggioni et al., 2001). It could follow that high concentrations of leptin may serve a protective role against damage by pro-inflammatory cytokines such as $TNF\alpha$.

18.7. POTENTIAL DISADVANTAGES OF RAISED LEPTIN IN PREECLAMPTIC PREGNANCIES

Preeclampsia is associated with an increase in sympathetic activity (Schobel et al., 1996), as elegantly demonstrated by measurement of increased postganglionic sympathetic-nerve activity in the blood vessels of patients' skeletal muscle. As a result, increased lipolysis may occur and it has been suggested that this provides an explanation for the often reported increase in plasma fatty acids in preeclampsia. No satisfactory explanation has been provided but it is possible that leptin may play a role as it is known to increases sympathetic activity via central pathways (Hall et al., 2001).

Leptin may also play a role in the acquired immune response. The absence of leptin in ob/ob mice is associated with reduced sensitivity to T cell activating stimuli. Leptin also polarizes T helper cells toward a Th1 phenotype (Martin-Romero et al., 2000). It may be relevant that preeclampsia is associated with a hitherto unexplained shift to a population of Th1 cells (Saito et al., 1999a) and that the T cells show evidence of activation (Saito et al., 1999b). It is interesting to speculate that the Th1 predominance and T cell activation in preeclampsia may be the consequence of elevation of plasma leptin.

The origin of oxidative stress in preeclampsia is likely to be multifactorial, with contributions from the placenta and from maternal leucocytes and endothelium. Leptin could also be implicated through increased synthesis of the free radicals at the level of the endothelial cell (Yamagishi et al., 2001). The repeated demonstration of oxidative stress in preeclampsia led to the carrying out of a randomized controlled trial of antioxidants vitamin C (1 gm/day) and vitamin E (400 IU/day) in a placebo controlled study in women at risk (Chappell et al., 1999) with risk being assessed on the basis of an abnormal uterine artery Doppler waveform analysis or because of previous preeclampsia. There was more than a 50 percent reduction in the incidence of preeclampsia in those women taking the active preparation. The concentrations of a number of markers of the disease process including plasma/serum concentrations of a lipid peroxides (assessed by evaluation of the isoprostane, 8-epi-$PGF_{2\alpha}$) and of leptin wave also assessed and compared these to concentrations in a simultaneously studied group of low risk women. The placebo group demonstrated high plasma concentrations of leptin from 16 weeks of gestation onwards but leptin concentrations fell in those women receiving antioxidants to values similar to those of controls (Chappell et al., 2002b). This was paralleled by a reduction in the plasma concentrations of 8 epi-prostaglandin $F_{2\alpha}$. It can be postulated that either (a) the antioxidants improved placental function and reduced placental hypoxia thereby reducing the hypoxic stimulus to leptin synthesis or (b) the antioxidants may have reduced cytokine synthesis from leucocytes (Erl et al., 1997), so reducing leptin synthesis from trophoblast and adipocyte by lowering the circulating cytokine concentrations.

18.8. LEPTIN CONCENTRATIONS IN UMBILICAL CORD BLOOD IN PREECLAMPSIA

Leptin has been implicated in fetal growth. In normal pregnancies, cord blood leptin concentrations are related to fetal weight or neonatal fat mass (Linneman et al., 2001). Higher concentrations in the umbilical vein than in umbilical arteries suggest the placenta as the source, although fetal tissues may also contribute. Few data are available from preeclamptic pregnancies but in a recent comprehensive study of cord plasma leptin from Norway in which 256 preeclamptic and 607 control pregnancies were investigated, Odegard et al. (2002) reported an increased concentration in blood from the preeclamptic group. Cord plasma leptin increased with gestational age in both groups, but at each gestational age values were higher in the preeclamptic pregnancies. The controls had a higher ponderal index, but adjustment for this, gestational age and gender did not affect the demonstration of increased concentrations in preeclampsia. Neither was there any difference between cord blood leptin concentrations from mild and severe cases of preeclampsia. The consistently higher concentrations of leptin in female newborns in both groups compared with males may reflect an increase in adipose tissue in females and also argued for a contribution of fetal adipocytes to cord blood leptin. These data from this large study contrast to the low concentrations in growth restricted pregnancies uncomplicated by preeclampsia (Linnemann et al., 2001). Two much smaller studies have found no difference between thecord blood leptin concentrations in preeclampsia and control groups (McCarthy et al., 1999; Diaz et al., 2002). McCarthy et al. (1999) reported a correlation with maternal plasma and cord leptin concentrations in the preeclamptic women, but not amongst control subjects and Laml et al. (2000) have reported positive associations between cord leptin blood concentrations and birth weight in 52 preeclamptic pregnancies. A positive relationship with birthweight was also found by Diaz et al. (2002) in 15 preeclamptic pregnancies. This relationship with birthweight would concur with the suggestion that an increase in leptin in both fetus and mother in preeclampsia could be a protective measure to maintain fetal growth. There is also indication in neonates that leptin may be raised when the mother has preeclampsia as one study of premature infants of gestational age 24- to 32-weeks (Hytinantti et al., 2000) has reported an independent association between the infants' plasma leptin concentrations and maternal preeclampsia in a multiple regression analysis when association between leptin with gestational age, preeclampsia and exposure to antenatal steroids were explored.

18.9. CONCLUSIONS

Most reports suggest an elevation of leptin in preeclamptic pregnancies. This review has attempted to explain why this may occur and what the consequences may be. However, much of what is written has to remain in the realms of speculation as very few mechanistic studies have as yet been attempted. Inferences are drawn almost entirely from a complex and often contradictory literature relating to the non-pregnant state and will only be substantiated by future investigation in pregnant animals and women.

REFERENCES

Anim-Nyame, N., Sooranna, S. R., Steer, P. J., and Johnson, M. R. (2000). Longitudinal analysis of maternal plasma leptin concentrations during normal pregnancy and preeclampsia. *Hum. Reprod.*, *15*, 2033–2036.

Bartha, J. L., Romero-Carmona, R., Escobar-Llompart, M., and Comino-Delgado, R. (2001). The relationships between leptin and inflammatory cytokines in women with preeclampsia. *Brit. J. Obstet. Gynaecol.*, *108*, 1272–1276.

Bi, S., Gavrilova, O., Gong, D. W., Mason, M. M., and Reitman, M. (1997). Identification of a placental enhancer for the human leptin gene. *J. Biol. Chem.*, *272*, 30583–30588.

Butte, N. F., Hopkinson, J. M., and Nicolson, M. A. (1997). Leptin in human reproduction: serum leptin levels in pregnant and lactating women. *J. Clin. Endocrinol. Metab.*, *82*, 585–589.

CESDI. (2001). Why mothers die 1997–1999. The confidential enquiries into maternal deaths in the United Kingdom. RCOG Press 2001.

Chappell, L. C., Seed, P. T., Briley, A. L., Kelly, K. J., Lee, R., Hunt, B. J., Parmer, K., Bewley, S. J., Shennan, A. H., Steer, P. J., and Poston, L. (1999). Effect of antioxidants on the occurrence of preeclampsia in women at increased risk: a randomized trial. *Lancet*, *354*, 810–816.

Chappell, L. C., Seed, P. T., Kelly, F. J., Hunt, B. J., Briley, B., Stock, M. J., Charnock-Jones, D. S., Smith, S. K., Mallet, A. I., and Poston, L. A. (2002a). Longitudinal study of biochemical parameters in women at risk of pre-eclampsia. *Am J Obstet Gynecol.*, *187*, 127–136.

Chappell, L. C., Seed, P. T., Kelly, F. J., Briley, A., Mallet, A. I., Hunt, B. J., Charnock-Jones, D. S., and Poston, L. (2002b). Vitamin E and C supplementation in women at risk of pre-eclampsia is associated with changes in indices of oxidative stress and placental function. *Am. J. Obstet. Gynecol., 187*, 777–784.

Clapp, J. F. and Kiess, W. (2000). Effects of pregnancy and exercise on concentrations of the metabolic markers tumor necrosis factor a and leptin. *Am. J. Obstet. Gynecol., 182*, 300–306.

Davey, D. A. and MacGillivray, I. (1988). The classification and definition of the hypertensive disorders of pregnancy. *Am. J. Obstet. Gynecol., 158*, 892–898.

Dekker, G. A. and Sibai, B. M. (1999). Etiology and pathogenesis of preeclampsia: current concepts. *Am. J. Obstet. Gynecol., 181*, 1037–1037.

Diaz, E., Halhali, A., Luna, C., Diaz, L., Avila, E., and Larrea, F. (2002). Newborn birth weight correlates with placental zinc, umbilical inslulin-like growth factor and leptin levels in preeclampsia. *Arch. Med. Res., 33*, 40–47.

Dötsch, J., Nüsken, K.-D., Knerr, I., Kirschbaum, M., Repp, R., and Rascher, W. (1999). Leptin and neuropeptide Y gene expression in human placenta: ontogeny and evidence for similarities to hypothalamic regulation. *J. Clin. Endocrinol. Metab., 84*, 2755–2758.

Erl, W., Weber, C., Wardemann, C., and Weber, P. C. (1997). Alpha-tocopheryl succinate inhibits monocytic cell adhesion to endothelial cells by suppressing NF-kB mobilization. *Am. J. Physiol., 273*, H634–H640.

Faggioni, R., Feinfold, K. R., and Grunfeld, C. (2001). Leptin regulation of the immune response and the immunodeficiency of malnutrition. *FASEB. J., 15*, 2566–2571.

Garcia, M. D., Casanueva, F. F., Dieguez, C., and Senaris, R. M. (2000). Gestational profile of leptin messenger ribonucleic acid (mRNA) content in the placenta and adipose tissue in the rat, and regulation of the mRNA levels of the leptin receptor subtypes in the hypothalamus during pregnancy and lactation. *Biol. Reprod., 62*, 698–703.

Geary, M., Pringle, P. J., Persaud, M., Wilshin, J., Hindmarsh, P. C., Rodeck, C. H., and Brook, C. G. (1999). Leptin concentrations in maternal serum and cord blood: relationship to maternal anthropometry and fetal growth. *Br. J. Obstet. Gynaecol., 106*, 1054–1060.

Green, E. D., Maffei, M., Braden, V. V., Proenca, R., DeSilva, U., Zhang, Y., Chua, S. C. Jr., Leibel, R. L., Weissenbach, J., and Friedman, J. M. (1995). The human obese (OB) gene: RNA expression pattern and mapping on the physical, cytogenic, and genetic maps of chromosome 7. *Genome Res., 5*, 5–12.

Grosfeld, A., Turban, S., Andre, J., Cauzac, M., Challier, J. C., Hauguel-de Mouzon, S., and Guerre-Millo, M. (2001). Transcriptional effect of hypoxia on placental leptin. *FEBBS. Lett., 502*, 122–126.

Grunfeld, C., Zhao, C., Fuller, J., Pollock, A., Moster, A., Friedman, J., and Feingold, K. R. (1996). Endotoxin and cytokines induce expression of leptin, the *ob* gene product, in hamsters. A role for leptin in the anorexia of infection. *J. Clin. Invest., 97*, 2152–2157.

Hall, J. E., Hildebrandt, D. A., and Kuo, J. (2001). Obesity hypertension: role of leptin and sympathetic nervous system. *Am. J. Hypertens., 6*, 103S–115S.

Hartmann, B. W., Wagenbichler, P., and Soregi, G. (1997). Maternal and umbilical-cord serum leptin concentrations in normal, full-term pregnancies. *N. Engl. J. Med., 337*, 863.

Higgins, J. R. and de Swiet, M. (2001). Blood pressure measurement and classification in pregnancy. *Lancet, 357*, 131–135.

Hoggard, N., Haggarty, P., Thomas, L., and Lea, R. G. (2001). Leptin expression in placental and fetal tissues: does leptin have a functional role? *Biochem. Soc. Trans., 29*, 57–63.

Hotamisligil, G. S. (1999). Mechanisms of TNF-alpha-induced insulin resistance. *Exp. Clin. Endocrinol. Diabetes, 107*, 119–125.

Hubel, C. A. and Roberts, J. M. (1999). Lipid metabolism and oxidative stress. In M. D. Lindheimer, J. M. Roberts, G. Cunningham, Appleton, and Lang (Eds.). *Chesley's Hypertensive Disorders in Pregnancy*, Connecticut, USA, pp. 453–486.

Hytinantti, T., Koistinen, H. A., Koivisto, V. A., Karonen, S.-L., Rutanen, E.-M., and Andersson, S. (2000). Increased leptin concentration in preterm infants of preeclamptic mothers. *Arch. Dis. Child Fetal Neonatal Ed., 83*, F13–F16.

Kimura, K., Tsuda, K., Baba, A., Kawabe, T., Boh-oka, S., Ibata, M., Moriwaki, C., Hano, T., and Nishio, I. (2000). Involvement of nitric oxide in endothelium-dependent arterial relaxation by leptin *Biochem. Biophys. Res. Commun., 273*, 745–749.

Laird, S. M., Quinton, N. D., Anstie, B., Li, T. C., and Blakemore, A. I. (2001). Leptin and leptin-binding activity in women with recurrent miscarriage: correlation with pregnancy outcome. *Hum. Reprod., 16*, 2009–2013.

Laivuori, H., Kaaja, R., Koistinen, H., Karonen, S. L., Andersson, S., Koivisto, V., and Ylikorkala, O. (2000). Leptin during and after preeclamptic or normal pregnancy: its relation to serum insulin and insulin sensitivity. *Metabolism*, *49*, 259–263.

Laivouri, H. and Roberts, J. M. (2002). Serum leptin concentration is higher in women destined to develop preeclampsia near term. *J. Soc. Gynecol. Invest.*, *9*(1, Suppl. 258A).

Laivouri, H., Gallaher, M. J., Powers, R. W., and Roberts, J. M. (2002). Leptin, fatty acid synthase and fatty acid translocase/CD36 mRNA are similar in adipocytes of preeclamptic and normal pregnant women. *J. Soc. Gynecol. Invest.*, *9*(1, Suppl. 180A).

Laml, T., Hartmann, B. W., Preyer, O., Ruecklinger, E., Soeregi, G., and Wagenbichler, P. (2000). Serum leptin concentration in cord blood: relationship to birthweight and gender in pregnancies complicated by preeclampsia. *Gynecol. Endocrinol.*, *14*, 442–447.

Laml, T., Preyer, O., Hartmann, B. W., Ruecklinger, E., Soeregi, G., and Wagenbichler, P. (2001). Decreased maternal serum leptin in pregnancies complicated by preeclampsia. *J. Soc. Gynecol. Invest.*, *8*, 89–93.

Lea, R. G., Howe, D., Hannah, L. T., Bonneau, O., Hunter, L., and Hoggard, N. (2000). Placental leptin in normal, diabetic and fetal growth-retarded pregnancies. *Mol. Hum. Reprod.*, *6*, 763–769.

Leung, D. N., Smith, S. C., To, K. F., Sahota, D. S., and Baker, P. N. (2001). Increased placental apoptosis in pregnancies complicated by preeclampsia. *Am. J. Obstet. Gynecol.*, *184*, 1249–1250.

Levine, R. J., Hauth, J. C., Curet, L. B., Sibai, B. M., Catalano, P. M., Morris, C. D., DerSimonian, R., Esterlitz, J. R., Raymond, E. G., Bild, D. E., Clemens, J. D., and Cutler, J. A.(1997). Trial of calcium to prevent preeclampsia. *N. Engl. J. Med.*, *337*, 69–76.

Linnemann, K., Malek, A., Schneider, H., and Fusch, C. (2001). Physiological and pathological regulation of feto/placento/maternal leptin expression. *Biochem. Soc. Trans.*, *29*, 86–90.

Martin-Romero, C., Santos-Alvarez, J., Goberna, R., and Sanchez-Margalet, V. (2000). Human leptin enhances activation and proliferation of human circulating T lymphocytes. *Cell Immunol.*, *199*, 15–24.

Masuzaki, H., Ogawa, Y., Sagawa, N., Hosoda, K., Matsumoto, T., Mise, H., Nishimura, H., Yoshimasa, Y., Tanaka, I., Mori, T., and Nakao, K. (1997). Nonadipose tissue production of leptin: Leptin as a novel placenta-derived hormone in humans. *Nat. Med.*, *3*, 1029–1033.

McCarthy, J. F., Misra, D. N., and Roberts, J. M. (1999). Maternal plasma leptin is increased in preeclampsia and positively correlates with fetal cord concentration. *Am. J. Obstet. Gynecol.*, *180*, 731–736.

Mise, H., Sagawa, N., Matsumoto, T., Yura, S., Nanno, H., Itoh, H., Mori, T., Masuzaki, H., Hosoda, K., Ogawa, Y., and Nakao, K. (1998). Augmented placental production of leptin in preeclampsia: Possible involvement of placental hypoxia. *J. Clin. Endocrinol. Metab.*, *83*, 3225–3229.

Mitchell, J. L., Morgan, D. A., Correia, M. L., Mark, A. L., Sivitz, W. I., and Haynes, W. G. (2001). Does leptin stimulate nitric oxide to oppose the effects of sympathetic activation? *Hypertension*, *238*, 1081–1086.

Odegard, R. A., Vatten, L. J., Nilsen, S. T., Salvesen, K. A., and Austgulen, R. (2002). Umbilical cord plasma leptin is increased in pre-eclampsia. *Am. J. Obstet. Gynecol.*, *186*, 427–432.

Park, H. Y., Kwon, H. M., Lim, H. J., Hong, B. K., Lee, J. Y., Park, B. E., Jang, Y., Cho, S. T., and Kim, H. S. (2001). Potential role of Leptin in angiogenesis: leptin induces endothelial cell proliferation and expression of matrix metalloproteinases in vivo and in vitro. *Exp. Mol. Med.*, *33*, 95–110.

Poston, L. (2001). Oxidative stress, pre-eclampsia, and antioxidants. *Reproductive Vascular Med.*, *1*, 107–113.

Redman, C. W. G., Sacks, G. P., and Sargent, I. L. (1999). Pre-eclampsia, an excessive maternal inflammatory response to pregnancy. *Am. J. Obstet. Gynecol.*, *180*, 499–506.

Roberts, J. M. and Cooper, D. W. (2001). Pathogenesis and genetics of pre-eclampsia. *Lancet*, *3576*: 53–56.

Roberts, J. M. and Hubel, C. A. (1999). Is oxidative stress the link in the two-stage model of pre-eclampsia? *Lancet*, *354*, 788–789.

Roberts, J. M. and Redman, C. W. (1993). Pre-eclampsia: more than pregnancy-induced hypertension. *Lancet*, *341*, 1447–1451.

Sacks, G. P., Studena, K., Sargent, K., and Redman, C. W. G. (1998). Normal pregnancy and preeclampsia both produce inflammatory changes in peripheral blood leucocytes akin to those of sepsis. *Am. J. Obstet. Gynecol.*, *179*, 80–86.

Saito, S., Sakai, M., Sasaki, Y., Tanebe, K., Tsuda, H., and Michimata, T. (1999a). Quantitative analysis of peripheral blood Th0, Th1, Th2 and the Th1: Th2 cell ratio during normal human pregnancy and preeclampsia. *Clin. Exp. Immunol.*, *117*, 550–555.

Saito, S., Umekage, H., Sakamoto, Y., Sakai, M., Tanebe, K., Sasaki, Y., and Morikawa, H. (1999b). Increased T-helper-1-type immunity and decreased T-helper-2-type immunity in patients with preeclampsia. *Am. J. Reprod. Immun.*, *41*, 297–306.

Sarraf, P., Frederich, R. C., Turner, E. M., Ma, G., Jaskowiak, N. T., Rivet, D. J. III, Flier, J. S., Lowell, B. B., Fraker, D. L., and Alexander, H. R. (1997). Multiple cytokines and acute inflammation raise mouse leptin levels: potential role in inflammatory anorexia. *J. Exp. Med.*, *185*, 171–175.

Sattar, N., Greer, I. A., Pirwani, I., Gibson, J., and Wallace, M. (1998). Leptin levels in pregnancy: Marker for fat accumulation and mobilization? *Acta Obstet. Gynecol. Scand.*, *77*, 278–283.

Schobel, H. P., Fischer, T., Heuszer, K., Geiger, H., and Schmieder, R. E. (1996). Preeclampsia—a state of sympathetic overactivity. *N. Engl. J. Med.*, *335*, 1480–1485.

Schubring, C., Englaro, P., Siebler, T., Blum, W. F., Demirakca, T., Kratzsch, J., and Kiess, W. (1998). Longitudinal analysis of maternal serum leptin levels during pregnancy, at birth and up to six weeks after birth: relation to body mass index, skinfolds, sex steroids and umbilical cord blood leptin levels. *Horm. Res.*, *50*, 276–283.

Schubring, C., Keiss, W., Englaro, P., Rascher, W., Dotsch, J., Hanitsch, S., Attanasio, A., and Blum, W. F. (1997). Levels of leptin in maternal serum, amniotic fluid and arterial and venous cord blood: relation to neonatal and placental weight. *J. Clin. Endocrinol. Metab.*, *82*, 1480–1483.

Sierra-Honigmann, M. R., Nath, A. K., Murakami, C., Garcia-Cardena, G., Papapetropoulos, A., and Sessa, W. C. (1998). Biological action of leptin as an angiogenic factor. *Science*, *281*, 1683–1686.

Solomon, C. G. and Seely, E. W. (2001). Hypertension in pregnancy. A manifestation of the insulin resistance syndrome? *Hypertension*, *37*, 232–239.

De Swiet, M. (2000). Maternal mortality: Confidential enquiries into maternal deaths in the United Kingdom. *Am. J. Obstet. Gynecol.*, *182*, 760–766.

Takahashi, N., Waelput, W., and Guisez, Y. (1999). Leptin is an endogenous protective protein against the toxicity exerted by tumor necrosis factor. *J. Exp. Med.*, *189*, 207–212.

Teppa, R. J., Ness, R. B., Crombelholme, W. R., and Roberts, J. M. (2000). Free leptin in increased in normal pregnancy and further increased in pre-eclampsia. *Metab.: Clin. Exp.*, *49*, 1043–1048.

Tschop, M., Strasburger, C. J., Hartmann, G., Biollaz, J., and Bartsch, P. (1998). Raised leptin concentrations at high altitude associated with loss of appetite. *Lancet*, *352*, 1119–1120.

Vitoratos, N., Chrystodoulacos, G., Kouskouni, E., Salamalekis, E., and Creatsas, G. (2001). Alterations of maternal and fetal leptin concentrations in hypertensive disorders of pregnancy. *Eur. J. Obstet. Gynecol. Reprod. Bio.*, *96* 59–62.

Williams, M. A., Havel, P. J., and Schwartz, M. W. (1999). Preeclampsia disrupts the normal relationship between serum leptin concentrations and adiposity in pregnant women. *Pead. Perinatal. Epidem.*, *13*, 190–204.

Yamagishi, S., Edelstein, D., Du, X., Kaneda, Y., Guzmán, M., and Brownlee, M. (2001). Leptin induces mitochondrial superoxide production and monocyte chemoattractant protein-1 expression in aortic endothelial cells by increasing fatty acid oxidation via protein kinase A. *J. Biol. Chem.*, *276*, 25096–25100.

Yang, S. Q., Lin, H. Z., Lane, M. D., Clemens, M., and Diehl, A. M. (1997). Obesity increases sensitivity to endotoxin liver injury: implications for the pathogenesis of steatohepatitis. *Proc. Natl. Acad. Sci.*, *94*, 2557–2562.

Yasumasu, T., Takahara, K., and Nakashima, Y. (2002). Hypoxia inhibits leptin production by cultured rat adipocytes. Letter. *Obesity Res.*, *10*, 128.

Zhang, F., Basinski, M. B., Beals, J. M., Briggs, S. L., Churgay, L. M., Clawson, D. K., DiMarchi, R. D., Furman, T. C., Hale, J. E., Hsiung, H. M., Schoner, B. E., Smith, D. P., Zhang, X. Y., Wery, J. P., and Schevitz, R. W. (1997). Crystal structure of the obese protein leptin-E100. *Nature*, *387*, 206–209.

19

Leptin and Hypothalamic Amenorrhea

MICHELLE P. WARREN AND JENNIFER E. DOMINGUEZ

19.1. DEFINITIONS

Amenorrhea is characterized by the lack of regular menstrual periods. It may be classified as primary or secondary in nature. Primary amenorrhea is a delay in menarche beyond the age of 14, if no secondary sexual characteristics have developed, and beyond the age of 16, if secondary sexual characteristics have developed (Sakala, 1997). Secondary amenorrhea is diagnosed when menstruation has been absent for more than 3 months, if menses were previously regular; or 6 months, if menses were irregular (Constantini and Warren, 1994). The causes of amenorrhea among premenopausal women are numerous, among them are premature ovarian failure, past contraceptive use, androgen excess, intensive physical training, nutritional restriction, and stress.

Hypothalamic amenorrhea (HA) results from a deficiency in the secretion of gonadotropin-releasing hormone (GnRH) from the hypothalamus. It is thought that the pulse generator in the arcuate nucleus of this region of the brain is especially vulnerable to stress and nutritional insult. These factors can disrupt the normal pulsatile secretion of GnRH, and subsequently of luteinizing hormone (LH) and follicle-stimulating hormone (FSH). In turn, this limits follicle stimulation and the production of estradiol.

Although the metabolic signals that inform the hypothalamus are still under investigation, considerable evidence suggests that leptin may be a significant mediator of reproductive function in women (Maffei et al., 1995; Ducy et al., 2000). A small polypeptide hormone encoded by the obesity (*ob*) gene, leptin is secreted primarily from adipocytes. It is related to the regulation of food intake and energy expenditure (Zhang et al., 1994), and is thought to inform the reproductive system of nutritional status. Leptin receptors have

---

MICHELLE P. WARREN AND JENNIFER E. DOMINGUEZ • Departments of Medicine and Obstetrics and Gynecology, Columbia University.

Leptin and Reproduction. Edited by Henson and Castracane, Kluwer Academic/Plenum Publishers, 2003.

been found on the hypothalamic neurons believed to be involved in the control of the GnRH pulse generator (Cheung et al., 1997), as well as on the ovary (Cioffi et al., 1966). Studies have shown a correlation between low leptin levels and amenorrhea (Miller et al., 1998; Warren et al., 1999), as well as disordered eating (Warren et al., 1999). They suggest that if leptin drops below a critical level, menstruation will not occur (Kopp et al., 1997), although the range of this level is broad.

We will focus here on the pathophysiology of HA, a disorder often found in women with low weight, low body fat, disordered eating, and/or stress and its relationship to leptin. The increased vulnerability of women with HA to osteoporosis will also be discussed.

19.2. THE HYPOTHALAMUS AND AMENORRHEA

19.2.1. The Reproductive Axis

To better understand the effects of disordered leptin levels on the reproductive axis, the normal functioning of the female reproductive system must be considered. Normal menstrual cyclicity is the result of a number of different endocrine systems working in concert. These include the hypothalamus region at the base of the brain and the adjacent anterior pituitary gland, as well as the ovaries and its related organs. The hypothalamus secretes GnRH, signaling the release of LH and FSH from the pituitary gland. These gonadotropins stimulate the ovaries to produce and secrete the sex hormones. The ovaries secrete primarily estrogens (mostly estradiol), progestins (mostly progesterone), and androgens, which in turn, regulate the hypothalamic–pituitary–ovarian axis via a feedback mechanism, acting directly on the pituitary, or indirectly on the hypothalamic GnRH pulses (Ferin et al., 1993).

19.2.1.1. GnRH Pulsatility. The pulse generator in the arcuate nucleus of the medial central area of the hypothalamus controls secretion of GnRH. Puberty is initiated when the hypothalamus is activated to secrete GnRH. Throughout the course of the female reproductive cycle, the hormone is released in varying spurts, normally occurring every 60–90 min. This rhythm of GnRH release causes gonadotropin secretion to occur in a likewise pulsatile manner. The gonadotropins thus affect the changes in the reproductive tract that characterize the menstrual cycle. GnRH release is regulated in response to changing levels of sex hormones, but the mechanism behind the pulse generator is not entirely understood. It is thought, however, to be highly sensitive to stress and metabolic factors. This may help explain the amenorrhea that often accompanies nutritional disturbances.

19.3. PHASES OF DYSFUNCTION IN HA

HA is common among women with low body weight and body fat due to excessive exercise or disordered eating (Locke and Warren, 2000). The disorder can be subcategorized into three conditions which describe the etiology of the amenorrhea. HA is common among athletes, a condition known as athletic, or exercise-induced amenorrhea, as well as among women with eating disorders. A third subtype, functional HA, describes normal weight women whose amenorrhea is of an unknown etiology.

19.3.1. Exercise-Induced

Menstrual irregularity is more prevalent among female athletes than in nonathletic women (Constantini and Warren, 1994) (Table 19.1). In the general adult population, menstrual irregularities have an estimated prevalence of 1.8–5 percent (Pettersson et al., 1973; Singh, 1981), while studies of adult athletes report a prevalence as high as 79 percent (Abraham et al., 1982). Many athletic women display what is known as the "female athletic triad," which consists of amenorrhea, osteoporosis, and eating disorders.

Up to 25 percent of female athletes develop exercise-induced amenorrhea. Athletes involved in sports requiring low body weight, such as ballet dancing, figure skating, gymnastics, long-distance running, or skiing are especially vulnerable. Often, these girls and women are pressured to keep their body fat below 10 percent of their total body weight, and body composition has been hypothesized to be a good predictor of amenorrhea (Warren and Shangold, 1997). Low body weight and disordered eating may be compounded by some combination of heavy exercise, mental stress, and the desire or pressure to maintain a certain body weight. While the use of performance enhancing drugs may also alter cycles, the mechanism by which it does so is probably different. In contrast, female athletes in sports that do not place such an emphasis on low body weight, such as swimming or basketball, tend to experience menstrual dysfunction in other forms, such as irregular cycles (Al-Othman and Warren, 1998).

Numerous studies of female athletes have confirmed the finding of depressed leptin levels in this population (Laughlin and Yen, 1997). One recent study found that leptin levels were significantly lower in elite athletes as compared to recreational athletes

Table 19.1. The prevalence of menstrual irregularities in different athletic disciplines (oligomenorrhea and amenorrhea)

Activity	Study	Number of subjects	Percentage with irregularities
General population			
	Petterson et al. (1973)	1,862	1.8
	Singh (1981)	900	5.0
Weight-bearing sports			
Ballet	Abraham et al. (1982)	29	79.0
	Brooks-Gunn et al. (1987)	53	59.0
Running	Feicht et al. (1978)	128	6–43
	Glass et al. (1987)	67	34.0
	Shangold and Levine (1982)	394	24.0
	Sanborn et al. (1982)	237	26.0
Non-weight bearing sports			
Cycling	Sanborn et al. (1982)	33	12.0
Swimming	Sanborn et al. (1982)	197	12.0

From Constantini, N. W. and Warren, M. P. (1994). Specific problems of the female athlete. In R. S. Panush, and N. E. Lane (Eds.). Clinical Rheumatology: Exercise and Rheumatic Disease, Baillere-Tindall, Philadelphia; with permission.

(Thong et al., 2000). Controlling for body fat explained the lower leptin levels seen in normally cycling athletes, but did not account for the significantly depressed leptin levels present in amenorrheic athletes.

19.3.2. Eating Disorder-Induced

The menstrual dysfunction seen in athletes in sports requiring low body weight is similar to that seen in women with eating disorders, particularly anorexia nervosa. In fact, having HA is a required *DSM-IV* criteria for the diagnosis of anorexia (Wilson and Walsh, 1991). In young women, this disorder is associated with severe weight loss and amenorrhea which often reverses with weight gain. In addition to the suppression of GnRH, anorexia nervosa produces other effects on the endocrine system, including changes in cortisol and thyroid hormones, which may also have effects on the reproductive system. Women suffering from bulimia may not necessarily have amenorrhea, but may experience other menstrual irregularities. Anorexic patients have been found to have low leptin levels, a finding which is predictive of amenorrhea in this population (Kopp et al., 1997).

19.3.3. Idiopathic/Functional HA

Functional HA (FHA) is a diagnosis of exclusion; HA of unknown etiology in normal weight, nonathletic women, has long been thought to be psychogenic, or caused by stress. However, mounting evidence suggests the possibility that this so-called idiopathic amenorrhea is actually caused by nutritional restriction and the resulting endocrine and metabolic response (Schweiger, 1991; Warren et al., 1994b; Laughlin, 1999; Warren et al., 1999). There is evidence of subclinical eating disorders in weight-stable, nonathletic women with FHA accompanied by a severe restriction of dietary fat intake. Unbalanced nutrient intake in psychogenic FHA has been associated with multiple endocrine-metabolic alterations (Laughlin et al., 1998).

In one study, women with FHA were compared with normally menstruating controls of similar Body Mass Index (BMI). The amenorrheic subjects had greater lean body mass and less body fat than the normal controls, and were found to consume less fat and carbohydrates than the BMI matched controls. Thus, while FHA is normally attributed to exercise or stress, in this population, it may actually be a response to nutritional disturbance (Couzinet et al., 1999).

Low leptin levels have been found in women with HA, even when weight is within the normal range. In one study, the difference in leptin levels between amenorrheic and eumenorrheic women remained significant when fat mass, body mass, and insulin were controlled (Miller et al., 1998).

19.4. ETIOLOGY OF HA

HA results from the disruption of the hypothalamus' pulsatile secretion of GnRH. The pulse generator appears to be sensitive to stress and metabolic factors and is highly vulnerable to environmental insults, particularly weight loss. GnRH secretion also appears to diminish when energy intake is decreased, particularly when expenditure of exercise exceeds the dietary intake, creating a negative energy balance (Warren and Shangold, 1997).

The inhibition of GnRH reduces the anterior pituitary's secretion of LH and FSH, thereby shutting down or limiting ovarian stimulation and estradiol production. A prolonged follicular phase, or the absence of a critical LH, or estradiol, surge mid-cycle results in the mild or intermittent suppression of menstrual cycles frequently observed in athletes. Lack of pituitary stimulation results in primary (manifested as delayed menarche) or secondary amenorrhea (Warren, 1980; Loucks and Heath, 1994). Frequently, the initial stages of HA are characterized by an inadequate luteal phase, a condition which is defined by low-peak progesterone levels and in which luteinization occurs without ovulation. Significantly greater LH pulse frequency is observed in women with luteal phase deficiency when compared with the frequency in normal controls (Soules et al., 1984).

19.4.1. Athletic

Early research into the causes of amenorrhea in athletes focused on body composition and the stress of exercise. Since then, however, growing evidence has suggested that the primary factor in exercise-induced reproductive dysfunction is the energy deficit athletic women incur when their daily caloric intake is insufficient for their activity level (Warren et al., 1999; Myerson et al., 1991; Loucks and Heath, 1994; Loucks et al., 1995). Moreover, a significant body of recent research suggests that a decrease in resting metabolic rate (RMR) is associated with a negative energy balance and with menstrual dysfunction (Tuschl et al., 1990; Salisbury et al., 1995; Kaufman et al., 2002) (Table 19.2). Metabolic rate may remain depressed in individuals who maintain a body weight lower

Table 19.2. Resting metabolic rate and bone mass density are depressed in ballet dancers with a history of amenorrhea

	Dancers (21) ± SD		Controls (27) ± SD	
	a. Ever had amenorrhea*	b. Normal	c. Ever had amenorrhea*	d. Normal
N^a	11	10	4	23
Wt (kg)	51.94 ± 8.37	52.90 ± 5.14	51.37 ± 2.38	52.79 ± 6.48
FFM (kg)	43.34 ± 5.71	41.17 ± 3.81	42.51 ± 2.75	40.29 ± 3.68
RMR (kcal/day)	1252.4 ± 198.0	1288.6 ± 141.0	1253.0 ± 105.7	1277.3 ± 100.2
RMR/FFM^b (kcal/kg/day)	28.7 ± 2.4	31.3 ± 1.1	29.1 ± 3.8	32.3 ± 2.6
EAT^c	23.4 ± 12.1	22.3 ± 8.4	4.5 ± 3.0	4.0 ± 2.4
BMD (g/cm^2)				
Spine^d	0.97 ± 0.11	1.11 ± 0.13	1.03 ± 0.09	1.06 ± 0.13
Leg	1.24 ± 0.09	1.24 ± 0.14	1.24 ± 0.10	1.20 ± 0.16
Arm^d	0.77 ± 0.07	0.82 ± 0.07	0.79 ± 0.06	0.82 ± 0.04

*Includes present and history of amenorrhea.
$^a p < 0.05$ ab versus cd
$^b p < 0.05$ a versus d
$^c p < 0.05$ b versus cd
$^d p < 0.05$ d versus ab

From Kaufman, B. A., Warren, M. P., Wang, J., Heymsfield, S. B., and Pierson, R. N. (2002). Bone density and amenorrhea in ballet dancers is related to a decreased resting metabolic rate. *J. Clin. Endocrinol. Metab., 87*(6); with permission.

than their normal or genetically predetermined weight (Leibel et al., 1995), as observed in recovered anorectics (Brooks-Gunn et al., 1987) and amenorrheic runners (Myerson et al., 1987). Even women who maintain a normal weight may suffer an energy deficit; their normal weight is likely maintained by a decrease in metabolic rate (Myerson et al., 1991).

The metabolic adaptation to energy deficit or starvation is evidenced by a number of factors including lowered T3 and occasionally T4 levels, as well as clinical signs of hypothyroidism (Fowler et al., 1972; Warren and Vande Wiele, 1973; Moshang and Utiger, 1977; Kiyohara et al., 1989). Metabolism of certain steroids, like testosterone, is similar to that seen in hypothyroidism (Boyar and Bradlow, 1977). Maximal aerobic power (VO_2) and heart rates are reduced, indicating an adaptation to low caloric intake (Fohlin, 1975). Basal growth hormone levels are elevated above normal, and insulin-like growth factor (IGF) is suppressed; both of these abnormalities resolve with nutritional therapy (Newman and Halmi, 1988). Twenty-four hour studies of women with HA have found that although episodic and circadian rhythms are normal, mean cortisol levels are elevated (Boyar et al., 1977), suggesting that a new set point has been determined by the hypothalamic-pituitary–adrenal (HPA) axis.

In addition to metabolic alterations, HA is associated with a constellation of neuro-endocrine secretory disturbances. Low body fat or body weight appears to inform the neuroendocrine controls of normal menstrual cycles in exercising women, producing the suppression of GnRH (Vigersky et al., 1977; Warren, 1983). Considerable research has sought to identify the signal that binds the metabolic aberrations to those seen in the repro-ductive system in women with HA. Increasing evidence suggests HA occurs as a result of an adaptation to an inadequate energy intake, and that changes in the HPA axis are sec-ondary. This hypothesis explains the similarity in the syndromes seen in female athletes, women with eating disorders, and in functional or idiopathic HA.

19.4.2. Eating Disorder-Induced

Women with eating disorder-induced HA are generally in a severe hypometabolic state which reverses with weight gain. Most women with anorexia nervosa regress to, or remain in, a prepubertal pattern of GnRH secretion, depending on the age of disease onset (Boyar et al., 1974; Pirke et al., 1985). When weight loss is severe, the GnRH response pat-tern is immature, with the FSH response being significantly greater than the LH response (Roth et al., 1972). Partially recovered anorexic patients tend to have exaggerated responses to GnRH (Mecklenburg et al., 1974; Warren et al., 1975; Beaumont et al., 1976). This hyperreactive phase is characterized by a greatly increased LH response. As weight increases, the response ratios return to normal; that is, the LH response is much greater than that of FSH (Fritz and Speroff, 1983). When an anorexia patient attains normal weight, GnRH secretion is released from inhibition by an unknown metabolic factor, and reproduc-tive hormones and ovarian morphology retrace their way through pubertal development (Warren et al., 1975; Kanders et al., 1988; van Binsbergen et al., 1990). However, HA and reproductive dysfunction may persist if disordered eating is present, suggesting the persist-ence of metabolic factors inhibiting reproduction which may be independent of weight.

One intriguing occurrence in anorexia nervosa is the persistence of the amenorrhea, despite a return to normal weight, which may occur in up to 10–30 percent of recovered anorectics (Falk and Halmi, 1982; Kohmura et al., 1986; Kreipe et al., 1989a). This persistence has been tied to continued disordered eating (Kreipe, Strauss et al., 1989b), and suggests that metabolic abnormalities may persist despite a return to normal weight. In fact, an immature pattern of LH secretion may occur (Katz et al., 1978) in this group despite a normal weight. Thus, some factor, other than weight, must inform the hypothalamus of the body's nutritional status.

Depressed Resting Metabolic Rate (RMR), which has been connected to high premorbid weights (Salisbury et al., 1995), bulimic behavior (Devlin et al., 1990), and restrained eating (Tuschl et al., 1990), have been described in recovered anorectics. In fact, decreased metabolic rates have been seen in other groups such as wrestlers who diet to "make weight" before a competition (Steen et al., 1988). In patients with eating disorders, there is a significant negative correlation between highest premorbid BMI and the number of calories required to maintain weight, suggesting a difference in energy metabolism which may relate to metabolic rate (Newman et al., 1987). The relationship of this alteration in the metabolism and its possible relationship to the reproductive dysfunction has never been studied in humans. However, one study investigated the causal role of energy deficit in amenorrhea in monkeys. The animals developed amenorrhea over a period of 7–24 months, as researchers gradually increased their levels of exercise, while keeping caloric intake constant (Williams et al., 2001). When additional calories were supplied, the animals regained normal menstrual function and increased reproductive hormone levels. These findings, along with the observation in women, that HA is reversible with weight gain, exercise reduction, or greater food intake, strongly support the negative energy balance theory. However, the mechanism by which these metabolic disturbances influence the HPA axis, remains to be elucidated, although evidence for the involvement of leptin is compelling.

19.4.3. Functional HA

The etiology of functional HA in weight-stable, nonathletic women is thought to be similar to the amenorrhea seen in athletic and eating disordered women. In all of these cases, a metabolic insufficiency likely underlies the reproductive dysfunction. Evidence of subclinical eating disorders has been found in weight-stable, nonathletic women with FHA. In one study, women with FHA matched to controls for age, BMI, and daily caloric intake, scored significantly higher on two measures of eating disorders, and consumed much less dietary fat, and a higher percentage of calories from carbohydrates than controls (Laughlin et al., 1998). Although the scores on the eating disorder assessments were subclinical, subjects with FHA showed multiple endocrine-metabolic abnormalities consistent with malnutrition. Another group of amenorrheic women who denied having an eating disorder and were within 10 percent of ideal body weight scored more than twice as high on a scale of disordered eating behavior and six times as high on a subscale of bulimic behavior than did normal controls without amenorrhea. The amenorrheic subjects also had higher fiber diets and expended more calories in aerobic activity per day. However, there was no difference between the groups in measured levels of emotional distress or depression (Warren et al., 1999). While stress may contribute to the disordered eating which may cause amenorrhea, it is unlikely that the stress itself is directly responsible for menstrual dysfunction in women with FHA. Rather, the low-energy availability is likely producing a decreased RMR which allows weight to remain stable in this population.

19.5. THE MECHANISM OF GnRH INHIBITION

Researchers have long sought the source of the metabolic signal to the GnRH pulse generator associated with low weight, and responsible for altered LH pulsatility (Dubey et al., 1986; Cameron, 1989; Bronson and Manning, 1991). The present major hypothesis for the etiology of energy deficit-induced HA is that the condition occurs when, an unknown metabolic signal suppresses GnRH, and therefore LH secretion, because caloric intake is insufficient for activity level. Another hypothesis suggests that HA occurs as a result of the activation of the HPA axis (Walsh et al., 1978; Gold et al., 1986; Hotta et al., 1986; Nappi et al., 1993). In this situation, corticotropin releasing hormone (CRH), the major neurohormonal stimulus to the pituitary–adrenal axis, is secreted in excess (due to stress) and may, in addition to other factors, suppress GnRH secretion.

Current evidence suggests that HA occurs as a result of the body's attempt to economize energy, in response to an inadequate energy intake to expenditure ratio. In this case, the changes to the HPA axis are secondary. There is a variety of evidence supporting this hypothesis. Women with HA exhibit a constellation of metabolic abnormalities, including lowered T_3 levels (Moshang et al., 1975; Kiyohara et al., 1989), increased GH, and decreased IGF-1 (Newman and Halmi, 1988), as well as decreased RMRs (Platte et al., 1994). In addition, abnormalities of LH secretion occur in normal women in response to caloric restriction, even before weight loss is seen (Olson et al., 1995). Further, normalization of menses appears to be associated with a weight gain (Kreipe et al., 1989a), and the persistence of amenorrhea with disordered eating (Kreipe et al., 1989b). Studies in primates show significant effects on LH pulsatility with caloric restriction and reversion by increasing food intake (Dubey et al., 1986; Cameron, 1989; Cameron et al., 1990). The suppression of LH pulses observed with acute fasting in primates reverses upon refeeding without effecting the HPA axis (Helmreich et al., 1993; Schreihofer et al., 1993).

In addition to the GnRH neurons, there are other neuronal networks in the arcuate nucleus which contain β-endorphin, dopamine, and norepinephrine, and these networks may also modulate GnRH secretion (Liu, 1998). Some of the neuropeptides involved in the control of GnRH release are opioids, neuropeptide Y, and CRF, and neurotransmitters that may play a role include noradrenaline, dopamine, serotonin, melatonin, and gamma-aminobutyric acid (GABA) (Genazzani et al., 1998). However, because neurohormones cannot be accurately measured in the peripheral circulation, their roles in the etiology of HA are not well-defined (Ferin et al., 1993).

Several observations suggest that energy deficit or an adaptation to large caloric needs may be fundamental to a prolonged HA state. First, low levels or an altered diurnal rhythm of the adipocyte hormone leptin seems to be part of the physiologic process of HA. Second is the intimate association of HA with osteopenia. Third, experiments in women runners suggest that a physical paradox exists whereby a large energy output is not compensated for by increased caloric intake, yet weight remains stable (Warren, 1999).

While GnRH inhibition clearly plays a major role in certain types of amenorrhea, the mechanism of HA onset is not entirely understood. While the roles of Dehydroepiandrosterone (DHEA), Insulin-like growth factor (IGF-1), and cortisol are also being explored, leptin has been highlighted in recent research, some of which suggests that leptin may be an independent regulator of metabolic function (Zhang et al., 1994).

19.5.1. Leptin

The 1994 discovery of leptin prompted new research into amenorrhea and body composition (Zhang et al., 1994; Ducy et al., 2000). Leptin levels appear to be controlled by total energy intake and fat stores, and correlate significantly with BMI in humans (Macut et al., 1998). Studies have shown that rodents with an inactive form of leptin tend to be amenorrheic and infertile, and fertility is restored with exogenous leptin administration (Legradi et al., 1997, 1998). In humans, low leptin levels have been associated with amenorrhea (Miller et al., 1998; Warren et al., 1999) and with disordered eating (Warren et al., 1999). In addition, it appears that if leptin drops below a critical level, menstruation will not occur (Kopp et al., 1997), although the range of this level is large. Because leptin is a regulator of the basal metabolic rate, it is thought to be a particularly important indicator of nutritional status (Maffei et al., 1995), and indeed, amenorrheic athletes typically have abnormal leptin secretion (Thong et al., 2000) and low metabolic rates (Kaufman et al., 2002). Leptin may be a significant mediator of reproductive function in that it responds to a negative energy balance found in women with exercise-induced HA (Laughlin and Yen, 1997). This response may trigger changes in GnRH pulsatility, as leptin receptors are found on hypothalamic neurons thought to be involved in the control of the GnRH pulse generator (Cheung et al., 1997).

Leptin also plays a role in thyroid function and the initiation of puberty; in the presence of nutritional deficiencies, alterations in leptin levels are directly associated with thyroid hormone changes (Warren et al., 1999). Studies on rodents have shown a direct correlation between leptin administration, and GnRH and gonadotropin secretion, especially during the time of sexual maturation (Ponzo et al., 2001; Tezuka et al., 2002). Thus, leptin may act not only as a regulator of metabolic rates, but also as a mediator of menstrual status, responding to starvation by slowing metabolism and possibly returning a woman to a prepubertal-like state.

19.6. HA AND OSTEOPOROSIS

In young women, HA is associated with reduced bone accretion or premature bone loss during adolescence (Drinkwater et al., 1984; Lindberg et al., 1984; Rigotti et al., 1984, 1991; Marcus et al., 1985; Biller et al., 1989; Dhuper et al., 1990; Warren et al., 1986, 1991; Jonnavithula et al., 1993), placing them at high risk for fractures, significant osteopenia, and severe osteoporosis at menopause. Studies have shown evidence of osteopenia in all of the above subtypes of HA (Grinspoon et al., 1999; Kaufman et al., 2002; Warren et al., 2002).

Two homeostatic mechanisms act on bone simultaneously: hormones and mechanical stress. Under normal circumstances, these mechanisms maintain both skeletal integrity and serum calcium levels. In fact, in postmenopausal women, exercise offsets the loss of bone mass associated with hypoestrogenism, and may even increase bone mineral density (BMD) above baseline. However, studies of hypoestrogenic ballet dancers suggest that despite high levels of weight-bearing exercise, their bones do not respond to mechanical stress (Warren et al., 1991, 2002) (Figures 19.1 and 19.2). It has also been shown that amenorrheic athletes consistently exhibit spinal BMD 10–20 percent lower than their

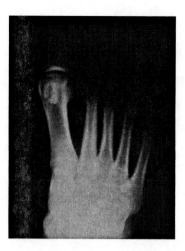

Figure 19.1. Bone density scan of the metatarsal cortical bones of a normally menstruating ballet dancer. Scan shows normal cortical bone accrual characteristic of ballet dancers, a group which sustains a heavy mechanical load on the foot. (Courtesy of Dr. William Hamilton.)

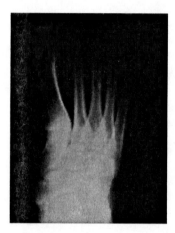

Figure 19.2. Bone density scan of the metatarsal cortical bones of an amenorrheic ballet dancer. Scan reveals the lack of cortical bone accrual in this amenorrheic dancer, as compared to a normally menstruating dancer. (Courtesy of Dr. William Hamilton.)

eumenorrheic counterparts (Lindburg et al., 1984; Fisher et al., 1986; Lloyd et al., 1986; Wolman et al., 1990; Bennell et al., 1997). In addition to the bone loss associated with menopause, menstrual disturbance, and factors such as diet, hormonal levels, and mechanical strain may cause bones to become more vulnerable to fracture and osteoporosis (Smith and Gilligan, 1994).

Because the period during adolescence and young adulthood is a critical window for bone formation, with at least 40 percent of bone mass formed in these years, amenorrhea has serious long-term ramifications for a woman's health (Riggs and Eastell, 1986; Kreipe and Forbes, 1990; Ott, 1990; Kreipe, 1992; Warren and Shangold, 1997). The bone mass attained during adolescence is a major determinant for fractures and osteoporosis later in life. As the skeleton enlarges in children and adolescents, bones are constantly growing. Bones model as bone mass is added to areas of high loading or stress, and remodel, as fatigue-damaged bone is resorbed and replaced by new bone (Barr and McKay, 1998). Women who fail to reach peak bone mass (the maximal amount of bone tissue accrued in individual bones and the whole skeleton) are at an increased risk for osteoporosis later in life, and the reduction of bone mass accretion during adolescence may be a major cause of low bone density and fracture in old age (Bass and Myburgh, 2000). This is a significant problem for women with HA. After the age of 40, bone mass begins to decline (Arnaud, 1996).

The relative osteopenia in weight-bearing bones when peak bone mass is not attained at a young age results in a higher risk for injury, particularly in athletes. A high incidence of stress fractures (61 percent), as directly related to delayed menarche, has been reported in young ballet dancers (Warren et al., 1986) and one study of ballet dancers found that the incidence of both scoliosis and stress fractures rose with each year of delayed menarche (Khan et al., 1999). High impact activity appears to have a beneficial effect on BMD primarily at the hip (Khan et al., 1999). Amenorrhea in athletic women affects trabecular and cortical bone, and studies have shown that spinal BMD is particularly affected in amenorrheic athletes (Rutherford, 1993); amenorrheic athletes' vertebral BMD has been found to be up to 25 percent lower than normally cycling athletes (Gibson et al., 1999). Weight-bearing physical activity does not completely compensate for the side effects of reduced estrogen levels even in weight-bearing bones in the lower extremities and the spine (Pettersson et al., 1999).

Early bone loss and osteoporosis are well-documented and common complications of anorexia nervosa (Rigotti et al., 1984; Brotman and Stern, 1985; Biller et al., 1989; Salisbury and Mitchell, 1991; Hay et al., 1992; Grinspoon et al., 1999). Up to one third of females who recovered from anorexia nervosa during adolescence were found to have persistent osteopenia (Bachrach et al., 1991). Young women with anorexia nervosa may suffer irreversible developmental and growth retardation; older chronic anorectics are at a higher risk for pathological fractures (Hartman et al., 1999).

The most recent research suggests that poor nutrition or an energy deficit with an adaptation to large caloric needs is fundamentally linked not only with a prolonged amenorrheic state, but with osteopenia as well (Tsafriri et al., 1982; Zanker and Swaine, 1998a, b; Zanker, 1999).

19.6.1. Mechanisms of Bone Loss in HA

It is unclear what changes in bone metabolism cause osteoporosis in women with HA. One study of women with anorexia nervosa suggested that bone formation is reduced while bone resorption remains normal (Soyka et al., 1999). Another suggested that bone formation is decreased and bone resorption increased (Bruni et al., 2000). Further analyses of bone turnover indices in women with HA and anorexia nervosa have also produced inconsistent results, with some data suggesting normal bone turnover, while others suggesting low bone formation that is uncoupled from bone resorption (Kreipe et al., 1993; Grinspoon et al., 1999). Results of a 1998 study of amenorrheic long-distance runners demonstrated reductions in bone turnover and in particular, bone formation (Zanker and

Table 19.3. Comparison of bone turnover markers in post-menopausal women and women with exercise-induced amenorrhea

		Bone		
Model	Study (n)	Formation	Resorption	Turnover
Post-menopausal estrogen deficiency	Manolagas, 1995 (*review*)	↑	↑	↑
Exercise-induced amenorrhea (runners)	Okano, 1995 (*8*)	↓ (osteocalcin)	N/A	?
Exercise-induced amenorrhea (runners)	Zanker, 1998 (*9*)	↓ (osteocalcin BAP*)	↓ (deoxypyridinoline)	↓

*Bone-specific alkaline phosphate.

From Warren, M. P. and Perlroth, N. A. (2001). The effects of intense exercise on the female reproductive system. *J. Endocrinol.*, *170*(1), 3–11; with permission.

Swaine, 1998a). The reduced bone formation was linked to a low BMI and an estrogen deficiency (Zanker and Swaine, 1998b).

Because estrogen deficiency is linked to eating disorders, amenorrhea, and can be associated with osteopenia, the hypothesis that this is the primary cause of bone deficiency in young women has been explored with largely negative results. Estrogen replacement does not normalize the low bone densities of dancers when it is given experimentally, as it does in postmenopausal women (Warren et al., 1994a). Furthermore, the metabolic profile of bone turnover characteristic of amenorrhea-associated osteopenia is distinct from the high turnover typical of hypoestrogenism (Table 19.3). Studies of biochemical markers of bone turnover in exercise-induced amenorrhea suggest a pattern different from that seen in cases of estrogen deficiency (Zanker and Swaine, 1998a,b; Zanker, 1999). Estrogen deficiency is marked by increased bone turnover with excessive resorption that responds to estrogen replacement. In contrast, researchers have reported low bone turnover in exercise-induced amenorrhea, and relate these depressed bone markers to an energy deficit (Zanker and Swaine, 1998a).

One recent study followed 111 subjects, 44 of whom had HA and 54 of whom were dancers. Both exercising and nonexercising amenorrheics showed reduced BMD in the spine, wrist, and foot. During the course of the study, seven of the amenorrheic women resumed menses, and while they showed an increase in spine and wrist BMD of 17 percent, they did not achieve normalization of BMD, suggesting that reduced bone accretion may result in a permanent failure to achieve peak bone mass. This underscores the importance of intervention in HA, either preventative or therapeutic (Warren et al., 2002). Another recent study of ballet dancers showed that in women with a history of amenorrhea, depressed RMR was associated with lower BMD and leptin levels (Kaufman et al., 2002). This suggests that these three factors may be related to the same underlying cause: inadequate nutrition as related to energy expenditure, resulting in a negative energy balance.

There are a number of nutritionally regulated processes which may underlie the osteopenia seen in women with HA. Hypercortisolemia (Ding et al., 1988; Loucks et al., 1989; Prior et al., 1990), hypoprogestinemia (Prior et al., 1990), IGF-1 deficiency (Zanker,

1999), or other metabolic adaptations seen in amenorrheic women may all contribute to bone loss. Chronic undernutrition depresses bone formation markers, triiodothyronine (T3), and IGF-1 in amenorrheic runners, but not in controls (Zanker, 1999). A study of ballet dancers showed normal vitamin D and parathyroid levels, suggesting that these may not be the causal factors in the bone loss (Warren et al., 1986). In past studies of women with anorexia nervosa, levels of IGF-1 and T3, two nutritionally dependent hormones, predicted the change in trabecular bone mass with estrogen–progestin therapy (Klibanski et al., 1995).

However, the discovery of leptin receptors on bone (Steppan et al., 1998; Thomas et al., 1998) suggests that leptin may function as a physiologic regulator of bone mass. This presents the possibility of a mechanism other than hypoestrogenism that could account for the low bone density and high stress fracture rates seen in women whose amenorrhea is associated with caloric deficiency, nutritional insults, and exercise (Grinspoon et al., 1999; Laughlin, 1999; Warren et al., 1999).

19.6.1.1. Leptin and Bone Accrual. Although the mechanism by which bone metabolism is altered in women with HA is still unclear, recent research supports a possible role for leptin as a biological signal which informs bone, as well as influences reproduction and metabolism depending on the body's nutritional status.

Mice that are genetically deficient in leptin (ob/ob) have provided an interesting model in which to study its various effects. Studies have shown that leptin administration in ob/ob mice has a positive effect on bone growth (Liu et al., 1997; Steppan et al., 2000). Liu et al. demonstrated osteoblast proliferation and pronounced endocortical bone formation in ob/ob mice after two weeks of leptin therapy, as well as normalization of the metabolic abnormalities seen in mice with this genetic condition (obesity, hypogonadism, hyperglycemia, hyperinsulinemia, and depressed IGF-1 levels) (Liu et al., 1997). Similarly, another study showed a reversal of decreased bone growth seen in ob/ob mice with 50 µg/day of leptin, as evidenced by increased femoral length, total body bone area, bone mineral content (BMC) and BMD (Steppan et al., 2000). Although one study did report an inhibitory effect of leptin on bone formation via a hypothalamic relay in ob/ob mice (Ducy et al., 2000), considerable evidence in both human and animal studies suggests a positive relationship between leptin levels and bone formation (Masuda et al., 1990; Thomas et al., 1998; Ogueh et al., 2000; Burguera et al., 2001; Reseland et al., 2001; Yamauchi et al., 2001). Still, important differences have been noted between the effects of leptin gene mutation in the mouse and the human, and further human studies are needed to clarify the issue.

Evidence from several in vitro studies of human tissues does, however, shed light on the role of leptin in osteoblast stimulation. Leptin gene expression has been reported in normal human osteoblasts (Reseland et al., 2001), as well as fetal mouse cartilage and bone (Hoggard et al., 1997). Studies have also showed the presence of the long form of the leptin receptor (OB-Rb) on primary osteoblasts (Steppan et al., 2000; Reseland et al., 2001). These factors also suggest that leptin may have local effects on bone formation, in addition to any indirect effects via the hypothalamic pathway. Additionally, Thomas et al. have shown that in human marrow stromal cell lines, leptin enhances osteoblast differentiation and inhibits the differentiation of common precursor cells into adipocytes (Thomas et al., 1998).

While it has been observed for some time that obesity causes a marked increase in BMD, and that conversely, low body weight is associated with decreased bone density, the precise mechanism by which the two factors are connected remains unknown (Lindsay et al., 1992). Leptin is secreted from adipocytes and is positively correlated with the percentage of fat. As has been shown in studies of ballet dancers, leptin levels, unlike weight, are an independent predictor of BMD. Studies have shown a positive correlation between total BMD values and plasma leptin concentrations in postmenopausal women (Yamauchi et al., 2001). In addition, Burguera et al. demonstrated that in ovariectomized rats, leptin administration reduced the usual bone loss seen with hypogonadism (Burguera et al., 2001). Despite this, results have been mixed (Goulding and Taylor, 1998), and some have suggested that leptin may play a greater role in bone formation in the growing skeleton, than in the mature one (Rauch et al., 1998). Leptin has been associated with bone development in both the human and mouse fetus (Hoggard et al., 1997; Ogueh et al., 2000). This may help explain the devastating effects of chronic energy depletion on the bones of young women in cases of HA.

19.7. TREATMENT OF HA

Ideally, women should not develop long-term amenorrhea. If amenorrhea is noted, a physical and pelvic examination is indicated to rule out androgenization, galactorrhea, and hyperprolactinemia, as well as pregnancy. A careful history of weight loss and eating disorders should also be taken. However, eating disorders can often go undetected, and a standardized scale or diagnostic interview may be helpful in making the diagnosis (Warren and Shangold, 1997). For women with nutritional imbalances or disordered eating, a crucial aspect of treatment is addressing and reversing unhealthy eating behaviors. A history of amenorrhea as an adolescent or young adult may be the only causal event in women presenting with osteoporosis or osteopenia later in life.

Management of HA includes restoring ovulatory cycles if possible, replacing estrogen when necessary, reassurance, and reevaluation. Gonadotropin deficiency may be reversible in women with HA upon the improvement of nutritional intake and body composition (Couzinet et al., 1999). Frequently, a weight gain of 1–2 kg, or a 10 percent decrease in the duration or intensity of exercise is enough to reverse reproductive dysfunction (Prior and Vigna, 1985; Drinkwater et al., 1990).

The most prescribed treatment for young amenorrheic women has traditionally been estrogen therapy and oral contraceptives. Because of estrogen's antiresorptive effects on bone modeling, exogenous estrogen is often given to prevent bone loss and/or increase overall bone accretion in women with HA, although clinical studies have not yet consistently demonstrated its efficacy in this group (Bruni et al., 2000; Robinson et al., 2000) (Table 19.4). Preliminary studies on small numbers of treated subjects ($n = 4$) suggest that oral contraceptives may be effective in athletic amenorrhea (DeCree et al., 1988), and another study showed that estrogen treatment produced increased spine and femoral BMD (Miller and Klibanski, 1999). However, another study of estrogens and oral contraceptives demonstrated no change in BMD with contraceptive use (Biller et al., 1993). Recent research on women with anorexia-induced osteopenia indicated that neither previous nor current estrogen use had any effect on BMD (Grinspoon et al., 2000). Research has not yet consistently demonstrated the efficacy of hormone replacement therapy or oral contraceptives in increasing the bone mass of women with HA (Robinson et al., 2000).

Table 19.4 Bone mineral density after treatment of amenorrhea with oral contraceptives and hormone replacement therapy (HRI)

Model	Study	Rx (n)	Bone mineral density		
			Spine	Hip	Total
Hypothalamic amenorrhea (including eating diorders)	Hergenroeder, 1997	Oral contraceptive (5) randomized, 12 months 0.035 mg ethinyl E2 0.5–1.0 mg norethindrone	↑	NS	↑
Exercise-induced amenorrhea	Cumming et al., 1996	HRT (8) observed, 24–30 months	↑	↑	—
Exercise-induced amenorrhea	Warren et al., 2003	HRT (13) randomized, 24 months	NS	NS	NS
Anorexia nervosa	Klibanski et al., 1995	HRT (22) randomized	NS*	—	—

*Patients with the lowest BMD showed some increase.

Adapted from Warren, M. P. and Perlroth, N. A. (2001). The effects of intense exercise on the female reproductive system. *J. Endocrinol.*, *170*(1), 3–11; with permission.

No response to estrogen–progestin therapy in replacement doses has also been reported when given to anorectic women with HA (Kreipe et al., 1993; Klibanski et al., 1995). Optimal treatment of osteoporosis in anorexia nervosa includes weight and diet normalization, and supplemental calcium and vitamin D. Unlike postmenopausal osteoporosis, estrogen replacement does not prevent or correct the osteoporosis that occurs in anorexia nervosa (Powers, 1999). Even after recovery, studies have found conflicting data about weight gain and changes in bone mineral density, and there is still uncertainty about the restoration of bone density with complete and sustained clinical recovery from anorexia nervosa (Bachrach et al., 1991; Klibanski et al., 1995; Hartman et al., 1999; Baker et al., 2000; Carruth and Skinner, 2000). The effect of estrogen supplementation is unique from patient to patient. Further studies of different hormone replacement therapy regimens are needed, and the possibilities of other therapies aimed at increasing bone formation such as recombinant human IGF, and recombinant human transforming growth factor, TGF-β are now under investigation (Bruni et al., 2000).

REFERENCES

Abraham, S. F., Beumont, P. J. V., Fraser, I. S., and Llewellyn-Jones, D. (1982). Body weight, exercise and menstrual status among ballet dancers in training. *Br. J. Obstet. Gynaecol.*, *89*, 507–510.

Al-Othman, F. N. and Warren, M. P. (1998). Exercise, the menstrual cycle, and reproduction. *Infertil. Reprod. Med. Clin. North Am.*, *9*, 667–687.

Arnaud, C. D. (1996). Osteoporosis: using "bone markers" for diagnosis and monitoring. *Geriatrics*, *51*, 24–30.

Bachrach, L. K., Katzman, D. K., Litt, I. F., Guido, D., and Marcus, R. (1991). Recovery from osteopenia in adolescent girls with anorexia nervosa. *J. Clin. Endocrinol. Metab.*, *72*, 602–606.

Baker, D., Roberts, R., and Towell, T. (2000). Factors predictive of bone mineral density in eating-disordered women: a longitudinal study. *Int. J. Eat. Disord.*, *27*, 29–35.

Barr, S. I. and McKay, H. A. (1998). Nutrition, exercise, and bone status in youth. *Int. J. Sports Med.*, *8*, 124–142.

Bass, S. L. and Myburgh, K. H. (2000). The role of exercise in the attainment of peak bone mass and bone strength. In M. P. Warren and N. W. Constantini (Eds.). *Contemporary Endocrinology: Sports Endocrinology*, Humana Press Inc., Totowa, NJ, pp. 253–280.

Beaumont, P. J. V., George, G. C. W., Pimstone, B. L., and Vinik, A. L. (1976). Body weight and pituitary response to hypothalamic releasing hormone in patients with anorexia nervosa. *J. Clin. Endocrinol. Metab.*, *43*, 487.

Bennell, K. L., Malcolm, S. A., Wark, J. D., et al. (1997). Skeletal effects of menstrual disturbances in athletes. *Scand. J. Med. Sci. Sports*, *7*, 261–273.

Biller, B. M. K., Saxe, V., Herzog, D. B., Rosenthal, D. I., Holzman, S., and Klibanski, A. (1989). Mechanisms of osteoporosis in adult and adolescent women with anorexia nervosa. *J. Clin. Endocrinol. Metab.*, *68*, 548–554.

Biller, B. M. K., Schoenfeld, D., and Klibanski, A. (1993). Premenopausal osteopenia: effects of estrogen administration (Abstract#1616). Endocrine Society Annual Meeting, *75*, 454.

Boyar, R. M. and Bradlow, H. L. (1977). Studies of testosterone metabolism in anorexia nervosa. In R. Vigersky (Ed.). *Anorexia Nervosa*. New York: Raven Press, p. 271.

Boyar, R. M., Hellman, L. D., Roffwarg, H., Katz, J., Zumoff, B., O'Connor, J., Bradlow, H. L., and Fukushima, D. K. (1977). Cortisol secretion and metabolism in anorexia nervosa. *N. Engl. J. Med.*, *296*, 190–193.

Boyar, R. M., Katz, J., Finkelstein, J. W., Kapen, S., Weiner, H., Weitzman, E. D., and Hellman, L. D. (1974). Anorexia nervosa: immaturity of the 24-hour luteinizing hormone secretory pattern. *N. Engl. J. Med.*, *291*, 861–865.

Bronson, F. H. and Manning, J. M. (1991). The energetic regulation of ovulation: a realistic role for body fat. *Biol. Reprod.*, *44*, 945–950.

Brooks-Gunn, J., Warren, M. P., and Hamilton, L. H. (1987). The relation of eating problems and amenorrhea in ballet dancers. *Med. Sci. Sports Exerc.*, *19*(1), 41–44.

Brotman, A. W. and Stern, T. A. (1985). Osteoporosis and pathological fractures in anorexia nervosa. *Am. J. Psychiatry.*, *142*, 495–496.

Bruni, V., Dei, M., Vicini, I., Beninato, L., and Magnani, L. (2000). Estrogen replacement therapy in the management of osteopenia related to eating disorders. *Ann. N. Y. Acad. Sci.*, Abstract, *900*, 416–421.

Burguera, B., Hofbauer, L., Thomas, T., Gori, F., Evans, G., Khosla, S., Riggs, B., and Turner, R. (2001). Leptin reduces ovariectomy-induced bone loss in rats. *Endocrinology*, *142*, 3546–3553.

Cameron, J. L. (1989). Influence of nutrition on the hypothalamic–pituitary–gonadal axis in primates. In K. M. Pirke, W. Wuttke, and U. Schweiger (Eds.). *The Menstrual Cycle and Its Disorders*, Springer-Verlag, Heidelberg, pp. 66–78.

Cameron, J. L., Nosbisch, C., Helmreich, D. L., and Parfitt, D. B. (1990). Reversal of exercise-induced amenorrhea in female cynomolgus monkeys (*Macaca fascicularis*) by increasing food intake (Abstract #1042). *Endocrine Society Annual Meeting*, *72*, 285.

Carruth, B. R. and Skinner, J. D. (2000). Bone mineral status in adolescent girls: effects of eating disorders and exercise. *J. Adolesc. Health.*, *26*, 322–329.

Cheung, C. C., Clifton, D. K., and Steiner, R. A. (1997). Proopiomelanocortin neurons are direct targets for leptin in the hypothalamus. *Endocrinology*, *138*, 4489–4492.

Cioffi, J., Shafer, A., and Zupancic, T. (1966). Novel B219/OB receptor isoforms: possible role of leptin in hematopoieses and reproduction. *Nature Med.*, *2*, 585–588.

Constantini, N. W. and Warren, M. P. (1994). Physical activity, fitness, and reproductive health in women: clinical observations. In C. Bouchard, R. I. Shephard, and T. Stephens (Eds.). *Physical Activity, Fitness, and Health: International Proceedings and Consensus Statement*, Human Kinetics, Champaign, pp. 955–966.

Couzinet, B., Young, J., Brailly, S., Le Bouc, Y., Chanson, P., and Schaison, G. (1999). Functional hypothalamic amenorrhoea: a partial and reversible gonadotrophin deficiency of nutritional origin. *Clin. Endocrinol. (Oxf.)*, *50*, 229–235.

Cumming, D. C. (1996). Exercise-associated amenorrhea, low bone density, and estrogen replacement therapy. *Archives of Internal Medicine*, *156*(19), 2193–2195.

DeCree, C., Lewin, R., and Ostyn, M. (1988). Suitability of cyproterone acetate in the treatment of osteoporosis associated with athletic amenorrhea. *Int. J. Sports Med.*, *9*, 187–192.

Devlin, M. J., Walsh, B. T., Kral, J. G., Heymsfield, S. B., Pi-Sunyer, F. X., and Dantzic, S. (1990). Metabolic abnormalities in bulimia nervosa. *Arch. Gen. Psychiatry 47*, 144–148.

Dhuper, S., Warren, M. P., Brooks-Gunn, J., and Fox, R. P. (1990). Effects of hormonal status on bone density in adolescent girls. *J. Clin. Endocrinol. Metab.*, *71*, 1083–1088.

Ding, J. H., Sheckter, C. B., Drinkwater, B. L., Soules, M. R., and Bremner, W. J. (1988). High serum cortisol levels in exercise-associated amenorrhea. *Ann. Intern. Med.*, *108*, 530–534.

Drinkwater, B. L., Bruemner, B., and Chesnut, C. H., III (1990). Menstrual history as a determinant of current bone density in young athletes. *JAMA*, *263*, 545–548.

Drinkwater, B. L., Nilson, K., Chesnut, C. H., III, Bremner, W. J., Shainholtz, S., and Southworth, M. B. (1984). Bone mineral content of amenorrheic and eumenorrheic athletes. *N. Engl. J. Med.*, *311*(5), 277–281.

Dubey, A. K., Cameron, J. L., Steiner, R. A., and Plant, T. M. (1986). Inhibition of gonadotropin secretion in castrated male rhesus monkeys (*Macaca mulatta*) induced by dietary restriction: analogy with the prepubertal hiatus of gonadotropin release. *Endocrinology*, *118*, 518–525.

Ducy, P., Amling, M., Takeda, S., Priemel, M., Schilling, A. F., Beil, F. T., Shen, J., Vinson, C., Rueger, J. M., and Karsenty, G. (2000). Leptin inhibits bone formation through a hypothalamic relay: a central control of bone mass. *Cell*, *100*, 197–207.

Falk, J. R. and Halmi, K. A. (1982). Amenorrhea in anorexia nervosa: examination of the critical body weight hypothesis. *Biol. Psychiatry*, *17*, 799–806.

Feicht, C. B., Johnson, T. S., and Martin, B. J. (1978). Secondary amenorrhea in athletes. *Lancet*, *2*, 1145–1146.

Ferin, M., Jewelewicz, R., and Warren, M. P. (1993) *The Menstrual Cycle.* Oxford University Press, New York.

Fisher, E. C., Nelson, M. E., Frontera, W. R., et al. (1986). Bone mineral content and levels of gonadotropins and estrogen in amenorrheic running women. *J. Clin. Endocrinol. Metab.*, *62*, 1232–1236.

Fohlin, L. (1975). Exercise, performance, and body dimensions in anorexia nervosa before and after rehabilitation. *Acta. Med. Scand.*, *204*, 61.

Fowler, P. B. S., Banim, S. O., and Ikram, H. (1972). Prolonged ankle reflex in anorexia nervosa. *Lancet*, *2*, 307.

Fritz, M. A. and Speroff, L. (1983). Current concepts of the endocrine characteristics of normal menstrual function: the key to diagnosis and management of menstrual disorders. *Clin. Obstet. Gynecol.*, *26*, 647–689.

Genazzani, A. R., Petraglia, F., Bernardi, F., Casarosa, E., Salvestroni, C., Tonetti, A., Nappi, R. E., Luisi, S., Palumbo, M., Purdy, R. H., and Luisi, M. (1998). Circulating levels of allopregnanolone in humans: gender, age, and endocrine influences. *J. Clin. Endocrinol. Metab.*, *83*, 2099–2103.

Gibson, J. H., Mitchell, A., Reeve, J., and Harries, M. G. (1999). Treatment of reduced bone mineral density in athletic amenorrhea: a pilot study. *Osteoporos. Int.*, *10*, 284–289.

Glass, A. R., Deuster, P. A., Kyle, S. B., Yahiro, J. A., Vigersky, R. A., and Schoomaker, E. B.(1987). Amenorrhea in Olympic marathon runners. *Fertility and Sterility*, *48*, 740–745.

Gold, P., Gwirtsman, H., Avgerinos, P., Nieman, L., Gallucci, W., Kaye, W., Jimerson, D., Ebert, M., Rittmaster, R., Loriaux, L., and Chrousos, G. (1986). Abnormal hypothalamic–pituitary–adrenal function in anorexia nervosa. *N. Engl. J. Med.*, *314*, 1335–1342.

Goulding, A. and Taylor, R. W. (1998). Plasma leptin values in relation to bone mass and density and to dynamic biochemical markers of bone resorption and formation in postmenopausal women. *Calcif. Tissue Int.*, *63*, 456–458.

Grinspoon, S., Miller, K. K., and Coyle, C. (1999). Severity of osteopenia in estrogen-deficient women with anorexia nervosa and hypothalamic amenorrhea. *J. Clin. Endcrinol. Metab.*, Abstract, *68*, 402–411.

Grinspoon, S., Thomas, E., Pitts, S., Gross, E., Mickley, D., Miller, K., Herzog, D., and Klibanski, A. (2000). Prevalence and predictive factors for regional osteopenia in women with anorexia nervosa. *Ann. Intern. Med.*, *133*, 790–794.

Hartman, D., Crisp, A., Rooney, B., Rackow, C., Atkinson, R., and Patel, S. (1999). Bone density of women who have recovered from anorexia nervosa. *Int. J. Eat. Dis.*, Abstract, *28*(1), 107–112.

Hay, P., Delahunt, J. W., Hall, A., Mitchell, A. W., Harper, G., and Salmond, C. (1992). Predictors of osteopenia in premenopausal women with anorexia nervosa. *Calcif. Tissue Int.*, *50*, 498–501.

Helmreich, D. L., Mattern, L. G., and Cameron, J. L. (1993). Lack of a role of the hypothalamic–pituitary–adrenal axis in the fasting-induced suppression of luteinizing hormone secretion in adult male rhesus monkeys (*Macaca mulatta*). *Endocrinology*, *132*, 2427–2437.

Hergenroeder, A. C., O'Brian Smith, E., Shypailo, R., et al. (1997). Bone mineral changes in young women with hypothalamic amenorrhea treated with oral contraceptives, medroxyprogesterone, or placebo over 12 months. *Am. J. Obstet. Gynecol.*, *176*, 1017–1025.

Hoggard, N., Hunter, L., Duncan, J. S., Williams, L. M., Trayhurn, P., and Mercer, J. G. (1997). Leptin and leptin receptor mRNA and protein expression in the murine fetus and placenta. *Proc. Natl. Acad. Sci. USA*, *94*, 11073–11078.

Hotta, I., Shebasoki, K., Masuda, A., Imaki, T., Hiroshi, D., Ling, N., and Shizume, K. (1986) The responses of plasma adrenocorticotropin and cortisol to corticotropin releasing hormone (CRH) and cerebrospinal fluid immunoreactive CRH in anorexia nervosa patients. *J. Clin. Endocrinol. Metab.*, 62, 319–324.

Jonnavithula, S., Warren, M. P., Fox, R. P., and Lazaro, M. I. (1993). Bone density is compromised in amenorrheic women despite return of menses: a 2-year study. *Obstet. Gynecol.*, 81, 669–674.

Kanders, B., Dempster, D. W., and Lindsay, R. (1988). Interaction of calcium nutrition and physical activity on bone mass in young women. *J. Bone Miner. Res.*, 3(2), 145–149.

Katz, J. L., Boyar, R. M., Roffwarg, H., Hellman, L., and Weiner, H. (1978). Weight and circadian luteinizing hormone secretory pattern in anorexia nervosa. *Psychosom. Med.*, 40, 549–567.

Kaufman, B. A., Warren, M. P., Dominguez, J. E., Wang, J., Heymsfield, S. B., and Pierson, R. N. (2002). Bone density and amenorrhea in ballet dancers are related to a decreased resting metabolic rate and lower leptin levels. *J. Clin. Endocrinol. Metab.* 2002; 87(6): 2777–2783.

Khan, K. M., Warren, M. P., Stiehl, A., McKay, H. A., and Wark, J. D. (1999). Bone mineral density in active and retired ballet dancers. *J. Dance Med. Sci.*, 3, 15–23.

Kiyohara, K., Tamai, H., Takaichi, Y., Nakagawa, T., and Kumagai, L. F. (1989). Decreased thyroidal triiodothyronine secretion in patients with anorexia nervosa: influence of weight recovery. *Am. J. Clin. Nutr.*, 50, 767–772.

Klibanski, A., Biller, B. M. K., Schoenfeld, D. A., Herzog, D. B., and Saxe, V. C. (1995). The effects of estrogen administration on trabecular bone loss in young women with anorexia nervosa. *J. Clin. Endocrinol. Metab.*, 80, 898–904.

Kohmura, H., Miyake, A., Aono, T., and Tanizawa, O. (1986). Recovery of reproductive function in patients with anorexia nervosa: a 10-year follow up study. *Eur. J. Obstet. Gynecol. Reprod. Biol.* 22, 293–296.

Kopp, W., Blum, W. F., von Prittwitz, S., Ziegler, A., Lubbert, H., Emons, G., Herzog, W., Herpertz, S., Deter, H. C., Remschmidt, H., and Hebebrand, J. (1997). Low leptin levels predict amenorrhea in underweight and eating disordered females. *Mol. Psychiatry*, 2, 335–340.

Kreipe, R. E. (1992). Bones of today, bones of tomorrow (editorial). *Am. J. Dis. Child*, 146, 22–25.

Kreipe, R. E. and Forbes, G. B. (1990). Osteoporosis: a "new morbidity" for dieting female adolescents? *Pediatrics* 86, 478–480.

Kreipe, R. E., Churchill, B. H., and Strauss, J. (1989a). Long-term outcome of adolescents with anorexia nervosa. *Am. J. Dis. Child*, 143, 1322–1327.

Kreipe, R. E., Hicks, D. G., Rosier, R. N., and Puzas, J. E. (1993). Preliminary findings on the effects of sex hormones on bone metabolism in anorexia nervosa. *J. Adolesc. Health*, 14, 319–324.

Kreipe, R. E., Strauss, J., Hodgeman, C. H., and Ryan, R. M. (1989b). Menstrual cycle abnormalities and subclinical eating disorders: a preliminary report. *Psychosom. Med.*, 51, 81–86.

Laughlin, G. A. (1999). The role of nutrition in the etiology of functional hypothalamic amenorrhea. *Curr. Opin. Endocrinol. Diabetes*, 6, 38–43.

Laughlin, G. A. and Yen, S. S. C. (1997). Hypoleptinemia in women athletes: absence of a diurnal rhythm with amenorrhea. *J. Clin. Endocrinol. Metab.*, 82, 318–321.

Laughlin, G. A., Dominguez, C. E., and Yen, S. S. C. (1998). Nutritional and endocrine-metabolic aberrations in women with functional hypothalamic amenorrhea. *J. Clin. Endcrinol. Metab.*, 83, 25–32.

Legradi, G., Emerson, C. H., Ahima, R. S., Flier, J. S., and Lechan, R. M. (1997). Leptin prevents fasting-induced suppression of prothyrotropin-releasing hormone messenger ribonucleic acid in neurons of the hypothalamic paraventricular nucleus. *Endocrinology*, 138, 2569–2576.

Legradi, G., Emerson, C. H., Ahima, R. S., Rand, W. M., Flier, J. S., and Lechan, R. M. (1998). Arcuate nucleus ablation prevents fasting-induced suppression of proTRH mRNA in the hypothalamic paraventricular nucleus. *Neuroendocrinology*, 68, 89–97.

Leibel, R. L., Rosenbaum, M., and Hirsch, J. (1995). Changes in energy expenditure resulting from altered body weight. *N. Engl. J. Med.*, 332, 621–628.

Lindberg, J. S., Fears, W. B., Hunt, M. M., Powell, M. R., Boll, D., and Wade, C. E. (1984). Exercise-induced amenorrhea and bone density. *Ann. Intern. Med.*, 101, 647–648.

Lindsay, R., Cosman, F., Herrington, B. S., and Himmelstein, S. (1992). Bone mass and body composition in normal women. *J. Bone Miner. Res.*, 7, 55–63.

Liu, C., Grossman, A., Bain, S., Strchan, M., Puermner, D., Bailey, C., Humes, J., Lenox, J., Yamamoto, G., Sprugel, K., Kuijper, J., Weigle, S., Dumam, D., and Moore, E. (1997). Leptin stimulates cortical bone formation in obese (ob/ob) mice. *J. Bone. Miner. Res.*, 12, S115.

Liu, J. H. (1998). Anovulation of CNS Origin. In B. R. Carr and R. E. Blackwell (Eds.). *Textbook of Reproductive Medicine*, Appleton and Lange, Stamford, CT, pp. 309–322.

Lloyd, S. J., Triantafyllou, S. J., Baker, E. R., Houts, P. S., Whiteside, J. A., Kalenak, A., and Stumpf, P. (1986). Women athletes with menstrual irregularity have increased musculoskeletal injuries. *Med. Sci. Sports Exerc.*, *18*(4), 374–379.

Locke, R. J. and Warren, M. P. (2000). How to prevent bone loss in women with hypothalamic amenorrhea. *Womens Health Primary Care*, *3*, 270–278.

Loucks, A. B. and Heath, E. M. (1994). Induction of low-T3 syndrome in exercising women occurs at a threshold of energy availability. *Am. J. Physiol.*, *266*, R817–R823.

Loucks, A. B., Brown, R., King, K., Thuma, J. R., and Verdun, M. (1995). A combined regimen of moderate dietary restriction and exercise training alters luteinizing hormone pulsatility in regularly menstruating young women. *Endocrine Society Annual Meeting*, Washington DC (June 14–17), Abstract, 558–558.

Loucks, A. B., Mortola, J. F., Girton, L., and Yen, S. S. C. (1989). Alterations in the hypothalamic–pituitary–ovarian and the hypothalamic–pituitary–adrenal axes in athletic women. *J. Clin. Endocrinol. Metab.*, *68*, 402–411.

Macut, D., Micic, D., Pralong, F. P., Bischof, P., and Campana, A. (1998). Is there a role for leptin in human reproduction? *Gynecol. Endocrinol.*, *12*, 321–326.

Maffei, M., Halaas, J., Ravussin, E., Pratley, R. E., Lee, G. H., Zhang, Y., Fei, H., Kim, S., Lallone, R., and Ranganathan, S. (1995). Leptin levels in human and rodent: measurement of plasma leptin and ob RNA in obese and weight-reduced subjects. *Nat. Med.*, *1*, 1155–1161.

Manolagas, S. C., and Jilka, R. L. (1995). Bone marrow, cytokines, and bone remodeling: Emerging insights into the pathophysiology of osteoporosis. *N. Eng. J. Med.*, *332*(5):305–311.

Marcus, R., Cann, C. E., Madvig, P., Minkoff, J., Goddard, M., Bayer, M., Martin, M. C., Gaudiani, L., Haskell, W., and Genant, H. K. (1985). Menstrual function and bone mass in elite women distance runners. *Ann. Intern. Med.*, *102*, 158–163.

Masuda, A., Shibasaki, T., Hotta, M., Yamauchi, N., Ling, N., Demura, H., and Shizume, K. (1990). Insulin-induced hypoglycemia, L-dopa and arginine stimulate GH secretion through different mechanisms in man. *Regul. Pept.*, *31*, 53–64.

Mecklenburg, R. S., Loriaux, D. L., and Thompson, R. H. (1974). Hypothalamic dysfunction in patients with anorexia nervosa. *Medicine*, *53*, 147.

Miller, K. K. and Klibanski, A. (1999). Amenorrheic bone loss. *J. Clin. Endocrinol. Metab.*, *84*, 1775–1783.

Miller, K. K., Parulekar, M. S., Schoenfeld, E., Anderson, E., Hubbard, J., Klibanski, A., and Grinspoon, S. K. (1998). Decreased leptin levels in normal weight women with hypothalamic amenorrhea: the effects of body composition and nutritional insults. *J. Clin. Endocrinol. Metab.*, *83*, 2309–2312.

Moshang, T., Jr. and Utiger, R. D. (1977). Low triiodothyronine euthyroidism in anorexia nervosa. In R. Vigersky (Ed.) *Anorexia Nervosa*. New York: Raven Press. p. 263.

Moshang, T., Jr., Parks, J. S., Baker, L., Vaidya, V., Utiger, R. D., Bongiovanni, A. M., and Snyder, P. J. (1975). Low serum triiodothyronine in patients with anorexia nervosa. *J. Clin. Endocrinol. Metab.*, *40*, 470–473.

Myerson, M., Gutin, B., Warren, M. P., May, M., Contento, I., Lee, M., Pierson, R. N., and Pi-Sunyer, F. X. (1987). Energy balance of amenorrhea and eumenorrheic runners. *Med. Sci. Sports. Exerc.*, Abstract, *19*, S37.

Myerson, M., Gutin, B., Warren, M. P., May, M., Contento, I., Lee, M., Pi-Sunyer, F. X., Pierson, R. N., and Brooks-Gunn, J. (1991). Resting metabolic rate and energy balance in amenorrheic and eumenorrheic runners. *Med. Sci. Sports. Exerc.*, *23*(1), 15–22.

Nappi, R. E., Petraglia, F., Genazzani, A. D., D'Ambrogio, G., Zara, C., and Genazzani, A. R. (1993). Hypothalamic amenorrhea: evidence for a central derangement of hypothalamic–pituitary–adrenal cortex axis activity. *Fertil. Steril.*, *59*, 571–576.

Newman, M. M. and Halmi, K. A. (1988). The endocrinology of anorexia nervosa and bulimia nervosa. *Neurol. Clin.*, *6*, 195–212.

Newman, M. M., Halmi, K. A., and Marchi, P. (1987). Relationship of clinical factors to caloric requirements in subtypes of eating disorders. *Biol. Psychiatry.*, *22*, 1253–1263.

Ogueh, O., Sooranna, S., Nicolaides, K. H., and Johnson, M. R. (2000). The relationship between leptin concentration and bone metabolism in the human fetus. *J. Clin. Endocrinol. Metab.*, *85*, 1997–1999.

Okano, H., Mizunuma, H., Soda, M. Matsui, H., Honjo, S., and Ibuki, Y. (1995). Effects of exercise and amenorrhea on bone mineral density in teenage runners. *Endocrine Journal*, *42*, 271–276.

Olson, B. R., Cartledge, T., Sebring, N., Defensor, R., and Nieman, L. (1995). Short-term fasting affects luteinizing hormone secretory dynamics but not reproductive function in normal-weight sedentary women. *J. Clin. Endocrinol. Metab.*, *80*, 1187–1193.

Ott, S. M. (1990). Editorial: attainment of peak bone mass. *J. Clin. Endocrinol. Metab.*, *71*, 1082A–1082C.

Pettersson, F., Fries, H., and Nillius, S. J. (1973). Epidemiology of secondary amenorrhea: incidence and prevalence rates. *Am. J. Obstet. Gynecol.*, *7*, 80–86.

Pettersson, U., Stalnacke, B., Ahlenius, G., Henriksson-Larsen, K., and Lorentzon, R. (1999). Low bone mass density at multiple skeletal sites, including the appendicular skeleton in amenorrheic runners. *Calcif. Tissue Int.*, *64*, 117–125.

Pirke, K. M., Pahl, J., Schweiger, U., and Warnhoff, M. (1985). Metabolic and endocrine indices of starvation in bulimia: a comparison with anorexia nervosa. *Psychiatry. Res.*, *15*, 33–39.

Platte, P., Pirke, K. M., Trimborn, P., Pietsch, K., Krieg, J. C., and Fichter, M. M. (1994). Resting metabolic rate and total energy expenditure in acute and weight recovered patients with anorexia nervosa and in healthy young women. *Int. J. Eating. Disord.*, *16*, 45–52.

Ponzo, O. J., Szwarcfarb, B., Rondina, D., Carbone, S., Reynoso, R., Scacchi, P., and Moguilevsky, J. A. (2001). Changes in the sensitivity of gonadotrophin axis to leptin during sexual maturation in female rats. *Neuroendocrinol. Lett.*, *22*, 427–431.

Powers, P. S. (1999). Osteoporosis and eating disorders. *J. Pediatr. Adolesc. Gynecol.*, *12*, 51–57.

Prior, J. C. and Vigna, Y. M. (1985). Gonadal steroids in athletic women: contraception, complications and performance. *Sports Med.*, *2*, 287–295.

Prior, J. C., Vigna, Y. M., Schechter, M. T., and Burgess, A. E. (1990). Spinal bone loss and ovulatory disturbances. *N. Engl. J. Med.*, *323*, 1221–1227.

Rauch, F., Blum, W. F., Klein, K., Allolio, B., and Schonau, E. (1998). Does leptin have an effect on bone in adult women? *Calcif. Tissue Int.*, *63*, 453–455.

Reseland, J. E., Syversen, U., Bakke, I., Qvigstad, G., Eide, L. G., Hjertner, O., Gordeladze, J. O., and Drevon, C. A. (2001). Leptin is expressed in and secreted from primary cultures of human osteoblasts and promotes bone mineralization. *J. Bone Miner. Res.*, *16*, 1426–1433.

Riggs, B. L. and Eastell, R. (1986). Exercise, hypogonadism and osteopenia. *JAMA*, *256*, 392–393.

Rigotti, N. A., Neer, R. M., Skates, S. J., Herzog, D. B., and Nussbaum, S. R. (1991). The clinical course of osteoporosis in anorexia nervosa: a longitudinal study of cortical bone mass. *JAMA*, *265*, 1133–1138.

Rigotti, N. A., Nussbaum, S. R., Herzog, D. B., and Neer, R. M. (1984). Osteoporosis in women with anorexia nervosa. *N. Engl. J. Med.*, *311*, 1601–1606.

Robinson, E., Bachrach, L. K., and Katzman, D. K. (2000). Use of hormone replacement therapy to reduce the risk of osteopenia in adolescent girls with anorexia nervosa. *J. Adolesc. Health*, *26*, 343–348.

Roth, J. C., Kelch, R. P., Kaplan, S. L., and Grumbach, M. M. (1972). FSH and LH response to luteinizing hormone-releasing factor in prepubertal and pubertal children, adult males and patients with hypogonadotropic and hypergonadotropic hypogonadism. *J. Clin. Endocrinol. Metab.*, *35*, 926–930.

Rutherford, O. M. (1993). Spine and total body bone mineral density in amenorrheic endurance athletes. *J. Appl. Physiol.*, *74*, 2904–2908.

Sakala, E. P. (1997). *Obstetrics & Gynecology*. Williams & Wilkins, Baltimore, 266–272.

Salisbury, J. J. and Mitchell, J. E. (1991). Bone mineral density and anorexia nervosa in women. *Am. J. Psychiatry*, *148*, 768–774.

Salisbury, J. J., Levine, A. S., Crow, S. J., and Mitchell, J. E. (1995). Refeeding, metabolic rate, and weight gain in anorexia nervosa: a review. *Int. J. Eating Disord.*, *17*, 337–345.

Sanborn, C. F., Martin, B. J., and Wagner, W. W., Jr. (1982) Is athletic amenorrhea specific to runners? *Am. J. Obstet. Gynecol.*, *143*, 859–861.

Schreihofer, D. A., Amico, J. A., and Cameron, J. L. (1993). Reversal of fasting-induced suppression of luteinizing hormone (LH) secretion in male rhesus monkeys by intragastric nutrient infusion: evidence for rapid stimulation of LH by nutritional signals. *Endocrinology*, *132*, 1890–1897.

Schweiger, U. (1991). Menstrual function and luteal-phase deficiency in relation to weight changes and dieting. *Clin. Obstet. Gynecol.*, *34*(1), 191–197.

Shangold, M. M., and Levine, H. S. The effect of marathon training upon menstrual function. *Am. J. Obstet. Gynecol.*, *143*, 862–869.

Singh, K. B. (1981). Menstrual disorders in college students. *Am. J. Obstet. Gynecol.*, *1210*, 299–302.

Smith, E. L. and Gilligan, C. (1994). Bone Concerns. In M. M. Shangold and G. Mirkin (Eds.). *Women and Exercise: Physiology and Sports Medicine*, F. A. Davis Company, Philadelphia, pp. 89–101.

Soules, M. R., Steiner, R. A., Clifton, D. K., and Bremner, W. J. (1984). Abnormal patterns of pulsatile luteinizing hormone in women with luteal phase deficiency. *Obstet. Gynecol.*, *63*, 626–629.

Soyka, L. A., Grinspoon, S., Levitsky, L. L., Herzog, D. B., and Klibanski, A. (1999). The effects of anorexia nervosa on bone metabolism in female adolescents. *J. Clin. Endcrinol. Metab.*, Abstract, *84*(12), 4489–4496.

Steen, S. N., Oppliger, R. A., and Brownell, K. D. (1988). Metabolic effects of repeated weight loss and regain in adolescent wrestlers. *JAMA*, *260*, 47–50.

Steppan, C. M., Crawford, D. T., Chidsey-Frink, K. L., Ke, H., and Swick, A. G. (2000). Leptin is a potent stimulator of bone growth in ob/ob mice. *Regul. Pept.*, *92*, 73–78.

Steppan, C. M., Chidsey-Frink, K. L., Crawford, D. T., et al. (1998). Leptin administration causes bone growth in ob/ob mice. *The Endocrine Society, 80th Annual Meeting*, New Orleans, LA, Abstract.

Tezuka, M., Irahara, M., Ogura, K., Kiyokawa, M., Tamura, T., Matsuzaki, T., Yasui, T., and Aono, T. (2002). Effects of leptin on gonadotropin secretion in juvenile female rat pituitary cells. *Eur. J. Endocrinol.*, *146*, 261–266.

Thomas, T., Gori, F., Burguera, B., et al. (1998). Leptin acts on human marrow stromal precursor cells to enhance osteoblast differentiation and to inhibit adipocyte differentiation: a potential mechanism for increased bone mass in obesity. *The Endocrine Society, 80th Annual Meeting*, New Orleans, LA, Abstract.

Thong, F. S., McLean, C., and Graham, T. E. (2000). Plasma leptin in female athletes: relationship with body fat, reproductive, nutritional, and endocrine factors. *J. Appl. Physiol.*, *88*, 2037–2044.

Tsafriri, A., Dekel, N., and Bar-Ami, S. (1982). The role of oocyte maturation inhibitor in follicular regulation of oocyte maturation. *J. Reprod. Fertil.*, *64*, 541–551.

Tuschl, R. J., Platte, P., Laessle, R. G., Stichler, W., and Pirke, K. M. (1990). Energy expenditure and every day eating behavior in healthy young women. *Am. J. Clin. Nutr.*, *52*, 81–86.

van Binsbergen, C. J. M., Coelingh Bennink, H. J. T., Odink, J., Haspels. A. A., and Koppeschaar. H. P. F. (1990). A comparative and longitudinal study on endocrine changes related to ovarian function in patients with anorexia nervosa. *J. Clin. Endocrinol. Metab.*, *71*, 705–711.

Vigersky, R. A., Andersen, A. E., Thompson, R. H., and Loriaux, D. L. (1977). Hypothalamic dysfunction in secondary amenorrhea associated with simple weight loss. *N. Engl. J. Med.*, *297*, 1141–1145.

Walsh, B. T., Katz, J. L., Levin, J., Kream, J., Fukushima, D. K., Hellman, L. D., Weiner, H., and Zumoff, B. (1978). Adrenal activity in anorexia nervosa. *Psychosom. Med.*, *40*, 499–506.

Warren, M. P. (1980). The effects of exercise on pubertal progression and reproductive function in girls. *J. Clin. Endocrinol. Metab.*, *51*(5), 1150–1157.

Warren, M. P. (1983). The effects of undernutrition on reproductive function in the human. *Endocr. Rev.*, *4*, 363–377.

Warren, M. P. (1999). Health issues for women athletes: exercise-induced amenorrhea [see comments]. *J. Clin. Endocrinol. Metab.*, *84*, 1,892–1,896.

Warren, M. P. and Shangold, M. M. (1997). *Sports Gynecology: Problems and Care of the Athletic Female.* Blackwell Science, Cambridge, MA.

Warren, M. P. and Vande Wiele, R. L. (1973). Clinical and metabolic features of anorexia nervosa. *Am. J. Obstet. Gynecol.*, *117*(3), 435–449.

Warren, M. P., Fox, R. P., DeRogatis, A. J., and Hamilton, W. G. (1994a). Osteopenia in hypothalamic amenorrhea: a 3-year longitudinal study. *The Endocrine Society Annual Meeting*, Anaheim, California, Abstract.

Warren, M. P., Brooks-Gunn, J., Hamilton, L. H., Warren, L. F., and Hamilton, W. G. (1986). Scoliosis and fractures in young ballet dancers: relation to delayed menarche and secondary amenorrhea. *N. Engl. J. Med.*, *314*, 1348–1353.

Warren, M. P., Jewelewicz, R., Dyrenfurth, I., Ans, R., Khalaf, S., and Vande Wiele, R. L. (1975). The significance of weight loss in the evaluation of pituitary response to LH–RH in women with secondary amenorrhea. *J. Clin. Endocrinol. Metab.*, *40*, 601–611.

Warren, M. P., Brooks-Gunn, J., Fox, R. P., Lancelot, C., Newman, D., and Hamilton, W. G. (1991). Lack of bone accretion and amenorrhea: evidence for a relative osteopenia in weight bearing bones. *J. Clin. Endocrinol. Metab.*, *72*, 847–853.

Warren, M. P., Holderness, C. C., Lesobre, V., Tzen, R., Vossoughian, F., and Brooks-Gunn, J. (1994b). Hypothalamic amenorrhea and hidden nutritional insults. *J. Soc. Gynecol. Invest.*, *1*, 84–88.

Warren, M. P., Voussoughian, F., Geer, E. B., Hyle, E. P., Adberg, C. L., and Ramos, R. H. (1999). Functional hypothalamic amenorrhea: hypoleptinemia and disordered eating. *J. Clin. Endocrinol. Metab.*, *84*, 873–877.

Warren, M. P., Brooks-Gunn, J., Fox, R. P., Holderness, C. C., Hyle, E. P., and Hamilton, W. G. (2002). Osteopenia in exercise associated amenorrhea using ballet dancers as a model: a longitudinal study. *J. Clin. Endocrinol. Metab.*, *87*(7): 3162–3168.

Williams, N. I., Helmreich, D. L., Parfitt, D. B., Caston-Balderrama, A., and Cameron, J. L. (2001). Evidence for a causal role of low energy availability in the induction of menstrual cycle disturbances during strenuous exercise training. *J. Clin. Endocrinol. Metab.*, *86*, 5184–5193.

Wilson, G. T. and Walsh, B. T. (1991). Eating disorders in the DSM-IV. *J. Abnorm. Psychol.*, *100*(3), 362–365.

Wolman, R. L., Clark, P., McNally, E., Harries, M., and Reeve, J. (1990). Menstrual state and exercise as determinants of spinal trabecular bone density in female athletes. *Brit. Med. J.*, *301*, 516–518.

Yamauchi, M., Sugimoto, T., Yamaguchi, T., Nakaoka, D., Kanzawa, M., Yano, S., Ozuru, R., Sugishita, T., and Chihara, K. (2001). Plasma leptin concentrations are associated with bone mineral density and the presence of vertebral fractures in postmenopausal women. *Clin. Endocrinol. (Oxf.)*, *55*, 341–347.

Zanker, C. L. (1999). Bone metabolism in exercise associated amenorrhoea: the importance of nutrition. *Br. J. Sports Med.*, *33*, 228–229.

Zanker, C. L. and Swaine, I. L. (1998a). Bone turnover in amenorrhoeic and eumenorrhoeic distance runners. *Scand. J. Med. Sci. Sports*, *8*, 20–26.

Zanker, C. L. and Swaine, I. L. (1998b). The relationship between bone turnover, oestradiol, and energy balance in women distance runners. *Br. J. Sports Med.*, *32*, 167–171.

Zhang, Y., Proenca, R., Maffei, M., Barone, M., Leopold, L., and Friedman, J. M. (1994). Positional cloning of the mouse *obese* gene and its human homologue. *Nature*, *372*, 425–432.

The Role of Leptin in Polycystic Ovary Syndrome

Denis A. Magoffin, Priya S. Duggal, and Robert J. Norman

20.1. THE POLYCYSTIC OVARY SYNDROME

Polycystic ovary syndrome (PCOS) is the most common reproductive endocrine disease in women of reproductive age. Within the infertile population, approximately three quarters of women with anovulatory infertility have PCOS, thus accounting for approximately one third of women with secondary amenorrhea and approximately 90 percent of women with oligomenorrhea (Franks and White, 1993). It is clear that PCOS affects a large number of women and accounts for a major proportion of women with menstrual irregularities.

The health consequences of PCOS are varied. Not only is there a high incidence of infertility, but approximately 60 percent of these women are hirsute (Franks and White, 1993). Other consequences of PCOS are a possible increased risk of recurrent early pregnancy loss (Franks and White, 1993) and increased risk for cardiovascular disease (Conway et al., 1992) with an estimated 11-fold increased risk of myocardial infarction between the ages of 50–61 years (Dahlgren et al., 1992). There are, however, no clinical data indicating increased death rates. There is also an increased risk of endometrial cancer at a young age. Cases of endometrial cancer have been reported in women with PCOS as young as 17 years (Smyczek-Gargya and Geppert, 1992). Thus, many serious health problems are associated with PCOS.

Despite the seriousness of PCOS and the high prevalence of this disorder, the etiology of PCOS remains a mystery. The metabolic profile of the PCOS patient is heterogeneous

Denis A. Magoffin • Cedars-Sinai Burns and Allen Research Institute, The David Geffen School of Medicine at UCLA.

Priya S. Duggal and Robert J. Norman • Reproductive Medicine Unit, Department of Obstetrics & Gynaecology, University of Adelaide, The Queen Elizabeth Hospital.

Leptin and Reproduction. Edited by Henson and Castracane, Kluwer Academic/Plenum Publishers, 2003.

and has led to much discussion and controversy regarding even the definition of PCOS (Lobo, 1995). The fact of having "polycystic" ovarian morphology is insufficient to diagnose PCOS. Several reports describe women with "polycystic ovaries" having normal menstrual cycles (Roberts and Haines, 1960; Goldzieher and Axelrod, 1963; Wong et al., 1995). Clearly ultrasound evidence or even histological evidence of multiple small antral follicles is not sufficient to conclude that a woman has PCOS. Indeed, other potential abnormalities such as elevated serum luteinizing hormone (LH) or LH:FSH ratio are not always present (Lobo et al., 1983) nor are physical changes such as hirsutism always present (Carmina et al., 1992; Norman et al., 1995). Two obligatory symptoms occur in PCOS, namely hyperandrogenism and chronic anovulation (Lobo, 1995). The common physiological link in all subtypes of PCOS is the chronic failure of follicles to develop beyond the early antral stage (Yen, 1980; Hughesdon, 1982; Erickson, 1994).

A consistent finding in women with PCOS is that the ovaries produce abnormally high amounts of androgens, but the role of hyperandrogenism in inhibiting selection of dominant follicles is unknown. There is good evidence, based on clinical observations, to conclude that elevated androgens are capable of interfering with selection of dominant follicles and causing PCOS. Women who have clinical conditions where circulating androgen levels are elevated due to non-ovarian causes develop polycystic ovaries. In conditions such as congenital adrenal hyperplasia wherein there is a defect of the 21-hydroxylase or the 11β-hydroxysteroid dehydrogenase enzymes, the adrenal gland is unable to make glucocorticoids. In this condition, there are excessive circulating androgen concentrations and polycystic ovaries develop (Erickson et al., 1989; Lobo, 1995). In the case of female to male transsexuals high concentrations of exogenous androgen are administered. These women also develop polycystic ovaries (Futterweit and Deligdisch, 1986). Female to male transsexuals have been used frequently as a model to study the role of hyperandrogenism in the genesis of PCOS. In both congenital adrenal hyperplasia and female to male transsexuals, when the excessive androgen concentrations are reduced to normal, the ovaries resume normal function. These observations provide strong and convincing evidence that hyperandrogenism is a reversible cause of polycystic ovaries.

20.2. INSULIN AND OBESITY IN PCOS

There has been a considerable amount of effort expended to study the mechanisms regulating androgen production by ovarian theca cells in order to understand the causes of hyperandrogenism in PCOS. From these studies we have learned that there are metabolic abnormalities in PCOS that appear to cause elevated androgen production. One of the most important among these metabolic abnormalities is insulin resistance (Dunaif, 1997). Approximately 50 percent of women with PCOS are insulin resistant and conversely, approximately 50 percent of women with insulin resistance develop polycystic ovaries (Conn et al., 2000).

Since the first published description of the association between insulin resistance and the androgen excess of PCOS (Burghen et al., 1980), researchers have sought to identify the mechanism by which insulin regulates thecal androgen biosynthesis. When a woman is insulin resistant there is a moderate compensatory increase in circulating insulin levels. Although blood glucose concentrations are controlled within normal levels, there is an increase in the fasting insulin to glucose ratio that has been shown to correlate well with the degree of insulin resistance in women with PCOS (Legro et al., 1998). Ample evidence supports a causal link between the hyperinsulinemia and the increased ovarian androgen synthesis characteristic of PCOS (Ehrmann et al., 1995; Dunaif, 1997). Receptors for insulin and insulin-like growth factor I (IGF-I) are abundant in the human ovary (Bergh et al., 1993; El-Roeiy et al., 1994; Barbieri et al., 1988; Franks et al., 1999) and the stimulatory effects of insulin on ovarian androgen production are mediated through the insulin receptor (Franks et al., 1999). These observations provide a basis for the hypothesis that insulin excess can contribute to excessive ovarian androgen production. Although basal androgen production in cultured theca cells is enhanced 20-fold in polycystic ovaries compared to controls (Gilling-Smith et al., 1994, 1997), the stimulatory effect of insulin on steroidogenesis is also augmented in polycystic ovaries (Barbieri et al., 1986) suggesting an ovarian "hypersensitivity" to insulin action in PCOS.

Insulin resistance and obesity are linked. Insulin secretion is proportional to adiposity and insulin provides a signal to the brain that regulates food intake and body weight (Woods and Seeley, 2001). Central obesity is associated with both insulin resistance and PCOS. Recent evidence demonstrated that intra-abdominal fat accumulation correlates with insulin resistance (Cnop et al., 2002), supporting the concept that insulin resistance, obesity and ovarian hyperandrogenism are linked.

20.3. LEPTIN AND OBESITY IN PCOS

The demonstration that leptin regulates fertility in the mouse led to speculation regarding its potential role in human infertility (Ezzell, 1996). It immediately became apparent that there was a strong positive correlation between serum leptin concentrations and body fat in humans (Maffei et al., 1995; Caro et al., 1996; Considine et al., 1996; Dagogo-Jack et al., 1996) demonstrating that obesity is associated with high circulating leptin levels. Women with a low percentage of body fat such as trained distance runners, ballet dancers, and women with anorexia nervosa often are infertile (Green et al., 1988; De Souza and Metzger, 1991). At the other extreme, obese women exhibit a high incidence of oligo- or amenorrhea and infertility (Stein and Leventhal, 1935; Goldzieher, 1981; Green et al., 1988). Thus, it was reasonable to propose that the link between obesity and infertility might involve leptin.

A few years after the initial discovery of leptin and its relationship to obesity in mice, the connection between this hormone and reproduction in humans was examined in relationship to women with PCOS. The first study indicated that leptin levels were disproportionately high in women who had PCOS (Brzechffa et al., 1996). Because of the well-known effects of leptin on ovarian steroidogenesis as described previously by this group, the authors speculated that leptin might have a role in the pathogenesis of PCOS. Several other articles subsequently confirmed the apparent association between hyperleptinemia and PCOS (Vicennati et al., 1998; Carmina et al., 1999; El Orabi et al., 1999). It was soon realized, however, that there should be a compensation for body mass and for fat distribution because of the increased weight in PCOS subjects compared to controls. Chapman et al. (1997) were the first to show that multiple regression analysis for a variety of factors led to the relationship between leptin and body mass index (BMI) being

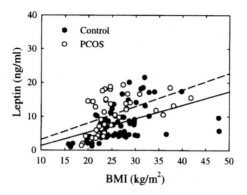

Figure 20.1. The relationship between serum leptin concentrations and obesity. Leptin was measured by RIA in fasting morning serum from regularly cycling control women or women with PCOS. The lines are the linear regressions of control (solid line) and PCOS (dashed line) data, respectively.

paramount rather than that of PCOS (Figure 20.1). In the next few months, following the publication of this article, another three studies showed no differences in leptin in PCOS compared to weight managed controls (Laughlin et al., 1997; Mantzoros et al., 1997; Rouru et al., 1997). One study has shown that girls with early puberty (a known feature of some women with PCOS) are more likely to have high leptin values (Cizza et al., 2001) but this may be explained on the basis of BMI alone rather than any other primary criteria. There is now a general agreement from a large number of studies that there is no peripheral hyperleptinemia in PCOS above that of obesity alone (Gennarelli et al., 1998; Kowalska et al., 1999; Pirwany et al., 2001; Takeuchi and Tsutsumi, 2000; Veldhuis et al., 2001). Several good reviews on leptin and PCOS have been written (Conway and Jacobs, 1997; Jacobs and Conway, 1999; Gonzalez et al., 2000).

20.4. PCOS AND LEPTIN IN THE HYPOTHALAMUS

Initial studies suggested that gonadotropin-releasing hormone (GnRH) injected into women led to an increased leptin concentration in circulation, followed by a decreased amount as down regulation of the receptor occurred. Several studies have looked at the concordance of LH pulsatility with that of leptin with and without the addition of GnRH (Sir-Petermann et al., 1999; Sir-Petermann et al., 2001; Spritzer et al., 2001). The well-known increased pulse amplitude of LH in PCOS was confirmed by all studies, but there has been controversy about the relationship with leptin. Studies of pulsatility in normal controls show a close relationship between LH and leptin in normal subjects but not in PCOS (Laughlin et al., 1997). Veldhuis et al. (2001) studied 11 patients with PCOS and 9 non-PCOS controls who had blood removed every 20 min for 12 hr. There was a positive correlation between LH and leptin, with LH lagging by 20 min in normal subjects, but this association was not found in PCOS. Similarly, there was a relationship between leptin and testosterone in normal subjects, but not in PCOS. PCOS non-obese adolescents with menstrual disturbance, while having normal serum leptin concentrations, show a disturbance of the regularity of leptin secretion patterns with loss of the synchrony between leptin and LH, and leptin and androgens. This disruption of the leptin gonadal axis integration is of unknown etiology and consequence, and it is not known whether this persists into adulthood. Leptin is known to modulate levels of neuropeptide Y in the brain and this peptide is elevated in obese and non-obese women with PCOS (Baranowska et al., 1999).

20.5. THE ROLE OF LEPTIN IN OVARIAN HYPERANDROGENISM

20.5.1. PCOS and Ovarian Leptin

Studies have shown that the concentration of leptin in human follicular fluid is equivalent to that in the plasma, although leptin binding in follicular fluid is less than that in the plasma (Fedorcsak et al., 2000). It has been speculated that some of the well known effects of obesity on reproduction and poor reproductive performance may be as a result of the increased leptin in follicular fluid reflecting the concentrations in the circulating blood. One study has shown that the lower the level of leptin in follicular fluid in relation to the BMI, the better the outcome when assisted reproductive technology is performed (Mantzoros et al., 2000).

20.5.2. Leptin Receptors in PCOS

Oksanen et al. (2000) analyzed the leptin gene by sequencing samples from 38 well-characterized patients with PCOS. There were no mutations in the coding exons studied. Using single stranded conformational polymorphism (SSCP) analysis and sequencing of the leptin receptor gene, several variants in Exon two, four and 12 were detected. However, allele frequencies of these polymorphisms were no different from those of the general population reflected in 122 controls. Some of the polymorphisms were related to serum insulin concentrations, which suggested that variations in the leptin receptor gene locus may influence insulin regulation, although there is no major defect in the leptin receptor itself.

20.5.3. Direct Effects of Leptin on Human Ovarian Steroidogenesis

The demonstration that multiple splice variants of leptin receptor mRNA are expressed in the human ovary raised the possibility of direct regulation of ovarian steroidogenesis by leptin (Ezzell, 1996; Karlsson et al., 1997; Agarwal et al., 1999).

In human theca cells leptin did not alter the LH stimulation of androstenedione production (Agarwal et al., 1999), but at physiological concentrations blocked the synergistic stimulation of LH-stimulated androstenedione production by IGF-I (Figure 20.2). A direct effect of leptin was also demonstrated using human granulosa cells in vitro (Agarwal et al., 1999). Leptin was without effect on basal estradiol production and similarly had no effect on the ability of FSH to stimulate aromatase activity. The physiological concentrations of leptin commonly found in obese women completely blocked the ability of

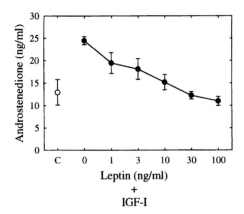

Figure 20.2. Leptin inhibition of IGF-I stimulation of androgen production by human theca cells. Human theca cells microdissected from healthy follicles of regularly cycling women were cultured in defined medium for 2 days with 10 ng/ml of LH with and without (control) 50 ng/ml of IGF-I. Increasing concentrations of recombinant human leptin caused a concentration-related inhibition of the IGF-I stimulation of androstenedione production. Data are the mean $\pm$ SEM of androstenedione concentrations measured by RIA.

IGF-I to augment FSH-stimulated estradiol production but had no effect on progesterone production (Agarwal et al., 1999). Thus, the direct effect of leptin was specific for estradiol production.

Taken together, the available data demonstrate direct inhibitory actions of leptin not only on granulosa cell aromatization of androstenedione, but also on the production of androstenedione by the theca cells. In obese women with PCOS this combination of effects could be an important contributor to infertility by preventing the dominant follicle from secreting adequate amounts of estradiol to sustain reproductive cyclicity and to prepare the endometrium for implantation.

20.6. LEPTIN AND INSULIN RESISTANCE IN PCOS

There is an ongoing uncertainty as to the relationship between leptin and the hyperinsulinemia that characterises PCOS (Brzcheffa et al., 1996; Laughlin et al., 1997; Rouru et al., 1997; Vicennati et al., 1998; El Orabi et al., 1999). Descriptions of no relationship with insulin or of a negative relationship with insulin sensitivity (Micic et al., 1997; Michelmore et al., 2001) are found in the literature. When obese women (Morin-Papunen et al., 1998; Kowalska et al., 2001) or adolescent girls (Freemark and Bursey, 2001) are treated with metformin to improve insulin sensitivity, there is a significant decrease in circulating leptin concentrations. Metformin therapy also causes a significant reduction in insulin concentrations and BMI. Therefore, it appears as if most of the relationship between high leptin concentrations and insulin sensitivity can be described on the basis of BMI alone, rather than a causative relationship between the two hormones (Figure 20.3).

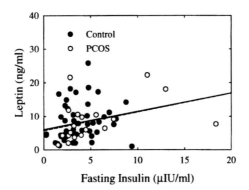

Fasting Insulin (μIU/ml)

Figure 20.3. The relationship between serum leptin concentrations and fasting insulin concentrations. Leptin and insulin concentrations were measured by RIA in fasting morning serum from regularly cycling control women and women with PCOS. The lines are the linear regressions of control (solid line) and PCOS (dashed line) data, respectively.

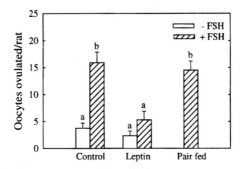

Figure 20.4. Leptin inhibition of ovulation in the rat. Rats were implanted with osmotic minipumps containing FSH or saline for 2 days. The morning of day three, the rats were injected with an ovulatory dose of hCG. Beginning at the time of hCG injection, the rats were injected with recombinant human leptin or saline (control) every 3 hr. On the morning of day four, the number of oocytes in the ampulla were counted. Pair fed rats were given the same amount of food consumed by the leptin-treated rats. Data are the mean ± SEM. Bars with different letters are significantly different. Redrawn with permission from Duggal et al. (2000).

20.7. LEPTIN AND OVULATION

Although extremes of leptin concentrations are associated with anovulation, there is little known regarding the effects of leptin on ovulation in women. In vivo studies using immature gonadotropin-primed rats, and studies using the in vitro perfused rat ovary model have yielded some intriguing results.

As shown in Figure 20.4, in vivo treatment of female rats with leptin significantly reduced the number of oocytes ovulated (Freemark and Bursey, 2001). Because leptin has central effects to decrease gonadotropin secretion, follicle development was stimulated by FSH infusion using osmotic minipumps and ovulation was induced by hCG treatment. To avoid the confounding factor of decreased food consumption in leptin-treated animals, the rats not receiving leptin treatment were given the same amount of food as the leptin-treated rats.

There was no effect of leptin on follicle development or steroid hormone production, indicating that leptin has a direct inhibitory effect on the process of ovulation (Duggal et al., 2000).

This concept was confirmed by studies using in vitro perfusion of whole rat ovaries. In vivo eCG treatment was used to promote follicle growth and maturation (van Cappellen et al., 1995) prior to perfusion of isolated rat ovaries in vitro (Brannstrom et al., 1987; Brannstrom, 1993). Leptin significantly inhibited LH-induced ovulation by the perfused ovaries (Duggal et al., 2000). The mechanism does not appear to involve inhibition of prostaglandin synthesis or recruitment of leukocytes into the theca interna (Duggal et al., 2002). Further studies will be required to determine if direct inhibition of ovulation plays a role in PCOS.

20.8. LEPTIN AND ASSISTED REPRODUCTION

20.8.1. PCOS, Leptin, and Ovulation Induction

Circulating leptin levels are not different between the two groups when BMI is taken into consideration (Pirwany et al., 2001). However, various interventions appear to alter the concentrations of leptin in women with PCOS attempting to become pregnant. Huber Buchholz et al. (1999) investigated women with PCOS who were undergoing a weight loss program and in which euglycemic glucose clamps were carried out over the course of 6 months. There was no difference in leptin levels between those who ovulated and those who did not, nor between patients who ovulated and those who remained anovulatory. Imani et al. (2000) used leptin to predict the success of clomiphene citrate in causing ovulation in women with PCOS. They had previously proposed a formula that BMI was one of the predicted variables that was most associated with a response to clomiphene (the higher the BMI the less the response). Incorporating leptin into the equation gave a marginally better result than the BMI alone, which implies that the lower the leptin in relation to the BMI the better the outcomes. Various studies have looked at the effect of diazoxide and metformin on leptin as part of a clinical program to induce ovulation. The use of 300 mg a day of diazoxide led to a reduction in insulin and in leptin, but all the correlations were related to BMI and leptin, and not to insulin alone (Krassas et al., 1998). It was, therefore, concluded that the changes were due to decreases in BMI rather than direct effects on leptin. Ng et al. (2001) using 1500 mg of metformin a day noted that despite a decrease in leptin levels, there was no change in ovulation. Others have shown a preferential decrease in leptin in women with PCOS treated with metformin compared to those control subjects who also received metformin (Pasquali et al., 2000; Freemark and Bursey, 2001; Kowalska et al., 2001). There is no apparent relationship between insulin and leptin.

20.8.2. PCOS, Leptin, and IVF

Assisted reproductive technology (ART) offers an approach to look at the effect of leptin on oocytes and embryo development. The basic biology surrounding the influence

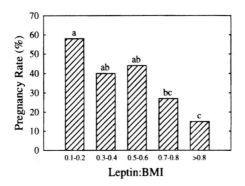

Figure 20.5. The effect of leptin and body mass on pregnancy rates. Pregnancy rates were determined from 172 IVF cycles in 139 women and plotted as a function of serum leptin : BMI ratio. Leptin was measured in morning samples of non-fasting serum by RIA. BMI was calculated from data obtained at the beginning of the IVF treatment cycles. Bars with different letters denote significant differences by χ^2 analysis. Redrawn with permission from Brannian et al. (2001).

of leptin on oocytes and embryo development was well described prior to any studies on human ART (Cioffi et al., 1997). Leptin levels are known to rise during treatment with FSH during ovarian hyperstimulation and values are much higher at the time of hCG injection and in the luteal phase compared to non-stimulated cycles (Lindheim et al., 2000; Mantzoros et al., 2000; Brannian et al., 2001; Unkila-Kallio et al., 2001). Claims have been made that leptin concentrations are higher in patients who become pregnant on IVF at 12 days following embryo transfer, but that there was no difference between singleton and twin pregnancies (Unkila-Kallio et al., 2001). Leptin concentrations do not appear to be altered in unexplained infertility, endometriosis, or tubal occlusion (Unkila-Kallio et al., 2001) and, therefore, the observed responses are dependent on gonadotrophic stimulation of the ovary and multiple follicle development. Lower levels of leptin in the follicular fluid appear to be a good marker of success, while a low leptin to BMI ratio (Figure 20.5) is predictive of better embryos and pregnancy success (Mantzoros et al., 2000; Unkila-Kallio et al., 2001). It is possible that lower intraovarian leptin concentrations are more helpful for ovarian steroidogenesis and oocyte maturation. Estradiol administered to non-pregnant women leads to an increase in leptin but there appears to be no particular effect on induction of leptin secretion other than that of a BMI effect (Panidis et al., 2000). It is, therefore, likely that the changes seen during assisted reproduction are due to FSH stimulation of leptin production rather than due to any estradiol influence on adipocyte secretion.

20.9. SUMMARY

The relationships between leptin, obesity and insulin resistance are complex and not fully understood. Although leptin is clearly elevated in obese individuals, the data do not support a causal role for hyperleptinemia in PCOS. In vitro studies support the potential for leptin to contribute to certain features of PCOS such as anovulation, but the importance of leptin independent of obesity per se has yet to be established in women.

REFERENCES

Agarwal, S. K., Vogel, K., Weitsman, S. R., and Magoffin, D. A. (1999). Leptin antagonizes the insulin-like growth factor-I augmentation of steroidogenesis in granulosa and theca cells of the human ovary. *J. Clin. Endocrinol. Metab.*, *84*, 1072–1076.

Baranowska, B., Radzikowska, M., Wasilewska-Dziubinska, E., Kaplinski, A., Roguski, K., and Plonowski, A. (1999). Neuropeptide Y, leptin, galanin and insulin in women with polycystic ovary syndrome. *Gynecol. Endocrinol.*, *13*, 344–351.

Barbieri, R. L., Makris, A., Randall, R. W., Daniels, G., Kistner, R. W., and Ryan, K. J. (1986). Insulin stimulates androgen accumulation in incubations of ovarian stroma obtained from women with hyperandrogenism. *J. Clin. Endocrinol. Metab.*, *62*, 904–910.

Barbieri, R. L., Smith, S., and Ryan, K. J. (1988). The role of hyperinsulinemia in the pathogenesis of ovarian hyperandrogenism. *Fertil. Steril.*, *50*, 197–212.

Bergh, C., Carlsson, B., Olsson, J.-H., Selleskog, U., and Hillensjo, T. (1993). Regulation of androgen production in cultured human thecal cells by insulin-like growth factor I and insulin. *Fertil. Steril.*, *59*, 323–331.

Brannian, J. D., Schmidt, S. M., Kreger, D. O., and Hansen, K. A. (2001). Baseline non-fasting serum leptin concentration to body mass index ratio is predictive of IVF outcomes. *Hum. Reprod.*, *16*, 1819–1826.

Brannstrom, M. (1993). In vitro perfused rat ovary. *Meth. Toxicol.*, *3B*, 160–169.

Brannstrom, M., Johansson, B., Sogn, J., and Janson, P. O. (1987). Characterisation of an in vitro perfused rat ovary model: ovulation rate, oocyte maturation, steroidogenesis and influence of PMSG priming. *Acta. Physiol. Scand.*, *130*, 107–114.

Brzechffa, P. R., Jakimiuk, A. J., Agarwal, S. K., Weitsman, S. R., Buyalos, R. P., and Magoffin, D. A. (1996). Serum immunoreactive leptin concentrations in women with polycystic ovary syndrome. *J. Clin. Endocrinol. Metab.*, *81*, 4166–4169.

Burghen, G. A., Givens, J. R., and Kitabchi, A. E. (1980). Correlation of hyperandrogenism with hyperinsulinism in polycystic ovarian disease. *J. Clin. Endocrinol. Metab.*, *50*, 113–116.

Carmina, E., Ferin, M., Gonzalez, F., and Lobo, R. A. (1999). Evidence that insulin and androgens may participate in the regulation of serum leptin levels in women. *Fertil. Steril.*, *72*, 926–931.

Carmina, E., Koyama, T., Chang, L., Stanczyk, F. Z., and Lobo, R. A. (1992). Does ethnicity influence the prevalence of adrenal hyperandrogenism and insulin resistance in polycystic ovary syndrome? *Am. J. Obstet. Gynecol.*, *167*, 1807–1812.

Caro, J. F., Sinha, M. K., Kolaczynski, J. W., Zhang, P. L., and Considine, R. V. (1996). Leptin: the tale of an obesity gene. *Diabetes*, *45*, 1455–1462.

Chapman, I. M., Wittert, G. A., and Norman, R. J. (1997). Circulating leptin concentrations in polycystic ovary syndrome: relation to anthropometric and metabolic parameters. *Clin. Endocrinol. (Oxf.)*, *46*, 175–181.

Cioffi, J. A., Van Blerkom, J., Antczak, M., Shafer, A., Wittmer, S., and Snodgrass, H. R. (1997). The expression of leptin and its receptors in pre-ovulatory human follicles. *Mol. Hum. Reprod.*, *3*, 467–472.

Cizza, G., Dorn, L. D., Lotsikas, A., Sereika, S., Rotenstein, D., and Chrousos, G. P. (2001). Circulating plasma leptin and IGF-1 levels in girls with premature adrenarche: potential implications of a preliminary study. *Horm. Metab. Res.*, *33*, 138–143.

Cnop, M., Landchild, M. J., Vidal, J., Havel, P. J., Knowles, N. G., Carr, D. R., Wang, F., Hull, R. L., Boyko, E. J., Retzlaff, B. M., Walden, C. E., Knopp, R. H., and Kahn, S. E. (2002). The concurrent accumulation of intra-abdominal and subcutaneous fat explains the association between insulin resistance and plasma leptin concentrations: distinct metabolic effects of two fat compartments. *Diabetes*, *51*, 1005–1015.

Conn, J. J., Jacobs, H. S., and Conway, G. S. (2000). The prevalence of polycystic ovaries in women with type 2 diabetes mellitus. *Clin. Endocrinol. (Oxf.)*, *52*, 81–86.

Considine, R. V., Sinha, M. K., Heiman, M. L., Kriauciunas, A., Stephens, T. W., Nyce, M. R., Ohannesian, J. P., Marco, C. C., McKee, L. J., Bauer, T. L., and Caro, J. F. (1996). Serum immunoreactive-leptin concentrations in normal-weight and obese humans. *N. Engl. J. Med.*, *334*, 292–295.

Conway, G. S., Agrawal, R., Betteridge, D. J., and Jacobs, H. S. (1992). Risk factors for coronary artery disease in lean and obese women with the polycystic ovary syndrome. *Clin. Endocrinol.*, *37*, 119–125.

Conway, G. S. and Jacobs, H. S. (1997). Leptin: a hormone of reproduction. *Hum. Reprod.*, *12*, 633–635.

Dagogo-Jack, S., Fanelli, C., Paramore, D., Brothers, J., and Landt, M. (1996). Plasma leptin and insulin relationships in obese and nonobese humans. *Diabetes*, *45*, 695–698.

Dahlgren, E., Janson, P. O., Johansson, S., Lapidus, L., and Oden, A. (1992). Polycystic ovary syndrome and risk for myocardial infarction: evaluated from a risk factor model based on a prospective population study of women. *Acta Obstet. Gynecol. Scand., 71,* 599–604.

De Souza, M. and Metzger, D. (1991). Reproductive dysfunction in amenorrheic athletes and anorexic patients: a review. *Med. Sci. Sports Exerc., 23,* 995–1007.

Duggal, P. S., Ryan, N. K., Van der Hoek, K. H., Ritter, L. J., Armstrong, D. T., Magoffin, D. A., and Norman, R. J. (2002). The effects of leptin administration and feed restriction on thecal leukocytes in the pre-ovulatory rat ovary and the effects of leptin on meiotic maturation, granulosa cell proliferation, steroid hormone and prostaglandin E_2 release using rat ovarian follicle culture. Reproduction, *123,* 899–905.

Duggal, P. S., Van der Hoek, K. H., Milner, C., Ryan, N. K., Armstrong, D. T., Magoffin, D. A., and Norman, R. J. (2000). The in vivo and in vitro effects of exogenous leptin on ovulation in the rat. *Endocrinology, 141,* 1971–1976.

Dunaif, A. (1997). Insulin resistance and the polycystic ovary syndrome: mechanism and implications for pathogenesis. *Endocrinol. Rev., 18,* 774–800.

Ehrmann, D. A., Sturis, J., Byrne, M. M., Karrison, T., Rosenfield, R. L., and Polonsky, K. S. (1995). Insulin secretory defects in polycystic ovary syndrome. Relationship to insulin sensitivity and family history of non-insulin-dependent diabetes mellitus. *J. Clin. Invest., 96,* 520–527.

El Orabi, H., Ghalia, A. A., Khalifa, A., Mahfouz, H., El Shalkani, A., and Shoieb, N. (1999). Serum leptin as an additional possible pathogenic factor in polycystic ovary syndrome. *Clin. Biochem., 32,* 71–75.

El-Roeiy, A., Chen, X., Roberts, V. J., Shimasaki, A., Ling, N., LeRoith, D., Roberts, C. T., and Yen, S. S. C. (1994). Expression of the genes encoding the insulin-like growth factors (IGF-I and II), the IGF and insulin receptors, and IGF-binding proteins-1-6 and the localization of their gene products in normal and polycystic ovary syndrome ovaries. *J. Clin. Endocrinol. Metab., 78,* 1488–1496.

Erickson, G. F. (1994). Polycystic ovary syndrome: normal and abnormal steroidogenesis. In R. Schats, and J. Schoemaker (Eds.). *Ovarian Endocrinopathies,* pp. 103–115. Parthenon Publishing Group: New York.

Erickson, G. F., Magoffin D. A., and Jones, K. L. (1989). Theca function in polycystic ovaries of a patient with virilizing congenital adrenal hyperplasia. *Fertil. Steril., 51,* 173–176.

Ezzell, C. (1996). The molecular link between fat and female fertility. *J. NIH Res., 8,* 24–25.

Fedorcsak, P., Storeng, R., Dale, P. O., Tanbo, T., Torjesen, P., Urbancsek, J., and Abyholm, T. (2000). Leptin and leptin binding activity in the preovulatory follicle of polycystic ovary syndrome patients. *Scand. J. Clin. Lab. Invest., 60,* 649–655.

Franks, S., Gilling-Smith, C., Watson, H., and Willis, D. (1999). Insulin action in the normal and polycystic ovary. *Endocrinol. Metab. Clin. N. Am., 28,* 361–378.

Franks, S. M. and White, D. M. (1993). Prevalence of and etiological factors in polycystic ovarian syndrome. *Ann. NY Acad. Sci., 687,* 112–114.

Freemark, M. and Bursey, D. (2001). The effects of metformin on body mass index and glucose tolerance in obese adolescents with fasting hyperinsulinemia and a family history of type 2 diabetes. *Pediatrics, 107,* E55.

Futterweit, W. and Deligdisch, L. (1986). Histopathological effects of exogenously administered testosterone in 19 female to male transsexuals. *J. Clin. Endocrinol. Metab., 62,* 16–21.

Gennarelli, G., Holte, J., Wide, L., Berne, C., and Lithell, H. (1998). Is there a role for leptin in the endocrine and metabolic aberrations of polycystic ovary syndrome? *Hum. Reprod., 13,* 535–541.

Gilling-Smith, C., Story, H., Rogers, V., and Franks, S. (1997). Evidence for a primary abnormality of thecal cell steroidogenesis in the polycystic ovary syndrome. *Clin. Endocrinol. (Oxf.), 47,* 93–99.

Gilling-Smith, C., Willis, D. S., Beard, R. W., and Franks, S. (1994). Hypersecretion of androstenedione by isolated thecal cells from polycystic ovaries. *J. Clin. Endocrinol. Metab., 79,* 1158–1165.

Goldzieher, J. W. (1981). Polycystic ovarian disease. *Fertil. Steril., 35,* 371–394.

Goldzieher, J. W. and Axelrod L. R. (1963). Clinical and biochemical features of polycystic ovarian disease. *Fertil. Steril., 14,* 631–653.

Gonzalez, R. R., Simon, C., Caballero-Campo, P., Norman, R., Chardonnens, D. Devoto, L., and Bischof, P. (2000). Leptin and reproduction. *Hum. Reprod. Update, 6,* 290–300.

Green, B. B., Weiss, N. S., and Daling, J. R. (1988). Risk of ovulatory infertility in relation to body weight. *Fertil. Steril., 50,* 721–726.

Huber-Buchholz, M. M., Carey, D. G., and Norman, R. J. (1999) Restoration of reproductive potential by lifestyle modification in obese polycystic ovary syndrome: role of insulin sensitivity and luteinizing hormone. *J. Clin. Endocrinol. Metab., 84,* 1470–1474.

Hughesdon, P. E. (1982). Morphology and morphogenesis of the Stein-Levinthal ovary and so called "hyperthecosis". *Obstet. Gynecol. Surv.*, *37*, 59–77.

Imani, B., Eijkemans, M. J., de Jong, F. H., Payne, N. N., Bouchard, P., Giudice, L. C., and Fauser, B. C. (2000). Free androgen index and leptin are the most prominent endocrine predictors of ovarian response during clomiphene citrate induction of ovulation in normogonadotropic oligoamenorrheic infertility. *J. Clin. Endocrinol. Metab.*, *85*, 676–682.

Jacobs, H. S. and Conway, G. S. (1999). Leptin, polycystic ovaries and polycystic ovary syndrome. *Hum. Reprod. Update*, *5*, 166–171.

Karlsson, C., Lindell, K., Svensson, E., Bergh, C., Lind, P., Billig, H., Carlsson, L. M., and Carlsson B. (1997). Expression of functional leptin receptors in the human ovary. *J. Clin. Endocrinol. Metab.*, *82*, 4144–4148.

Kowalska, I., Kinalski, M., Straczkowski, M., Wolczyski, S., and Kinalska, I. (2001). Insulin, leptin, IGF-I and insulin-dependent protein concentrations after insulin-sensitizing therapy in obese women with polycystic ovary syndrome. *Eur. J. Endocrinol.*, *144*, 509–515.

Kowalska, I., Kinalski, M., Wolczynski, S., Straczkowski, M., Kinalska, I., and Szamatowicz, M. (1999). The influence of obesity on ovarian function. II. Plasma leptin concentration in women with polycystic ovary syndrome. *Ginekol. Pol.*, *70*, 428–432.

Krassas, G. E., Kaltsas, T. T., Pontikides, N., Jacobs, H., Blum, W., and Messinis, I. (1998). Leptin levels in women with polycystic ovary syndrome before and after treatment with diazoxide. *Eur. J. Endocrinol.*, *139*, 184–189.

Laughlin, G. A., Morales, A. J., and Yen, S. S. C. (1997). Serum leptin levels in women with polycystic ovary syndrome: the role of insulin resistance/hyperinsulinemia. *J. Clin. Endocrinol. Metab.*, *82*, 1692–1696.

Legro, R. S., Finegood, D., and Dunaif, A. (1998). A fasting glucose to insulin ratio is a useful measure of insulin sensitivity in women with polycystic ovary syndrome. *J. Clin. Endocrinol. Metab.*, *83*, 2694–2698.

Lindheim, S. R., Sauer, M. V., Carmina, E., Chang, P. L., Zimmerman, R., and Lobo, R. A. (2000). Circulating leptin levels during ovulation induction: relation to adiposity and ovarian morphology. *Fertil. Steril.*, *73*, 493–498.

Lobo, R. A. (1995). A disorder without identity: "HCA," "PCO," "PCOD," "PCOS," "SLS." What are we to call it?! *Fertil. Steril.*, *63*, 1158–1160.

Lobo, R. A., Kletzky, O. A., Campeau, J. D., and diZerega, G. S. (1983). Elevated bioactive luteinizing hormone in women with the polycystic ovary syndrome (PCO). *Fertil. Steril.*, *39*, 674–678.

Maffei, M., Halaas, J., Ravussin, E., Pratley, R. E., Lee, G. H., Zhang, Y., Fei, H., Kim, S., Lallone, R., Ranganathan, S., Kern, P. A., and Friedman, J. M. (1995). Leptin levels in human and rodent: measurement of plasma leptin and ob RNA in obese and weight-reduced subjects. *Nat. Med.*, *1*, 1155–1161.

Mantzoros, C. S., Cramer, D. W., Liberman, R. F., and Barbieri, R. L. (2000). Predictive value of serum and follicular fluid leptin concentrations during assisted reproductive cycles in normal women and in women with the polycystic ovarian syndrome. *Hum. Reprod.*, *15*, 539–544.

Mantzoros, C. S., Dunaif, A., and Flier, J. S. (1997). Leptin concentrations in the polycystic ovary syndrome. *J. Clin. Endocrinol. Metab.*, *82*, 1687–1691.

Michelmore, K., Ong, K., Mason, S., Bennett, S., Perry, L., Vessey, M., Balen, A., and Dunger, D. (2001). Clinical features in women with polycystic ovaries: relationships to insulin sensitivity, insulin gene VNTR and birth weight. *Clin. Endocrinol.*, *55*, 439–446.

Micic, D., Macut, D., Popovic, V., Sumarac-Dumanovic, M., Kendereski, A., Colic, M., Dieguez, C., and Casanueva, F. F. (1997). Leptin levels and insulin sensitivity in obese and non-obese patients with polycystic ovary syndrome. *Gynecol. Endocrinol.*, *11*, 315–320.

Morin-Papunen, L. C., Koivunen, R. M., Tomas, C., Ruokonen, A., and Martikainen, H. K. (1998). Decreased serum leptin concentrations druting metformin therapy in obese women with polycystic ovary syndrome. *J. Clin. Endocrinol. Metab.*, *83*, 2566–2568.

Ng, E. H., Wat, N. M., and Ho, P. C. (2001). Effects of metformin on ovulation rate, hormonal and metabolic profiles in women with clomiphene-resistant polycystic ovaries: a randomized, double-blinded placebo-controlled trial. *Hum. Reprod.*, *16*, 1625–1631.

Norman, R. J., Masters, S. C., Hague, W., Beng, C., Pannall, P., and Wang, J. X. (1995). Metabolic approaches to the subclassification of polycystic ovary syndrome. *Fertil. Steril.*, *63*, 329–335.

Oksanen, L., Tiitinen, A., Kaprio, J., Koistinen, H. A., Karonen, S., and Kontula, K. (2000). No evidence for mutations of the leptin or leptin receptor genes in women with polycystic ovary syndrome. *Mol. Hum. Reprod.*, *6*, 873–876.

Panidis, D. K., Rousso, D. H., Matalliotakis, I. M., Kourtis, A. I., Stamatopoulos, P., and Koumantakis, E. (2000). The influence of long-term administration of conjugated estrogens and antiandrogens to serum leptin levels in women with polycystic ovary syndrome. *Gynecol. Endocrinol.*, *14*, 169–172.

Pasquali, R., Gambineri, A., Biscotti, D., Vicennati, V., Gagliardi, L., Colitta, D., Fiorini, S., Cognigni, G. E., Filicori, M., and Morselli-Labate, A. M. (2000). Effect of long-term treatment with metformin added to hypocaloric diet on body composition, fat distribution, and androgen and insulin levels in abdominally obese women with and without the polycystic ovary syndrome. *J. Clin. Endocrinol. Metab.*, *85*, 2767–2774.

Pirwany, I. R., Fleming, R., Sattar, N., Greer, I. A., and Wallace, A. M. (2001). Circulating leptin concentrations and ovarian function in polycystic ovary syndrome. *Eur. J. Endocrinol.*, *145*, 289–294.

Roberts, D. W. and Haines, M. (1960). Is there a Stein-Leventhal syndrome? *Br. Med. J.*, *5187*, 1709–1711.

Rouru, J., Anttila, L., Koskinen, P., Penttila, T. A., Irjala, K., Huupponen, R., and Koulu, M. (1997). Serum leptin concentrations in women with polycystic ovary syndrome. *J. Clin. Endocrinol. Metab.*, *82*, 1697–1700.

Sir-Petermann, T., Piwonka, V., Perez, F., Maliqueo, M., Recabarren, S. E., and Wildt, L. (1999). Are circulating leptin and luteinizing hormone synchronized in patients with polycystic ovary syndrome? *Hum. Reprod.*, *14*, 1435–1439.

Sir-Petermann, T., Recabarren, S. E., Lobos, A., Maliqueo, M., and Wildt, L. (2001). Secretory pattern of leptin and LH during lactational amenorrhoea in breastfeeding normal and polycystic ovarian syndrome women. *Hum. Reprod.*, *16*, 244–249.

Smyczek-Gargya, B. and Geppert, M. (1992). Endometrial cancer associated with polycystic ovaries in young women. *Path. Res. Pract.*, *188*, 946–948.

Spritzer, P. M., Poy, M., Wiltgen, D., Mylius, L. S., and Capp, E. (2001). Leptin concentrations in hirsute women with polycystic ovary syndrome or idiopathic hirsutism: influence on LH and relationship with hormonal, metabolic, and anthropometric measurements. *Hum. Reprod.*, *16*, 1340–1346.

Stein, I. F. and Leventhal, M. L. (1935). Amenorrhea associated with bilateral polycystic ovaries. *Am. J. Obstet. Gynecol.*, *29*, 181–186.

Takeuchi, T. and Tsutsumi, O. (2000). Basal leptin concentrations in women with normal and dysfunctional ovarian conditions. *Int. J. Gynaecol. Obstet.*, *69*, 127–133.

Unkila-Kallio, L., Andersson, S., Koistinen, H. A., Karonen, S. L., Ylikorkala, O., and Tiitinen, A. (2001). Leptin during assisted reproductive cycles: the effect of ovarian stimulation and of very early pregnancy. *Hum. Reprod.*, *16*, 657–662.

van Cappellen, W. A., Meijs-Roelofs, H. M. A., Kramer, P., van Leeuwen, E. C. M., de Leeuw, R., and de Jong, F. H. (1995). Recombinant FSH (Org32489) induces follicle growth and ovulation in the adult cyclic rat. *J. Endocrinol.*, *144*, 39–47.

Veldhuis, J. D., Pincus, S. M., Garcia-Rudaz, M. C., Ropelato, M. G., Escobar, M. E., and Barontini, M. (2001). Disruption of the synchronous secretion of leptin, LH, and ovarian androgens in nonobese adolescents with the polycystic ovarian syndrome. *J. Clin. Endocrinol. Metab.*, *86*, 3772–3778.

Vicennati, V., Gambineri, A., Calzoni, F., Casimirri, F., Macor, C., Vettor, R., and Pasquali, R. (1998). Serum leptin in obese women with polycystic ovary syndrome is correlated with body weight and fat distribution but not with androgen and insulin levels. *Metabolism*, *47*, 988–992.

Wong, I. L., Morris, R. S., Lobo, R. A., Paulson, R. J., and Sauer, M. V. (1995). Isolated polycystic morphology in ovum donors predicts response to controlled hyperstimulation. *Hum. Reprod.*, *10*, 524–528.

Woods, S. C. and Seeley R. J. (2001). Insulin as an adiposity signal. *Int. J. Obes. Relat. Metab. Disord.*, *25*(Suppl. 5), S35–S38.

Yen, S. S. C. (1980). The polycystic ovary syndrome. *Clin. Endocrinol.*, *12*, 177–208.

Index

Adenomyosis, 66
Adipocyte(s), 6, 16, 25, 26, 28–30, 33, 39, 40, 43, 44, 46, 64, 80, 173, 177, 189, 206, 207, 221, 232, 240, 287, 300, 303, 304, 306, 311, 323, 324, 341
AdipoQ, 56
Adipose tissue, 3, 30, 31, 33, 39, 41, 44–47, 53, 57, 64, 151, 156, 159, 191, 193, 202, 205, 246, 247, 253, 268–271, 281, 288
  omental, 40, 58, 67, 243, 245
  subcutaneous, 40, 57, 58, 67, 125, 126, 156, 243, 245, 247
  retroperitoneal, 57
  visceral, 41, 125, 156, 335
Adrenocorticotropic hormone (ACTH), 43, 152, 208
Agouti-related peptide, 292, 293
Amenorrhea, 55, 81, 160, 191, 290, 312–314, 317–320, 320, 322–325
  athletic/exercise induced, 312–315, 319, 321, 322, 324
  hypothalamic, 8, 159, 312, 316, 318, 322, 323, 325
  idiopathic/functional, 313, 314, 316, 317
  primary, 311, 315
  oligo-, 8, 120, 333, 335
  secondary, 118, 311, 315, 333
Amniochorion, 205, 207, 243, 252
Amnion, 190, 201, 202, 207
Amniotic fluid, 155, 201, 208
Androgen(s), 47, 89, 126, 141, 156, 182, 211, 246, 249, 250, 253, 269, 290, 311, 334, 335, 337, 338

Androstenedione, 102, 103, 105, 337, 338
Angiogenesis, 5, 8, 84, 89, 114, 152, 159, 194, 202, 210, 227, 231,233, 251, 253, 305
Anorexia, 26, 55, 86, 159, 160, 191, 304, 314, 316, 321, 325, 335
Anti-obesity hormone, 5, 152
Antioxidant(s), 306
Apoptosis, 8, 305
Arcuate nuclei, 23, 29, 56, 152, 157, 293, 311
Aromatase, 61, 62, 99, 100–103, 105, 337
Assisted reproductive technologies (ART), 6, 77, 79, 81, 84, 89, 90, 340
Autocrine regulation, 201, 210
Autocrine-paracrine regulation, 7, 46, 47, 53, 67, 79, 190, 201, 208, 214, 239, 243; see also Paracrine regulation

Baboon(s), 104, 140, 241, 244–246, 248, 252; see also Primates
BALB/cJ mice, 118, 275–277
Beals syndrome,145, 146
Beta adrenergic(s), 45
  receptors, 43
Blastocyst(s), 85, 86, 112, 113, 192, 252
Blood-brain barrier, 4, 87, 137
Blood-testis barrier, 137
Body mass index (BMI), 40, 55, 61,64–67, 80–83, 88, 138, 143, 145, 155, 156, 159, 161, 206, 208, 303, 304, 314, 317, 319, 322, 337, 338, 340, 341
Bone, 5, 46, 195, 196, 251, 319, 321–324, 335

Bone (*cont.*)
  mass density (BMD), 315, 320–325
Bombesin, 5
Bovine, 98, 100, 101, 118
Brain, 30, 68, 87, 161, 172, 177, 182, 196, 208, 241, 279, 311, 312, 335
Bulimia, 159, 160, 314, 317

Calmodulin 33
Caloric intake, 41, 47, 55, 183, 315, 318, 323
Cancer, 114, 248, 249, 333
Capillary endothelium, 16
Carbenoxolone 229
Catecholamine(s), 43, 47; *see also* Dopamine; Epinephrine; Norepinephrine
Cerebrospinal fluid, 23
CHO cells, 230
Cholecystokinin (CCK), 5, 41, 46
Cholocystectomy, 57
Chorio-allantoic placenta, 222
Choriocarcinoma cell(s), 190, 203, 212, 247–249
Chorionic tissue, 206, 207
Circadian rhythm, 27, 31, 52, 157, 191, 316; *see also* Sleep cycle
Clomiphene citrate, 80, 81, 340
Cocaine and amphetamine regulated transcript (CART), 5, 153, 293
Conceptus growth/development, 5
Congenital adrenal hyperplasia, 334
Corpus luteum, 56, 79, 98, 101, 104, 221, 225, 243
Corticosterone, 118
Corticotroph(s), 153
Corticotropin-releasing hormone (CRH), 5, 23, 26, 55, 119, 228, 318, 319
Cortisol, 43, 47, 58, 61, 119, 121–124, 126, 152, 158, 252, 314, 316, 318
"critical weight" hypothesis, 25, 39, 40, 89, 104–105, 118, 134, 154, 161, 172, 184, 269, 288, 312, 319
Cumulus cells, 56, 65, 78, 153
Cyclic adenosine monophosphate (cAMP), 20, 43, 45, 102–104
Cyclic guanosine monophophate (cGMP), 15, 20, 21, 29, 30, 31
Cytotrophoblast, 7, 67, 113, 190, 192, 201, 203, 211–214, 242–244, 248, 252, 253; *see also* Placenta; Trophoblast
CYP17, 101
C57BL/6J mice, 118, 272, 275–279

Db/db mice, 195, 196, 231, 269–271, 275, 288, 289
Decidua, 7, 112, 202, 205, 231, 233, 243, 252
Dehydroepiandrosterone (DHEA), 63, 250, 318
Dexamethasone, 43, 45, 229, 230
Diabetes, 5, 7, 54, 118, 152, 191, 194, 196, 210, 211, 252, 276–278, 302

Dihydrotestosterone (DHT), 41, 126
Diurnal profile/cyclicity, 41, 143, 177, 180
Dopamine, 28, 318
Dorsomedial nuclei, 152
DSM-IV criteria, 314

Embryo, 67, 82, 83, 85, 86, 194, 210, 213
  transfer, 66, 81, 82, 83, 341
Embryonic development, 6, 66, 85, 192, 340
Endometriosis, 6, 67, 113, 114, 341
Endometrium, 6, 66, 67, 83, 112–114, 192, 213, 222, 225, 231, 252, 333, 338
Endothelial
  cell(s), 33, 299, 300, 304, 306
  dysfunction, 8, 299, 300, 305
Energy, 121, 123–126, 161, 177, 179, 183, 184, 189, 191, 193, 196, 202, 204, 208, 224, 232, 239, 268, 288–291, 311, 314–318, 322
Epididymal fat pad(s), 16
Epinephrine, 28, 29, 43; *see also* Catecholamine
Estradiol, 26, 41, 56–60, 62–69, 79, 80, 99–101, 104, 105, 112, 137, 145, 152, 157, 175, 176, 178, 191, 192, 240, 244–246, 273, 280, 311, 312, 315, 337, 338, 341
Estrogen, 6, 23, 41, 47, 54, 68, 80, 155, 156, 160, 173, 205, 209, 211, 224, 225, 239, 240, 243, 245, 246, 248, 249, 251, 253, 254, 269, 293, 312, 322, 324
  receptor, 56, 244, 245, 248, 249
    antagonist(s), 248
  replacement, 244; *see also* Hormone replacement therapy (HRT)
  response element, 244, 248

Fat
  cell volume, 40
  depot/stores, 40, 123, 140, 146, 156, 157, 161, 189, 268, 279, 283, 289, 316
  mass/adiposity, 40, 46, 64, 67, 68, 87, 88, 118, 139, 140, 154, 160, 161, 172, 193, 201, 202, 205, 211, 222, 268, 270, 271, 276, 277, 279, 288, 290, 291, 303, 306, 314, 335
Fatty acid synthase/translocase, 303
Female athletic triad, 313
Fertility, 5, 21, 77, 81, 83, 89, 90, 117, 119, 123, 126, 171, 172, 230, 277, 278, 281, 289, 291, 319, 335; *see also* Infertility; Sterility
Fetal
  adipose tissue, 7, 155, 240, 253, 307
  adrenal, 245, 253
  birth weight, 190, 193, 207, 208, 240, 250, 306, 307
  development/growth, 9, 135, 155, 189–194, 196, 202, 204, 207–211, 221, 228, 230–233, 240, 241, 250, 251, 253, 254, 281, 306

Fetal (cont.)
  morbidity/mortality, 299
  tissue(s), 155, 189, 190, 193, 195, 196, 204, 212, 226, 229, 241, 243, 245, 250, 253, 306
Fetectomy, 245–248
Fibroblast growth factor(s), 84, 251
Follicle(s), 56, 79, 81, 83–86, 97, 99, 100, 102, 103, 153, 159, 192, 210, 292, 311, 334, 337–340
Follicle stimulating hormone (FSH), 15, 17–19, 21, 26, 30, 54, 61, 63, 65, 79, 81, 82, 97, 99–101, 104, 119, 135, 141, 142, 153, 155, 157–160, 191, 269, 273, 290–292, 311, 312, 315, 316, 334, 337, 338, 341
  releasing hormone (FSHRH) 15, 17–21, 23
Forskolin, 206, 212

Galanin, 25
  -like peptide (GALP), 25, 153
Gamete intra-fallopian transfer (GIFT), 82
Gamma amino butyric acid (GABA), 293, 318
Gender differences , 6, 40, 46, 58, 67, 125, 126, 137, 142–144, 146, 153–155, 157, 158, 169, 173, 179, 181, 182, 209, 243, 268, 306
Gestational age/ontogeny, 201, 205, 206, 222, 226, 232, 240, 241, 243, 245, 246, 252, 253, 281, 282, 300, 301, 306, 307
Glucose, 6, 42, 44, 46, 119, 124, 125, 126, 178, 194, 232, 303, 334, 340
Glucagon-like peptide (GRP-1), 5, 41
Glucocorticoid(s), 5, 43–45, 47, 119, 229–232, 252
Glutamate, 293
Glycogen, 123
Gonadotropin(s), 5, 7, 15, 21, 23, 24, 53–56, 77, 79, 81, 86, 89, 101, 104, 111, 119, 120, 123, 126, 135, 137, 138, 145, 152, 154, 157, 159, 191, 208, 269, 270, 288, 290–292, 312, 314, 339
  releasing hormone (GnRH), 7, 8, 15, 17–19, 25, 53–56, 65, 69, 118, 119, 123, 133–135, 137, 140, 141, 144–146, 153, 155, 158–160, 171–173, 177–179, 181–184, 202, 268, 269, 273, 279, 281, 289–293, 311, 312, 314–316, 318, 319, 336; see also LHRH
  -pulse generator, 55, 119, 120, 123, 134, 135, 137, 139, 157, 171, 175, 191, 312, 318, 319
Gonad(s), 9, 137, 142, 145, 153, 155, 171, 173
Gonadectomy, 174–176, 178
Gonadotrope(s), 19, 20, 80, 135, 153
Granulosa cells, 6, 56, 61, 62, 65, 77–79, 83, 87, 90, 98–105, 137, 153, 225
Growth hormone (GH), 26, 121,123, 124, 126, 136, 144, 155, 158, 196, 208, 210, 250, 251, 279, 318
GT-1 cells, 281
Guanosine triphosphate (GTP), 15, 30, 31

Hematopoiesis, 5, 46, 152, 159, 195, 202, 210

Hemochorial placenta, 242, 243
HepG2 cells, 102
Hexosamine(s), 6, 42, 44, 47
Hirsuitism, 334
Hormone replacement therapy (HRT), 68, 69, 324, 325; see also Estrogen replacement
Human chorionic gonadotropin (hCG), 7, 61, 79, 84, 99, 104, 125, 201, 202, 206, 207, 212, 240, 251, 339, 341
Human placental lactogen (hPL), 207, 211, 240
Human serum amyloid P component (SAP) promoter, 289
Hyaditiform mole, 212
Hydrocortisone, 43
Hyperandrogenism, 334, 335
Hypercortisolemia, 173, 322
Hyperglycemia 118, 119, 173, 323
Hyperinsulinemia, 335
Hyperinsulinemic-euglycemic clamp technique, 42
Hyperphagia, 144, 232, 233, 288, 290
Hyperprolactinemia, 324
Hypertension, 252, 299, 300, 302
Hypoestrogenism, 319, 323
Hypogonad(al/ism), 41, 118, 120, 144, 159, 269, 287, 288, 290, 323
Hypogonadotropism, 172–174, 176, 177, 191, 287, 288, 290, 324
Hypoinsulinemia, 173, 194, 211, 323, 338
Hypoleptin(a)emia, 55, 161, 205, 211, 241, 243, 252, 253, 289, 290, 304, 335, 341
Hypoprogestinemia, 322
Hypothalamic
  level, 6, 17, 55, 68, 87, 104, 105, 134, 135, 153, 189, 205, 225, 233, 243, 269, 312, 323
  -pituitary axis, 21, 23, 55, 56, 89, 152, 269, 270
  -pituitary-adrenal axis, 26, 251, 316, 318
  -pituitary-gonadal axis, 133, 134, 137, 146, 151, 152, 154, 156–158, 160, 161, 273
  -pituitary-ovarian axis, 53, 69, 312
  -pituitary-testicular axis, 143
Hypothalamus, 3, 4, 9, 15, 16, 18, 20, 21, 24–26, 29, 30, 53, 56, 77, 117–119, 137, 151, 202, 221, 224, 228, 241, 242, 248, 275, 288, 293, 311, 317, 336
Hypothyroidism, 316
Hypoxia, 8, 66, 85, 89, 90, 192, 210, 252, 304, 305
Hysterectomy, 57, 59

Immune
  cells, 30, 305, 306
  response, 5, 46, 152, 195, 305
Inflammation, 152, 210, 299, 304, 305
Implantation, 7, 66, 67, 69, 82, 83, 89, 111, 112, 192, 201, 202, 204, 212–214, 230, 231, 233, 252, 269, 281, 338; see also Placental invasion

Infection, 26; *see also* Sepsis

Infertility, 54, 67, 89, 114, 117–120, 172, 269, 275, 288, 294, 319, 335; *see also* Fertility
  anovulatory, 8, 333, 334, 338, 339, 341

Insulin, 5, 42, 44, 45, 47, 55, 56, 61, 68, 81, 86, 87, 89, 90, 99–105, 118, 177, 191, 193, 202, 209–211, 232, 276, 289, 314, 334, 338–340
  -like growth factor(s), 61, 81, 86, 87, 101, 104, 136, 153–155, 158, 177, 193, 196, 213, 228, 316, 318, 322, 323, 325, 335, 337, 338
  receptor substrate protein(s), 86, 101
  resistance, 87, 118, 119, 232, 300, 303, 304, 334, 335, 338, 341

Integrin(s), 7, 67, 113, 192, 201, 212, 213

Interleukin(s), 4, 23, 26, 30, 43, 67, 113, 192, 195, 201, 202, 212, 213, 304

Intrauterine growth retardation/restriction (IUGR), 7, 9, 194, 196, 209, 231, 232, 250, 252, 301, 302, 307

Isoproterenol, 16, 28, 43

In vitro fertilization (IVF), 56, 65, 66, 78, 81, 82–85, 88, 89, 101, 192, 340, 341

Janus-activated kinase/signal transducers and activators of transcription (JAK/STAT), 4, 102, 103, 192, 224, 241

Kidney/renal, 30, 202, 204, 243, 281, 299, 301

Labyrinth cell(s)/zone, 222, 226, 228; *see also* Placenta; Trophoblast

Lactation/nursing, 201, 207–209, 223, 233, 268, 269, 272, 273, 281, 288

Leiomyoma(s), 67

Lep
  gene, 3, 4, 39, 40, 44, 45, 159; *see also* obese (*ob*) gene
  promoter, 42, 44, 45, 248

Leptin
  binding protein(s)/activity, 4, 156, 204–206, 222, 224
  chromosomal position, 4
  clearance, 4, 204, 206, 211, 223, 228, 240, 252, 281, 301
  half-life, 4
  negative feedback, 43, 47
  non-adipose production in
    fetal cartilage, 4
    hypothalamus, 151
    mammary tissue, 4
    pituitary, 4, 151
    mammary gland, 151
    muscle, 151
    placenta, 4, 46, 151, 189–194, 201, 202, 205, 207, 211, 224, 232, 240, 250, 253, 281, 300, 303

Leptin (*cont.*)
  non-adipose production in (*cont.*)
    stomach, 4, 46, 151
  receptor, 4, 7, 23, 53, 61, 66, 67, 77, 78, 83–88, 98, 99, 104, 111–113, 117, 119, 125, 135–137, 144, 151, 153, 178, 189, 192, 194–196, 208, 231, 239, 250–253, 270, 287, 289, 300, 305, 311, 323, 337
    isoforms, 4, 33, 56, 66, 135, 136, 151, 152, 156, 190, 193, 194, 201, 202, 204, 206, 213, 223–228, 230–233, 241–243, 271, 275, 281, 288, 292, 309, 323, 337; *see also* Soluble leptin receptor
    family, 4, 135, 151, 202, 223
  resistance, 5, 77, 87–90, 193, 205, 224, 225, 239, 254, 305

Leukocyte activation, 304, 340

Leydig cells, 103, 104, 125, 153, 154, 158, 271, 273, 274

Lipodystrophy/atrophy, 40, 120, 145, 146

Lipopolysaccharide (LPS), 16, 26, 30, 305

Luciferase reporter, 248, 249

Luteinizing hormone (LH), 7, 17–22, 24–26, 30, 54–56, 60, 63, 69, 80, 97, 99, 100, 102, 103, 119–125, 135, 138, 141, 142, 144, 152, 153, 155, 157–160, 173–179, 181, 182, 191, 232, 269, 270, 273, 280, 289–292, 311, 312, 315–318, 334, 336, 337, 340
  -releasing hormone (LHRH), 15, 16, 18, 20–26, 30, 114; *see also* GnRH

Macrosomia, 211

Maternal-fetoplacental unit, 159, 194, 202, 212, 239, 245

MCF-7 cells, 248, 249

MC4R knockout mice, 153, 281

Median eminence, 17, 19, 24

Medroxyprogesterone acetate, 66, 112

Melanin concentrating hormone (MCH), 5, 153

Melanocortin 4 receptor, 293

Melatonin, 152, 182, 318

Menarche, 118, 157, 170, 243, 311, 315, 321

Menopause, 5, 6, 58, 67–69, 79, 98, 243, 244, 319, 320, 322, 324

Menses/menstruation, 55, 170, 279, 311–314

Menstrual cycle, 5, 6, 53, 55, 59, 62–64, 66, 67, 69, 86, 113, 114, 120, 145, 160, 244, 313, 315–317, 319, 320, 333, 334, 336
  follicular phase, 57, 60, 62–65, 98, 113, 191, 315
  luteal phase, 5, 57, 59, 62–64, 69, 80, 113, 241, 245, 315, 341

Metabolic gate, 8, 287

Metalloproteinase(s), 7, 67, 113, 192, 201, 212, 213, 231, 252

Metformin, 338, 340

Metyrapone, 229

Mitogen activated protein (MAP) kinase, 4, 86, 103, 194, 241

Modifier gene(s), 8, 275, 278

Monkey(s), 20, 54, 120, 133–135, 137, 138, 141, 142, 145, 146, 280, 292, 317; see also Primate; Rhesus macaque

N-acetylglucosamine, 40, 42

Neonate, 201, 207–209, 240, 243, 251, 299, 306

Neuron specific enolase promoter, 275

Neuropeptide Y (NPY), 5, 23, 24, 55, 56, 68, 119, 135, 137, 153, 233, 279, 292, 318, 337

Nitric oxide (NO), 5, 15, 22, 30–33, 66, 85

synthase (NOS), 5, 15, 16, 20, 22–24, 26, 29, 30, 32, 33, 153, 157

synthesis, 8, 29, 305

release, 16, 20, 21, 24, 30, 32, 89

Norepinephrine, 28, 29, 55, 318

Norethisterone, 68

Obese (ob) gene, 3, 39, 54, 56, 57, 62, 117, 134, 151, 201, 211, 221, 239, 270, 288, 311; see also Lep, gene

Obesity, 3, 8, 26, 39, 53, 61, 65, 77, 80, 81, 87–89, 99, 104, 118, 119, 125, 138, 144, 154, 159, 161, 211, 239, 270, 275, 278, 280, 283, 287–290, 293, 300, 323, 334–338, 341

Ob/ob mouse, 3, 7, 21, 25, 39, 88, 97, 104, 118, 119, 194–196, 204, 231, 233, 239, 254, 267, 269–279, 281, 287–289, 291, 292, 305, 323

O-glycosylation, 42

Oocyte, 81, 84, 192, 339, 340

function, 77, 341

maturation, 6, 62, 66, 78, 79, 83, 85, 86, 89, 90

quality, 6, 83, 85, 88, 89

Oophorectomy, 59

Orexin(s), 5, 177

Osteopenia, 318, 319, 321, 324

Osteoporosis, 312, 313, 319, 320, 321, 324, 325

Ovary, 5, 56, 57, 61, 65, 69, 79, 89, 90, 102, 137, 225, 226, 232, 268, 288, 290, 312, 316, 335

(dis)function, 8, 62, 86, 88, 98, 100, 104, 111, 269, 270, 289, 292, 311

stimulation, 80, 81, 101, 315, 340

Ovariectomy, 54, 56–60, 68, 324

Ovulation, 65, 79, 84, 86, 98, 120, 125, 153, 180, 181, 189, 221, 279, 288, 315, 324, 339, 340

Parabiosis experiments, 3, 117, 269

Paracrine regulation, 7, 153, 193; see also Autocrine-paracrine regulation

Paraventricular nucleus (PVN), 23

Pars distalis, 153

Pars tuberalis, 153

Parturition, 201, 204, 206, 211, 222, 240, 251, 269, 281, 282, 288, 301, 307

Permissive mechanism(s), 184

Phentolamine, 16, 28

Photoperiod, 173, 177, 179–183

Phytoestrogen(s), 69

Pig(s)/porcine/swine, 98, 104, 190, 225, 244

Pituitary, 9, 16, 18–20, 24, 26, 28–30, 56, 135, 137, 153, 155, 157, 171, 208, 221, 250, 251, 268, 269, 312, 315

Placenta(l), 7, 159, 190, 192, 201, 204, 206, 210, 213, 221–223, 226–228, 230, 231, 239, 241–243, 246, 247, 251, 252, 299, 305, 306; see also Trophoblast

formation, 7, 46, 233, 299

growth/development, 8, 202, 208, 210, 231, 240

growth factor, 301

invasion, 9, 201, 212, 213, 231, 252; see also Implantation

leptin promoter/enhancer, 46, 192, 206, 207, 212, 247, 300, 304

spiral arteries, 299

sufficiency, 9, 240, 304

villi, 201, 206, 207, 240–243, 246, 247, 252

plasminogen activator inhibitor-1, 301, 302

Polycystic ovarian (PCO) disease/syndrome, 8, 65, 81, 87, 88, 90, 104, 120, 333–341

Post-transcriptional mechanisms, 246

PPAR gamma, 44, 45, 47

Pre-eclampsia, 7, 192, 196, 210, 252, 299–307

Pregnancy, 5, 65, 81–83, 89, 90, 111, 155, 159, 191–194, 202, 204, 205, 210, 211, 221–226, 228–230, 232, 239–241, 243–245, 250, 252–254, 268, 272, 273, 278, 281, 282, 288, 299–306, 324, 340

maintenance, 207, 231, 239, 251, 281, 282, 299

Primate(s), 7, 134, 135, 137, 139–141, 144, 155, 159, 222, 239, 241, 242, 244, 250, 251, 253, 318; see also Baboon; Monkey; Rhesus macaque

Progesterone, 6, 57–61, 64, 66, 65, 69, 98, 99, 101, 104, 112, 221, 232, 251, 269, 288, 312, 315

Prolactin, 26, 28, 29, 32, 55, 152, 182, 208, 221, 232, 233

Proopiomelanocortin (POMC), 5, 137, 292, 293

Propranolol, 16, 28

Proteinuria, 299

Puberty, 5, 6, 41, 104, 121, 123, 133–135, 145, 151, 154, 157,-159, 161, 170–172, 175, 180, 183, 191, 280, 281, 292, 316

induction/activation, 15, 120, 137, 143, 159, 169, 173, 177, 179, 184, 268, 289, 291, 319

onset, 9, 21, 25, 111, 125, 134, 138–146, 153–156, 177, 181, 182, 189, 279, 287, 288, 290, 336

Relaxin, 269

Resting metabolic rate (RMR), 315, 317, 318

Rhesus macaque, 25, 120–124, 138–140, 144,146, 158; *see also* Monkey; Primate
R105X mutation, 270

Satiety factor/appetite control, 3, 117, 202, 207, 232, 239, 252, 254, 305
Scoliosis, 321
Seminiferous tubule(s), 274
Sepsis, 43; *see also* Infection
Serotonin, 5
Sertoli cell(s), 125
Sheep/ovine, 7, 98, 120, 169, 172, 173, 177, 180, 181, 190, 209, 251, 287
Side-chain cleavage, 97, 101–104, 125
Sleep cycle, 42; *see also* Circadian rhythm
Sodium nitroprusside, 20
Soluble leptin receptor, 4, 7, 88, 190, 204, 205, 222, 228, 232, 241, 252, 281; *see also* Leptin receptor isoforms
Sp1, 42, 44
Species dependency/specificity, 6, 173, 241
Sperm/Spermatogenesis, 182, 273, 274, 279
Spontaneous abortion/miscarriage, 66, 81, 207, 251, 300
STAT transcription factor(s), 62, 63, 85
Sterility, 8, 97, 269–275, 279, 281, 288; *see also* Fertility
Steroid(s), 5, 21, 120, 173, 174, 179, 181, 243, 307, 316
  produced by the fetus, 7, 135; *see also* Androgens; Dehydroepiandrosterone (DHEA)
  produced by the gonad, 41, 46, 62, 125, 290; *see also* Steroid(s), produced by the ovary; Steroid(s), produced by the testis
  produced by the ovary, 6, 8, 9, 64, 66, 69, 83, 86, 89, 90,97–99, 101–105, 112, 154, 159, 335, 337, 340, 341
  produced by the placenta, 7, 135, 211
  produced by the testis, 104, 138
Steroidogenesis, 6, 61, 62, 134, 335
Steroidogenic acute regulatory (STAR) protein, 101, 103, 104, 111, 125
Stress, 26, 27, 30, 306, 311–315, 317
Suppressors of cytokine signaling, (SOCS-3) 5, 88, 205

Sympathetic nervous system, 42, 47
Syncytiotrophoblast, 190, 201, 202, 206, 208, 242, 243, 245, 248, 251, 254, 300; *see also* Placenta; Trophoblast

Tanner stages, 143, 154, 156, 158, 170
Testis, 5, 99, 125, 135, 137, 140, 141, 154, 182, 270
Testosterone, 6, 41, 58, 63, 65, 67, 103, 104, 119, 121, 125, 126, 138–140, 142, 143, 154, 156–158, 160, 182, 209, 249, 273, 316, 336
Thecal cell(s), 6, 56, 61, 77, 78, 79, 83, 89, 90, 98, 99, 102, 103, 105, 137, 153, 334, 335, 337, 338
Thermogenesis/thermoregulation, 288, 289
Thiazolidinediones, 44
Thyroid stimulating hormone/thyrotropin (TSH), 26, 152
Tissue specificity, 6, 211, 239–241, 244, 247, 248
Transforming growth factor(s), 102, 135, 213, 276, 325
Troglitazone, 44, 45
Tumor necrosis factor(s), 16, 26, 30, 32, 43, 195, 206, 210, 212, 304, 305
Trophoblast, 7, 9, 190, 201, 202, 205–207, 212, 213, 222, 227, 239, 240, 242–244, 247, 251–254, 304, 306; *see also* Cytotrophoblast; Placenta; Syncytiotrophoblast
Twins, 250

Umbilical cord, 155, 190–193, 207–209, 211, 240, 250, 252, 253, 306, 307
Urocortin, 5
Uterus, 53, 111, 202, 204, 224–226, 268, 289, 290
Uterine myoma(s), 114; *see also* Leiomyoma

Vascular endothelial growth factor (VEGF) 66, 81, 84, 85, 89, 194, 210, 231, 251
Vascular system/tissue 30, 190, 227, 251, 299, 305
Ventromedial nucleus 56, 152

11 beta-hydroxylase, 229, 334
11 beta-hydroxysteroid dehydrogenase, 43, 229, 334
17 alpha-hydroxylase, 97
17–20 lyase, 102, 138
3T3-L1 cells 43, 44, 206